TROUBLESHOOTING ELECTRICAL/ELECTRONIC SYSTEMS

Third Edition

AMERICAN TECHNICAL PUBLISHERS, INC.
ORLAND PARK, ILLINOIS 60467-5756

Glen A. Mazur

Thomas E. Proctor

Troubleshooting Electrical/Electronic Systems contains procedures commonly practiced in industry and the trade. Specific procedures vary with each task and must be performed by a qualified person. For maximum safety, always refer to specific manufacturer recommendations, insurance regulations, specific job site and plant procedures, applicable federal, state, and local regulations, and any authority having jurisdiction. The material contained is intended to be an educational resource for the user. American Technical Publishers, Inc. assumes no responsibility or liability in connection with this material or its use by any individual or organization.

American Technical Publishers, Inc., Editorial Staff

Editor in Chief:
    Jonathan F. Gosse
Vice President—Production:
    Peter A. Zurlis
Art Manager:
    James M. Clarke
Technical Editor:
    Peter A. Zurlis
    Daniel M. Johnson
Copy Editor:
    Diane J. Weidner
Cover Design:
    Jennifer M. Hines
Illustration/Layout:
    Jeana M. Platz
    Jennifer M. Hines
    Mark S. Maxwell
Multimedia Coordinator:
    Carl R. Hansen
CD-ROM Development:
    Gretje Dahl
    Daniel Kundrat
    Nicole S. Polak

Acknowledgments

DOE/NREL, Robb Williamson
Fluke Corporation

Microsoft, Windows, Windows Vista, Windows XP, Windows NT, PowerPoint, and Internet Explorer are either registered trademarks or trademarks of Microsoft Corporation in the United States and/or other countries. Adobe, Acrobat, and Reader are either registered trademarks of Adobe Systems Incorporated in the United States and/or other countries. Pentium is a registered trademark of Intel Corporation in the United States and/or other countries. Firefox is a registered trademark of Mozilla Corporation in the United States and other countries. Quick Quiz, Quick Quizzes, and Master Math are either registered trademarks or trademarks of American Technical Publishers, Inc. National Electrical Code and NEC are registered trademarks of the National Fire Protection Association, Inc.

3 4 5 6 7 8 9 – 10 – 9 8 7 6

Printed in the United States of America

ISBN 978-0-8269-1791-1

 This book is printed on 30% recycled paper.

Contents

Interactive CD-ROM Contents

- **Quick Quizzes®**
- **Illustrated Glossary**
- **Flash Cards**

- **Virtual Meters**
- **Media Clips**
- **ATPeResources.com**

Troubleshooting Electrical/Electronic Systems, 3rd Edition, covers all aspects of troubleshooting electrical and electronic systems. The book is designed for use in the electrical industry, electrical training programs, and related fields. Each chapter in this text-workbook contains electrical and electronic system applications, step-by-step troubleshooting procedures, and realistic troubleshooting activities that reinforce the concepts presented. Topics covered range from electrical theory to troubleshooting industrial circuits and components. Detailed illustrations provide a visual, systematic approach to troubleshooting electrical/electronic systems. This latest edition has been updated to include expanded coverage of alternative energy systems, NFPA 70E® requirements, and motor nameplate interpretation.

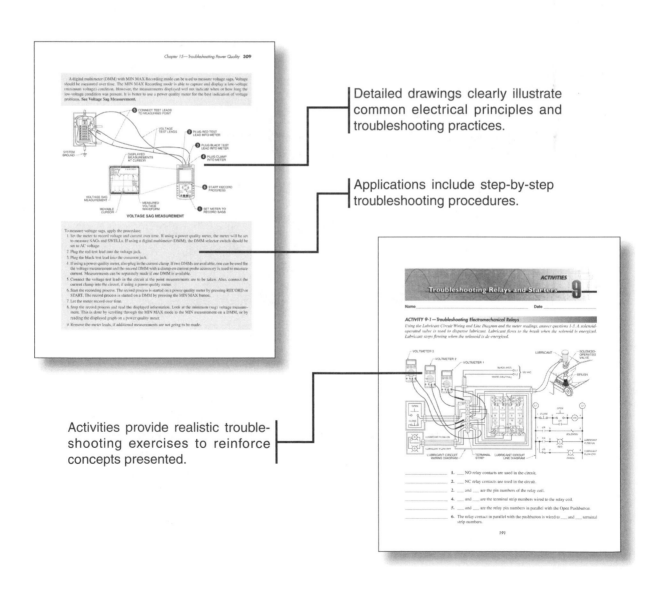

Detailed drawings clearly illustrate common electrical principles and troubleshooting practices.

Applications include step-by-step troubleshooting procedures.

Activities provide realistic troubleshooting exercises to reinforce concepts presented.

The interactive CD-ROM included with the book is a study aid with the following features:

- Quick Quizzes® that reinforce fundamental concepts, with 10 questions per chapter
- An Illustrated Glossary of industry terms, with links to illustrations, video clips, and animated graphics
- Flash Cards that enable a review of common electrical system troubleshooting terms and definitions
- Virtual Meters that facilitate interactive demonstrations of meter functions
- Media Clips that depict electrical troubleshooting concepts through video clips and animated graphics
- Access to ATPeResources.com, which provides a comprehensive array of instructional resources

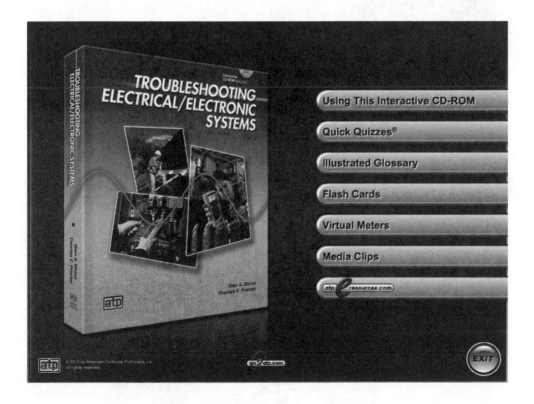

To obtain information on related training products, visit the American Tech web site at www.go2atp.com.

The Publisher

Also Available from American Tech...

Troubleshooting Electrical/Electronic Systems Workbook reinforces the textbook's content, provides applications, and tests student understanding of the concepts presented in *Troubleshooting Electrical/Electronic Systems.*

The Workbook includes nine sets of prints of common electrical/electronic systems, emphasizing the content of *Troubleshooting Electrical/Electronic Systems.* Tech-Cheks contain questions based on the corresponding chapter of *Troubleshooting Electrical/Electronic Systems.* Trade Tests are composed of worksheets developed from the nine sets of prints and are also based on the content of individual and preceding chapters in *Troubleshooting Electrical/Electronic Systems.*

The Publisher

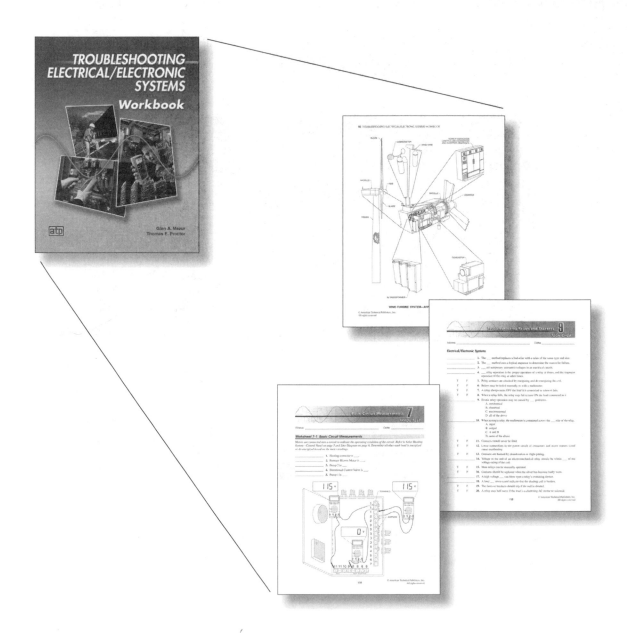

APPLICATION 1-1—Troubleshooting Levels

Troubleshooting is the systematic elimination of the various parts of a system or process to locate a malfunctioning part. A *system* is a combination of components, units, or modules that are connected to perform work or meet a specific need. A *process* is a sequence of operations that accomplishes desired results. A *malfunction* is the failure of a system, equipment, or part to operate properly.

To locate and correct a malfunction quickly, troubleshooting is performed at different levels. The different levels of troubleshooting electrical and electronic systems are the system, equipment/unit, board/module, and component levels. **See Troubleshooting Levels.**

TROUBLESHOOTING LEVELS

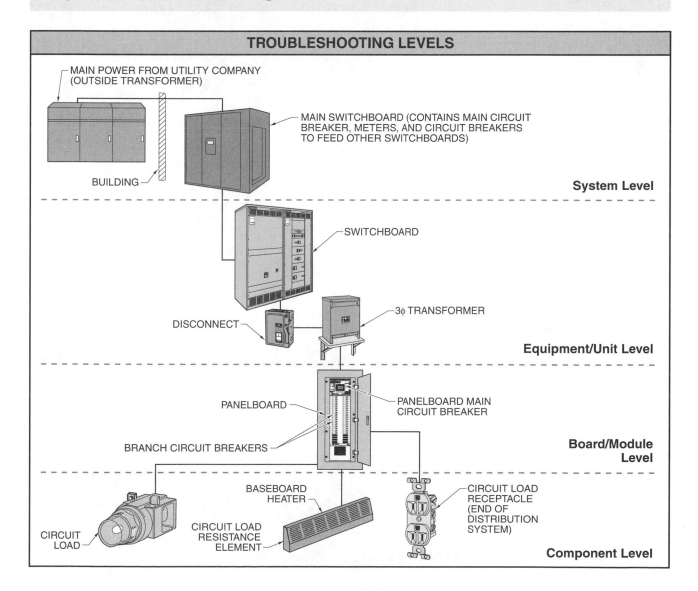

MAIN POWER FROM UTILITY COMPANY (OUTSIDE TRANSFORMER)

MAIN SWITCHBOARD (CONTAINS MAIN CIRCUIT BREAKER, METERS, AND CIRCUIT BREAKERS TO FEED OTHER SWITCHBOARDS)

BUILDING

System Level

SWITCHBOARD

DISCONNECT

3φ TRANSFORMER

Equipment/Unit Level

PANELBOARD

PANELBOARD MAIN CIRCUIT BREAKER

BRANCH CIRCUIT BREAKERS

Board/Module Level

BASEBOARD HEATER

CIRCUIT LOAD RECEPTACLE (END OF DISTRIBUTION SYSTEM)

CIRCUIT LOAD

CIRCUIT LOAD RESISTANCE ELEMENT

Component Level

System Level

A system is a combination of interconnected individual units, modules, and components. In a system, the combination of units, modules, and components works together to produce the desired results.

A system may include main or hand-held programming terminals, fixed or modular controllers, operator terminals, displays, and interfaces. **See Electrical/Electronic System.**

The programming terminals are usually located at a central location. Hand-held programming terminals for local programming changes and troubleshooting are connected at any point in the system. The controllers are located at individual machines or processes throughout the plant.

Troubleshooting at the system level requires knowledge of the operation of the hardware, software, and interfaces in the system. *Hardware* is the physical components in a system. Hardware malfunctions occur when two or more pieces of equipment are not properly sending or receiving data.

Software is the computer programs and procedures that operate the hardware. Software malfunctions are problems with the computer program(s).

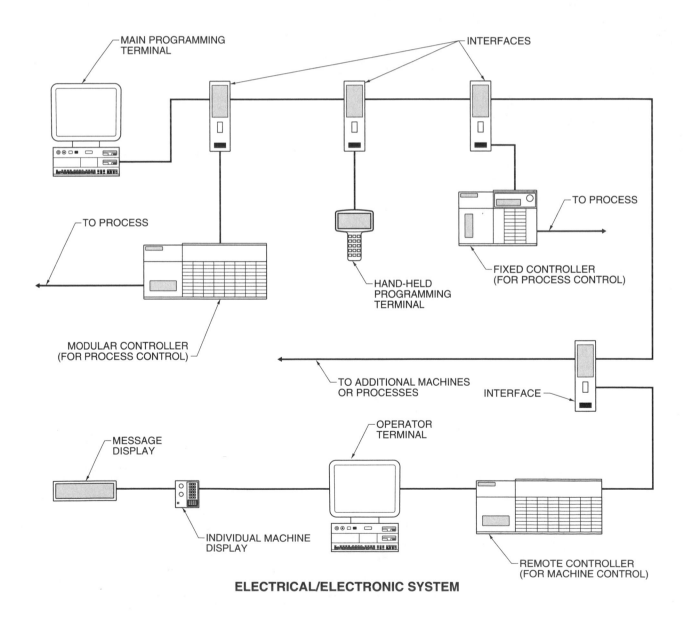

ELECTRICAL/ELECTRONIC SYSTEM

An *interface* is an electronic device that allows different levels or types of components to be used together in the same circuit. Interface problems occur when two or more pieces of equipment are working properly but are not able to communicate with each other.

A *systems analyst* is an individual who troubleshoots at the system level. *System analysis* is the breakdown of system requirements and components performed when designing, implementing, maintaining, or troubleshooting a system. A systems analyst observes each interface, connecting cable, program, piece of equipment, and language used.

System troubleshooting is performed on-site or off-site. *On-site troubleshooting* is troubleshooting at the location where the hardware is installed. *Off-site troubleshooting* is troubleshooting at a location other than where the hardware is installed. Off-site troubleshooting is performed when an on-site computer sends information about a malfunction to an off-site computer. The off-site computer is programmed to solve malfunctions in systems. When the malfunction is found, the off-site computer sends information to the on-site computer. **See Off-site Troubleshooting.**

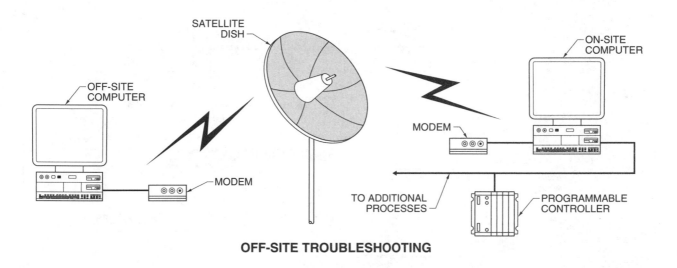

OFF-SITE TROUBLESHOOTING

Equipment/Unit Level

A *unit* is an individual component that performs a specific task by itself. Systems are made when units are connected. For example, a TV, Blu-ray disc player, DVD/CD player, and receiver each are separate units. When connected, an entertainment system is developed. **See Entertainment System.**

Troubleshooting at the equipment/unit level requires identification of the equipment or unit that is malfunctioning. When the malfunctioning equipment or unit is identified, substitution or testing is performed.

Substitution is the replacement of malfunctioning equipment or units. Substitution is performed when the equipment or unit is small, easily accessible, and a replacement is available.

Testing is performed when the equipment or unit is a hard-wired unit or is large in size, or when no replacement is available. A *hard-wired unit* has conductors connected to terminal screws. A programmable controller may have hundreds of hard-wired connections. Testing requires knowledge of using test equipment and ability to correctly interpret the test results.

For example, if a main computer cannot operate a machine connected to a programmable controller, a malfunction exists in the main programming terminal, interface, connecting cables, or programmable controller. A hand-held programming terminal is plugged into the programmable controller to troubleshoot at the equipment/unit level without disconnecting the equipment. **See Hand-held Programming Terminal.**

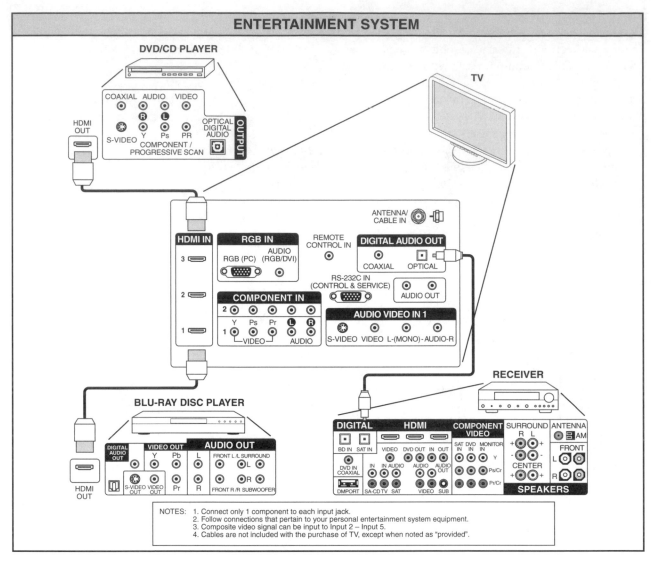

ENTERTAINMENT SYSTEM

NOTES: 1. Connect only 1 component to each input jack.
2. Follow connections that pertain to your personal entertainment system equipment.
3. Composite video signal can be input to Input 2 – Input 5.
4. Cables are not included with the purchase of TV, except when noted as "provided".

A hand-held programming terminal monitors information from and sends information to the programmable controller. Monitoring indicates whether information is received from the main computer. If no information is received, a malfunction exists between the programmable controller and the main computer. If information is received, a malfunction exists between the programmable controller and the machine. When a malfunctioning piece of equipment is identified, it is replaced or serviced at the board/module or component level.

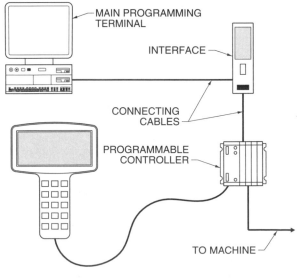

HAND-HELD PROGRAMMING TERMINAL

Board/Module Level

A *board* is a group of electronic components placed on a printed circuit (PC) board that performs a set task. A *module* is a group of electronic or electrical components housed in an enclosure that performs a set task.

Troubleshooting at the board/module level requires finding the malfunctioning board or module and replacing it. When troubleshooting electronic equipment at the board/module level, the entire board or module, such as a timer, counter, limit switch, photoelectric control, motor starter, or motor, that contains the malfunction is replaced. For example, a plug-in module on a programmable controller connects inputs (pushbuttons, limit switches, etc.) or outputs (lights, motor starters, solenoids, etc.). If a problem develops with the module, the module is removed, serviced, or replaced. **See Programmable Controller.**

Component Level

A *component* is an individual device used in a board or module. Components include springs, resistors, diodes, contacts, transistors, and capacitors. Troubleshooting at the component level requires finding the exact component that is malfunctioning. Examples of malfunctioning components include bad transistors or capacitors inside a timer or photoelectric control, or bad contacts or springs inside a magnetic motor starter or pushbutton station.

Troubleshooting at the component level is time consuming. It is often economical to replace an entire board or module. A voltmeter tests components or a group of components on a board. **See Component Checking.**

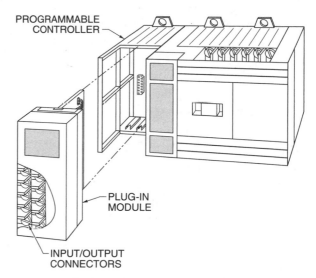

PROGRAMMABLE CONTROLLER

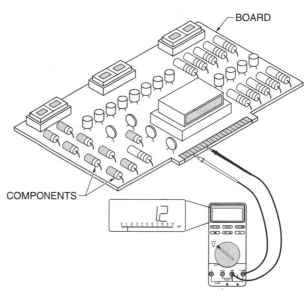

COMPONENT CHECKING

 Refer to **Activity 1-1—Troubleshooting Levels** ....

APPLICATION 1-2—Troubleshooting Meters

A skilled troubleshooter follows a logical plan that finds a malfunction quickly and efficiently. When a malfunctioning part is found, the part is repaired. Preventive maintenance is then performed to prevent future problems.

Preventive maintenance is maintenance performed to keep machines, assembly lines, production operations, and plant operations running with little or no downtime. Preventive maintenance keeps equipment in good condition. Troubleshooting methods include troubleshooting by knowledge and experience, troubleshooting by plant procedures, troubleshooting by manufacturer's procedures, or a combination of these.

Troubleshooting by Knowledge and Experience

Troubleshooting by knowledge and experience is a method of finding a malfunction in a system or process through applying information learned from past malfunctions. Troubleshooting by knowledge and experience is only temporarily effective because the primary malfunction is not corrected. For example, a fuse may blow on a machine, stopping production as a result. Changing fuses (or resetting breakers) starts production, but the reason the fuse blew is not clear.

Troubleshooting by Plant Procedures

In most plants, plant procedures are followed when troubleshooting equipment. Plant procedures are developed by supervisors or operators to ensure safe and efficient operation of equipment by plant personnel. Plant procedures are specific to the system or process used. **See Plant Procedure.** A plant procedure for replacing a defective module on a programmable controller includes the steps:

MIDWESTERN MANUFACTURING CO.
Defective Module Replacement

1. Inform the machine operator that power will be turned OFF.
2. Turn OFF, lock out, and tag the disconnect switch feeding power to the machine.
3. Using a voltmeter, test to ensure that no voltage is present at the power terminals of the programmable controller.
4. Remove the conductor connected to each terminal screw. Mark each conductor with the same number as the terminal screw.
5. Pull the module locking lever out and down until it is perpendicular to the face of the module.
6. Slide the module out and away from the receptacle.
7. Insert the replacement module into the receptacle and lock it in place using the locking lever.
8. Reconnect the conductors to the terminal screws.
9. Inform the machine operator that power will be turned ON.
10. Place all machine selector switches in the manual position.
11. Remove the lock and tag from the disconnect switch. Make sure the machine is clear of all employees.
12. Turn power ON.
13. Restart the machine.
14. Cycle the machine through one operation using the manual switches.
15. If the machine operates properly, place the selector switches in the automatic position.
16. Cycle the machine automatically several times.
17. Inform the supervisor if the machine is not operating properly.
18. Fill out a repair report if the machine is operating properly. Place one copy of the report in the maintenance folder inside the machine cabinet and return the other copy to the supervisor.
19. Remain with the operator until the machine is back in production.
20. Clean the area of all tools and any litter resulting from the maintenance call.

Troubleshooting by Manufacturer's Procedures

Troubleshooting by manufacturer's procedures is a method of finding malfunctioning equipment by using the procedures and recommendations provided by the manufacturer. Manufacturer's procedures differ from plant procedures in that manufacturer's procedures are shorter and more general than plant procedures. **See Manufacturer's Procedure.** A manufacturer's procedure for troubleshooting and replacing the power supply on a programmable controller includes the steps:

MIDCO TOOL DESIGN INC.
Programmable Controller Troubleshooting & Replacement

1. If there is no indication of power (status lights OFF) on the programmable controller, measure the voltage at the incoming power terminals on the power supply module.
2. If voltage is present and correct, replace power supply on programmable controller.
3. Remove power from the programmable controller.
4. Disconnect the power lines from the power supply terminals.
5. Disconnect the processor power cable from the power supply output terminal.
6. Remove the four mounting screws on power supply to free power supply from main panel.
7. Grasp the power supply firmly and pull out.
8. Press the replacement power supply into the main panel.
9. Replace and tighten the four mounting screws.
10. Connect the processor power cable.
11. Connect the power lines.
12. Turn power ON.

Plant or manufacturer's troubleshooting procedures may be in the form of a flow chart. A *flow chart* is a diagram that shows a logical sequence of steps for a given set of conditions. Flow charts help a troubleshooter to follow a logical path when trying to solve a problem. Flow charts use symbols and interconnecting lines to provide direction.

Symbols used with flow charts include an ellipse, rectangle, diamond, and arrow. An ellipse indicates the beginning and end of a flow chart or section of a flow chart. A rectangle contains a set of instructions. A diamond contains a question stated so that a yes or no answer is achieved. The yes or no answer determines the direction to follow through the flow chart. An arrow indicates the direction to follow based on the answers. **See Motor Troubleshooting Flow Chart.**

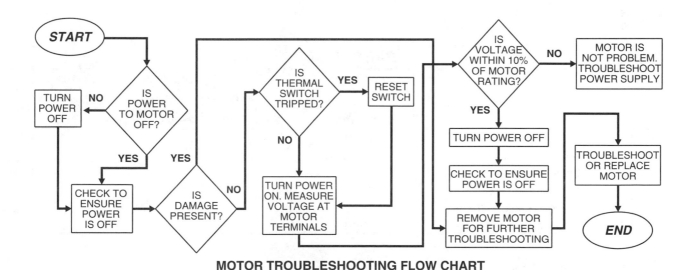

MOTOR TROUBLESHOOTING FLOW CHART

 Refer to **Activity 1-2 — Troubleshooting Meters** ....

APPLICATION 1-3—Troubleshooting Procedure

A *troubleshooting procedure* is a logical step-by-step process used to find a malfunction in a system or process as quickly and easily as possible. When troubleshooting a system, unit, or component, the following steps are performed: obtain information, observe system operation, isolate problems, test faults, and document findings.

Obtain Information

Obtain information from the manufacturer about the system, components, and potential malfunctions before a malfunction arises. Information is obtained by:

- Gathering technical information from the manufacturer, including service manuals, prints, and recommended spare parts lists. For example, when servicing a manual motor starter, contacts are often burnt and require replacement. By obtaining service bulletins, the exact part number is easily found. **See Service Bulletin.**
- Consulting listings of local dealer addresses. This provides several places to obtain help concerning a part or malfunction.
- Asking local dealers and manufacturers about training programs on the equipment or product line. Most major manufacturers offer technical training.

SERVICE BULLETIN

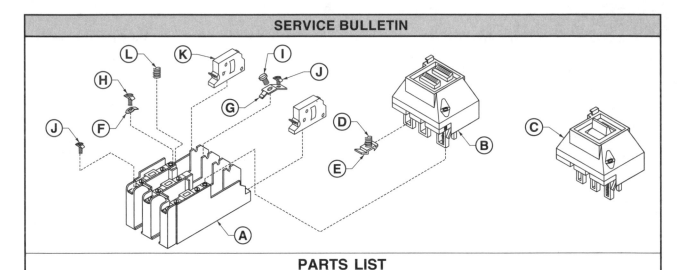

PARTS LIST

Item	Description	Part number	Size M-0 2-pole	Size M-0 3-pole	Size M-1 2-pole	Size M-1 3-pole	Size M-1P 2-pole
			Quantity				
A	Contact Block	00401	1	1	1	1	1
B	2-pole Contact actuator	01753	1	—	1	—	1
	3-pole Contact actuator	01751	—	1	—	1	—
K	Internal interlock (NO)	2065	—	—	—	—	—
	Internal interlock (NC)	2066	—	—	—	—	—
L	Return spring	17901	2	2	2	2	2

Observe System Operation

Observe a normally operating system or process during startup, slow operation, fast operation, and shutdown. Determine what part of the system is or is not working, and also the condition of the outputs (energized or de-energized), circuit, equipment, and operator. Look for problems, such as jammed, burnt, or broken material. Check fuses, circuit breakers, and overload resets. Also, check for any recent modifications or unauthorized changes.

Typical questions asked when troubleshooting include:

- What effect does a long downtime have on the process or product?
- Is it economical to repair or replace?
- Can the defective part be replaced with a spare or be fixed in the maintenance shop where time is not important to production?
- Is the repair going to last as long as a new component would?
- Is this a recurring problem?
- Are the required tools, instruments, and prints available to aid in troubleshooting?
- Are spare parts available?

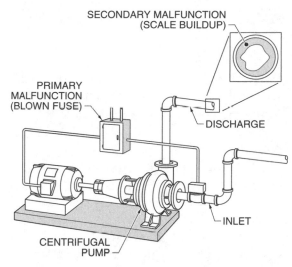

PIPING SYSTEM

Isolate Problems

Isolate the area or section that is probably malfunctioning. Narrow down the sources of the malfunction based on information obtained from observation and questioning. Look for potential secondary malfunction(s). For example, a motor drives a pump that moves product through a piping system. **See Piping System.**

The pump develops a higher-than-normal pressure if the pipe that moves the product is partially blocked. To develop the higher pressure, the motor draws additional current. The additional current blows a fuse. The blown fuse stops the motor. The primary malfunction is that the blown fuse stopped the motor. The secondary malfunction is the scale buildup in the pipe. Changing the fuse solves the problem only until the fuse blows again.

Test Malfunctions

Use test equipment to test and check the components that may be malfunctioning. Testing components help to determine primary and secondary malfunctions. For example, after changing fuses in a pump application, an ammeter is used to test the current draw of the motor. The motor is disconnected from the pump and rechecked if the current draw is too high. There is a motor problem if the current is still too high. There is a pump and/or piping system problem if the current draw is normal.

Document Findings

Record malfunctions found and service performed. List all findings. List components checked if a malfunction is not corrected. List any suggestions that may help to prevent the malfunction from recurring. Suggestions include replacing parts at a certain time, adding additional fuses, or changing the circuit design or operation. **See Motor Repair and Service Record.**

MOTOR REPAIR AND SERVICE RECORD

Motor File #: _004632_   Serial #: _____

Date Installed: _1/9_   Motor Location: _Equipment Room_

MFR: _Dayton_   Type: _D_   Frame: _145T_

HP: _10_   Volts: _230_   Amps: _15_

RPM: _1725_   Filter Sizes: _____

Date	Operation	Mechanic
2/7	Checked Current Drain - - OK	
3-14	Checked Belts - Realigned Belts	
5-16	Cleaned Motor	

SEMIANNUAL MOTOR MAINTENANCE CHECKLIST

Motor File #: _004632_   Serial #: _____

Date Installed: _1/9_   Motor Location: _Equipment Room_

MFR: _Dayton_   Type: _D_   Frame: _145T_

HP: _10_   Volts: _230_   Amps: _15_

RPM: _1725_   Date Serviced: _7-11_

Step	Operation	Mechanic
1	Turn OFF and lock out all power to the motor and its control circuit.	RNS
2	Clean motor exterior and all ventilation ducts.	
3	Check motor's wire raceway.	
4	Check and lubricate bearings as needed.	
5	Check drive mechanism.	
6	Check brushes and commutator.	
7	Check slip rings.	
8	Check motor terminations.	
9	Check capacitors.	
10	Check all mounting bolts.	
11	Check and record line-to-line resistance.	
12	Check and record megohmmeter resistance from L1 to ground.	
13	Check motor controls.	
14	Reconnect motor and control circuit power supplies.	
15	Check line-to-line voltage for balance and level.	
16	Check line current draw against nameplate rating.	
17	Check and record inboard and outboard bearing temperatures.	

 Refer to **Activity 1-3—Troubleshooting Procedure** ....

APPLICATION 1-4—Equipment Reliability

When troubleshooting or servicing a piece of equipment, the age of the equipment should be considered. Most equipment follows a typical life expectancy curve. **See Typical Equipment Life Expectancy Curve.**

Break-in Period

A *break-in period* is the time just after the installation of a new piece of equipment in which defects resulting from defective parts, poor manufacturing quality, contamination, or environmental stress appear. The highest failure rate for most equipment occurs during the break-in period. Troubleshooting a malfunctioning piece of equipment during the break-in period includes:

- Looking for signs of improper operator use.
- Determining if the equipment is correctly installed. The equipment may be connected to an improper voltage level, improperly grounded, or installed in an environment for which it was not designed.

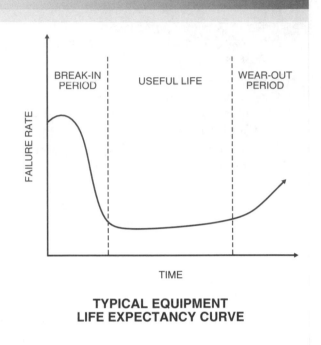

**TYPICAL EQUIPMENT
LIFE EXPECTANCY CURVE**

Useful Life

Useful life is the period of time after the break-in period when most equipment operates properly. Equipment may fail during its useful life. Troubleshooting a malfunctioning piece of equipment during its useful life includes:

- Checking for signs of improper maintenance, such as dirty or clogged filters. Most equipment needs periodic maintenance. Periodic maintenance usually includes cleaning and lubricating.

- Determining if the equipment is properly protected from improper environmental conditions. Higher-than-normal temperatures, high humidity, salt, corrosive chemicals, and dirt contribute to early failures.

- Looking for signs of damage caused by misuse or accidents. Accidents often occur because of careless operator use or improper environmental conditions. Such environmental conditions include voltage surges and flooding.

Wear-out Period

Wear-out period is the period of time after a piece of equipment's useful life terminates when normal equipment failures occur. Troubleshooting a piece of equipment in the wear-out period includes:

- Determining if the malfunction is worth repairing, and if the equipment will last long enough to justify the expense of the repair. For example, if a key on an old computer keyboard fails, it is likely that more keys will soon fail.

- Determining if the equipment is safe. Deterioration occurs in older equipment. If conductor insulation has deteriorated or if safety guards are not present, the equipment will not be safe even after a malfunction is repaired.

 *Refer to **Activity 1-4—Equipment Reliability** ....*

APPLICATION 1-5—*Troubleshooter Responsibilities*

All industries have systems and/or processes that must operate properly in order to produce products and services. To keep systems and/or processes operating properly, trained personnel are required to troubleshoot and repair any malfunction that occurs. Personnel who troubleshoot and repair systems or processes are part of the maintenance/service department of a company. The responsibilities of the maintenance/service personnel include:

- Installing new equipment according to prints and codes. Prints may be paper or electronic. Paper prints are the most common and are usually provided with each component or machine when purchased. Paper prints include diagrams that show power connections, assembly or connection procedures, and replacement parts. All prints use standard symbols and abbreviations to show circuit operation. **See Print.** Prints that include the circuit schematic and troubleshooting procedures are also available from most manufacturers.

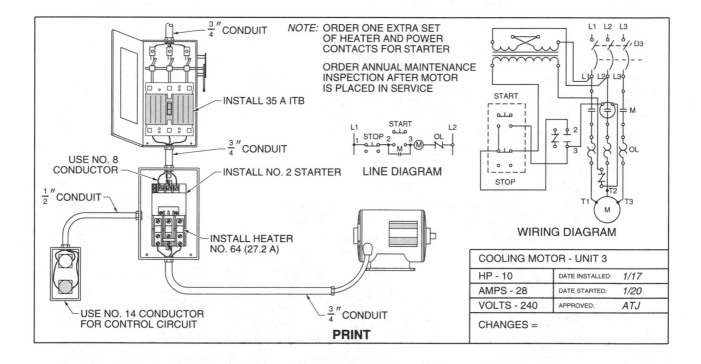

Electronic prints are prints that are captured electronically and stored on a computer disk. Electronic prints are used for large or complicated equipment. Electronic prints often include setup, operating, troubleshooting, and preventive maintenance procedures. Electronic prints are usually sold as an extra package.

- Following a logical startup procedure for a new system or process. Startup procedures include the following:

    Perform general checks of fluid levels, guard positions, belt tightness, and loose parts before turning power ON.

    Turn ON and test one part of a machine or system at a time.

    Test all manual operations first.

    Determine if all safety equipment and features are working properly.

    Anticipate component failure. Operate a machine long enough to allow pressure to build, belts to slip, and oil to heat up. The majority of component failure on new equipment occurs during the first few hours or days of operation.

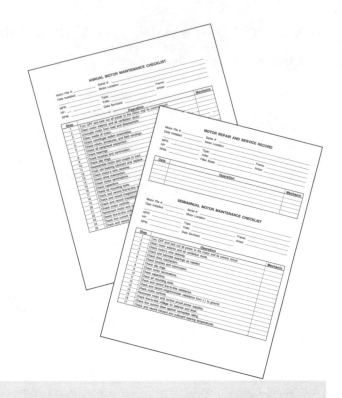

- Applying a preventive maintenance program. Preventive maintenance is action taken to maintain equipment in good operating condition to ensure uninterrupted operation. Preventive maintenance includes tightening belts, adding lubricant, cleaning, changing filters, calibrating, general inspecting, and setting control limits. Preventive maintenance is usually performed according to a set schedule. Monthly, semiannual, and annual preventive maintenance schedules are common. **See Motor Maintenance Checklists.**

- Preventing malfunctions. Every process or operation has potential problems that may cause malfunctions. Preventing malfunctions requires close attention to the entire process. This includes ensuring that the correct size and type of parts are fed into a machine, that the machine is operated properly, and that improvements, modifications, or design changes are made.

- Identifying malfunctions that occur. A skilled troubleshooter tries to identify the cause of a malfunction and correct it before the malfunction occurs again. Note all malfunctions that occur. The true cause of a malfunction may not be clear. For example, a part that jams in a machine may be caused by a part that is not within specifications, a machine that is not within specifications, operator misuse, improper lubrication, or a control malfunction. Most malfunctions are identified after they have appeared several times.

- Repairing equipment in an efficient manner. Repairing equipment in an efficient manner includes the steps:

    Anticipate and have the required tools and test equipment for a troubleshooting call.

    Troubleshoot from the simple to the complex.

    Ask the opinions of operators and other personnel that are familiar with the system.

    Determine if problems can be fixed temporarily to allow production to continue until the end of the operation.

- Maintaining good records of maintenance, equipment condition, and downtime. Provide justification for overtime, new tools and equipment, additional personnel, and improved safety to improve the maintenance operation.

- Handling emergencies. Emergencies that occur in a plant may be handled by the maintenance department until additional help arrives. Emergencies that may require trained maintenance personnel include:

    General emergencies. Examples of general emergencies include fires, floods, accidents, and power outages. Personnel should know plant evacuation procedures, emergency telephone numbers, the meaning of different alarms, and basic first aid procedures.

    Trade-related emergencies. Examples of trade-related emergencies include electrical shock, burns, and exposure to specific hazardous conditions. Maintenance personnel should know power disconnection procedures, shock treatment, CPR, and first aid.

    Operational emergencies. Operational emergencies are caused by hazardous materials, fire hazards, and safety problems. Personnel should know specific chemical, fire, and physical safety procedures.

- Remain informed about the latest equipment, procedures, and technologies. Trade and technical magazines provide information on new equipment, techniques, and technologies. Manufacturers and local suppliers offer training seminars about various product lines. The latest equipment, procedures, and technologies are often taught at local trade schools and in-plant training courses.

 Refer to **Activity 1-5 — Troubleshooter Responsibilities** ....

APPLICATION 1-6 — Electrical Shock

According to the National Safety Council, over 1000 people are killed each year in the United States from electrical shock. Electricity is the number one cause of fires. More than 100,000 people are killed in electrical fires each year.

Improper electrical wiring or misuse of electricity causes destruction of equipment and fire damage to property. Safe working habits are required when troubleshooting an electrical circuit or component because the electric parts that are normally enclosed are exposed.

An electrical shock results anytime a body becomes part of an electrical circuit. Electrical shock varies from a mild shock to fatal current. The severity of an electrical shock depends on the amount of electric current (in mA) that flows through the body, the length of time the body is exposed to the current flow, the path the current takes through the body, the physical size and condition of the body through which the current passes, and the amount of body area exposed to the electric contact.

The amount of current that passes through a circuit depends on the voltage and resistance of the circuit. During an electrical shock, a person's body becomes part of the circuit. The resistance a body offers to the flow of current varies. Sweaty hands have less resistance than dry hands. A wet floor has less resistance than a dry floor. The lower the resistance, the greater the current flow. The greater the current flow, the greater the severity of shock. **See Effect of Electric Current.**

EFFECT OF ELECTRIC CURRENT	
Current (in mA)	**Effect On Body**
8 or less	Sensation of shock but probably not painful
8 to 15	Painful shock Removal from contact point by natural reflexes
15 to 20	Painful shock May be frozen or locked to point of electric contact until circuit is de-energized
over 20	Causes severe muscular contractions, paralysis of breathing, heart convulsions

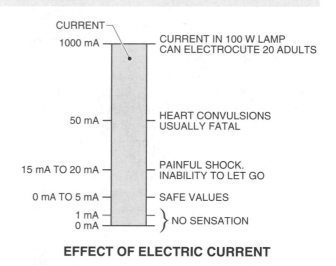

EFFECT OF ELECTRIC CURRENT

When handling an accident caused by electrical shock, apply the procedure:
1. Break the circuit to free the person immediately and safely. Never touch any part of a victim's body when in contact with the circuit. If the circuit cannot be turned OFF, use any nonconducting device to free the person. Resist the temptation to touch a person if the power is not turned OFF.

2. After the person is free from the circuit, send for help and determine if the person is breathing. If there is no breathing or pulse, start CPR. Always get medical attention for a victim of electrical shock.

3. If a person is breathing and has a pulse, check for burns and cuts. Burns are caused by contact with the live circuit, and are found at the points that the electricity entered and exited the body. Treat the entrance and exit burns as thermal burns and get medical help.

 Refer to **Activity 1-6—Electrical Shock**...

APPLICATION 1-7—Electrical Safety

Personal Protective Equipment

Personal protective equipment (PPE) is clothing and/or equipment worn by a technician to reduce the possibility of injury in the work area. The use of personal protective equipment is required whenever work may occur on or near energized exposed electrical circuits. Personal protective equipment includes flame-resistant clothing, head protection, eye protection, ear protection, hand and foot protection, and rubber insulated matting. **See Personal Protective Equipment.**

Clothing made of durable, fire-resistant material provides protection from contact with hot and sharp objects. Protective helmets or hard hats protect electrical workers from the impact of falling objects and electrical shock. Safety glasses, respirators, ear plugs, gloves, and foot protection may be used depending on the task. For example, insulated tools and gloves made from rubber can be used to provide maximum insulation from electrical shock hazards and should be worn when taking high-voltage and current measurements. Safety shoes with steel toes provide protection from falling objects. Insulated rubber boots and rubber mats provide insulation to prevent electrical shock.

PERSONAL PROTECTIVE EQUIPMENT

All personal protective equipment must meet Occupational Safety and Health Administration (OSHA) 29 Code of Federal Regulations (CFR) 1910, Subpart I—*Personal Protective Equipment* (sections 1910.132 to 1910.138), applicable American National Standards Institute (ANSI) standards, and other safety standards. The National Fire Protection Association standard *NFPA 70E, Standard for Electrical Safety in the Workplace,* addresses "electrical safety requirements for employee workplaces that are necessary for the safeguarding of employees in pursuit of gainful employment."

NFPA 70E

Per NFPA 70E, "Only qualified persons shall perform testing on or near live parts operating at 50 V or more." All personal protective equipment and tools are selected to be appropriate to the operating voltage (or higher) of the equipment or circuits being worked on or near. Equipment, devices, and tools must be suited for the work to be performed. In most cases, voltage-rated gloves and tools are required. Voltage-rated gloves and tools are rated and tested for the maximum line-to-line voltage on which work is to be performed. Protective gloves must be inspected or tested as required for maximum safety before each task.

Rubber Insulating Gloves

Rubber insulating gloves are important articles of personal protective equipment for electrical workers. Safety requirements call for the usage of rubber insulating gloves as well as cover gloves (commonly leather). The primary purpose of rubber insulating gloves and cover gloves is to insulate the hands and the lower arms from possible contact with live conductors.

Rubber insulating gloves offer a high resistance to current flow to help prevent an electrical shock, and the leather protectors protect the rubber insulating gloves and provide additional insulation. Rubber insulating gloves are rated and labeled for maximum allowed voltage.

Any substance that can physically damage rubber gloves must be removed before testing. Insulating gloves and protector gloves found to be defective shall not be discarded or destroyed in the field, but shall be tagged "unsafe" and returned to a supervisor.

Flame-Resistant (FR) Clothing

Sparks from an electrical circuit can cause a fire. Approved flame-resistant (FR) clothing must be worn in conjunction with rubber insulating gloves for protection from electrical arcs when performing certain operations on or near energized equipment or circuits. FR clothing must be kept as clean and sanitary as practical and must be inspected prior to each use. Defective clothing must be removed from service immediately and replaced. Defective FR clothing must be tagged "unsafe" and returned to a supervisor.

Eye Protection

Eye protection must be worn to prevent eye or face injuries caused by flying particles, contact arcing, and radiant energy. Eye protection must comply with OSHA 29 CFR 1910.133, *Eye and Face Protection.* Eye protection standards are specified in ANSI Z87.1, *Occupational and Educational Eye and Face Protection.* Eye protection includes safety glasses, face shields, and arc blast hoods. **See Eye Protection.**

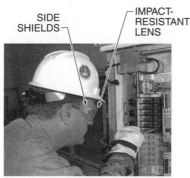
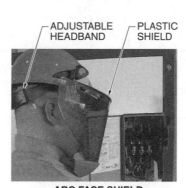
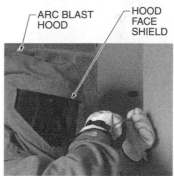

SIDE SHIELDS — IMPACT-RESISTANT LENS

ADJUSTABLE HEADBAND — PLASTIC SHIELD

ARC BLAST HOOD — HOOD FACE SHIELD

SAFETY GLASSES **ARC FACE SHIELD** **ARC BLAST HOOD**

EYE PROTECTION

Safety glasses are an eye protection device with special impact-resistant glass or plastic lenses, reinforced frames, and side shields. Plastic frames are designed to keep the lenses secured in the frame when an impact occurs in order to minimize the shock hazard when working with electrical equipment. Side shields provide additional protection from flying objects. Tinted-lens safety glasses protect against low-voltage arc hazards. Non-tinted safety glasses with side shields can be worn when working on low-voltage (50 V or less) PLC systems that are operating in a normally safe environment.

A *face shield* is an eye and face protection device that covers the entire face with a plastic shield, and is used for protection from flying objects. Tinted face shields protect against low-voltage arc hazards.

An *arc blast hood* is an eye and face protection device that covers the entire head with plastic and material. Arc blast hoods are used to protect against high-voltage arc blasts. Technicians working with energized high-voltage equipment must wear arc blast hoods.

Safety glasses, face shields, and arc blast hood lenses must be properly maintained to provide protection and clear visibility. Lens cleaners are available that clean without risk of lens damage. Pitted or scratched lenses reduce vision and may cause lenses to fail on impact.

Rubber Insulating Matting

Rubber insulating matting is a personal protective device placed on the floor to protect technicians from electrical shock when working on energized electrical circuits. Dielectric black fluted rubber matting is specifically designed for use in front of open cabinets or high-voltage equipment. Rubber insulating matting is used to protect technicians when voltages over 50 V are present. Rubber insulating matting is available in different sizes and is rated for maximum working voltage. **See Rubber Insulating Matting Ratings.**

RUBBER INSULATING MATTING RATINGS					
Safety Standard	**Material Thickness**		**Material Width (in.)**	**Test Voltage**	**Maximum Working Voltage**
	Inches	**Millimeters**			
BS921*	0.236	6	36	11,000	450
BS921*	0.236	6	48	11,000	450
BS921*	0.354	9	36	15,000	650
BS921*	0.354	9	48	15,000	650
VDE0680[†]	0.118	3	39	10,000	1000
ASTM D178[‡]	0.236	6	24	25,000	17,000
ASTM D178[‡]	0.236	6	30	25,000	17,000
ASTM D178[‡]	0.236	6	36	25,000	17,000
ASTM D178[‡]	0.236	6	48	25,000	17,000

* BSI–British Standards Institute
[†] VDE–Verband Deutscher Elektrotechniker Testing and Certification Institute
[‡] ASTM International

A troubleshooter must work safely at all times. Following are some basic safety rules that must be followed when working with electrical equipment.

General Electrical Safety Rules

- Always comply with the NEC®, OSHA, and NFPA 70E regulations.
- Use UL-approved appliances, components, and equipment.
- Keep electrical grounding circuits in good condition. Ground any conductive component or element that does not have to be energized. The grounding connection must be a low-resistance conductor heavy enough to carry the largest fault current that may occur.

- Turn OFF, lock out, and tag disconnect switches when working on any electrical circuit or equipment. Test all circuits after they are turned OFF. Insulators may not insulate, grounding circuits may not ground, and switches may not open the circuit.

- Use double-insulated power tools or power tools that include a third conductor grounding terminal that provides a path for fault current. Never use a power tool that has the third conductor grounding terminal removed. **See Normal Current and Fault Current.**

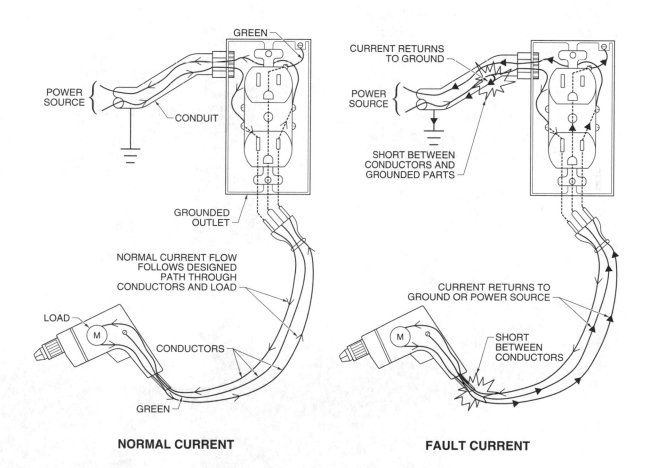

NORMAL CURRENT   **FAULT CURRENT**

- Always use personal protective equipment (PPE) and safety equipment.

- Know what to do in an emergency.

- Check conductors, cords, components, and equipment for signs of wear or damage. Replace any equipment that is not safe.

- Never throw water on an electrical fire. Turn OFF the power and use a Class C rated fire extinguisher. **See Fire Extinguisher Classes.** The four classes of fires are Class A, Class B, Class C, Class D, and Class K. Class A fires include burning wood, paper, textiles, and other ordinary combustible materials containing carbon. Class B fires include burning oil, gas, grease, paint, and other liquids that convert to a gas when heated. Class C fires include burning electrical devices, motors, and transformers. Class D is a specialized class of fires including burning metals such as zirconium, titanium, magnesium, sodium, and potassium. Class K fires include grease fires in commercial cooking equipment. Fire extinguishers are selected for the class of fire based on the combustibility of the material.

- When working in a dangerous area or with dangerous equipment, work with another individual.
- Learn CPR and first aid.
- Do not work when tired or taking medicine that causes drowsiness.
- Do not work in poorly lighted areas.
- Always use nonconductive ladders. Never use a metal ladder when working on or around electrical equipment.
- Ensure there are no atmospheric hazards such as flammable dust or vapor in the area. A live electrical circuit can emit a spark at any time.
- Use one hand when working on a live circuit to reduce the chance of an electrical shock passing through the heart and lungs.
- Never bypass or disable fuses or circuit breakers.
- Extra care must be taken in an electrical fire because burning insulation produces toxic fumes.
- Always fill out accident forms and report any electrical shock.

Electric Motor Safety

Two areas requiring attention when working with electric motors are the electrical circuit and rotating shaft. Basic electric motor safety rules include:

- Connecting a motor to the correct grounding system.
- Ensuring that guards or housings are connected to the rotating parts of a motor or anything connected to the motor. **See Motor Guard.**
- Using the correct motor type for the location. For example, a DC or universal motor shall never be used in a hazardous location that contains flammable materials because the sparking at the brushes can ignite the material.

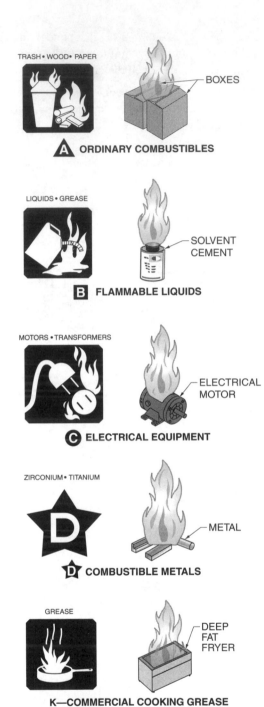

TRASH • WOOD • PAPER — BOXES
A ORDINARY COMBUSTIBLES

LIQUIDS • GREASE — SOLVENT CEMENT
B FLAMMABLE LIQUIDS

MOTORS • TRANSFORMERS — ELECTRICAL MOTOR
C ELECTRICAL EQUIPMENT

ZIRCONIUM • TITANIUM — METAL
D COMBUSTIBLE METALS

GREASE — DEEP FAT FRYER
K—COMMERCIAL COOKING GREASE

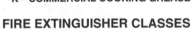

FIRE EXTINGUISHER CLASSES

- Connecting a motor to the correct voltage and power source.

- Providing the motor with the correct overload and overcurrent protection to protect the motor when starting or shorted (overcurrent protection) and when running (overload protection).

Grounding

Electrical circuits are grounded to safeguard equipment and personnel against the hazards of electrical shock. Proper grounding of electrical tools, machines, equipment, and delivery systems is one of the most important factors in preventing hazardous conditions.

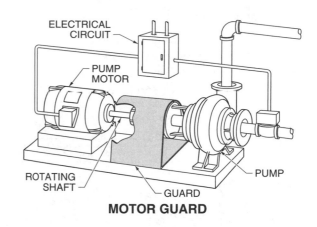

MOTOR GUARD

Grounding is the connection of all exposed non-current-carrying metal parts to the earth. Grounding provides a direct path for unwanted (fault) current to the earth without causing harm to persons or equipment. Grounding is accomplished by connecting the circuit to a metal underground pipe, a metal frame of a building, a concrete-encased electrode, or a ground ring. **See Grounding Methods.**

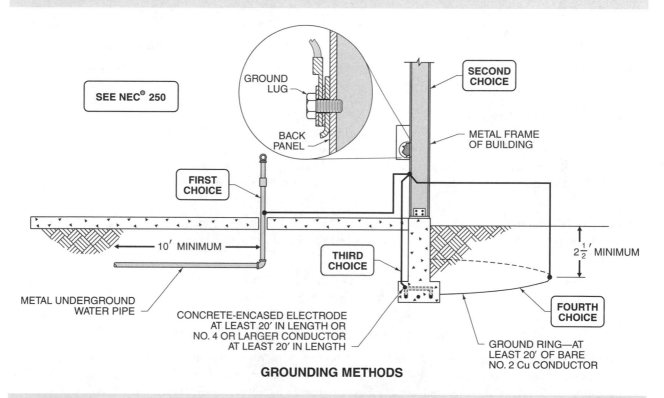

GROUNDING METHODS

Non-current-carrying metal parts that are connected to ground include all metal boxes, raceways, enclosures, and equipment. Unwanted current exists because of insulation failure or if a current-carrying conductor makes contact with a non-current-carrying part of the system. In a properly grounded system, the unwanted current flow trips fuses or circuit breakers. Once the fuse or circuit breaker is tripped, the circuit is opened and no additional current flows.

Codes and Standards

To protect people and property from electrical dangers, national, state, and local codes and standards are used. **See Codes and Standards.** A *code* is a regulation or minimum requirement. A *standard* is an accepted reference or practice. Codes and standards ensure electrical equipment is built and installed safely and every effort is made to protect people from electrical shock.

CODES AND STANDARDS

NFPA
Batterymarch Park
Quincy, MA 02269

NEMA
2101 L Street, N.W.
Washington, D.C. 20037

ANSI
11 West 42nd Street
New York, NY 10036

UL
333 Pfingsten Rd.
Northbrook, IL 60062

OSHA
230 South Dearborn Street
32nd Floor, Room 3244
Chicago, IL 60604

CSA
178 Rexdale Blvd.
Rexdale, ON M9W 1R3

National Fire Protection Association (NFPA). NFPA is a national organization that provides guidance in assessing the hazards of the products of combustion. The NFPA publishes the National Electrical Code (NEC®). The NEC's purpose is the practical safeguarding of persons and property from the hazards arising from the use of electricity. The NEC® is updated every three years. Many city, county, state, and federal agencies use the NEC® to set requirements for electrical installations. The NEC® is primarily concerned with protecting lives and property.

American National Standards Institute (ANSI). ANSI is a national organization that helps identify industrial and public needs for standards. ANSI coordinates and encourages activities in national standards development.

Occupational Safety and Health Administration (OSHA). OSHA is a federal organization that requires all employers to provide a safe environment for their employees. All work areas must be free from hazards likely to cause serious harm. The provisions of this act are enforced by federal inspection. With few exceptions, OSHA uses the NEC® to help assure a safe electrical environment.

National Electrical Manufacturers Association (NEMA). NEMA is a national organization that assists with information and standards concerning proper selection, ratings, construction, testing, and performance of electrical equipment. NEMA standards are used as guidelines for the manufacture and use of electrical equipment.

Underwriters' Laboratories, Inc. (UL). UL is an independent organization that tests equipment and products to see if they conform to national codes and standards. Equipment tested and approved by UL carries the UL label. UL approved equipment and products are listed in its annual publication.

Canadian Standards Association (CSA). CSA is a Canadian organization similar to UL that tests equipment and products to ensure they meet national standards.

Customer Relations

Troubleshooting is performed on broken or malfunctioning equipment. When performing a troubleshooting call, the troubleshooter should be prepared to communicate with upset individuals. Communicating with an upset individual can be accomplished by:

- Relaxing

    Relaxing takes the tension out of one's voice. Approach the individual as a friend. The upset individual is primarily mad at the situation or equipment.

- Listening

    Ask for an explanation of the problem even if it is already known. Let the individual explain without interruption.

- Agreeing

    Find some point of agreement. Agreeing demonstrates willingness to help and places the problem on the equipment.

- Finding the problem

    Find the problem and its cause.

- Making a decision

    Determine if the problem can be fixed and at what cost. State what is required to return the equipment to service.

- Coming to an agreement

    Agree to fix the problem, give the individual time to decide, call in a second opinion, or do nothing at the time.

 Refer to **Activity 1-7—Electrical Safety**

 Refer to *Quick Quiz® on CD-ROM*

Refer to Chapter 1 in the **Troubleshooting Electrical/Electronic Systems Workbook** for additional questions.

Name_____   Date _____

ACTIVITY 1-1—Troubleshooting Levels

State the level at which the malfunction occurred and was serviced (system, equipment, etc.).

1. After a bulk railcar is emptied halfway, no product moves to a storage silo.

_____ **A.** Malfunction occurred
at ___ level.

_____ **B.** Malfunction serviced
at ___ level.

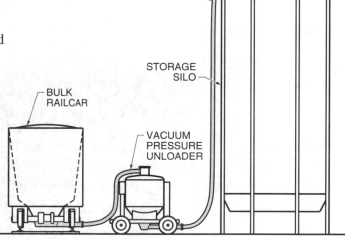

UNLOADER C

TROUBLESHOOTING CHART

UNLOADER C

Date ___3-21___   **Time** ___8:30___ **to** ___2:30___   **Service Person** _G.H._

Symptoms Observed: The vacuum pressure unloader pump is ON. Hoses from railcar to unloader and unloader to silo intake pipe are not kinked.

Probable Area of Malfunction: vacuum pressure unloader

Probable Malfunction: obstruction in hose or vacuum pump

Potential Secondary Malfunction(s):   shaft from pump motor to vacuum pressure unloader broken, hose or silo conveyor clogged

Findings/Service Performed: Turned vacuum unloader OFF and disconnected hose from silo intake pipe. Placed hose in portable barrel. Turned unloader ON, no product was pumped out. Turned unloader OFF and connected hose to silo. Disconnected hose at railcar. Turned unloader ON. Vacuum was felt at disconnected hose leading to unloader. Turned unloader OFF and removed discharge valve from railcar. Valve was clogged with a paper bag. Removed bag and placed unit back in service.

2. The letter *m* on a keyboard sticks and produces a line of *m*'s when pressed.

_____ **A.** Malfunction occurred
at ___ level.

_____ **B.** Malfunction serviced
at ___ level.

PC TERMINAL 0065

TROUBLESHOOTING CHART   PC TERMINAL 0065

Date 9-20	**Time** 10:00	**to** 10:45	**Service Person** J.P.	

Symptoms Observed:	The letters *b*, *n*, and *m* have a sticky feeling when pressed. The letter m reproduces several *m*'s at times
Probable Area of Malfunction:	keyboard, letter *m*
Probable Malfunction:	defective *m* key
Potential Secondary Malfunction(s):	customer misuse
Findings/Service Performed:	Found sticky substance in keyboard. Cleaned and tested keyboard. No further problems.
Suggestions:	Ask operators not to eat or drink around keyboard. Keep keyboard covered when not in use.

---

3. One button on a beverage gun does not dispense product when pressed.

_____ **A.** Malfunction occurred
at ___ level.

_____ **B.** Malfunction serviced
at ___ level.

PRINTED
CIRCUIT
BOARD

BEVERAGE GUN 01456

TROUBLESHOOTING CHART   BEVERAGE GUN 01456

Date 7-13	**Time** 9:00	**to** 10:00	**Service Person** J.R.	

Symptoms Observed:	All buttons except one dispense product.
Probable Area of Malfunction:	beverage gun unit
Probable Malfunction:	bad button or malfunctioning component in printed circuit board of beverage gun
Potential Secondary Malfunction(s):	cable leading from gun to solenoid valves that dispense product, solenoid valve
Findings/Service Performed:	Opened beverage gun and found cracked printed circuit board under malfunctioning button. Replaced entire beverage gun.
Suggestions:	Buttons do not have to be pressed hard.

4. A service door does not open when the up pushbutton is pressed.

_____ **A.** Malfunction occurred at ___ level.

_____ **B.** Malfunction serviced at ___ level.

MOTOR DRIVE UNIT

OVERHEAD DOOR B

TROUBLESHOOTING CHART   OVERHEAD DOOR B

Date 1-22	**Time** 8:15 **to** 10:00	**Service Person** T.R.

Symptoms Observed: Door does not open and motor makes no sound when the up pushbutton is pressed. Door has ice buildup along the bottom. Heat is turned down at night. Ice melts when heat is ON for some time.

Probable Area of Malfunction: fused disconnect unit, motor starter, motor/drive unit

Probable Malfunction: blown disconnect fuses, overloads tripped at motor starter, motor burned out.

Potential Secondary Malfunction(s): door jammed from misalignment, ice buildup caused fuses or overloads to trip

Findings/Service Performed: Motor overloads on reversing starter were tripped. Reset overloads and tested door. Door moves freely with no apparent problems. Current draw on motor in operation is within limits.

5. A counter is not counting the parts that are moving down a production line.

_____ **A.** Malfunction occurred at ___ level.

_____ **B.** Malfunction occurred at ___ level.

MAGNETS

LINE 2, COUNTER 4

TROUBLESHOOTING CHART   LINE 2, COUNTER 4

Date 5-18	**Time** 2:30 **to** 3:30	**Service Person** M.J.

Symptoms Observed: Counter is lit and displaying 0000. No counts are registering as product moves through the sensor.

Probable Area of Malfunction: counter or proximity switch

Probable Malfunction: proximity switch because counter is lit

Potential Secondary Malfunction(s): open or shorted cable from proximity switch to counter

Findings/Service Performed: Cable in good shape. Replaced proximity switch, no counts registering. Replaced counter, new counter working. Left new proximity switch to permit production.

ACTIVITY 1-2—Troubleshooting Methods

Complete the flow chart by placing the correct letter in the diamond or rectangle.

Problem: No heat, furnace fails to start

A. Is thermostat switch in OFF or cool position?

B. Is thermostat set too low?

C. Are breakers in tripped position?

D. Are fuses blown?

E. Is pilot out?

F. Is 24 V ± 2 V present on transformer secondary?

G. Is 115 V ± 5 V present on transformer primary?

H. Does furnace start?

I. Set thermostat higher.

J. Close switch.

K. Reset breakers.

L. Replace blown fuses.

M. Turn valve ON and relight pilot.

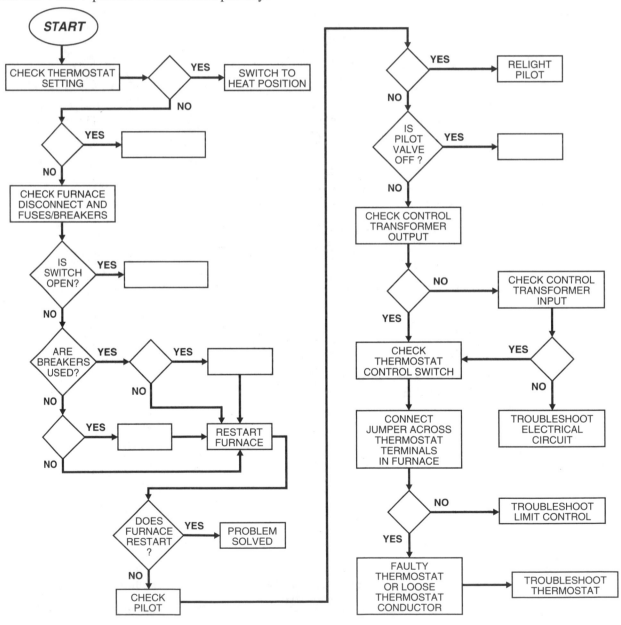

ACTIVITY 1-3—Troubleshooting Procedure

Using Annual Motor Maintenance Checklist and Motor Repair and Service Record, answer the following questions.

ANNUAL MOTOR MAINTENANCE CHECKLIST

Motor File #: *DC-10-12*   Serial #: *1000112B*   Date Installed: *11-18*

Motor Location: *Plant 3, Building A, Level 2, Fill Line 3, Infeed Motor*

_____   MFR: *North Co.*

Type: *DC Shunt*   Frame: *256J* HP: *10*   Volts: *500*

Amps: *18*   RPM: *650* Date Serviced: *2-4*   Class: *F*

Duty: *Cont*   SF: *1.0*   Starting Torque (ib-ft) *105*

Step	Operation	Mechanic
1	Turn OFF and lock out all power to the motor and its control circuit.	*RJ*
2	Clean motor exterior and all ventilation ducts. *Motor very dirty*	
3	Uncouple motor from load and disassemble.	
4	Clean inside of motor.	
5	Check rotors, armatures, and field windings. *Rotor has some burning*	
6	Check all peripheral equipment. *Belt OK (no wear)*	
7	Check bearings. *Added SAE #10 to sleeve bearings*	
8	Check brushes and commutator. *Brushes 1/2 worn.*	
9	Reassemble motor and couple to load.	
10	Check motor's wire raceway. *OK* Note: *Motor current measured*	
11	Check drive mechanism. *OK* *at 22 A and motor seems to*	
12	Check motor terminations. *OK* *be running hot.* *RJ*	
13	Check all mounting bolts. *OK*	
14	Check and record line-to-line resistance. *OK*	
15	Check and record megohmmeter resistance from T1 to ground. *OK*	
16	Check and record insulation polarization index. *OK*	
17	Check motor controls. *OK*	
18	Reconnect motor and control circuit power supplies. *OK*	
19	Check line-to-line voltage for balance and level. *Voltage low (460 V)*	

MOTOR REPAIR AND SERVICE RECORD

Date	Operation	Mechanic
10-29	*Fuses blown — changed w/same size*	*MA*
11-5	*Fuses blown — changed 20 A to 25 A*	*DF*
2-4	*Annual service — see checklist*	*RJ*

_____ **1.** The motor was installed on ___.

_____ **2.** The motor was in service for ___ months.

_____ **3.** The first sign of a problem occurred on ___.

_____ **4.** The first problem was ___.

_____ **5.** Was the motor serviced before the annual checkup was completed?

_____ 6. The current rating of the motor is ___ A.

_____ 7. The measured current draw of the motor during the annual checkup was ___ A.

_____ 8. The starting torque of the motor is ___ lb-ft.

_____ 9. The recommendation made by the maintenance person during the annual checkup was ___.

_____ 10. Is there any record that the recommendations made on the annual checkup report were followed?

_____ 11. The motor failed because of ___.

_____ 12. Was the motor undersized for the application?

_____ 13. Was the motor properly protected from contaminants?

_____ 14. Could a power line voltage lower than the voltage on the motor nameplate cause the problems?

_____ 15. Could the motor that replaced this motor have the same problems if preventive maintenance is not performed?

ACTIVITY 1-4—Equipment Reliability

For each maintenance procedure, list each period during which the procedure is generally performed. A = Break-in period, B = Useful life period, and C = Wear-out period.

_____ 1. Add new updates to a system or equipment.

_____ 2. Fill out and return manufacturer's warranty card.

_____ 3. Check the price of replacement equipment.

_____ 4. Read instruction manual.

_____ 5. Check that the incoming voltage level is correct.

_____ 6. Clean equipment components and change filters.

_____ 7. Look for signs of damage caused by misuse or accidents.

_____ 8. Determine if operators need training on the equipment.

_____ 9. Offer a service contract on the equipment.

_____ 10. Check that the equipment is properly grounded and safety features are working.

_____ 11. Check wire insulation and electrical connections for signs of wear.

_____ 12. File equipment manuals so that they are easily referenced.

ACTIVITY 1-5—Troubleshooter Responsibilities

_____ 1. Four checks that should be taken before starting new equipment include checking fluid levels, belt tightness, loose parts, and ___.

_____ 2. Four general emergencies that a trained maintenance person should be prepared to handle include fires, floods, accidents, and ___.

_____ **3.** Two electrical trade-related emergencies that a trained maintenance person should be prepared to handle are electrical shock and ___.

_____ **4.** In addition to fire hazards, two hazards that can cause an operational emergency are safety problems and ___.

_____ **5.** Six steps that a troubleshooter should take when communicating with an upset customer include relaxing, agreeing, finding the problem, making a decision, coming to an agreement, and ___.

ACTIVITY 1-6—Electrical Shock

1. Complete the table and plot the current for each resistance value on the graph. Mark the following points on the graph: A = Sensation of shock, B = Painful shock, able to remove oneself from contact point, C = Painful shock, may be locked to contact point, D = Fatal electrical shock.

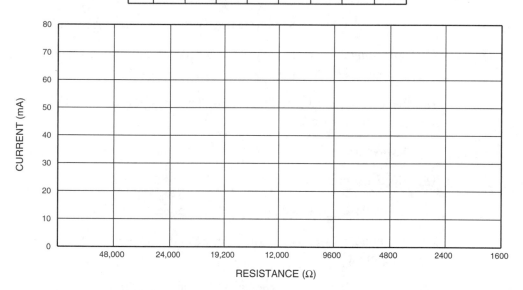

$$I = \frac{E}{R} \times 1000$$

I = CURRENT (IN mA)
E = VOLTAGE (IN VOLTS)
R = RESISTANCE (IN OHMS)

R	48,000	24,000	19,200	12,000	9600	4800	2400	1600
E	120	120	120	120	120	120	120	120
I								

ACTIVITY 1-7—Electrical Safety

T　F **1.** Flammable vapors or dust in the atmosphere can cause an explosion if there is an electrical spark.

T　F **2.** Only power tools that are double-insulated or include a third-wire power cord grounding conductor should be used.

T　F **3.** The grounding post on a plug can be removed when a three-prong plug does not fit into a socket.

T   F   **4.** A wood, metal, or plastic ladder can be used when working on an electrical circuit.

T   F   **5.** When working on a live electrical circuit, only one hand should be used so that the change of electric current will be reduced from flowing through the body.

T   F   **6.** When troubleshooting a problem, it is safe to bypass or disable a fuse or circuit breaker.

T   F   **7.** Always notify a supervisor and file an accident report after an electrical shock is received.

T   F   **8.** Using bare hands to free a shock victim may result in one's receiving a severe shock.

T   F   **9.** Ground any conductive component that does not have to be energized.

T   F   **10.** Before working on any electrical equipment, the circuit should be tested to make sure power is OFF, even if the disconnect switch is turned OFF and locked out.

T   F   **11.** It is safe to work on an electrical circuit when tired or drowsy.

T   F   **12.** Proper grounding helps prevent electrical shock.

T   F   **13.** To ground a system, all exposed non-current-carrying metal parts are connected to the earth ground.

T   F   **14.** Motors must be protected from overload and overcurrent conditions.

T   F   **15.** Grounding conductors carry only fault current and may be smaller and have a higher resistance than normal current-carrying conductors.

_____ **16.** Electricity can take multiple paths, but the largest current flows through the path of ___ resistance.

_____ **17.** Always use a Class ___ fire extinguisher on an electrical fire.

_____ **18.** Electrical fires give off ___.

_____ **19.** Always use ___ listed appliances, components, and equipment.

_____ **20.** Disconnect switches should be turned OFF, locked, and ___ before working on equipment.

_____ **21.** A(n) ___ protection device is used to protect a motor when starting.

_____ **22.** A(n) ___ protection device is used to protect a motor when running.

_____ **23.** A grounding system is designed to carry all ___ current to ground.

_____ **24.** A fuse or circuit breaker ___ when an overcurrent condition occurs in a circuit.

_____ **25.** ___ emergencies are caused by hazardous materials, fire hazards, and safety problems.

Basic Electrical Theory

APPLICATION 2-1 — Electrical Prefixes

Prefixes

Prefixes are used to avoid long expressions of units that are smaller and larger than the base unit. **See Common Prefixes.** For example, sentences 1 and 2 do not use prefixes. Sentences 3 and 4 use prefixes.

1. A solid-state device draws 0.000001 amperes (A).
2. A generator produces 100,000 watts (W).
3. A solid-state device draws 1 microampere (µA).
4. A generator produces 100 kilowatts (kW).

Converting Units

To convert between different units, the decimal point is moved to the left or right, depending on the unit. **See Conversion Table.** For example, an electronic circuit has a current flow of .000001 A. The current value is converted to simplest terms by moving the decimal point six places to the right to obtain 1.0 µA (from Conversion Table).

$$.000001.\ A\ =\ 1.0\,\mu A$$

MOVE DECIMAL POINT
6 PLACES TO RIGHT

Common Electrical Quantities

Abbreviations are used to simplify the expression of common electrical quantities. **See Common Electrical Quantities.** For example, milliwatt is abbreviated mW, kilovolt is abbreviated kV, and ampere is abbreviated A.

COMMON PREFIXES

Symbol	Prefix	Equivalent
G	giga	1,000,000,000
M	mega	1,000,000
k	kilo	1000
base unit	—	1
m	milli	.001
µ	micro	.000001
n	nano	.000000001

CONVERSION TABLE

Initial Units	Final Units						
	giga	mega	kilo	base unit	milli	micro	nano
giga		3R	6R	9R	12R	15R	18R
mega	3L		3R	6R	9R	12R	15R
kilo	6L	3L		3R	6R	9R	12R
base unit	9L	6L	3L		3R	6R	9R
milli	12L	9L	6L	3L		3R	6R
micro	15L	12L	9L	6L	3L		3R
nano	18L	15L	12L	9L	6L	3L	

R = move the decimal point to the right.
L = move the decimal point to the left.

COMMON ELECTRICAL QUANTITIES

Variable	Name	Unit of Measure and Abbreviation
E	voltage	volt — V
I	current	ampere — A
R	resistance	ohm — Ω
P	power	watt — W
PA	power (apparent)	volt-amp — VA
C	capacitance	farad — F
L	inductance	henry — H
Z	impedance	ohm — Ω
G	conductance	siemens — S
f	frequency	hertz — Hz
T	period	second — s

 Refer to Activity 2-1 — Electrical Prefixes ....

APPLICATION 2-2—Using Ohm's Law

Ohm's Law

Ohm's law is the relationship between the voltage, current, and resistance in an electrical circuit. Ohm's law states that current in a circuit is proportional to the voltage and inversely proportional to the resistance. Any value in these relationships is found using Ohm's law, which is written $I = \dfrac{E}{R}$, $E = R \times I$, and $R = \dfrac{E}{I}$. A commonly used variation of Ohm's law is $I = \dfrac{V}{R}$. **See Ohm's Law and Power Formula.**

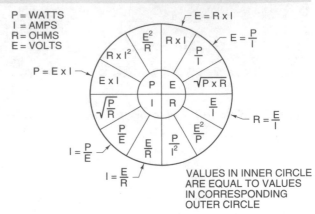

P = WATTS
I = AMPS
R = OHMS
E = VOLTS

VALUES IN INNER CIRCLE ARE EQUAL TO VALUES IN CORRESPONDING OUTER CIRCLE

OHM'S LAW AND POWER FORMULA

Current Calculation

Current (I) is the amount of electrons flowing through an electrical circuit. Current is measured in amperes (A). Current may be direct or alternating. *Direct current (DC)* is current that flows in one direction. *Alternating current (AC)* is current that reverses its direction of flow at regular intervals. To calculate current, apply the formula:

$I = \dfrac{E}{R}$
where
I = current (in amps)
E = voltage (in volts)
R = resistance (in ohms)

also used as $I = \dfrac{V}{R}$
where
I = current (in amps)
V = voltage (in volts)
R = resistance (in ohms)

Example: Calculating Current

An electrical circuit has a resistance of 80 Ω and is connected to a 120 V supply. Calculate the amount of current in the circuit. **See Current Calculation.**

$I = \dfrac{E}{R}$

$I = \dfrac{120}{80}$

$I = \mathbf{1.5\ A}$

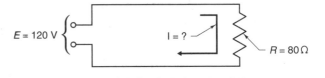

E = 120 V   I = ? R = 80 Ω

CURRENT CALCULATION

Voltage Calculation

Voltage (E) is the amount of electrical pressure in a circuit. Voltage is measured in volts (V). Voltage is produced by the conversion of chemical energy (battery), light (photocell), electromagnetic energy (generator), heat (thermocouple), pressure (piezoelectricity), or friction (static electricity) to electrical energy. Voltage is produced by generating an excess of electrons at one terminal of a voltage source and a deficiency of electrons at the other terminal. The greater the difference in electrons between the terminals, the higher the voltage. To calculate voltage, apply the formula:

$E = I \times R$
where
E = voltage (in volts)

I = current (in amps)
R = resistance (in ohms)

Example: Calculating Voltage

An electrical circuit has 5 A of current and a resistance of 46 W. Calculate the voltage in the circuit.
See Voltage Calculation.

$$E = I \times R$$
$$E = 5 \times 46$$
$$E = \textbf{230 V}$$

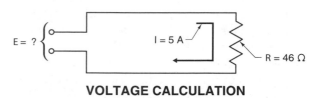

VOLTAGE CALCULATION

Resistance Calculation

Resistance (R) is opposition to the flow of electrons. Resistance is measured in ohms (Ω). A *conductor* is any material through which current flows easily. Examples of conductors include copper, aluminum, brass, gold, and other metals. An *insulator* is a material through which current cannot flow easily. Examples of insulators include rubber, glass, plastic, paper, varnish, and dry wood. To calculate resistance, apply the formula:

$$R = \frac{E}{I}$$

where
R = resistance (in ohms)
E = voltage (in volts)
I = current (in amps)

Example: Calculating Resistance

An electrical circuit draws 8 A when connected to a 460 V supply. Calculate the resistance of the circuit.
See Resistance Calculation.

$$R = \frac{E}{I}$$

$$R = \frac{460}{8}$$

$$R = \textbf{57.5 } \Omega$$

RESISTANCE CALCULATION

Refer to **Activity 2-2—Using Ohm's Law**...

APPLICATION 2-3—Using the Power Formula

The power formula is the relationship between the voltage, current, and power in an electrical circuit. The power formula states that the power in a circuit is equal to the voltage times the current. Any value in these relationships is found using the power formula, which is written $P = E \times I$, $E = \frac{P}{I}$, and $I = \frac{P}{E}$.

Power Calculation

Power (P) is the rate of doing work or using energy. Power may be expressed as true or apparent. *True power* is the actual power used in an electrical circuit. True power is measured using a wattmeter and is expressed in watts (W). *Apparent power* is the product of voltage and current in a circuit calculated without considering the phase shift that may be present between total voltage and current in the circuit. Apparent power is measured in volt-amperes (VA). A phase shift exists in most AC circuits that contain devices causing capacitance or inductance.

Capacitance is the property of an electric device that permits the storage of electrically separated charges when potential differences exist between the conductors. *Inductance* is the property of a circuit that causes it to oppose a change in current due to energy stored in a magnetic field. All coils (motor windings, transformers, solenoids, etc.) create inductance in an electrical circuit. True power equals apparent power in an electrical circuit containing only resistance. True power is less than apparent power in a circuit containing inductance or capacitance. To calculate apparent power, apply the formula:

$$P_A = E \times I$$
where
P_A = apparent power (in volt-amps)
E = measured voltage (in volts)
I = measured current (in amps)

To calculate true power, apply the formula:

$$P_T = I^2 \times R$$
where
P_T = true power (in watts)
I = total circuit current (in amps)
R = total resistive component of the circuit (in ohms)

Power factor (PF) is the ratio of true power used in an AC circuit to apparent power delivered to the circuit. Power factor is expressed as a percentage. True power equals apparent power when the power factor is 100%. When the power factor is less than 100%, the circuit is less efficient and has higher operating costs. To calculate power factor, apply the formula:

$$PF = \frac{P_T}{P_A}$$
where
PF = power factor (percentage)
P_T = true power (in watts)
P_A = apparent power (in volt-amps)

Example: Calculating Apparent Power

An electrical circuit has a measured current of 25 A and a measured voltage of 440 V. Calculate the apparent power of the circuit. **See Apparent Power Calculation.**

$$P_A = E \times I$$
$$P_A = 440 \times 25$$
$$P_A = \textbf{11,000 VA}$$

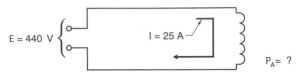

APPARENT POWER CALCULATION

Example: Calculating True Power

An electrical circuit has a current of 25 A and a total resistive component of 21 Ω. Calculate the true power of the circuit. **See True Power Calculation.**

$$P_T = I^2 \times R$$
$$P_T = (25)^2 \times 21$$
$$P_T = 625 \times 21$$
$$P_T = \textbf{13,125 W}$$

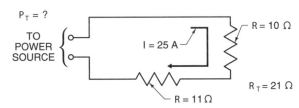

TRUE POWER CALCULATION

Calculating Power Factor

The apparent power of an electrical circuit is 13,125 VA and the true power is 11,000 W. Calculate the power factor of the circuit.

$$PF = \frac{P_T}{P_A}$$
$$PF = \frac{11,000}{13,125} \times 100$$
$$PF = .838 \times 100$$
$$PF = \textbf{83.80\%}$$

*Refer to **Activity 2-3—Using the Power Formula** ....*

APPLICATION 2-4—Temperature Conversion

Temperature is a measurement of the intensity of heat. *Ambient temperature* is the temperature of the air surrounding a device. Current flowing in a conductor produces heat. The heat produced in an electrical circuit may be by design, such as from a heating element, or unintentional, as from a bad conductor splice. Unintentional heat deteriorates insulation and lubrication, and can destroy electric devices.

Temperature rise is an increase in temperature above ambient temperature. All electric and electronic devices produce heat and are designed to function correctly within a given temperature rise. **See Temperature Measurement.** Temperature is usually measured in degrees Fahrenheit (°F) or degrees Celsius (°C). **See Fahrenheit and Celsius Scales.**

Converting Fahrenheit to Celsius

To convert a Fahrenheit temperature reading to Celsius, subtract 32 from the Fahrenheit reading and divide by 1.8. To convert Fahrenheit to Celsius, apply the formula:

$$°C = \frac{°F - 32}{1.8}$$

where
°C = degrees Celsius
°F = degrees Fahrenheit
32 = difference between bases
1.8 = ratio between bases

Example: Converting Fahrenheit to Celsius

A digital thermometer indicates a reading of 185°F. Convert the Fahrenheit temperature to Celsius.

$$°C = \frac{°F - 32}{1.8}$$

$$°C = \frac{185 - 32}{1.8}$$

$$°C = \frac{153}{1.8}$$

$$°C = \mathbf{85°C}$$

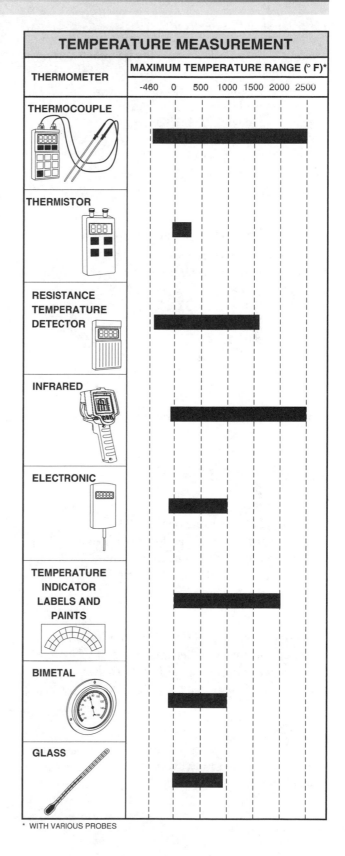

DIGITAL DISPLAY

PROBE

TEMPERATURE MEASUREMENT

THERMOMETER	MAXIMUM TEMPERATURE RANGE (° F)*
	-460 0 500 1000 1500 2000 2500
THERMOCOUPLE	
THERMISTOR	
RESISTANCE TEMPERATURE DETECTOR	
INFRARED	
ELECTRONIC	
TEMPERATURE INDICATOR LABELS AND PAINTS	
BIMETAL	
GLASS	

* WITH VARIOUS PROBES

Converting Celsius to Fahrenheit

To convert a Celsius temperature reading to Fahrenheit, multiply 1.8 by the Celsius reading and add 32. To convert Celsius to Fahrenheit, apply the formula:

$$°F = (1.8 × °C) + 32$$

where
$°F$ = degrees Fahrenheit
1.8 = ratio between bases
$°C$ = degrees Celsius
32 = difference between bases

Example: Converting Celsius to Fahrenheit

A temperature monitor indicates 55°C. Convert the Celsius temperature to Fahrenheit.

$$°F = (1.8 × °C) + 32$$
$$°F = (1.8 × 55) + 32$$
$$°F = 99 + 32$$
$$°F = \mathbf{131° F}$$

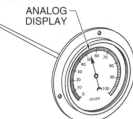
ANALOG DISPLAY

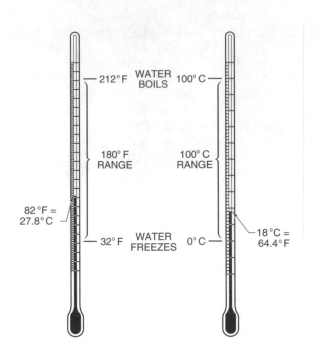

FAHRENHEIT SCALE   CELSIUS SCALE

 Refer to **Activity 2-4—Temperature Conversion**..

APPLICATION 2-5—AC Sine Wave

An *AC sine wave* is a symmetrical waveform that contains 360 electrical degrees. The wave reaches its peak positive value at 90°, returns to 0 at 180°, increases to its peak negative value a 270°, and returns to 0 at 360°. **See AC Sine Wave.**

Cycle

A *cycle* is one complete wave of alternating voltage or current. An *alternation* is half of a cycle. A sine wave has one positive alternation and one negative alternation per cycle.

Period

Period (T) is the time required to produce one complete cycle of a waveform. Period is usually measured in seconds.

1 CYCLE   1 POSITIVE ALTERNATION   1 NEGATIVE ALTERNATION

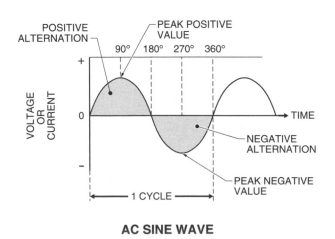

AC SINE WAVE

Frequency

Frequency (f) is the number of cycles per second (cps) in an AC sine wave. Frequency is measured in hertz (Hz). One hertz equals one cycle per second. **See Frequency.** Standard power is distributed at a frequency of 60 Hz. An audio amplifier amplifies all frequencies between 20 Hz and 20 kHz. To calculate frequency, apply the formula:

$$f = \frac{1}{T}$$

where
f = frequency (in hertz)

1 = constant
T = period (in seconds)

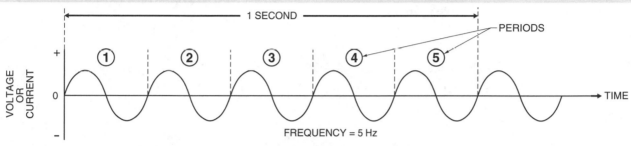

FREQUENCY

Example: Calculating Frequency

A waveform requires .02 seconds to complete one cycle. Calculate the frequency of the waveform. **See Frequency Calculation.**

$$f = \frac{1}{T}$$

$$f = \frac{1}{.02}$$

$$f = \textbf{50 Hz}$$

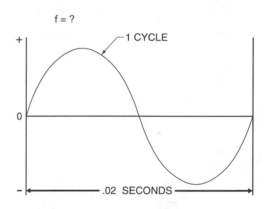

FREQUENCY CALCULATION

Period is calculated by applying the formula:

$$T = \frac{1}{f}$$

where
T = period (in seconds)
1 = constant
f = frequency (in hertz)

Example: Calculating Period

The frequency of an electrical signal is the standard 60 Hz power distribution frequency. Calculate the period required to complete one cycle. **See Period Calculation.**

$$T = \frac{1}{f}$$

$$T = \frac{1}{60}$$

$$T = \textbf{.0167 sec}$$

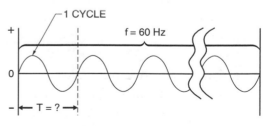

PERIOD CALCULATION

Amplitude

The *amplitude* of a sine wave is the height of the waveform. Voltmeters may specify a sine wave as peak, average, or root-mean-square.

Peak Value

The *peak value* (V_{max}) of a sine wave is the maximum value of either the positive or negative alternation. The positive and negative alternations are equal in a sine wave. **See Sine Wave Peak Values.**

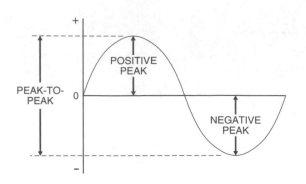

SINE WAVE PEAK VALUES

Peak-to-Peak

The *peak-to-peak value* ($V_{p\text{-}p}$) of a sine wave is the value measured from the maximum positive alternation to the maximum negative alternation. Peak-to-peak value is twice the peak value. To calculate peak-to-peak value, apply the formula:

$$V_{p\text{-}p} = 2 \times V_{max}$$

where
2 = constant

$V_{p\text{-}p}$ = peak-to-peak value
V_{max} = peak value

Example: Calculating Peak-to-Peak Voltage

A voltage source has a positive peak alternation of 165 V. Calculate the peak-to-peak voltage. **See Peak-to-Peak Calculation.**

$$V_{p\text{-}p} = 2 \times V_{max}$$
$$V_{p\text{-}p} = 2 \times 165$$
$$V_{p\text{-}p} = \textbf{330 V}$$

PEAK-TO-PEAK CALCULATION

Average Value

The *average value* (V_{avg}) of a sine wave is the mathematical mean of all instantaneous voltage values in the sine wave. Average value is equal to .637 of the peak value of a sine wave. **See Sine Wave Average Values.** To calculate average value, apply the formula

$$V_{avg} = V_{max} \times .637$$

where
V_{avg} = average value (in volts)
V_{max} = peak value (in volts)
.637 = constant

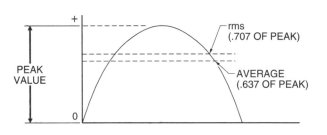

SINE WAVE AVERAGE VALUES

Example: Calculating Average Value

A voltage source has a positive peak alternation of 165 V. Calculate the average value. **See Average Calculation.**

$$V_{avg} = V_{max} \times .673$$
$$V_{avg} = 165 \times .673$$
$$V_{avg} = \textbf{111.045 V}$$

AVERAGE CALCULATION

Root-Mean-Square (Effective) Value

The *root-mean-square* (V_{rms}) *value*, or effective value, of a sine wave is the value that produces the same amount of heat in a pure resistive circuit as a DC of the same value. The rms value is obtained by squaring the instantaneous voltages along a waveform, averaging the squared values, and taking the square root of that number. The rms value is indicated by most AC voltmeters and ammeters. The rms value is equal to .707 of the peak in a sine wave. To calculate rms value, apply the formula:

$$V_{rms} = V_{max} \times .707$$
where
V_{rms} = rms value (in volts)

V_{max} = peak value (in volts)
.707 = constant

Example: Calculating Root-Mean-Square Value

A voltage source has a positive peak alternation of 165 V. Calculate the rms value. **See rms Calculation.**

$$V_{rms} = V_{max} \times .707$$
$$V_{rms} = 165 \times .707$$
$$V_{rms} = \textbf{116.655 V}$$

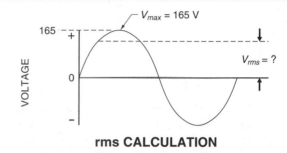

rms CALCULATION

Voltage conversion tables are available to allow easy conversion from one value to another. Any value can be converted to another value by applying a conversion factor. For example, to convert an rms to a peak value, multiply the rms value by 1.414. **See Voltage Conversions.**

VOLTAGE CONVERSIONS

To Convert	To	Multiply By
rms	Average	.9
rms	Peak	1.414
Average	rms	1.111
Average	Peak	1.567
Peak	rms	.707
Peak	Average	.637
Peak	Peak-to-peak	2

Phase Shift

Phase shift is the state when voltage and current in a circuit do not reach their maximum amplitude and zero level simultaneously. Phase shift occurs in inductive or capacitive circuits.

Resistive Circuits

Alternating voltage and current are in-phase in resistive circuits. *In-phase* is the state when voltage and current reach their maximum amplitude and zero level simultaneously. A *resistive circuit* is a circuit that contains only resistive components, such as heating elements and incandescent lamps. **See In-Phase AC Sine Wave.**

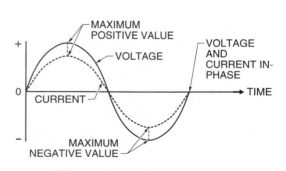

IN-PHASE AC SINE WAVE

True power is the actual power in a circuit. Apparent power is the product of voltage (in volts) and current (in amps) in a circuit. Apparent power equals true power when voltage and current are in-phase. Apparent power equals true power in a pure resistive circuit. True power is less than apparent power in circuits containing inductance or capacitance.

Inductive Circuits

A phase shift exists between alternating voltage and current in an inductive circuit. An *inductive circuit* is a circuit in which current lags voltage. The greater the inductance in a circuit, the larger the phase shift. True power in a circuit decreases as phase shift increases. Inductance reduces true power in a circuit. **See Inductive Circuit AC Sine Wave.**

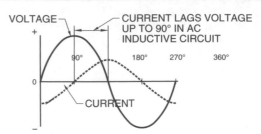

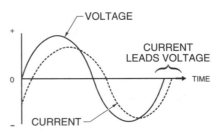

INDUCTIVE CIRCUIT AC SINE WAVE

Capacitive Circuits

A phase shift exists between voltage and current in a capacitive circuit. A *capacitive circuit* is a circuit in which current leads voltage. The greater the capacitance in a circuit, the larger the phase shift. Capacitance (like inductance) reduces the true power in a circuit. Capacitance is added to a circuit to increase true power when a circuit includes resistive and inductive loads. **See Capacitive Circuit AC Sine Wave.**

CAPACITIVE CIRCUIT AC SINE WAVE

 Refer to **Activity 2-5—AC Sine Wave** ....

APPLICATION 2-6—Drawings and Diagrams

Drawings and diagrams are used in the electrical/electronic field to convey facts, ideas, information and directions, and to provide information on the operation of devices and circuits. Standard practices for drawing electrical and electronic diagrams are detailed in ANSI Standard Y14.15, Reaffirmed 1988, *Electrical and Electronic Diagrams*. This standard gives definitions, general information, and recommended drafting practices. **See Drawing Standards.**

General rules applicable to all diagrams include:
- Combined forms of diagrams may be used when the net result is helpful to the user.
- Line thickness and lettering shall conform with ANSI 14.2. Uppercase letters are generally used. Lowercase letters may be used to avoid ambiguity.
- Graphic symbols shall conform with ANSI Y32.2. Special symbols shall be explained by a note on the diagram. The terminal symbol (o) may be added as required to any basic symbol.
- Switch symbols and relay contacts shall be shown in the OFF position.
- Abbreviations shall conform with ANSI Z32.13.
- Diagrams should be drawn so that the main features are prominent. Crowding of parts should be avoided.
- Grouping of parts may be shown by a phantom line enclosing the parts.
- Wire colors may be indicated by abbreviations or numerical codes.

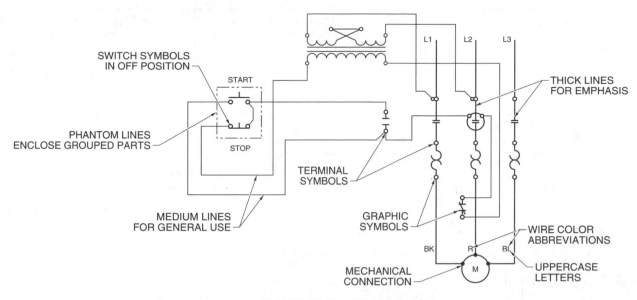

DRAWING STANDARDS

WIRE COLORS		
Color	**Abbreviation**	**Numerical Code**
Black	BK	0
Brown	BR	1
Red	R	2
Orange	O	3
Yellow	Y	4
Green	G	5
Blue	BL	6
Violet	V	7
Gray	GY	8
White	W	9

DIAGRAM LINE CONVENTIONS		
Line Application	**Thickness**	**Line**
General use	Medium	———————
Mechanical connection	Medium	- - - - - - -
Bracket-connecting dash line	Medium	— — — — —
Brackets, leaders, etc.*	Thin	———————
Mechanical-grouping boundary line*	Thin	—·—·—·—
For emphasis*	Thick	━━━━━━

* use of line thicknesses is optional

   Standard types of electrical diagrams include line (one-line), schematic (elementary), wiring (connection), interconnection, and terminal diagrams. Other drawings and diagrams include pictorial, sectional view, charts and graphs, application, operational, and block.

Line (One-Line) Diagrams

A *line (one-line) diagram* is a diagram that shows, with single lines and graphic symbols, the logic of an electrical circuit or system of circuits and components. Line diagrams show the connection of major equipment, protective relays, meters, and instruments. Line diagrams show the relationship between circuits and their components but do not show the actual location of the components. Line diagrams provide fast, easy understanding of the connections and use of each component. **See Line Diagram.**

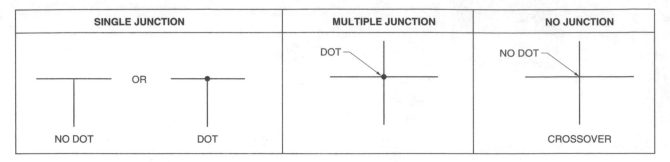

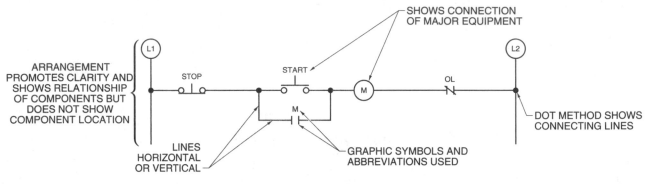

LINE DIAGRAM

The arrangement of a line diagram should promote clarity. Graphic symbols, abbreviations, and device designations are drawn per standards. Connecting lines may be shown by the dot or no dot method. The main circuit is shown in the most direct path and logical sequence. Lines between symbols should be horizontal or vertical. Line crossing should be minimized.

Schematic (Elementary) Diagrams

A *schematic (elementary) diagram* is a diagram that shows with graphic symbols the electrical connections and functions of a specific circuit arrangement. A *function* is a mathematical, graphical, or tabular statement of the influence one device or component has on another device or component.

A schematic diagram does not show the physical relationship of the components in a circuit. A schematic diagram is used to trace a circuit and its functions without regard to the actual size, shape, or location of the component device or parts. **See Schematic Diagram.**

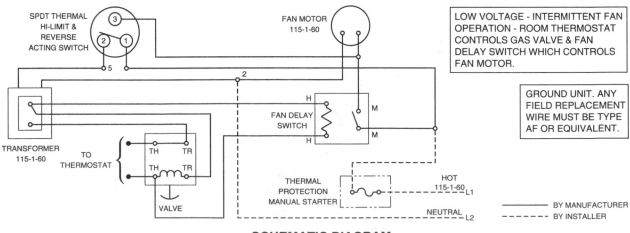

SCHEMATIC DIAGRAM

Wiring (Connection) Diagrams

A *wiring (connection) diagram* is a diagram that shows the connection of an installation or its component devices or parts. A wiring diagram shows, as closely as possible, the actual location of each component in a circuit, including the control circuit and the power circuit. Internal and external connections may be shown in sufficient detail to make or trace connections. **See Wiring Diagram.**

Wiring diagrams are used in troubleshooting because they show the actual component layout and connections. Wiring diagrams are limited in quickly showing circuit logic because the conductor connections are often hard to follow. When a wiring diagram is used with its line diagram, a clear understanding of the circuit wiring and operation is given.

Symbols used with wiring diagrams may be the symbols used with schematic diagrams or may be simple rectangles and circles. No attempt is made to show exact sizes of component devices or parts. Terminals should be shown as circles even when their physical appearances differ. Three types of wiring diagrams include the continuous-line type, interrupted-line type, and tabular type.

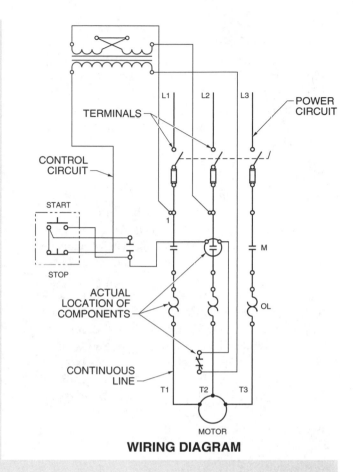

WIRING DIAGRAM

Interconnection Diagrams

An *interconnection diagram* is a diagram that shows the external connection between unit assemblies or equipment. Interconnection diagrams may be either the wiring type or cable type. The wiring type shows each wire. The cable type only shows cables. Combinations of the two types may also be used. Connection lines representing wires or cables shall be drawn horizontally or vertically and shall run parallel in a direct and logical path between terminals. For wiring, interconnection diagrams shall include wire color information, wire size, and wire type. For cables, interconnection diagrams shall include the part number, reference or designation, and cable type designation. **See Interconnection Diagram.**

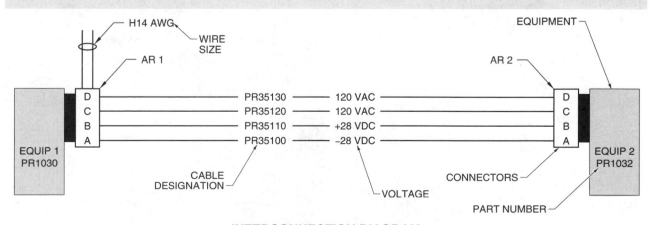

INTERCONNECTION DIAGRAM

Terminal Diagrams

A *terminal diagram* is a diagram that relates the internal circuit of a piece of equipment to its terminal physical configuration. Terminal diagrams may be orthographic or pictorial. Reference letters and/or numbers are used to associate internal circuits with external connectors. **See Terminal Diagram.**

Pictorial Drawings

A *pictorial drawing* is a drawing that shows the length, height, and depth of an object in one view. Pictorial drawings are used as working drawings to show the general location, function, and appearance of parts and/or assemblies. Pictorial drawings may be isometric, oblique, or perspective.

An *isometric drawing* is a pictorial drawing with a 120° included axis. Isometric drawings are used in promotional literature showing and describing products for sale. Isometrics are also used to show steps of assembly. Exploded isometrics are often used in service bulletins to show the relationship of individual components. The components are keyed to a parts list that provides statistical information. **See Isometric Drawing.**

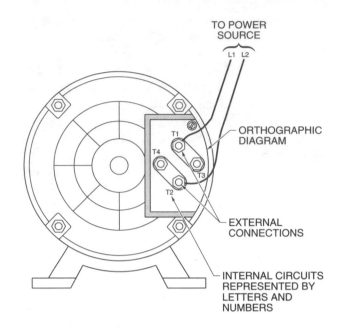

TERMINAL DIAGRAM

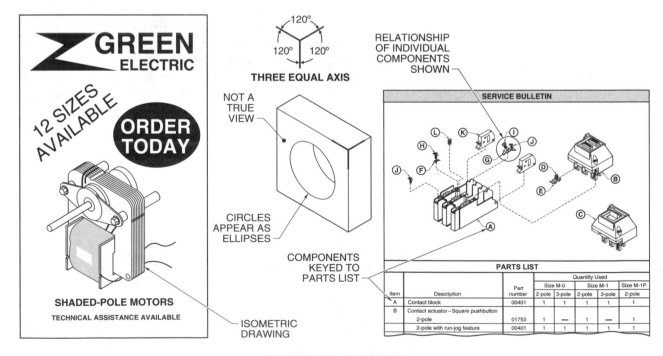

ISOMETRIC DRAWING

Sectional View Drawings

A *sectional view drawing* is a drawing that shows the internal features of an object. An imaginary cutting plane is passed through the object perpendicular to the line of sight. The portion of the object between the cutting plane and the observer is removed, revealing the internal features. Sectional view drawings may be pictorial or orthographic.

Section lines are drawn on all surfaces cut by the cutting plane. General purpose section lines are commonly used. Whenever adjacent parts are shown in sectional views, the direction of the lines is changed. Specific section lines representing various materials, such as metals, soil, insulation, etc., may also be used. **See Sectional Views.**

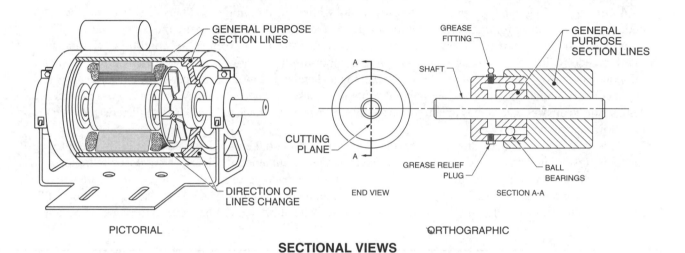

SECTIONAL VIEWS

SELECTED SYMBOLS FOR SECTION LINING

GENERAL PURPOSE — CORRECT, INCORRECT, 45° 2.5 mm (.10), IRREGULAR SPACING, LINES TOO CLOSE, LINE WIDTHS VARY, LINES TOO THICK, LINES TOO SHORT OR TOO LONG

METAL — STEEL; BRONZE, BRASS, AND COPPER; ALUMINUM; WHITE METAL, ZINC, LEAD, BABBITT, AND ALLOYS; MAGNESIUM, ALUMINUM, AND ALUMINUM ALLOYS

WOOD — FINISH, ROUGH, BLOCKING

MASONRY — COMMON/FACE, CONCRETE BLOCK

EARTH/CONCRETE — EARTH/COMPACT FILL, CAST-IN-PLACE/PRECAST, LIGHTWEIGHT

ELECTRICAL — RUBBER, PLASTIC, AND ELECTRICAL; ELECTRIC WINDINGS, ELECTROMAGNETS, RESISTANCES, ETC.

Charts and Graphs

Charts and graphs are synonymous terms for graphical (pictorial) representations of data. Charts and graphs are used to show relationships between two or more variables. Charts and graphs include tabular, line, bar, flow, pie, and alignment. **See Charts and Graphs.**

A *tabular chart* is a chart that presents data in columnar form. Tabular charts are the least graphical of the various charts and graphs. Pictorial representations are often referenced to show the tabulated data.

A *line graph* is a graph that shows two related variables plotted on coordinate paper. The plotted data is joined by a continuous line. On a line graph, one known variable is plotted horizontally and another is plotted vertically. The relationship between the two variables is represented by a straight or curved line. The point at which either variable line intersects the straight or curved line represents the value of the unknown variable.

A *bar graph* is a graph that compares statistical data with vertical or horizontal bars. The bars may be color-coded or cross-hatched for comparative purposes. If so, a legend should accompany the bar graph.

A *flow chart* is a chart that shows a logical sequence of steps for a given set of conditions. Symbols and interconnecting lines are used to provide direction. Symbols used with flow charts include an ellipse, rectangle, diamond, and arrow. An ellipse indicates the beginning and end of a flow chart or section of a flow chart. A rectangle contains a set of instructions. A diamond contains a question stated so that a yes or no answer is achieved. The yes or no answer determines the direction to follow through the flow chart. An arrow indicates the direction to follow based on the answers.

A *pie chart* is a chart that compares parts of a whole in relation to their total use. A circle is divided into segments representing data on a percentage basis. Pie charts represent amounts clearly, as opposed to trends.

An *alignment chart* is a chart that shows the relationship among three variables. An alignment chart is used to solve three-variable equations. A straight line is drawn joining known values. The line is extended to intersect the scale of the third variable. The point where the line intersects the third scale is the value that solves the equation.

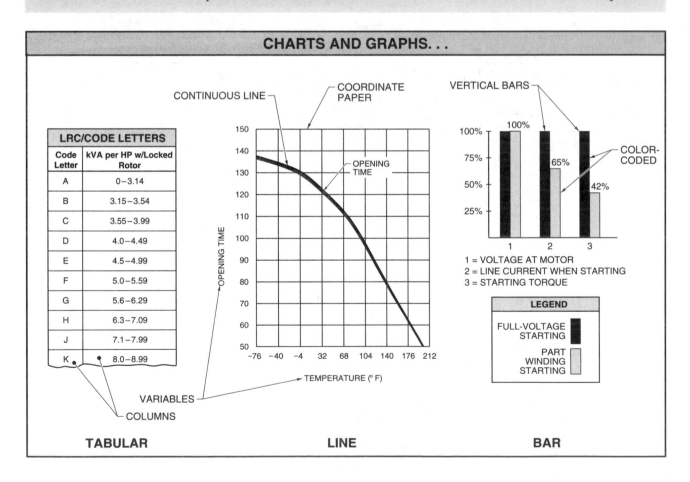

CHARTS AND GRAPHS. . .

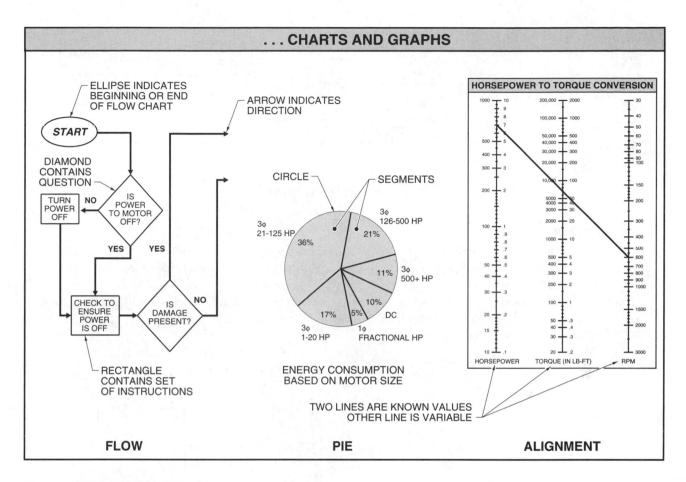

... CHARTS AND GRAPHS

ELLIPSE INDICATES BEGINNING OR END OF FLOW CHART

ARROW INDICATES DIRECTION

START

DIAMOND CONTAINS QUESTION

TURN POWER OFF — NO — IS POWER TO MOTOR OFF?

YES   YES

CHECK TO ENSURE POWER IS OFF — IS DAMAGE PRESENT? — NO

RECTANGLE CONTAINS SET OF INSTRUCTIONS

FLOW

CIRCLE   SEGMENTS

3φ 21-125 HP   36%
3φ 126-500 HP   21%
11%   3φ 500+ HP
10%
3φ 1-20 HP   17%   5%   DC
1φ FRACTIONAL HP

ENERGY CONSUMPTION BASED ON MOTOR SIZE

PIE

HORSEPOWER TO TORQUE CONVERSION

HORSEPOWER   TORQUE (IN LB-FT)   RPM

TWO LINES ARE KNOWN VALUES OTHER LINE IS VARIABLE

ALIGNMENT

Application Drawings

An *application drawing* is a drawing that shows the use of a particular piece of equipment or product in an application. Application drawings show product use and are not intended to show component connection or wiring. Application drawings present ideas on the use of a product in problem solving and are used by manufacturers to promote products. **See Application Drawing.**

Application drawings are used in troubleshooting to present ideas for the use of new or different components to replace older or troublesome components.

Operational Diagrams

An *operational diagram* is a diagram that shows the operation of individual components or circuits. Operational diagrams show the relationship between the input and output of a component or circuit. Operational diagrams provide an easily understood graphic representation of circuit operation. **See Operational Diagram.**

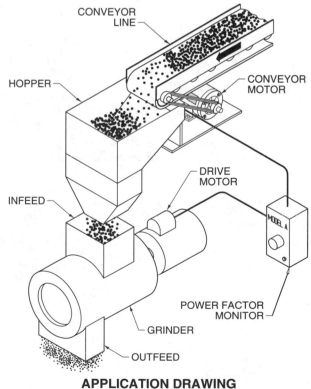

CONVEYOR LINE

HOPPER

CONVEYOR MOTOR

DRIVE MOTOR

INFEED

MODEL A

POWER FACTOR MONITOR

GRINDER

OUTFEED

APPLICATION DRAWING

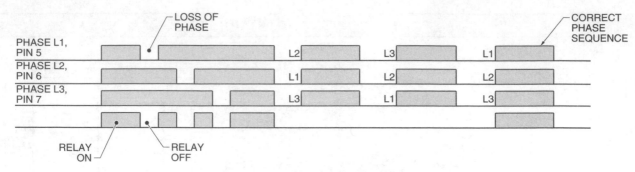

OPERATIONAL DIAGRAM

Block Diagrams

A *block diagram* is a diagram that shows the relationship between individual sections (blocks) of circuits or components. The circuit or component inside each block is not shown. Text describes the function of the block. Complicated circuits are not shown. **See Block Diagram.**

Block diagrams are used in troubleshooting to show the contents provided in a circuit or component under test. Block diagrams are also used to help isolate problems by checking individual blocks that may have failed.

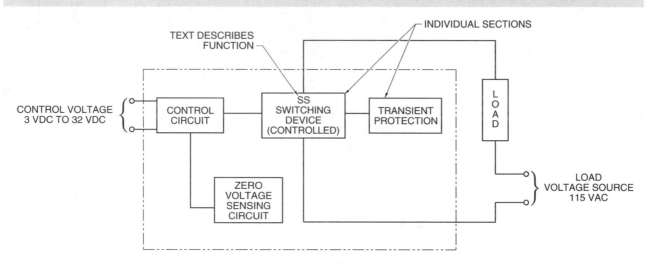

BLOCK DIAGRAM

 Refer to **Activity 2-6—Drawings and Diagrams** ....

Refer to Quick Quiz® on CD-ROM

 Refer to Chapter 2 in the **Troubleshooting Electrical/Electronic Systems Workbook** *for additional questions.*

Basic Electrical Theory

Name_____ **Date**_____

ACTIVITY 2-1—Electrical Prefixes

State the value for each meter reading in the units given.

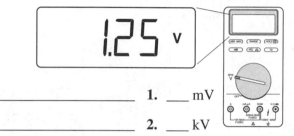

_____ 1. ___ mV

_____ 2. ___ kV

_____ 3. ___ MV

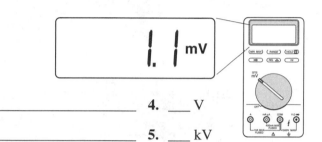

_____ 4. ___ V

_____ 5. ___ kV

_____ 6. ___ GV

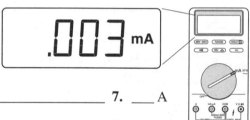

_____ 7. ___ A

_____ 8. ___ μA

_____ 9. ___ kA

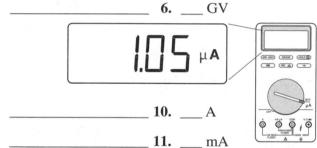

_____ 10. ___ A

_____ 11. ___ mA

_____ 12. ___ kA

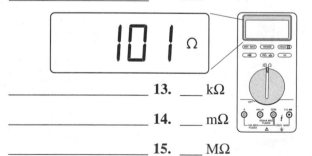

_____ 13. ___ kΩ

_____ 14. ___ mΩ

_____ 15. ___ MΩ

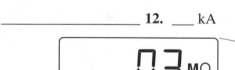

_____ 16. ___ Ω

_____ 17. ___ kΩ

_____ 18. ___ GΩ

_____ 19. ___ mV

_____ 20. ___ kV

_____ 21. ___ μV

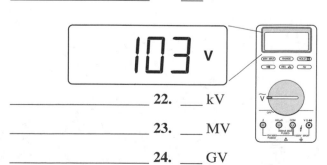

_____ 22. ___ kV

_____ 23. ___ MV

_____ 24. ___ GV

ACTIVITY 2-2—Using Ohm's Law

1. A meter is normally set on the lowest range that produces a reading. Using the resistance and voltage values, determine the lowest meter range that produces a reading in each circuit.

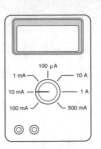

_____ **A.** Range = ___

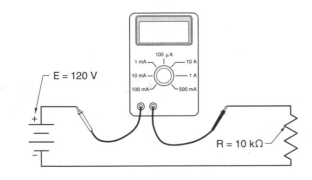

_____ **B.** Range = ___

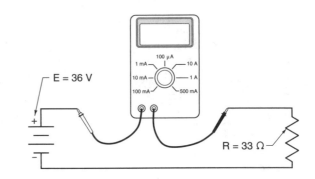

_____ **C.** Range = ___

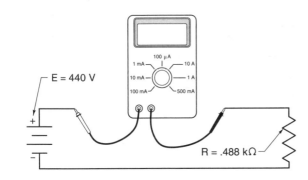

_____ **D.** Range = ___

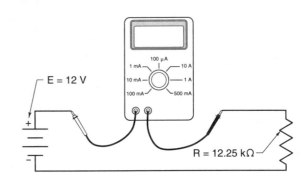

_____ **E.** Range = ___

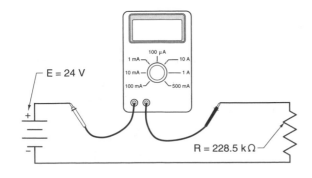

_____ **F.** Range = ___

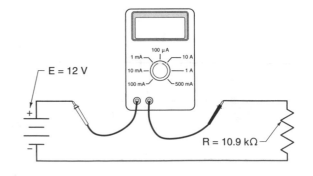

2. Using the resistance values and ammeter readings, determine the voltage.

_____ **A.** ___ V

_____ **B.** ___ kV

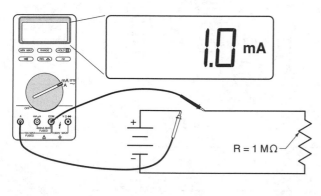

_____ **C.** ___ V

_____ **D.** ___ mV

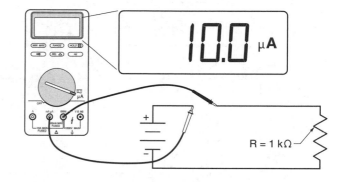

_____ **E.** ___ V

_____ **F.** ___ kV

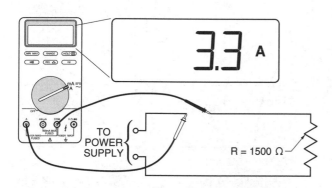

_____ **G.** ___ V

_____ **H.** ___ kV

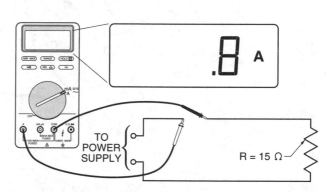

3. Using the ammeter and voltmeter readings, state the resistance values.

_____ **A.** R_1 = ___ Ω

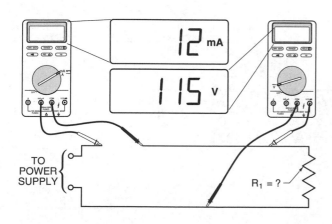

_____ **B.** R_1 = ___ Ω

_____ **C.** R_1 = ___ kΩ

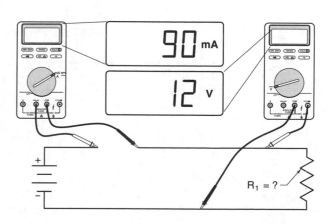

_____ **D.** $R_1 = $ ___ Ω

_____ **E.** $R_2 = $ ___ Ω

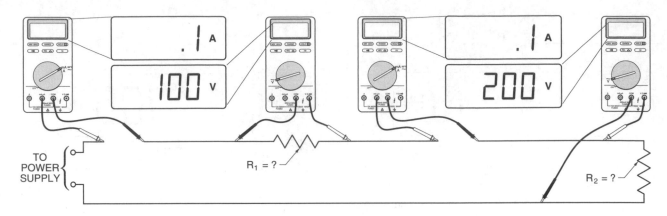

_____ **F.** $R_1 = $ ___ Ω _____ **G.** $R_1 = $ ___ W

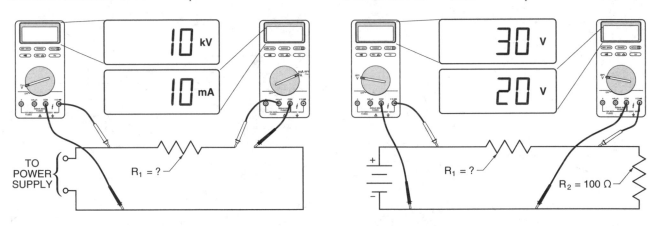

ACTIVITY 2-3—Using the Power Formula

1. Determine the power that each heating unit produces when the control switch is closed.

_____ **A.** $P = $ ___ W

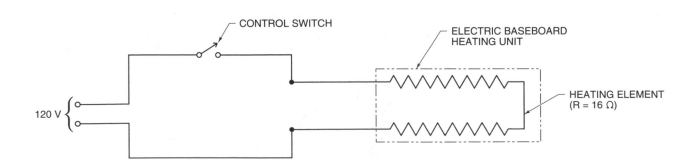

_____ **B.** P = ___ W

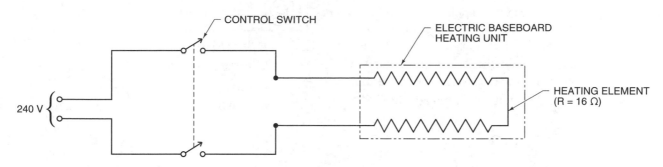

2. Determine the power that the coffeemaker produces when the thermostat is closed (brew) and open (warm). *Note:* When the thermostat is closed, the warm heating element is removed from the circuit.

_____ **A.** P = ___ W

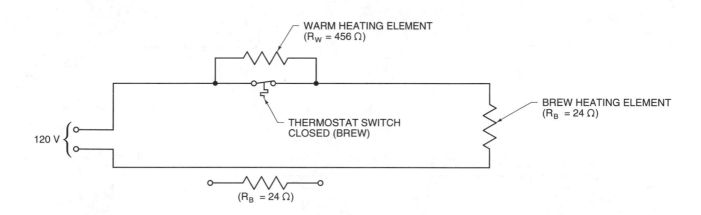

_____ **B.** P = ___ W

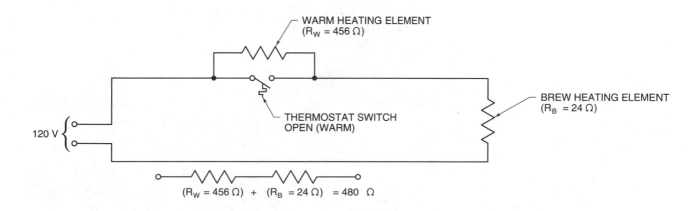

3. A typical electric range includes two coils per heating element. The two coils are connected to a multiposition switch that connects the coils to the supply voltage. The switch applies 120 V or 230 V to each coil or any series or parallel combination of the coils. Ten different wattages are possible using two coils and two voltages. The wattage output of the coils varies from 160 W to 2880 W. Determine the power for each switching arrangement.

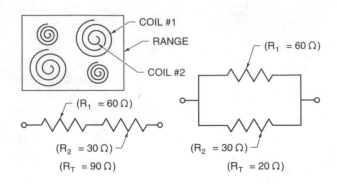

COIL #1
RANGE
COIL #2

$(R_1 = 60\,\Omega)$
$(R_2 = 30\,\Omega)$
$(R_T = 90\,\Omega)$

$(R_1 = 60\,\Omega)$
$(R_2 = 30\,\Omega)$
$(R_T = 20\,\Omega)$

_____ **A.** P = ___ W

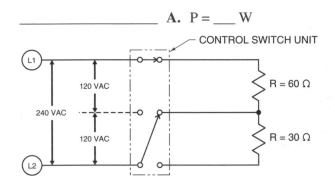

_____ **B.** P = ___ W

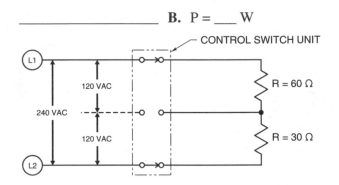

_____ **C.** P = ___ W

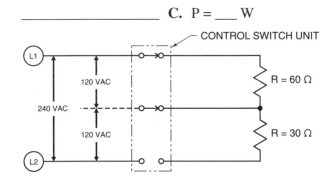

_____ **D.** P = ___ W

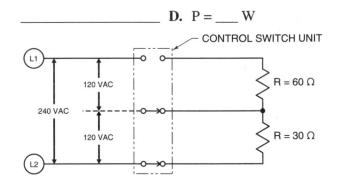

_____ **E.** P = ___ W

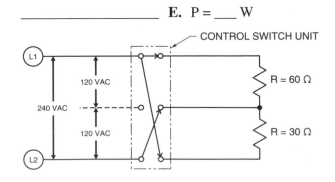

_____ **F.** P = ___ W

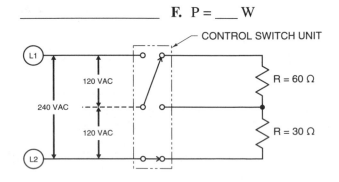

_____ **G.** P = ___ W

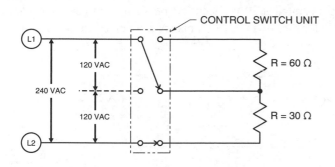

_____ **H.** P = ___ W

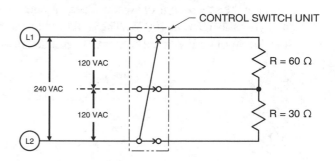

_____ **I.** P = ___ W

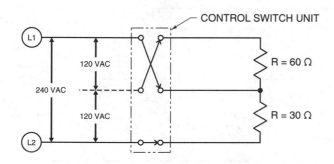

_____ **J.** P = ___ W

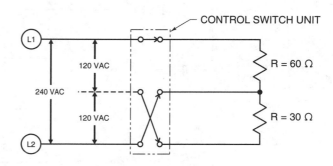

4. Determine the low, medium, and high heating position of the control switch, and also the power in each switch position.

_____ **A.** Position 1 = ___

_____ **B.** Power in circuit when switch is in Position 1 = ___ W

_____ **C.** Position 2 = ___

_____ **D.** Power in circuit when switch is in Position 2 = ___ W

_____ **E.** Position 3 = ___

_____ **F.** Power in circuit when switch is in Position 3 = ___ W

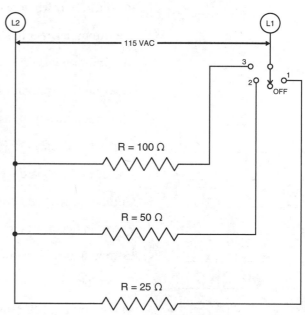

5. An automobile electrical system has a 12 VDC supply. The rear window defrost circuit has a resistance of 8 Ω. The starter motor draws between 20 A and 25 A. The fan motor draws between 2.5 A and 6 A, depending on the position of the speed-control switch. Each dashboard light is rated at 30 W. Each low beam and high beam headlight element is rated at 100 W. The dim/bright switch determines which elements are connected in the circuit. Using Automobile Electrical System, determine the maximum current and power in each circuit.

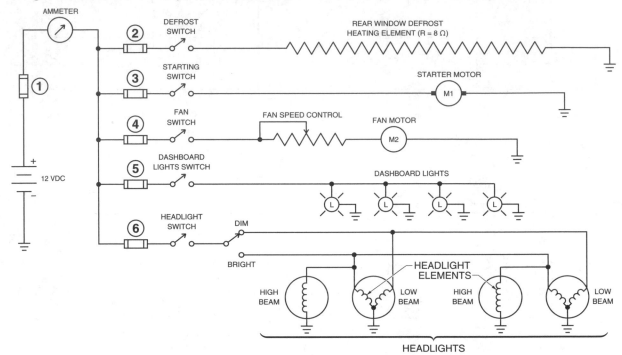

AUTOMOBILE ELECTRICAL SYSTEM

_____ **A.** Current in rear window defrost circuit = ___ A

_____ **B.** Power in rear window defrost circuit = ___ W

_____ **C.** Maximum current in starter motor circuit = ___ A

_____ **D.** Maximum power in starter motor circuit = ___ W

_____ **E.** Maximum current in fan motor circuit = ___ A

_____ **F.** Maximum power in fan motor circuit = ___ W

_____ **G.** Current in dashboard light circuit = ___ A

_____ **H.** Power in dashboard light circuit = ___ W

_____ **I.** Current in headlight circuit when headlight switch is in the dim position = ___ A

_____ **J.** Power in headlight circuit when headlight switch is in the dim position = ___ W

_____ **K.** Current in headlight circuit when headlight switch is in the bright position = ___ A

_____ **L.** Power in headlight circuit when headlight switch is in the bright position = ___ W

_____ **M.** Maximum total current in the circuit when all loads are energized = ___ A

_____ **N.** Maximum total power in the circuit when all loads are energized = ___ W

6. Using Standard Fuse Sizes and the calculated current values from Automobile Electrical System, determine the maximum standard fuse size for each circuit. *Note:* Round all fuse sizes to the next largest size.

_____ **A.** The size of Fuse 1 using a safety factor of 115% is ___ A.

STANDARD FUSE SIZES
¼, ½, ¾, 1, 1¼, 1½, 2, 2½, 3, 3½, 4, 4½, 5, 6, 7, 8, 9, 10, 12, 15, 20, 25, 30, 35, 40, 45, 50, 60, 70, 80, 90, 100, 110, 120, 130...

_____ **B.** The size of Fuse 2 using a safety factor of 125% is ___ A.

_____ **C.** The size of Fuse 3 using a safety factor of 135% is ___ A.

_____ **D.** The size of Fuse 4 using a safety factor of 125% is ___ A.

_____ **E.** The size of Fuse 5 using a safety factor of 125% is ___ A.

_____ **F.** The size of Fuse 6 using a safety factor of 125% is ___ A.

ACTIVITY 2-4—Temperature Conversion

Calculate the equivalent temperature readings for the applications.

_____ **1.** A temperature of 0°C is ___°F.

_____ **2.** The maximum operating temperature for a motor is ___°C (140°F).

_____ **3.** Ambient earth temperature is ___°F (20°C).

_____ **4.** Armored cable installed in thermal insulation shall have conductors rated at ___°C (194°F).

_____ **5.** Sealing compound shall provide a seal against passage of gas or vapors through the seal fitting, shall not have a melting point of less than ___°F (93°C), and shall not be affected by the surrounding atmosphere or liquids.

_____ **6.** Most motors have a temperature rise of ___°F (40°C) listed on the motor nameplate.

_____ **7.** When cleaning a water-soaked motor, the water temperature should not exceed ___°C (200°F).

_____ **8.** The melting point of most solder used in the electrical field is ___°F (200°C).

ACTIVITY 2-5—AC Sine Wave

1. Using the sine waves, determine the values.

_____ **A.** Number of cycles = ___

_____ **B.** Number of alternations = ___

_____ **C.** Period of one cycle = ___

_____ **D.** Frequency = ___ Hz

TIME = .25 SECONDS

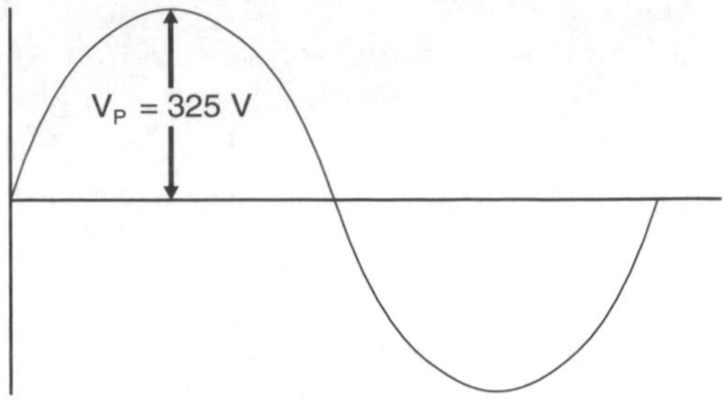

_____ E. Peak-to-peak voltage = ___ V

_____ F. Average voltage = ___ V

_____ G. Root-mean-square voltage = ___ V

$V_P = 325$ V

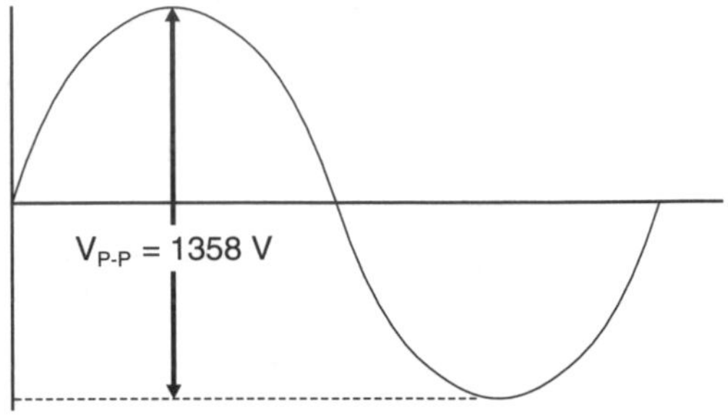

_____ H. Peak voltage = ___ V

_____ I. Average voltage = ___ V

_____ J. Root-mean-square voltage = ___ V

$V_{P-P} = 1358$ V

2. Identify if the current is in-phase, leading, or lagging.

_____ A. Current is ___.

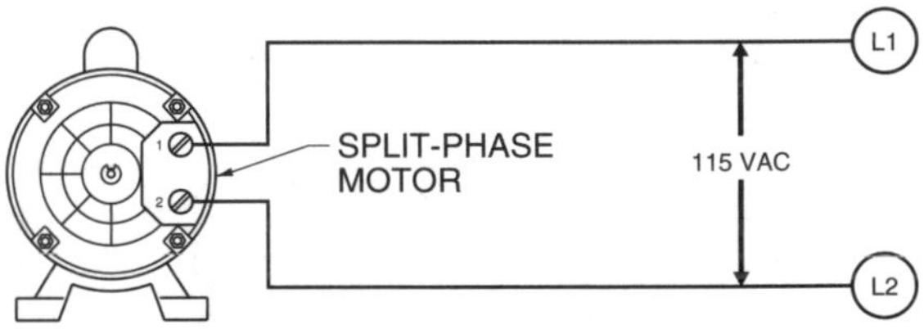

SPLIT-PHASE MOTOR

115 VAC

L1

L2

_____ B. Current is ___.

L1

115 VAC

L2

RESISTANCE HEATING ELEMENT

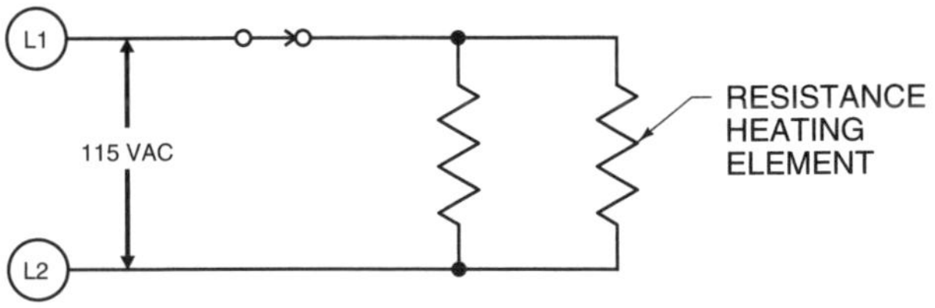

_____ C. Current is ___.

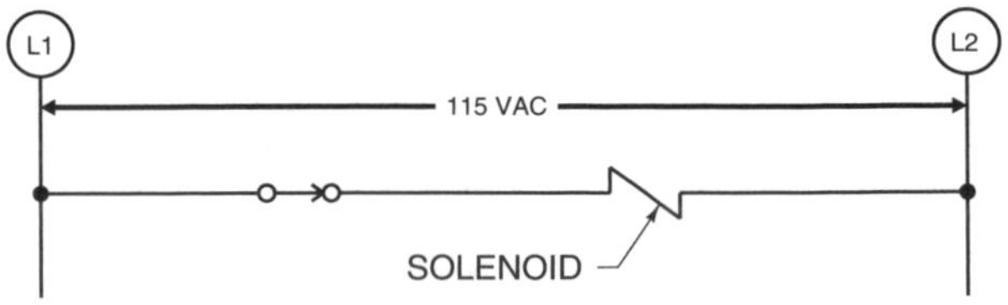

L1

L2

115 VAC

SOLENOID

ACTIVITY 2-6—Drawings and Diagrams

1. Identify the drawings, diagrams, and types of symbols.

_____ **A.**

_____ **B.**

_____ **C.**

_____ **D.**

_____ **E.**

_____ **F.**

_____ **G.**

_____ **H.**

_____ **I.**

_____ **J.**

_____ **K.**

_____ **L.**

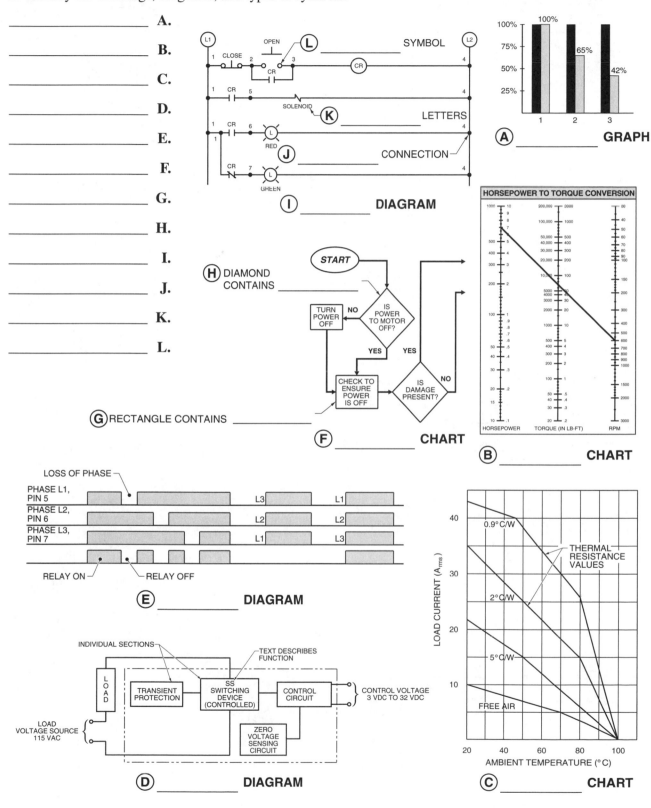

2. Identify the drawings, diagrams, and types of symbols.

_____ **A.**

_____ **B.**

_____ **C.**

_____ **D.**

_____ **E.**

_____ **F.**

_____ **G.**

_____ **H.**

_____ **I.**

_____ **J.**

_____ **K.**

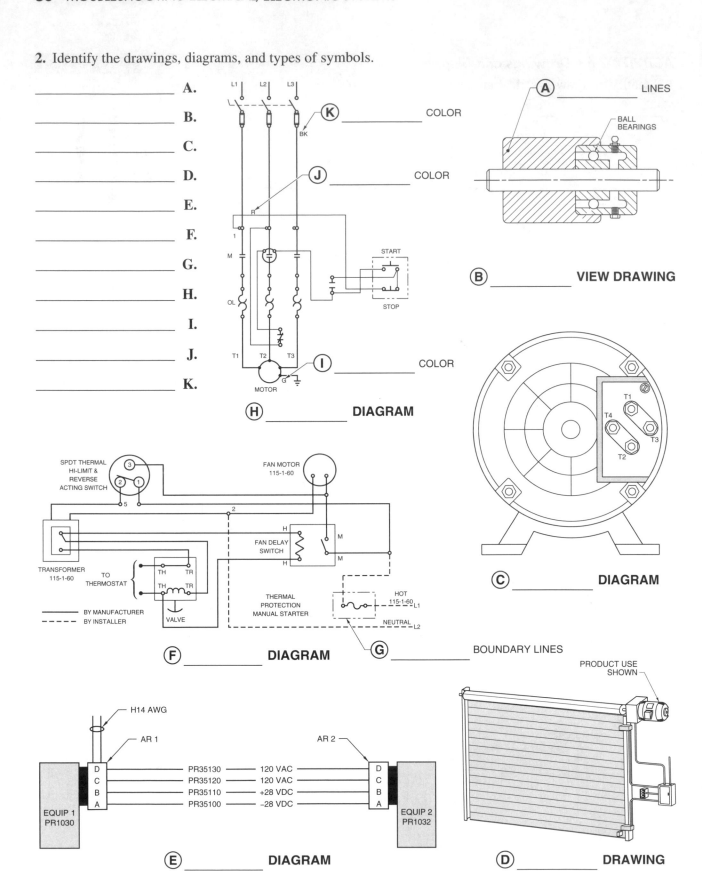

APPLICATION 3-1—Electrical Symbols and Abbreviations

Electrical prints are used when troubleshooting, servicing, or repairing electrical systems or equipment. Electrical prints use standard symbols and abbreviations to show circuit operation and device use. **See Selected Symbols and Selected Abbreviations. See Electrical Symbols and Abbreviations in Appendix.**

SELECTED SYMBOLS

LIMIT SWITCHES		FULL-WAVE RECTIFIER	BATTERY	SILICON CONTROLLED RECTIFIER	CONTROL TRANSFORMER SINGLE VOLTAGE	FUSE POWER OR CONTROL
NO	NC				H1 H2	
HELD CLOSED	HELD OPEN	AC / +DC −DC / AC	+ ‖‖‖ −		X2 X1	

SELECTED ABBREVIATIONS

Abbr/ Acronym	Meaning	Abbr/ Acronym	Meaning	Abbr/ Acronym	Meaning
A	Ammeter; Ampere; Anode; Armature	FU	Fuse	PNP	Positive-Negative-Positive
AC	Alternating Current	FWD	Forward	POS	Positive
AC/DC	Alternating Current; Direct Current	G	Gate; Giga; Green; Conductance	POT.	Potentiometer
A/D	Analog to Digital	GEN	Generator	P-P	Peak-to-Peak
AF	Audio Frequency	GRD	Ground	PRI	Primary Switch
AFC	Automatic Frequency Control	GY	Gray	PS	Pressure Switch
Ag	Silver	H	Henry; High Side of Transformer; Magnetic Flux	PSI	Pounds Per Square Inch
ALM	Alarm			PUT	Pull-Up Torque
AM	Ammeter; Amplitude Modulation	HF	High Frequency	Q	Transistor
AM/FM	Amplitude Modulation; Frequency Modulation	HP	Horsepower	R	Radius; Red; Resistance; Reverse
		Hz	Hertz	RAM	Random-Access Memory
ARM.	Armature	I	Current	RC	Resistance-Capacitiance
Au	Gold	IC	Integrated Circuit	RCL	Resistance-Inductance-Capacitance

 Refer to **Activity 3-1—Electrical Symbols and Abbreviations**...

APPLICATION 3-2—Series, Parallel, and Series/Parallel Connections

Switches, loads, meters, and other components in electrical circuits are connected in series, parallel, or series/parallel combinations. A *series* connection has two or more components connected so that there is one path for current flow. A *parallel* connection has two or more components connected so that there is more than one path for current flow. A *series/parallel* connection is a combination of series- and parallel-connected components.

Switches and loads are connected in series or parallel. Ammeters are connected in series. Voltmeters are connected in parallel. **See Series Connection and Parallel Connection.**

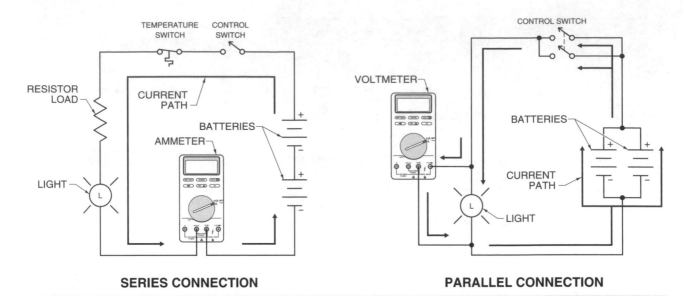

SERIES CONNECTION   **PARALLEL CONNECTION**

Switch Connections

When connected in series, all switches in a circuit must be closed before current flows. Opening any one or more of the switches stops current flow. When connected in parallel, one or more switches must be closed before current flows. All closed switches must be opened to stop current flow. When connected in series/parallel, two or more switches must be closed before current flows. Any one or more series-connected switches, or all parallel-connected switches, must be opened to stop current flow. **See Switch Connections.**

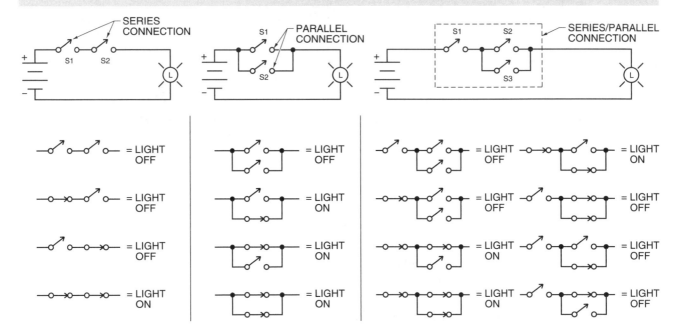

SWITCH CONNECTIONS

Series Load Connections

The total voltage in a circuit containing series-connected loads equals the sum of the voltages across each load. Total voltage is divided across each load. The voltage across each load decreases if loads are added in series and increases if loads are removed.

The current in a circuit containing series-connected loads is the same throughout the circuit. The current in the circuit decreases if loads are added in series and increases if loads are removed. The total resistance in a circuit containing series-connected loads equals the sum of the resistances of all loads. The resistance in the circuit increases if loads are added in series and decreases if loads are removed. **See Series Load Connections.**

Parallel Load Connections

The total voltage in a circuit containing parallel-connected loads is the same as the voltage across each load. The voltage across each load remains the same if parallel loads are added or removed. The total current in a circuit containing parallel-connected loads equals the sum of the current through all loads.

The total current increases if loads are added in parallel and decreases if loads are removed. The total resistance in a circuit containing parallel-connected loads is less than the smallest resistance value. The total resistance decreases if loads are added in parallel and increases if loads are removed. **See Parallel Load Connections.**

Series/Parallel Load Connections

The total voltage in a circuit containing series/parallel-connected loads equals the sum of the voltage across all series-connected loads and the voltage through the parallel-connected load combinations.

The total current in a circuit containing series/parallel-connected loads equals the total current flowing through all series loads or the sum of the current flowing through all parallel combinations. The total resistance in a circuit containing series/parallel-connected loads equals the sum of the series loads and the equivalent resistance of the parallel combinations. **See Series/Parallel Load Connections.**

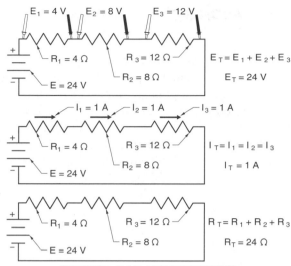

SERIES LOAD CONNECTIONS

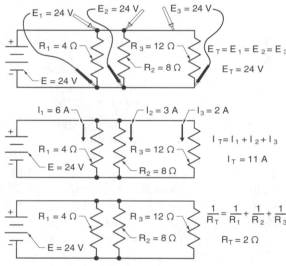

PARALLEL LOAD CONNECTIONS

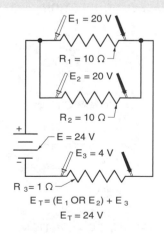

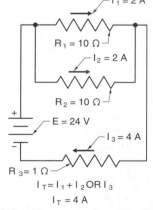

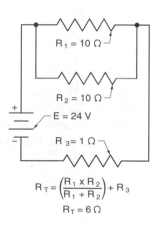

SERIES/PARALLEL LOAD CONNECTIONS

 Refer to **Activity 3-2—Series, Parallel, and Series/Parallel Connections** ....

APPLICATION 3-3—Circuit Polarity

All DC voltage sources have a positive and a negative terminal. The positive and negative terminals establish polarity in a circuit. All points in a DC circuit have polarity. *Polarity* is the positive (+) or negative (–) state of an object. The polarity of a component must be known before connecting it to a DC voltmeter or ammeter. **See Circuit Polarity.**

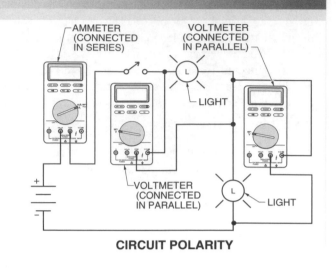

CIRCUIT POLARITY

 Refer to **Activity 3-3—Circuit Polarity** ....

APPLICATION 3-4—Batteries and Solarcells

A *battery* is a DC voltage source that converts chemical energy to electrical energy. Batteries are constructed by connecting individual cells in series, parallel, or series/parallel combinations.

A *cell* is a unit that produces electricity at a fixed voltage and current level. The voltage potential of a battery is increased by connecting cells in series. The current capacity of a battery is increased by connecting cells in parallel. Cells must have the same voltage and current rating when connected. **See Battery Connections.**

A *solarcell* is a voltage source that converts light energy to electrical energy. Solarcells are used in place of batteries in low-current devices, such as calculators, or are used to recharge battery-operated devices. Solarcells are used individually or in combinations to produce high voltage and current.

Solarcells are connected in series, parallel, or series/parallel. The voltage potential is increased when solarcells are connected in series. The current capacity is increased when solarcells are connected in parallel. **See Solarcell Connections.**

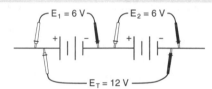

BATTERIES—SERIES CONNECTION

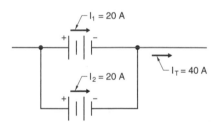

BATTERIES—PARALLEL CONNECTION

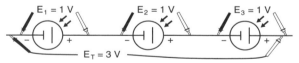

SOLARCELLS— SERIES CONNECTION

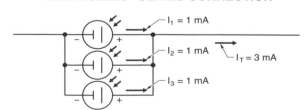

SOLARCELLS— PARALLEL CONNECTION

 Refer to **Activity 3-3—Batteries and Solarcells** ....

APPLICATION 3-5—Resistance in a Series Circuit

A *series circuit* is a circuit that contains two or more loads and one path through which current flows. The total resistance of the loads (resistors) connected in series is equal to the sum of the individual resistances. To find the total resistance of a series circuit, apply the formula:

$$R_T = R_1 + R_2 + R_3 + ...$$

where
R_T = total resistance (in ohms)
R_1 = resistance 1 (in ohms)
R_2 = resistance 2 (in ohms)
R_3 = resistance 3 (in ohms)

Example: Finding Total Resistance— Series Circuit

A series circuit has resistances of 500 Ω, 700 Ω, and 100 Ω. Calculate the total resistance in the circuit. **See Total Resistance—Series Circuit.**

$$R_T = R_1 + R_2 + R_3$$
$$R_T = 500 + 700 + 100$$
$$R_T = \mathbf{1300\ \Omega}$$

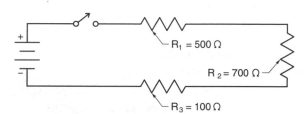

TOTAL RESISTANCE — SERIES CIRCUIT

 Refer to **Activity 3-5—Resistance in a Series Circuit** ....

APPLICATION 3-6—Resistance in a Parallel Circuit

A *parallel circuit* is a circuit that contains two or more loads and has more than one path through which current flows. To find the total resistance of a parallel circuit with two resistors, apply the formula:

$$R_T = \frac{R_1 \times R_2}{R_1 + R_2}$$

To find the total resistance of a parallel circuit with three or more resistors, apply the formula:

$$R_T = \frac{1}{\dfrac{1}{R_1} + \dfrac{1}{R_2} + \dfrac{1}{R_3} + ...}$$

Example: Finding Total Resistance— Parallel Circuit

A parallel circuit has resistances of 2 Ω and 4 Ω. Calculate the total resistance in the circuit. **See Total Resistance—Parallel Circuit.**

$$R_T = \frac{R_1 \times R_2}{R_1 + R_2}$$
$$R_T = \frac{2 \times 4}{2 + 4}$$
$$R_T = \mathbf{1.33\ \Omega}$$

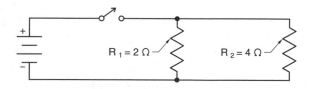

TOTAL RESISTANCE — PARALLEL CIRCUIT

 Refer to **Activity 3-6—Resistance in a Parallel Circuit** ....

APPLICATION 3-7—Resistance in a Series/Parallel Circuit

Most electrical circuits are combinations of series and parallel circuits. The total resistance of a series/parallel circuit is found by calculating the equivalent resistance of the parallel circuit(s) and adding the value to the resistance of the loads connected in series. To find the total resistance in a series/parallel circuit, apply the formula:

$$R_T = \frac{R_{P1} \times R_{P2}}{R_{P1} + R_{P2}} + R_{S1} + R_{S2} + \ldots$$

where

R_T = total resistance (in ohms)
R_{P1} = parallel resistance 1 (in ohms)
R_{P2} = parallel resistance 2 (in ohms)
R_{S1} = series resistance 1 (in ohms)
R_{S2} = series resistance 2 (in ohms)

Example: Finding Total Resistance— Series/Parallel Circuit

A series/parallel circuit has resistances of 150 Ω, 75 Ω, and 125 Ω connected in series and resistances of 60 Ω and 30 Ω connected in parallel. Calculate the total resistance.
See Total Resistance—Series/Parallel Circuit.

$$R_T = \frac{R_{P1} \times R_{P2}}{R_{P1} + R_{P2}} + R_{S1} + R_{S2} + \ldots$$

$$R_T = \frac{60 \times 30}{60 + 30} + 150 + 75 + 125$$

$$R_T = \frac{1800}{90} + 350$$

$$R_T = 20 + 350$$

$$R_T = \mathbf{370 \ \Omega}$$

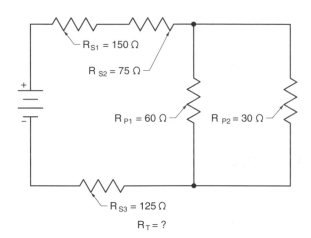

TOTAL RESISTANCE—SERIES/PARALLEL CIRCUIT

 Refer to **Activity 3-7—Resistance in a Series/Parallel Circuit** ....

APPLICATION 3-8—Capacitor Connection

Capacitors Connected in Series

Capacitors with different voltage ratings should not be connected. The applied voltage is divided over each capacitor for capacitors connected in series. The applied voltage is divided evenly between two capacitors of the same value. For example, two 110 V capacitors connected in series have 110 V across them when connected to a 220 V power source. **See Capacitors Connected in Series.**

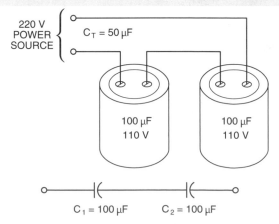

CAPACITORS CONNECTED IN SERIES

The total capacitance of two capacitors connected in series is less than the lowest-value capacitor. To find total capacitance of two capacitors connected in series, apply the formula:

$$C_T = \frac{C_1 \times C_2}{C_1 + C_2}$$

where
C_T = total capacitance (in μF)

C_1 = capacitance of capacitor 1 (in μF)
C_2 = capacitance of capacitor 2 (in μF)

To find the total capacitance of three or more capacitors connected in series, apply the formula:

$$\frac{1}{C_T} = \frac{1}{C_1} + \frac{1}{C_2} + ...$$

Example: Finding Total Capacitance—Capacitors Connected in Series

Two 100 μF capacitors are connected in series. Find the total capacitance. **See Two Capacitors Connected in Series.**

$$C_T = \frac{C_1 \times C_2}{C_1 + C_2}$$

$$C_T = \frac{100 \times 100}{100 + 100}$$

$$C_T = \frac{10,000}{200}$$

$$C_T = \textbf{50 μF}$$

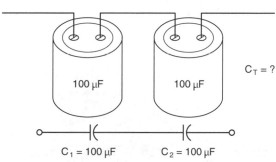

TWO CAPACITORS CONNECTED IN SERIES

100 μF   100 μF C_T = ?

C_1 = 100 μF C_2 = 100 μF

Example: Finding Total Capacitance—Three Capacitors Connected in Series

Capacitors of 10 μF, 50 μF, and 100 μF are connected in series. Find the total capacitance. **See Three Capacitors Connected in Series.**

$$\frac{1}{C_T} = \frac{1}{C_1} + \frac{1}{C_2} + ...$$

$$\frac{1}{C_T} = \frac{1}{10} + \frac{1}{50} + \frac{1}{100}$$

$$\frac{1}{C_T} = .1 + .02 + .01$$

$$\frac{1}{C_T} = .13$$

$$C_T = \textbf{7.69 μF}$$

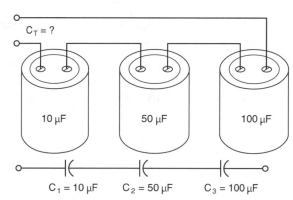

C_T = ?

10 μF   50 μF   100 μF

C_1 = 10 μF C_2 = 50 μF C_3 = 100 μF

THREE CAPACITORS CONNECTED IN SERIES

Capacitors Connected in Parallel

Capacitors connected in parallel have the same voltage across them. The voltage rating of each capacitor must be the same or greater than the supply voltage. For example, two 220 V capacitors connected in parallel have 220 V across them when connected to a 220 V power source. **See Capacitors Connected in Parallel.**

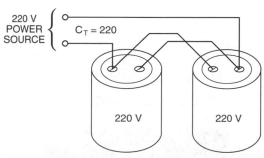

220 V POWER SOURCE C_T = 220

220 V   220 V

CAPACITORS CONNECTED IN PARALLEL

The equivalent capacitance of capacitors connected in parallel is equal to the sum of the individual capacitors. To find total capacitance of capacitors connected in parallel, apply the formula:

$$C_T = C_1 + C_2 + \ldots$$

Example: Finding Total Capacitance—Capacitors Connected in Parallel

Two 50 μF capacitors are connected in parallel. Find the total capacitance. **See Two Capacitors Connected in Parallel.**

$$C_T = C_1 + C_2 + \ldots$$
$$C_T = 50 + 50$$
$$C_T = \mathbf{100\ \mu F}$$

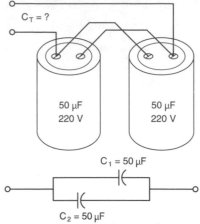

TWO CAPACITORS CONNECTED IN PARALLEL

Capacitors Connected in Series/Parallel

To calculate the total capacitance of capacitors connected in series/parallel, apply the procedure:

1. Calculate the capacitance of each parallel branch.

2. Calculate the capacitance of each series branch.

Example: Finding Total Capacitance—Capacitors Connected in Series/Parallel

Two 30 μF capacitors are connected in parallel with a 50 μF capacitor connected in series. Find the total capacitance. **See Capacitors Connected in Series/Parallel.**

1. Calculate the capacitance of the parallel branch.

$$C_T = C_1 + C_2$$
$$C_T = 30 + 30$$
$$C_T = \mathbf{60\ \mu F}$$

2. Calculate the capacitance of the series combination.

$$C_T = \frac{C_1 \times C_2}{C_1 + C_2}$$
$$C_T = \frac{60 \times 50}{60 + 50}$$
$$C_T = \frac{3000}{110}$$
$$C_T = \mathbf{27.27\ \mu F}$$

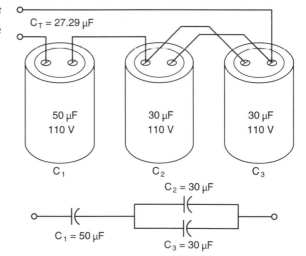

CAPACITORS CONNECTED IN SERIES/PARALLEL

 Refer to **Activity 3-8—Capacitor Connection** ....

 *Refer to Quick Quiz® on CD-ROM*

Refer to Chapter 3 in the **Troubleshooting Electrical/Electronic Systems Workbook** *for additional questions.*

Symbols and Circuits

Name_____ **Date** _____

ACTIVITY 3-1—Electrical Symbols and Abbreviations

Using the line and wiring diagrams, identify the symbols.

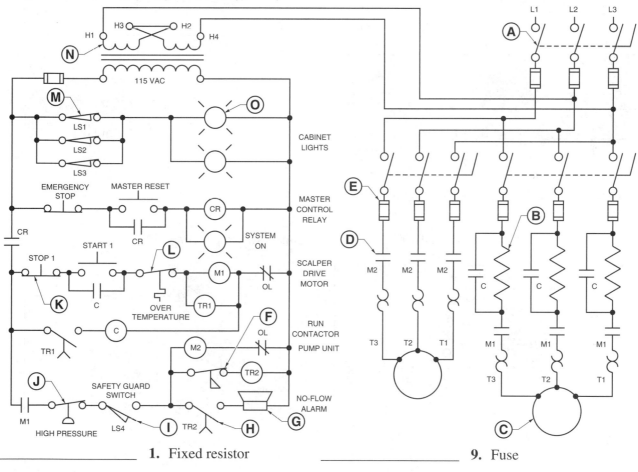

_____ **1.** Fixed resistor

_____ **2.** Control transformer

_____ **3.** Disconnect switch

_____ **4.** NC limit switch

_____ **5.** NO limit switch

_____ **6.** Three-phase motor

_____ **7.** NC flow switch

_____ **8.** NC pressure switch

_____ **9.** Fuse

_____ **10.** Pilot light

_____ **11.** Horn

_____ **12.** NO contact

_____ **13.** NC temperature switch

_____ **14.** NO, timed closed contact

_____ **15.** NC pushbutton

ACTIVITY 3-2—Series, Parallel, and Series/Parallel Connections

1. Determine if the light is ON or OFF for each combination of switch positions.

SWITCH COMBINATIONS				
	Switch			
	1	2	3	4
A	X	X	X	X
B	O	O	O	O
C	X	O	O	X
D	X	O	X	X
E	O	X	X	O
F	X	X	X	O
G	X	X	O	X

X = switch closed
O = switch open

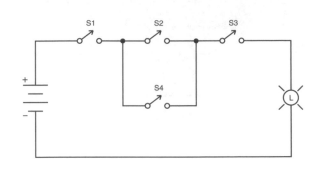

_____ **A.** Light is ___. _____ **E.** Light is ___.

_____ **B.** Light is ___. _____ **F.** Light is ___.

_____ **C.** Light is ___. _____ **G.** Light is ___.

_____ **D.** Light is ___.

2. Determine if the light is ON or OFF for each combination of switch positions.

SWITCH COMBINATIONS								
	Switch							
	1	2	3	4	5	6	7	8
A	O	X	X	X	X	X	X	X
B	X	X	O	O	O	X	O	O
C	X	O	O	O	X	O	O	X
D	X	X	X	O	X	O	O	O
E	O	O	O	X	X	X	O	X
F	X	X	O	O	O	O	X	X
G	X	X	X	O	O	O	O	X
H	X	X	X	O	O	O	X	O
I	X	O	X	X	X	X	X	O

X = switch closed
O = switch open

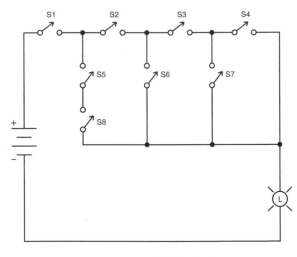

_____ **A.** Light is ___. _____ **F.** Light is ___.

_____ **B.** Light is ___. _____ **G.** Light is ___.

_____ **C.** Light is ___. _____ **H.** Light is ___.

_____ **D.** Light is ___. _____ **I.** Light is ___.

_____ **E.** Light is ___.

3. Connect Meter 1 in parallel with the battery to measure the voltage at the battery. Connect Meter 2 in parallel with the collision stop lights to test the voltage to all three lights. Connect Meter 3 so that it reads the voltage at the left turn signal light. Connect Meter 4 so that it reads the voltage from the fuse that protects the running lights.

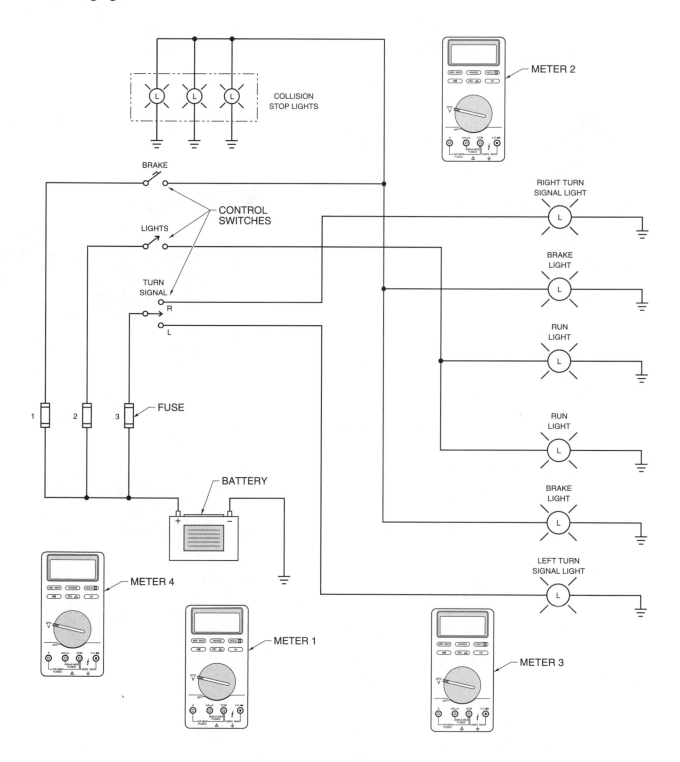

AUTO LIGHTING SYSTEM

4. Using Auto Lighting System, determine whether the devices are connected in series or parallel.

_____ **A.** The collision stop lights are connected in ___.

_____ **B.** The two brake lights are connected in ___.

_____ **C.** The fuses are connected in ___ with the lights in the circuit.

_____ **D.** The control switches are connected in ___ with the lights.

_____ **E.** The battery is connected in ___ with the lights.

ACTIVITY 3-3—Circuit Polarity

1. Using One-Station Power Window Circuit, complete the following.

    A. Mark each pushbutton as up or down.

    B. Connect the voltmeter to measure the voltage at the motor when the Up Pushbutton is pressed.

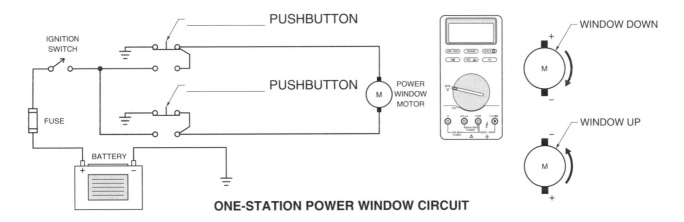

ONE-STATION POWER WINDOW CIRCUIT

2. Using Two-Station Power Window Circuit, complete the following.

    A. Mark each pushbutton as up or down.

    B. Connect the voltmeter to measure the voltage at the motor when the Down Pushbutton is pressed.

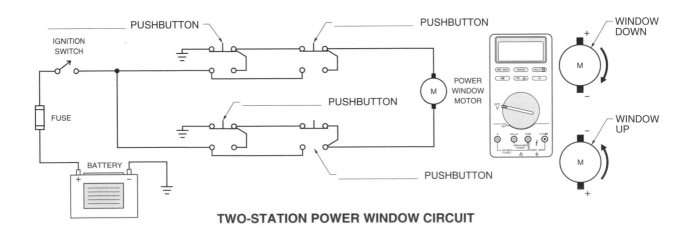

TWO-STATION POWER WINDOW CIRCUIT

ACTIVITY 3-4—Batteries and Solarcells

1. Determine the output voltage and available current for each battery connection.

_____ **A.** E = ___ V   _____ **C.** E = ___ V

_____ **B.** I = ___ A   _____ **D.** I = ___ A

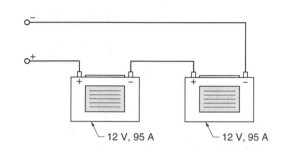

_____ **E.** E = ___ V

_____ **F.** I = ___ A

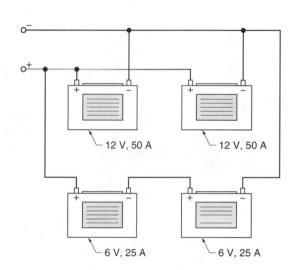

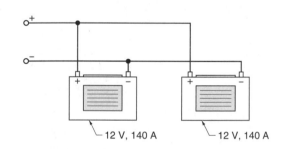

_____ **G.** E = ___ V   _____ **I.** E = ___ V

_____ **H.** I = ___ A   _____ **J.** I = ___ A

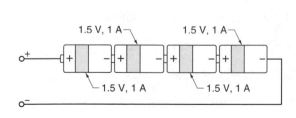

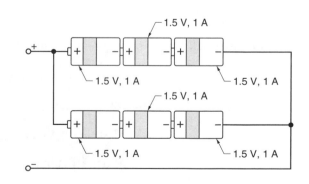

2. Using Solarcell Battery Charging Circuit, mark the correct polarity (+) or (−) for each meter lead.

_____ **A.** ___

_____ **B.** ___

_____ **C.** ___

_____ **D.** ___

_____ **E.** ___

_____ **F.** ___

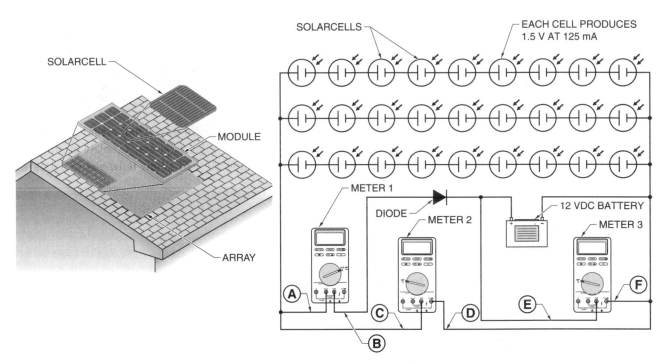

SOLARCELL BATTERY CHARGING CIRCUIT

3. Using Solarcell Battery Charging Circuit, list the meter readings for the light conditions.

_____ **A.** Meter 1 reading is ___ mA when light shines on solarcells.

_____ **B.** Meter 2 reading is ___ V when light shines on solarcells.

_____ **C.** Meter 3 reading is ___ V when light shines on solarcells.

_____ **D.** Meter 1 reading is ___ mA when no light shines on solarcells.

_____ **E.** Meter 2 reading is ___ V when no light shines on solarcells.

_____ **F.** Meter 3 reading is ___ V when no light shines on solarcells.

ACTIVITY 3-5—Resistance in a Series Circuit

Determine the total resistance and unknown electrical quantity.

_____ **1.** $R_T =$ ___ Ω

_____ **2.** $I_T =$ ___ mA

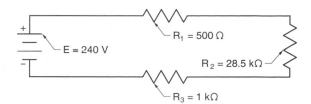

_____ **3.** $R_T =$ ___ Ω

_____ **4.** $P_T =$ ___ W

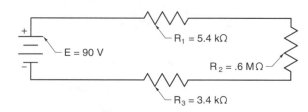

_____ **5.** $R_T =$ ___ kΩ

_____ **6.** $P_T =$ ___ mW

ACTIVITY 3-6—Resistance in a Parallel Circuit

Determine the total resistance and unknown electrical quantity.

_____ **1.** $R_T =$ ___ Ω

_____ **2.** $I_T =$ ___ mA

_____ **3.** $R_T =$ ___ Ω

_____ **4.** $I_T =$ ___ mA

_____ **5.** $R_T =$ ___ Ω

_____ **6.** $P_T =$ ___ W

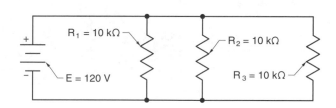

ACTIVITY 3-7—Resistance in a Series/Parallel Circuit

1. Determine the total resistance in the circuit by solving each resistance combination.

_____ **A.** $R_T =$ ___ Ω

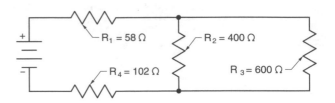

_____ **B.** $R_T =$ ___ Ω

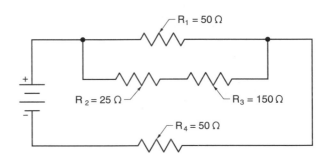

_____ **C.** $R_T =$ ___ Ω

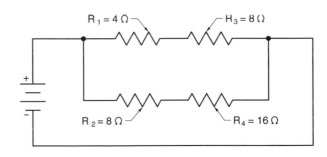

_____ **D.** $R_T =$ ___ Ω

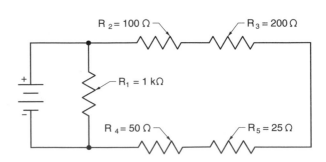

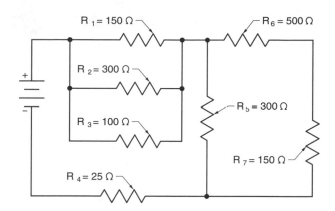

_____ **E.** $R_T =$ ___ Ω

2. Determine the total resistance in the circuit for each switch combination. Round the answer to the nearest whole number.

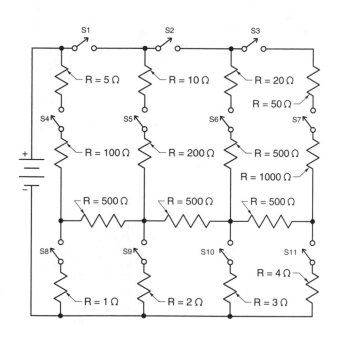

	SWITCH COMBINATIONS										
	Switch										
	1	2	3	4	5	6	7	8	9	10	11
A	X	O	O	X	O	O	O	X	O	O	O
B	X	X	X	O	O	O	X	O	O	O	X
C	X	O	O	X	X	O	O	O	O	X	O
D	O	O	O	X	O	O	O	O	O	O	X
E	X	X	X	O	O	X	X	X	X	O	O
F	X	X	O	O	O	X	O	X	O	O	O
G	X	O	O	O	X	O	O	O	O	X	X
H	O	X	X	X	X	X	X	O	X	O	O
I	X	O	X	X	O	X	X	O	O	O	X
J	X	X	O	O	O	X	X	O	X	O	O
K	X	X	X	O	O	O	X	X	O	O	O
L	X	X	O	O	X	X	O	O	O	O	X
M	X	X	X	X	O	O	X	O	X	O	O
N	X	X	X	O	X	O	X	O	O	X	O

X = switch closed
O = switch open

_____ **A.** $R_T =$ ___ Ω _____ **H.** $R_T =$ ___ Ω

_____ **B.** $R_T =$ ___ Ω _____ **I.** $R_T =$ ___ Ω

_____ **C.** $R_T =$ ___ Ω _____ **J.** $R_T =$ ___ Ω

_____ **D.** $R_T =$ ___ Ω _____ **K.** $R_T =$ ___ Ω

_____ **E.** $R_T =$ ___ Ω _____ **L.** $R_T =$ ___ Ω

_____ **F.** $R_T =$ ___ Ω _____ **M.** $R_T =$ ___ Ω

_____ **G.** $R_T =$ ___ Ω _____ **N.** $R_T =$ ___ Ω

ACTIVITY 3-8—Capacitor Connection

Determine the total capacitance in each circuit.

_____ **1.** $C_T =$ ___ μF

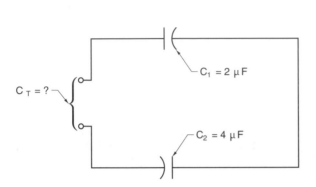

_____ **2.** $C_T =$ ___ μF

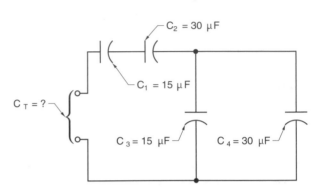

_____ **3.** $C_T =$ ___ μF

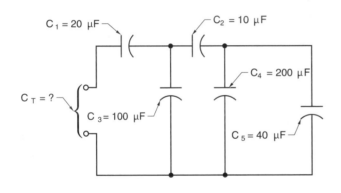

_____ **4.** $C_T =$ ___ μF

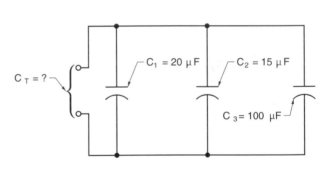

_____ **5.** $C_T =$ ___ μF

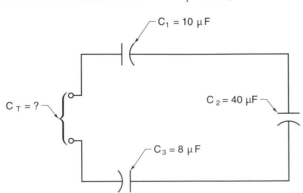

_____ **6.** $C_T =$ ___ μF

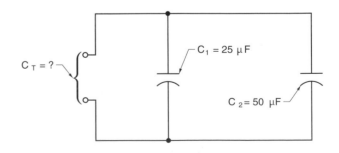

Meter Symbols and Terminology

APPLICATION 4-1—Measurement Precautions

A variety of analog and digital meters measure electrical quantities such as voltage, current, resistance, power, speed, temperature, etc. Each meter includes an instruction manual that shows and explains individual meter specifications and features, proper operating procedure and safety, and specific warnings and applications. The manufacturer's instruction manual should be read before using any test equipment. Always refer to the instruction manual for information concerning specific use, safety, and meter capacity. Safety precautions are required when using meters because of the high voltage and currents in some circuits. Safety precautions for using meters include:

- Ensuring that the meter leads are connected to the correct meter jacks.
- Ensuring that the selector switch is set to the proper range and function before applying test leads to a circuit.
- Knowing that any reading taken varies from the true reading by the accuracy limits of the meter.
- Averaging several readings when accuracy is critical to reduce the effect of stray alternating fields.
- Starting with the highest range when measuring unknown values.
- Checking meter leads for frayed or broken insulation.
- Avoiding taking measurements under humid or damp conditions.
- Ensuring that hands, shoes, and work areas are dry.

 Refer to **Activity 4-1—Measurement Precautions** ....

APPLICATION 4-2—Meter Abbreviations and Symbols

Meter switches, scales, displays, and manuals use abbreviations and symbols to convey information. An *abbreviation* is a letter or combination of letters that represent a word. Abbreviations are dependent on a particular language. A *symbol* is a graphic element that represents a quantity or unit. Symbols are independent of language because a symbol can be recognized regardless of the language a person speaks. **See Selected Meter Abbreviations and Selected Meter Symbols.**

SELECTED METER ABBREVIATIONS

AC	Alternating current or voltage		RPM	Revolutions per minute
DC	Direct current or voltage		COM	Common
V	Volts		OL	Overload
mV	Millivolts (0.001 V)		T	Time
kV	Kilovolts (1000 V)		LSD	Least significant digit
A	Amperes		MAX	Maximum
mA	Milliamperes		MIN	Minimum
μA	Microamperes		AVG	Average
W	Watts		V_{ave}	Average voltage
kΩ	Kilohms		V_p	Peak voltage
MΩ	Megohms	LOG	Readings are being recorded	V_{p-p}	Peak-to-peak voltage
Hz	Hertz	LO	Low	V_{rms}	Root-mean-square (rms) voltage
kHz	Kilohertz	nS	Nanosiemens (1×10⁻⁹) or	Hi-Z	High input impedance
μF	Microfarads		0.000000001 siemens	dB	Decibel
nF	Nanofarads	MEM	Memory	dBV	Decibel volts
°F	Degrees Fahrenheit	MS	Time display in minutes: seconds	dBW	Decibel watts
°C	Degrees Celsius	HM	Time display in hours: minutes		

SELECTED METER SYMBOLS

$\sim$	AC	⚲	See service manual	◯	Switch position OFF (power)
$===$	DC	▢	Double insulation	|	Switch position ON (power)
$\equiv$	AC or DC	⊏▭⊐	Fuse	⊙	Manual Range mode
$+$	Positive	⊟ ⊟ }	Battery	⚠	Warning: Dangerous or high voltage that could result in personal injury
$-$	Negative	H	Hold	⚠	Caution: Hazard that could result in equipment damage or personal injury
⏚	Ground	🔒	Lock	⏚1000 V MAX	Terminals must not be connected to a circuit with higher than listed voltage
$\pm$	Plus or minus	)))))	Audio beeper	△	Relative mode — displayed value is difference between present measurement and previous stored measurement
▸▪	Diode	⊣⊢	Capacitor	Ω	Ohms resistance
▸▪)))))	Diode Test	%	Percent	☼	Meter display light
<	Less than	▷	Move right	⚡	> 30VAC or VDC present
>	Greater than	◁	Move left	⎍	Trigger on positive slope
△	Increase setting	⊘	No (do not use)	⎍	Trigger on negative slope
▽	Decrease setting				

Abbreviations

All meters use standard abbreviations to represent a quantity or term. Abbreviations can be used individually (100 V) or in combinations (100 kV).

Symbols

Meters use symbols to display information. Most meters use standard international symbols that are designed to represent an electrical component (battery, diode, etc.), term (ground, positive, etc.), or message to the user (warning, see manual, etc.).

A component symbol used on a meter may differ from the same symbol used in wiring or schematic diagrams. A *component* is an individual device. For example, the symbol for a battery on a meter is not the same as the symbol for a battery in a schematic diagram. **See Battery Symbols**.

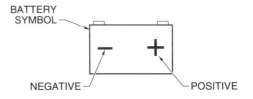

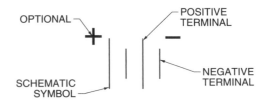

BATTERY SYMBOLS

 Refer to **Activity 4-2—Meter Abbreviations and Symbols** ....

APPLICATION 4-3—*Meter Terminology*

Meter Terminology

All meters use terms to describe displayed information. Understanding of meter terminology is required for proper meter use. For example, a meter may display an electrical quantity as an average, effective, peak, or rms value. The difference between each value must be understood to correctly interpret the measurements. **See Meter Terminology.**

METER TERMINOLOGY		
TERM	**SYMBOL**	**DEFINITION**
AC		Continually changing current that reverses direction at regular intervals; standard U.S. frequency is 60 Hz
AMBIENT TEMPERATURE		Temperature of air surrounding a meter or equipment to which the meter is connected
AC/DC		Indicates ability to read or operate on alternating and direct current
ACCURACY ANALOG METER		Largest allowable error (in percent of full scale) made under normal operating conditions; the reading of a meter set on the 250 V range with an accuracy rating of ±2% could vary ±5 V; analog meters have greater accuracy when readings are taken on the upper half of the scale
DIODE		Semiconductor that allows current to flow in only one direction
DISCHARGE		Removal of an electric charge
DUAL TRACE		Feature that allows two separate waveforms to be displayed simultaneously
EARTH GROUND		Reference point that is directly connected to ground
FREQUENCY		Number of complete cycles occurring per second
FUNCTION SWITCH		Switch that selects the function (AC voltage, DC voltage, etc.) that a meter is to measure
GLITCH		Momentary spike in a waveform
GLITCH DETECT		Function that increases the meter sampling rate to maximize the detection of the glitch(es)
TRIGGER		Device which determines the starting point of a measurement
WAVEFORM		Pattern defined by an electrical signal
ZOOM		Function that allows a waveform (or part of waveform) to be magnified

 Refer to **Activity 4-3—*Meter Terminology*** ....

APPLICATION 4-4—*Reading Analog Displays*

Analog Displays

An *analog display* is an electromechanical device that indicates readings by the mechanical motion of a pointer. Analog displays use scales to display measured values. Analog scales may be linear or nonlinear. A *linear scale* is a scale that is divided into equally spaced segments. A *nonlinear scale* is a scale that is divided into unequally spaced segments. **See Analog Displays.**

Analog scales are divided using primary divisions, secondary divisions, and subdivisions. A *primary division* is a division with a listed value. A *secondary division* is a division that divides primary divisions in halves, thirds, fourths, fifths, etc. A *subdivision* is a division that divides secondary divisions in halves, thirds, fourths, fifths, etc. Secondary divisions and subdivisions do not have listed numerical values.

When reading an analog scale, add the primary, secondary, and subdivision readings. **See Reading Analog Scales.**

To read an analog scale, apply the procedure:
1. Read the primary division.
2. Read the secondary division if the pointer moves past a secondary division. *Note:* This may not occur with very low readings.
3. Read the subdivision if the pointer is not directly on a primary or secondary division. Round the reading to the nearest subdivision if the pointer is not directly on a subdivision. Round the reading to the next highest subdivision if rounding to the nearest subdivision is unclear.

Add the primary division, secondary division, and subdivision readings to obtain the analog reading.

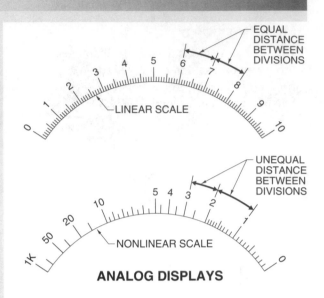

ANALOG DISPLAYS

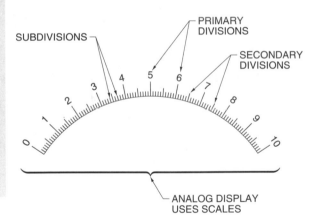

ANALOG DISPLAY USES SCALES

	READING
❶ PRIMARY DIVISION	2
❷ SECONDARY DIVISION	.5
❸ SUBDIVISION	.3
METER READING	2.8

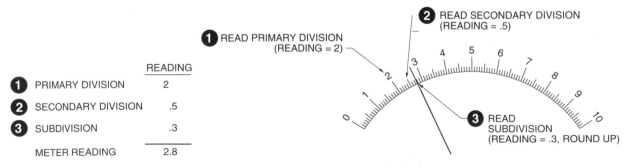

READING ANALOG SCALES

 Refer to **Activity 4-4—*Reading Analog Displays*** ....

APPLICATION 4-5—Reading Digital Displays

Digital Displays

A *digital display* is an electronic device that displays readings as numerical values. Digital displays help eliminate human error when taking readings by displaying exact values measured. Errors occur when reading a digital display if the displayed prefixes, symbols, and decimal point are not properly applied.

Digital displays display values using either a light-emitting diode (LED) or a liquid crystal display (LCD). LED displays are easier to read and use more power than LCD displays. Most portable digital meters use an LCD.

The exact value on a digital display is determined from the numbers displayed and the position of the decimal point. A range switch determines the placement of the decimal point.

Typical voltage ranges on a digital display are 3 V, 30 V, and 300 V. The highest possible reading with the range switch on 3 V is 2.999 V. The highest possible reading with the range switch on 30 V is 29.99 V. The highest possible reading with the range switch on 300 V is 299.9 V. Accurate readings are obtained by using the range that gives the best resolution without overloading. **See Digital Display.**

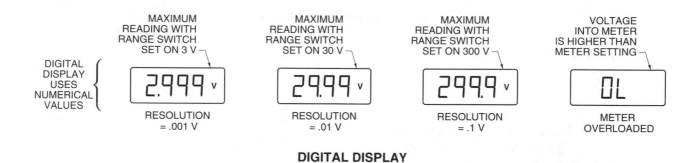

DIGITAL DISPLAY

Bar Graph

Most digital displays include a bar graph to show changes and trends in a circuit. A *bar graph* is a graph composed of segments that function as an analog pointer. The displayed bar graph segments increase as the measured value increases and decrease as the measured value decreases. Reverse the polarity of test leads if a negative sign is displayed at the beginning of the bar graph. **See Bar Graph Display.**

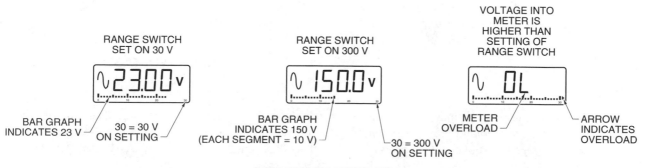

BAR GRAPH DISPLAY

Wrap-Around Bar Graph

A *wrap-around bar graph* is a bar graph that displays a fraction of the full range on the graph. The pointer wraps around and starts over when the limit of the bar graph is reached. **See Wrap-Around Bar Graph.**

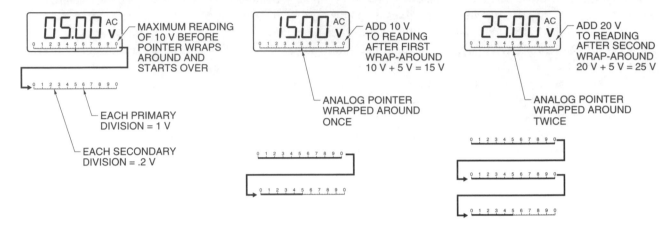

WRAP-AROUND BAR GRAPH

Using a Bar Graph

A bar graph reading is updated 30 times per second. A digital reading is updated 4 times per second. The bar graph is used when quickly changing signals cause the digital display to flash or when there is a change in the circuit that is too rapid for the digital display to detect.

For example, mechanical relay contacts may bounce open when exposed to vibration. Contact bounce causes intermittent problems in electrical equipment. Frequency and severity of contact bounce increase as a relay ages.

A contact's resistance changes momentarily from zero to infinity and back when a contact bounces open. A digital display cannot indicate contact bounce because most digital displays require more than 250 milliseconds (ms) to update their displays. The quick response of bar graphs enable detection of most contact bounce problems. The contact bounce is displayed by the movement of one or more segments the moment the contact opens.

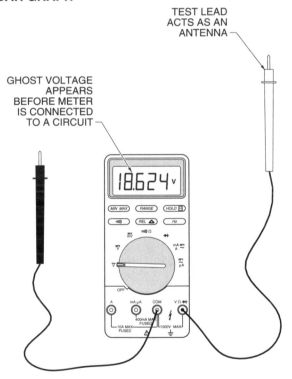

GHOST VOLTAGES

Ghost Voltages

A meter set to measure voltage may display a reading before the meter is connected to a powered circuit. The displayed voltage is a ghost voltage that appears as changing numbers on a digital display or as a vibrating analog display. A *ghost voltage* is a voltage that appears on a meter that is not connected to a circuit. **See Ghost Voltages.**

Ghost voltages are produced by the magnetic fields generated by current-carrying conductors, fluorescent lighting, and operating electrical equipment. Ghost voltages enter a meter through the test leads because test leads not connected to a circuit act as antennae for stray voltages.

Ghost voltages do not damage a meter. Ghost voltages may be misread as circuit voltages when a meter is connected to a circuit that is believed to be powered. A circuit that is not powered can also act as an antenna for stray voltages. To ensure true circuit voltage readings, connect a meter to a circuit for a long enough time so that the meter displays a constant reading.

 Refer to **Activity 4-5—Reading Digital Displays** ....

 Refer to Quick Quiz® on CD-ROM

 Refer to Chapter 4 in the **Troubleshooting Electrical/Electronic Systems Workbook** for additional questions.

Name_____ Date _____

ACTIVITY 4-1—Measurement Precautions

Answer the questions using Model 01 Meter Specifications.

_____DC_____ **1.** The meter has greater accuracy when measuring ___ voltage.

_____.1mVac_____ **2.** The meter has greater accuracy when measuring ___ current.

_____.1mVac_____ **3.** ___ VAC is the minimum value that can be displayed when the meter is set to autoranging.

_____999.9Vac_____ **4.** ___ VAC is the maximum value that can be displayed when the meter is set to autoranging.

_____ **5.** ___ mA is the minimum value of continuous AC current that can be displayed when the meter is set to autoranging.

_____ **6.** ___ A is the maximum value of continuous AC current that can be displayed when the meter is set to autoranging.

_____ **7.** ___ VDC is the minimum value that can be displayed when the meter is set to autoranging.

MODEL 01 METER SPECIFICATIONS		
Meter Function	**Range**	**Accuracy**
$\widetilde{V}$	400.0 mV	±.7%
	4.000 V	±.7%
	40.00 V	±.7%
	400.0 V	±.7%
	1000 V	±.7%
$\overline{\overline{V}}$	400.0 mV	±.1%
	4.000 V	±.1%
	40.00 V	±.1%
	400.0 V	±.1%
	1000 V	±.1%
Ω	400.0 W	±.2%
	4.000 kΩ	±.2%
	40.00 kΩ	±.2%
	400.0 kΩ	±.2%
	4.000 MΩ	±.2%
	40.00 MΩ	±1%
$\widetilde{mA}$	40.00 mA	±1%
	400.0 mA	±1%
	4000 mA	±1%
A~	10.00 A*	±1%
$\overline{\overline{mA}}$	40.00 mA	±.2%
	400.0 mA	±.2%
	4000 mA	±.2%
A ===	10.00 A*	±.2%
Frequency	199.99 Hz	±.005%
	1999.9 Hz	±.005%
	19.999 kHz	±.005%
	199.99 kHz	±.005%

* 10 A continuous, 20 A for 30 seconds maximum

_____ **8.** ___ VDC is the maximum value that can be displayed when the meter is set to autoranging.

_____ **9.** ___ mA is the minimum value of continuous DC current that can be displayed when the meter is set to autoranging.

_____ 10. ___ A is the maximum value of continuous DC current that can be displayed when the meter is set to autoranging.

_____ 11. ___ Ω is the minimum resistance value that can be displayed when the meter is set to autoranging.

_____ 12. ___ MΩ is the maximum resistance value that can be displayed when the meter is set to autoranging.

_____ 13. ___ Hz is the minimum frequency value that can be displayed when the meter is set to autoranging.

_____ 14. ___ kHz is the maximum DC frequency value that can be displayed when the meter is set to autoranging.

_____ 15. The meter measures ___ most accurately when measuring different electrical quantities.

ACTIVITY 4-2—Meter Abbreviations and Symbols

Identify the abbreviations and symbols.

_____ 1. Direct current

_____ 2. Audio beeper

_____ 3. See manual

_____ 4. DC or AC

_____ 5. Diode

_____ 6. DC voltage setting

_____ 7. AC voltage setting

_____ 8. Millisecond

_____ 9. Maximum

_____ 10. Relative mode

_____ 11. Average

_____ 12. Alternating current

_____ 13. Hold

_____ 14. Battery

_____ 15. Minimum

_____ 16. Capacitor

_____ 17. Danger

_____ 18. Hertz

_____ 19. Ohms

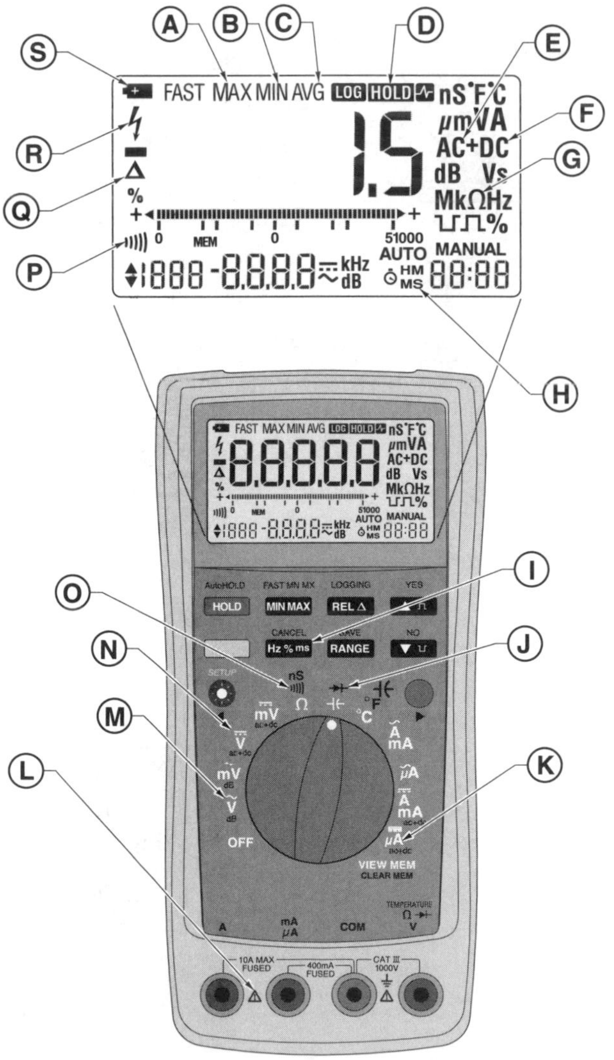

ACTIVITY 4-3—Meter Terminology

Using the analog meter readings and specifications, determine the actual value range and accuracy of the readings.

1. A meter has a range setting of 50 V and an accuracy of ±2%.

_____ **A.** The actual value is between ___ V and ___ V when a measured reading of 5 V is obtained.

_____ **B.** The accuracy of the reading is ___%.

_____ **C.** The actual value is between ___ V and ___ V when a measured reading of 12 V is obtained.

_____ **D.** The accuracy of the reading is ___%.

_____ **E.** The actual value is between ___ V and ___V when a measured reading of 18 V is obtained.

_____ **F.** The accuracy of the reading is ___%.

_____ **G.** The actual value is between ___ V and ___ V when a measured reading of 25 V is obtained.

_____ **H.** The accuracy of the reading is ___%.

2. A meter has a range setting of 500 mA and an accuracy of ±1.5%.

_____ **A.** The actual value is between ___ mA and ___ mA when a measured reading of 16 mA is obtained.

_____ **B.** The accuracy of the reading is ___%.

_____ **C.** The actual value is between ___ mA and ___ mA when a measured reading of 48 mA is obtained.

_____ **D.** The accuracy of the reading is ___%.

_____ **E.** The actual value is between ___ mA and ___ mA when a measured reading of 90 mA is obtained.

_____ **F.** The accuracy of the reading is ___%.

_____ **G.** The actual value is between ___ mA and ___ mA when a measured reading of 260 mA is obtained.

_____ **H.** The accuracy of the reading is ___%.

Using the digital meter readings and specifications, determine the actual value range of the readings.

3. A meter has a range setting of 40.00 V and an accuracy of ±.5%.

_____ **A.** The actual value is between ___ V and ___ V when a measured reading of 3.00 V is obtained.

_____ **B.** The actual value is between ___ V and ___ V when a measured reading of 3.50 V is obtained.

_____ **C.** The actual value is between ___ V and ___ V when a measured reading of 3.55 V is obtained.

_____ **D.** The actual value is between ___ V and ___ V when a measured reading of 12.50 V is obtained.

_____ **E.** The actual value is between ___ V and ___ V when a measured reading of 18.09 V is obtained.

4. A meter has a range setting of 400.0 V and an accuracy of ±.5%.

_____ **A.** The actual value is between ___ V and ___ V when a measured reading of 30.0 V is obtained.

_____ **B.** The actual value is between ___ V and ___ V when a measured reading of 35.0 V is obtained.

_____ **C.** The actual value is between ___ V and ___ V when a measured reading of 35.5 V is obtained.

_____ **D.** The actual value is between ___ V and ___ V when a measured reading of 125.0 V is obtained.

_____ **E.** The actual value is between ___ V and ___ V when a measured reading of 180.9 V is obtained.

5. A meter has a range setting of 32.00 mA and an accuracy of ±.75%.

_____ **A.** The actual value is between ___ mA and ___ mA when a measured reading of 00.32 mA is obtained.

_____ **B.** The actual value is between ___ mA and ___ mA when a measured reading of 00.99 mA is obtained.

_____ **C.** The actual value is between ___ mA and ___ mA when a measured reading of 01.01 mA is obtained.

_____ **D.** The actual value is between ___ mA and ___ mA when a measured reading of 12.50 mA is obtained.

_____ **E.** The actual value is between ___ mA and ___ mA when a measured reading of 18.66 mA is obtained.

6. A meter has a range setting of 320.0 Ω and an accuracy of ±.3%.

_____ **A.** The actual value is between ___ Ω and ___ Ω when a measured reading of 001.2 Ω is obtained.

_____ **B.** The actual value is between ___ Ω and ___ Ω when a measured reading of 015.0 Ω is obtained.

_____ **C.** The actual value is between ___ Ω and ___ Ω when a measured reading of 025.6 Ω is obtained.

_____ **D.** The actual value is between ___ Ω and ___ Ω when a measured reading of 100.1 Ω is obtained.

_____ **E.** The actual value is between ___ Ω and ___ Ω when a measured reading of 180.9 Ω is obtained.

ACTIVITY 4-4—Reading Analog Displays

List the correct reading for each range setting.

_____ 1. The reading equals ___ V with the range switch set on 1 V.

_____ 2. The reading equals ___ V with the range switch set on 10 V.

_____ 3. The reading equals ___ V with the range switch set on 100 V.

_____ 4. The reading equals ___ Ω with the range switch set on R × 1.

_____ 5. The reading equals ___ Ω with the range switch set on R × 100.

_____ 6. The reading equals ___ Ω with the range switch set on R × 1k.

_____ 7. The reading equals ___ V with the range switch set on 1 V.

_____ 8. The reading equals ___ V with the range switch set on 10 V.

_____ 9. The reading equals ___ V with the range switch set on 100 V.

_____ 10. The reading equals ___ V with the range switch set on 1 V.

_____ 11. The reading equals ___ V with the range switch set on 10 V.

_____ 12. The reading equals ___ V with the range switch set on 100 V.

_____ 13. The reading equals ___ Ω with the range switch set on R × 1.

_____ 14. The reading equals ___ Ω with the range switch set on R × 100.

_____ 15. The reading equals ___ Ω with the range switch set on R × 1k.

ACTIVITY 4-5—Reading Digital Displays

1. Determine the resolution of the meter readings.

_____ **A.** The resolution equals
___ V.

$$15 \text{ v}$$

_____ **B.** The resolution equals
___ V.

$$1.555 \text{ v}$$

_____ **C.** The resolution equals
___ V.

$$30.1 \text{ v}$$

_____ **D.** The resolution equals
___ V.

$$12.25 \text{ v}$$

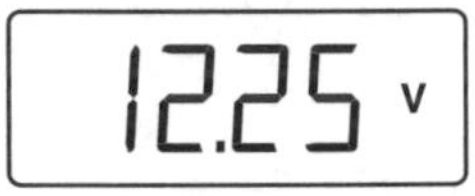

_____ **E.** The resolution equals
___ V.

$$230.8 \text{ v}$$

_____ **F.** The resolution equals
___ V.

$$115 \text{ v}$$

2. Determine the value of the individual segments on the bar graphs.

_____ **A.** Each segment equals
___ V.

RANGE SWITCH
SET ON 30 V

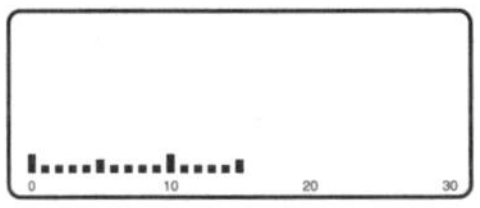

_____ **B.** Each segment equals
___ V.

RANGE SWITCH
SET ON 300 V

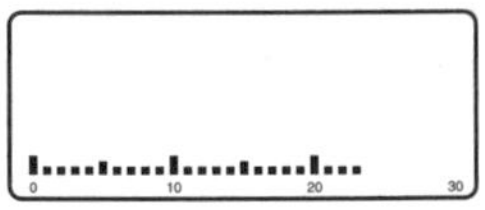

_____ **C.** Each segment equals
___ V.

RANGE SWITCH
SET ON 3 V

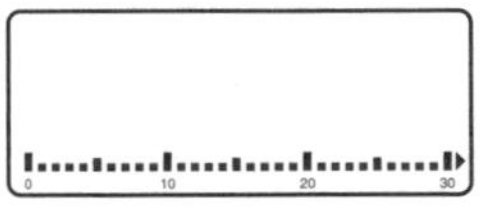

_____ **D.** Each segment equals
___ V.

RANGE SWITCH
SET ON 300 mV

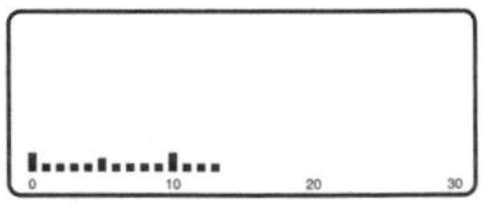

_____ **E.** Each segment equals
___ V.

RANGE SWITCH
SET ON 3000 V

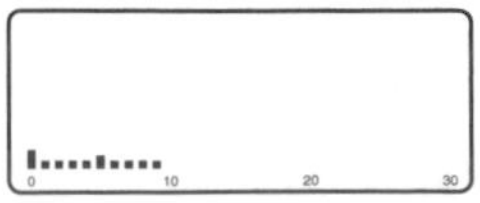

_____ **F.** Each segment equals
___ V.

RANGE SWITCH
SET ON 30 mV

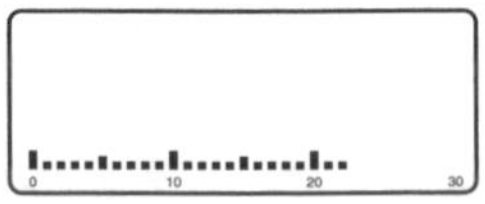

APPLICATION 5-1—Multimeters

A *multimeter* is a meter that is capable of measuring two or more electrical quantities. Multimeters may be analog or digital. The most common multimeter is a volt-ohm-milliammeter (VOM). A *VOM* is a meter that measures voltage, resistance, and current. The function switch is set on AC when measuring alternating voltage (VAC) and is set on DC when measuring direct voltage or current (DC).

If a multimeter does not have a separate setting for resistance, the positive DC setting is used. The function and range switches must be set at the correct quantity and the correct scale must be read when taking electrical measurements using a multimeter. **See Multimeter.**

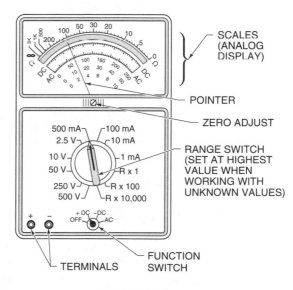

MULTIMETER

Multimeter Settings

The normal setting for measuring DC voltage or current is +DC. This setting makes the red (+) lead positive. The alternative setting for measuring DC voltage or current is –DC. This setting makes the red (+) lead negative. This setting is used when it is preferable to have the test lead with the alligator clip (black) positive and the test lead with the pointer (red) negative.

Multimeter Use

Multimeters measure many different electrical quantities. Care must be taken to ensure that a multimeter is set on the correct settings, connected to a circuit correctly, and the scale is read accurately. Ensure that a multimeter is properly set before connecting it to a circuit. **See Multimeter Use.**

To use a multimeter, apply the procedure:
1. Determine the required function (voltage, current, resistance, etc.) and set the range and/or function switch to the electrical quantity and function required.

Caution: A multimeter is damaged if it is set to measure current but is connected as a voltmeter.
2. Set the meter to the highest quantity expected. Select the highest range for unknown readings.
3. Connect the meter to the circuit per the manufacturer's recommendations. Ensure that the polarity is correct when measuring DC or voltage. Connect one lead (black lead) when connecting a meter to an unknown circuit. Slowly connect the other lead (red lead), observing the meter. Remove the lead immediately if the meter is overloaded.
4. Read the value on the meter.
5. Disconnect the meter from the circuit.

④ READ METER

SET METER TO
HIGHEST QUANTITY
EXPECTED ②

① DETERMINE
REQUIRED FUNCTION

③ CONNECT METER
TO CIRCUIT

⑤ DISCONNECT
METER

MULTIMETER USE

 Refer to **Activity 5-1—Multimeters** ....

APPLICATION 5-2—Test Lights

A *test light* is a light that is connected to two test leads and gives a visual indication when voltage is present in a circuit. The most common test light is a neon light. A *neon light* is a light that is filled with neon gas and uses two electrodes to ignite the gas. Neon lights with multiple bulbs can be used to determine the type of voltage and approximate amount of voltage.

The lamp glows if voltage is present in a circuit. If both sides of the bulb glow, the voltage in the circuit is AC. If one side glows, the voltage in the circuit is DC. The glowing side is the negative side of the DC circuit. The lamp glows brightly when high voltage is present. **See Neon Test Lights.**

Two or more test lights are connected in series to measure high-voltage circuits. Connecting test lights in series divides the voltage in the circuit between the lamps, allowing for a larger total applied voltage. When testing a circuit, one side of the neon lamp is touched to the circuit. Voltage is present if the lamp glows. High voltage makes the neon lamp glow before it is fully connected to the circuit.

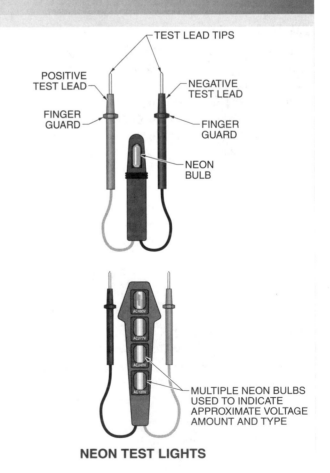

NEON TEST LIGHTS

Test Light Use

Exercise caution when testing voltages over 24 V. **See Test Light Use.**

Warning: Ensure that no body part contacts any part of the live circuit, including the metal contact points at the tip of the test light.

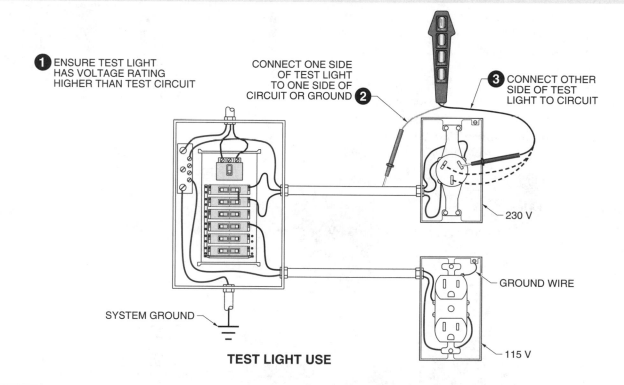

① ENSURE TEST LIGHT
HAS VOLTAGE RATING
HIGHER THAN TEST CIRCUIT

CONNECT ONE SIDE
OF TEST LIGHT
TO ONE SIDE OF
CIRCUIT OR GROUND ②

③ CONNECT OTHER
SIDE OF TEST
LIGHT TO CIRCUIT

230 V

GROUND WIRE

SYSTEM GROUND

115 V

TEST LIGHT USE

To use a test light, apply the procedure:

1. Ensure that the test light has a voltage rating higher than the highest potential voltage in the circuit. Connect two or more test lights in series if the voltage rating of the test light is less than the highest potential voltage in the circuit.

2. Connect one side of the test light to one side of the circuit or ground.

3. Connect the other side of the test light to the other side of the circuit. Voltage is present if the test light lights. Voltage is less than the rating of the test light if the test light lights dimly. Voltage is not present or is present at a very low level if the test light does not light.

Warning: If the test light does not light, a small voltage that can cause an electrical shock may still be present.

 Refer to **Activity 5-2—Test Lights** ....

APPLICATION 5-3—Voltage Testers

A *voltage tester* is a device that indicates approximate voltage level and type (AC or DC) by the movement and vibration of a pointer on a scale. Voltage testers contain a scale marked 120 VAC, 240 VAC, 480 VAC, 600 VAC, 120 VDC, 240 VDC, and 600 VDC.

Some voltage testers include a colored plunger to indicate the polarity of the test leads. If the red indicator is up, the red test lead is connected to the positive DC voltage. If the black indicator is up, the black test lead is connected to the positive DC voltage. **See Voltage Tester.**

Most voltage testers contain solenoids that vibrate when connected to AC. The solenoids do not vibrate when connected to DC. Most voltage testers include a neon light that determines if voltage is AC or DC. Both sides of the neon light glow when connected to AC. One side of the neon light glows when connected to DC. Voltage testers are designed for intermittent duty and should not be connected to a power supply for more than 15 seconds.

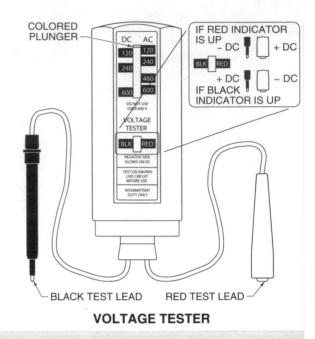

VOLTAGE TESTER

Voltage Tester Use

Exercise caution when testing AC voltages over 24 V. **See Voltage Tester Use.**

Warning: Ensure that no body part contacts any part of the live circuit, including the metal contact points at the tip of the tester.

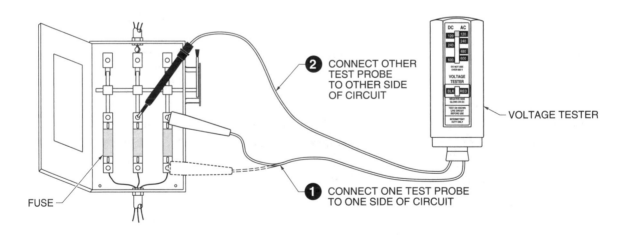

VOLTAGE TESTER USE

To use a voltage tester, apply the procedure:
1. Connect one test probe to one side of the circuit.
2. Connect the other test probe to the other side of the circuit. The indicator shows a reading and vibrates if the current is AC. The indicator shows a reading and does not vibrate if the current is DC.

 Refer to **Activity 5-3—Voltage Testers**..

APPLICATION 5-4—*Voltmeters*

A *voltmeter* is a meter that measures voltage. Voltmeters produce a visual display (analog or digital) of the amount of voltage present in a circuit. A voltmeter is used when exact voltage values are required. A voltmeter is connected directly across the part of a circuit requiring measurement. Voltage measurements are taken while equipment is operating. **See Voltmeter.**

High-voltage test leads extend the range of a voltmeter. High-voltage test leads measure voltage in TV receivers, oscilloscopes, and ignition systems. Typical high-voltage test leads available with most standard voltmeters are not intended for measuring high-voltage power circuits, such as substations, broadcast transmitters, and X-ray equipment. High-voltage power circuit measurements must be performed only by well-trained personnel.

All electric devices are designed to operate at a certain voltage level or range. Most manufacturers list the acceptable voltage range in which their equipment operates properly. Problems occur if the supply voltage is not within the equipment's acceptable range. **See Acceptable AC Voltage Ranges.**

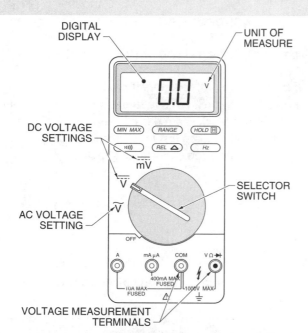

VOLTMETER

ACCEPTABLE AC VOLTAGE RANGES*		
Voltage Rating (60 Hz Power)	Minimum Voltage	Maximum Voltage
24	22	26
115	109	121
120	114	126
208	197	218
230	218	242
240	228	252
277	263	291
460	436	484
480	456	504

* in volts

Low voltage occurs due to loose connections, corrosion at a connection, undersized feeder conductors, damaged conductors, or when additional loads are added to a circuit.

Voltmeter—DC Voltage Measurement.

Exercise caution when measuring DC voltages over 60 V. **See Voltmeter — DC Voltage Measurement.**

Warning: Ensure that no body part contacts any part of a live circuit, including the metal contact points at the tip of the test leads.

To measure DC voltages with a voltmeter, apply the procedure:
1. Set the selector switch to DC voltage. Select a setting high enough to measure the highest possible circuit voltage if the meter has more than one voltage position. Set the meter on the highest voltage setting if the circuit voltage is unknown.
2. Plug the black test lead into the common jack. The common jack may be marked com (common), – (negative), lo (low), or have a black collar ring.

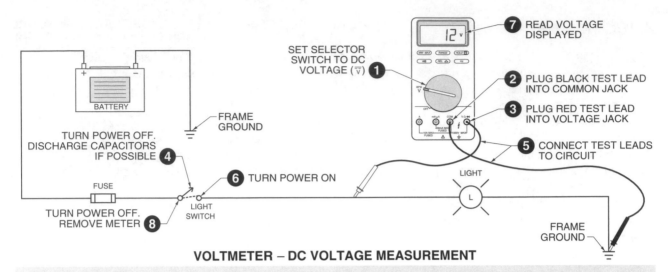

VOLTMETER – DC VOLTAGE MEASUREMENT

3. Plug the red test lead into the voltage jack. The voltage jack may be marked V (voltage), + (positive), hi (high), or have a red collar ring.

4. Turn the power to the circuit or device being tested OFF. Discharge all capacitors if possible.

5. Connect the meter test leads in the circuit. Connect the black (–) test lead to circuit ground and the red (+) test lead to the point at which the voltage is under test. Reverse the black and red test leads if a negative sign appears in front of the reading on a digital meter. Reverse the black and red test leads or change the polarity switch (if applicable) to – DC if the needle moves in the reverse direction on an analog meter.

6. Turn the power to the circuit ON.

7. Read the voltage displayed on the meter.

8. Turn the power to the circuit OFF and remove the meter.

Voltmeter—AC Voltage Measurement

Exercise caution when measuring AC voltages over 24 V. **See Voltmeter—AC Voltage Measurement.**

Warning: Ensure that no body part contacts any part of the live circuit, including the metal contact points at the tip of the test leads.

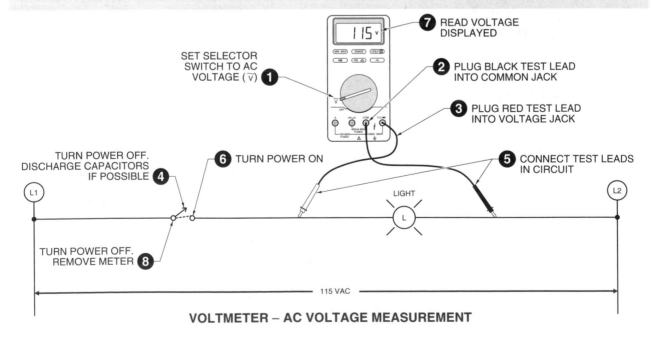

VOLTMETER – AC VOLTAGE MEASUREMENT

To measure AC voltages with a voltmeter, apply the procedure:

1. Set the selector switch to AC voltage. Select the highest setting if the meter has more than one voltage position. Set the meter on the highest voltage setting if the voltage in the circuit is unknown.

2. Plug the black test lead into the common jack. The common jack may be marked com (common), – (negative), lo (low), or have a black collar ring.

3. Plug the red test lead into the voltage jack. The voltage jack may be marked V (voltage), + (positive), hi (high), or have a red collar ring.

4. Turn the power to the circuit or device OFF. Discharge all capacitors if possible.

5. Connect the test leads in the circuit. The position of the test leads is arbitrary. Safe industrial practice is to connect the black test lead to the grounded (neutral) side of the AC voltage first and then the red lead. Remove the red lead first and then the black lead.

6. Turn the power to the circuit ON, if not already on.

7. Read the voltage displayed on the meter.

8. Turn the power to the circuit OFF (if required) and remove the meter.

 Refer to **Activity 5-4—Voltmeters** ....

APPLICATION 5-5—In-Line Ammeters

An *in-line ammeter* is a meter that measures current in a circuit by inserting the meter in series with the component under test. In-line ammeters measure either DC or AC. The ammeter leads must be connected with the correct polarity at the test opening when measuring DC with an in-line ammeter. An in-line ammeter is typically limited to measuring AC (and/or DC) current of 10 A or less.

When a motor is started, its current draw is 6 to 8 times the normal running current of the motor. An in-line ammeter is kept on a high scale to prevent damage to the meter if loads are switched ON when the ammeter is connected to the circuit.

In-Line Ammeter—DC Measurement

Care must be taken to protect the meter, circuit, and person using the meter when measuring DC with an in-line ammeter. Always apply the following rules when using an in-line ammeter:

- Check that the power to the test circuit is OFF before connecting and disconnecting test leads.

- Do not change the position of any switches or settings on the meter while the circuit under test is energized.

- Turn the power to the meter and circuit OFF before any meter settings are changed.

- Connect the ammeter in series with the components to be tested.

- Do not take current readings from any circuit in which the current may exceed the limit of the meter.

Many meters include a fuse in the low ampere range to prevent meter damage caused by excessive current. Before using a meter, check to see if the meter is fused on the current range being used. The meter is marked as fused or not fused at the test lead terminals. If the meter is not fused, an external fuse may be connected in series with the meter test leads. To protect the meter, the fuse rating should not exceed the current range of the meter. **See In-Line Ammeter — DC Measurement.**

Warning: Ensure that no body parts contact any part of the live circuit, including the metal contact points at the tip of the test leads.

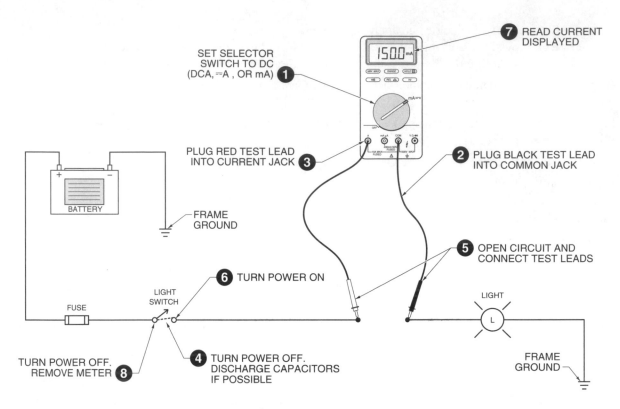

IN-LINE AMMETER – DC MEASUREMENT

To measure DC using an in-line ammeter, apply the procedure:
1. Set the selector switch to DC. Select a setting high enough to measure the highest possible circuit current if the meter has more than one DC position.
2. Plug the black test lead into the common jack. The common jack may be marked com (common), − (negative), lo (low), or have a black collar ring.
3. Plug the red test lead into the current jack. The current jack may be marked + (positive), mA (milliamps), hi (high), or have a red collar ring.
4. Turn the power to the circuit or device under test OFF and discharge all capacitors if possible.
5. Open the circuit and connect the test leads to each side of the opening. The black (negative) test lead is connected to the negative side of the opening and the red (positive) test lead is connected to the positive side of the opening. Reverse the black and red test leads if a negative sign appears in front of the displayed reading.
6. Turn the power to the circuit under test ON.
7. Read the current displayed on the meter.
8. Turn the power OFF and remove the meter from the circuit.

The same procedure is used to measure AC with an in-line ammeter, except that the selector switch is set on AC current.

 Refer to **Activity 5-5—In-Line Ammeters** ....

APPLICATION 5-6—Clamp-On Ammeters

A *clamp-on ammeter* is a meter that measures current in a circuit by measuring the strength of the magnetic field around a conductor. Clamp-on ammeters measure currents from .01 A or less to 1000 A or more. Most clamp-on ammeters measure AC, but DC clamp-on ammeters (or AC/DC) are also available. **See Clamp-On Ammeter.**

A clamp-on ammeter operates on the magnetic field present around a conductor as current flows through a circuit. Clamp-on ammeters cannot be connected incorrectly because there is no direct connection to the live circuit.

Clamp-on ammeters take current readings without opening a circuit. The jaws of a clamp-on ammeter are opened and encircled around the conductor under test. The clamp must be placed around only one conductor at a time. Readings are taken on bare or insulated conductors. When the jaws close, a current reading is indicated on the meter's scale. The reading indicates the amount of current drawn by loads connected to the conductor.

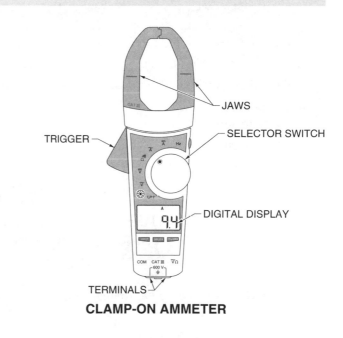

CLAMP-ON AMMETER

Clamp-On Ammeter — AC Measurement

Clamp-on ammeters measure the current in a circuit by measuring the strength of the magnetic field around a conductor. Care must be taken to ensure that the meter does not pick up stray magnetic fields. Whenever possible, separate the conductors under test from other surrounding conductors by a few inches. **See Clamp-On Ammeter—AC Measurement.**

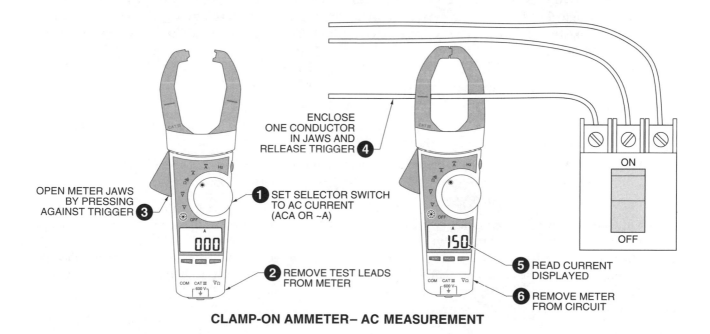

CLAMP-ON AMMETER— AC MEASUREMENT

To measure AC using a clamp-on ammeter, apply the procedure:

1. Set the selector switch to AC current. Select a setting high enough (10 A or 100 A) to measure the highest possible circuit current if the meter has more than one current position. Set the clamp-on ammeter on the highest current setting if the circuit current is unknown.

2. Remove the test leads from the meter to prevent the leads from accidentally becoming part of the circuit.

3. Open the meter's jaws by pressing against the trigger.

4. Enclose one conductor in the jaws and release the trigger. Ensure that the jaws are completely closed before taking readings.

5. Read the current displayed on the meter.

6. Remove the meter from the circuit.

The same procedure is used to measure DC with a clamp-on ammeter, except that the selector switch is set to the DC position.

The jaws are opened and closed immediately when taking a DC measurement to reduce the possibility of error. To further reduce the possibility of error, a reading is taken, the jaws are reversed, a second reading is taken, and the two readings are averaged.

 Refer to **Activity 5-6—Clamp-On Ammeters** ....

APPLICATION 5-7—Ohmmeters

An *ohmmeter* is an instrument that measures resistance. The higher the resistance reading, the lower the current flow. The lower the resistance reading, the higher the current flow.

Ohmmeter leads are connected across the device under test with all power removed from the circuit. An ohmmeter supplies power through batteries inside the meter. A standard ohmmeter cannot measure high resistance values, such as several million ohms. A high applied voltage is required to measure high resistance values.

Ohmmeter Resistance Measurement

Ensure that no voltage is present in the circuit or component under test before taking resistance measurements. Low voltage applied to a meter set to measure resistance causes inaccurate readings. High voltage applied to a meter set to measure resistance causes meter damage. Check for voltage using a voltmeter. **See Ohmmeter Resistance Measurement.**

To measure resistance using an ohmmeter, apply the procedure:

1. Check that all power is OFF in the circuit or component under test.

2. Set the selector switch to the resistance position, which is marked R × 1, R × 100, R × 10,000, etc., on analog meters. The resistance position is marked Ω on digital meters.

3. Plug the black test lead into the common jack. The common jack may be marked com (common), – (negative), or lo (low).

4. Plug the red test lead into the Ω jack.

5. Ensure that the meter batteries are in good condition and then zero the meter. On analog meters, the condition of the batteries is determined by the movement of the pointer. An analog meter is zeroed by touching the two test leads together and setting the zero adjustment to place the pointer on the "0" resistance mark. On digital meters, the battery symbol is displayed when the batteries are low. Digital meters are zeroed by the manufacturer.

6. Connect the meter test leads across the component under test. Ensure that contact between the test leads and the circuit is good. Dirt, solder flux, oil, and other foreign substances greatly affect resistance readings.

7. Read the resistance displayed on the meter. Check the circuit schematic for parallel paths. Paths in parallel with the resistance under test cause reading errors. Remove one end of the component under test from the circuit if parallel paths exist.

8. Turn the meter OFF after measurements are taken to save battery life.

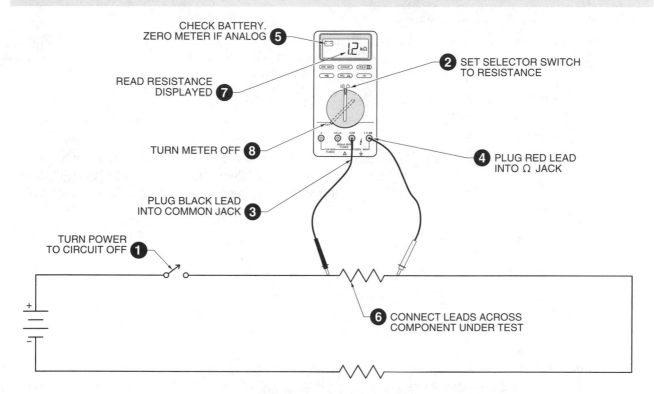

CHECK BATTERY.
ZERO METER IF ANALOG ❺

❷ SET SELECTOR SWITCH TO RESISTANCE

READ RESISTANCE DISPLAYED ❼

TURN METER OFF ❽

❹ PLUG RED LEAD INTO Ω JACK

PLUG BLACK LEAD INTO COMMON JACK ❸

TURN POWER TO CIRCUIT OFF ❶

❻ CONNECT LEADS ACROSS COMPONENT UNDER TEST

OHMMETER RESISTANCE MEASUREMENT

 Refer to **Activity 5-7—Ohmmeters** ....

APPLICATION 5-8—Continuity Checkers

A *continuity checker* is an instrument that indicates an open or closed circuit in a circuit in which all power is OFF. A continuity checker is an economical tester that is used to test switches, fuses, grounds, and is also used for identifying individual conductors in a multiwire cable.

A bulb on the continuity checker lights when there is a path for current to flow. Expensive models also give an audible indication when there is a complete path. Most multimeters have a built in continuity checker. **See Continuity Test.**

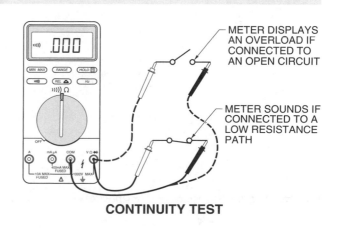

METER DISPLAYS AN OVERLOAD IF CONNECTED TO AN OPEN CIRCUIT

METER SOUNDS IF CONNECTED TO A LOW RESISTANCE PATH

CONTINUITY TEST

Short or Open Conductor Testing

Multiconductor cable is used in communication, entertainment, automotive, and security applications. A short or open may develop in a conductor if the cable is abused. A continuity checker tests for a shorted or open conductor. **See Short Conductor Testing** and **Open Conductor Testing.**

To test a multiconductor cable for a short between two conductors, apply the procedure:
1. Set selector switch to continuity setting.
2. Connect one test lead to one conductor.
3. Connect the other test lead to each of the remaining conductors. The meter sounds if any two conductors are shorted. Connect the first test lead to a different conductor and the other test lead to each of the remaining conductors until all are checked.

To test a multiconductor cable for an open conductor, apply the procedure:
1. Set selector switch to continuity setting.
2. Connect one pair of conductors.
3. Connect the meter test leads to the other ends of the conductors. The meter sounds if a conductor is continuous. The meter does not sound if a conductor is open. Connect each conductor to a different conductor in the cable to determine which conductor is open. Connect two different conductors until all are checked.

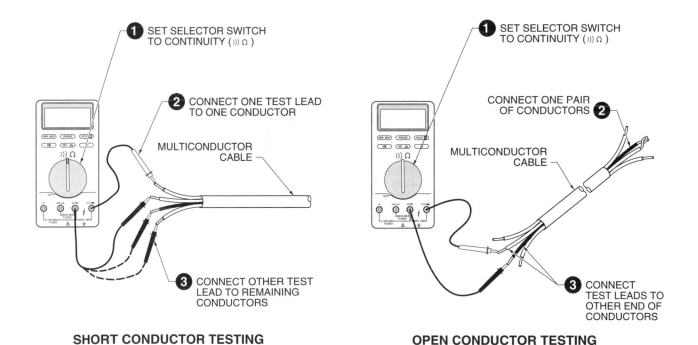

SHORT CONDUCTOR TESTING

OPEN CONDUCTOR TESTING

 *Refer to **Activity 5-8—Continuity Checkers** ....*

APPLICATION 5-9—Wattmeters

A *wattmeter* is an instrument that measures true power in a circuit. A wattmeter accounts for the phase difference between current and voltage. **See Wattmeter.** Wattmeters are designed for measuring power in single-phase and three-phase circuits. Apparent power is calculated after voltage and current are measured separately.

Wattmeter Power Measurement

Most wattmeters have more than one power range. Always start with the highest range if a wattmeter has more than one power range. **See Wattmeter Power Measurement.**

To measure power with a wattmeter, apply the procedure:

1. Ensure that the voltage in the circuit does not exceed the voltage limit of the meter. Check the circuit voltage using a voltmeter. A voltmeter is used on high-voltage circuits because it has a higher voltage range than a wattmeter.

2. Ensure that the current in the circuit does not exceed the current limit of the meter. Check the current using a clamp-on ammeter. A clamp-on ammeter is used on high-current circuits because it has a higher current range than a wattmeter.

3. Set the wattmeter at the highest wattage range.

4. Connect the circuit under test to the wattmeter with the power OFF. Check that the power is OFF by using a voltmeter.

5. Turn the power to the test circuit ON after the meter is properly connected.

6. Read the voltage, current, and power displayed on the meter. *Note:* The displayed power measurement is true power.

7. Turn the power OFF and remove the meter from the circuit.

CLAMP-ON CURRENT TRANSFORMER FOR CURRENT MEASUREMENT

VOLTAGE MEASUREMENT PROBES

OFF   W   kW   Vrms   Arms

WATTMETER

Multiply the voltage reading by the current reading to determine the apparent power in the circuit. Divide true power by apparent power and multiply by 100 to determine the efficiency of the circuit.

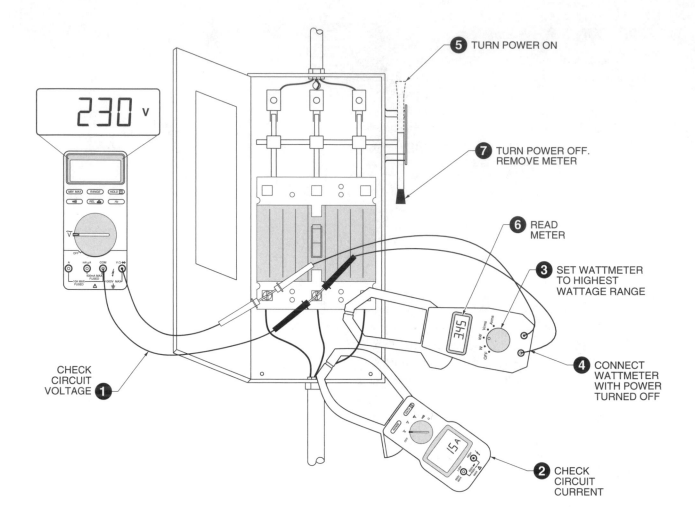

5 TURN POWER ON

7 TURN POWER OFF. REMOVE METER

6 READ METER

3 SET WATTMETER TO HIGHEST WATTAGE RANGE

4 CONNECT WATTMETER WITH POWER TURNED OFF

CHECK CIRCUIT VOLTAGE **1**

2 CHECK CIRCUIT CURRENT

WATTMETER POWER MEASUREMENT

Refer to **Activity 5-9—Wattmeters** ....

 Refer to Quick Quiz® on CD-ROM

 Refer to Chapter 5 in the **Troubleshooting Electrical/Electronic Systems Workbook** for additional questions.

Meters

Name _____   **Date** _____

ACTIVITY 5-1—Multimeters

Set the selector switches to measure the electrical quantities required. Draw the position on each meter. Select the setting that places the pointer in the center of the scale if more than one setting is possible.

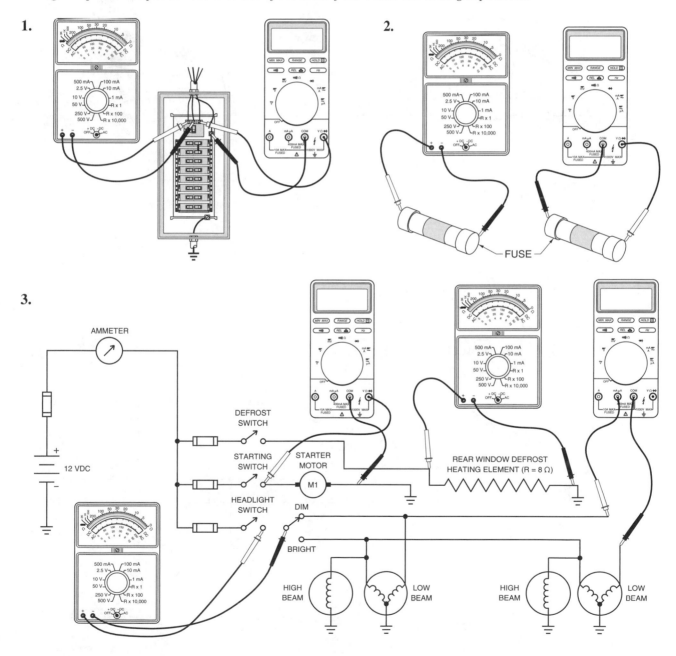

ACTIVITY 5-2—Test Lights

Determine the problem. Select the number of the problem from the list of possible problems.

1. The rear window defroster does not melt ice.

_____ **A.** The problem is ___.

_____ **B.** Potential problem(s) are ___.

Number	Possible Problems
1	Open fuse
2	Loose or corroded connection
3	Load has an internal open circuit
4	Broken wire
5	Faulty control switch
6	Open ground wire

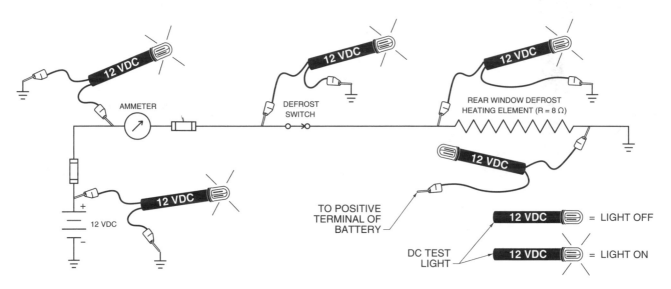

2. The high beams do not work on both headlights.

_____ **A.** The problem is ___.

_____ **B.** Potential problem(s) are ___.

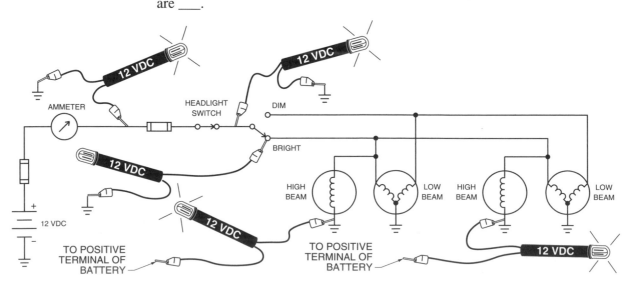

3. The fan motor operates only in high speed.

_____ **A.** The problem is ___.

_____ **B.** Potential problem(s) are ___.

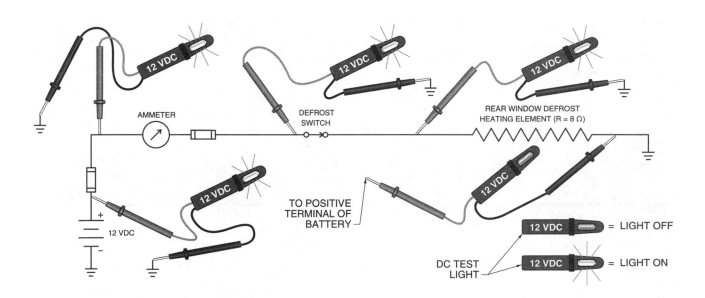

4. One dashboard light does not work even after a new bulb is installed.

_____ **A.** The problem is ___.

_____ **B.** Potential problem(s) are ___.

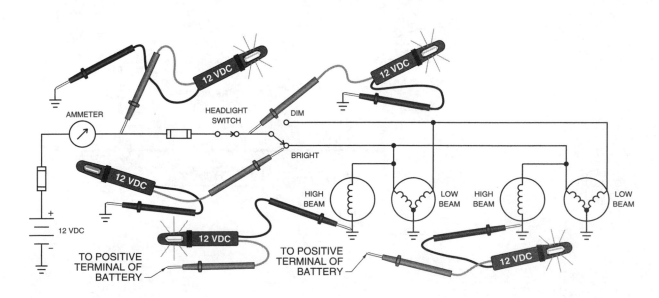

ACTIVITY 5-3—Voltage Testers

Use the Voltage Test tables to determine the problems.

1. A voltage tester is used to test a split-wired receptacle circuit. In the split-wired circuit, the top half of the receptacle is controlled by the switch. The bottom half of the receptacle has power at all times.

_____ **A.** The problem is ___.   _____ **B.** The problem is ___.

VOLTAGE TEST 1		
Voltage Tester	Test Position	Voltage Tester Reading
1	1	115 V (at all times)
1	2	0 V (switch in ON positon)
1	2	0 V (switch in OFF position)
2	1	0 V (switch in ON position)
2	1	0 V (switch in OFF position)
2	2	115 V (switch in ON position)
2	2	115 V (switch in OFF position)

VOLTAGE TEST 2		
Voltage Tester	Test Position	Voltage Tester Reading
1	1	115 V (at all times)
1	2	0 V (switch in ON position)
1	2	0 V (switch in OFF position)
2	1	115 V (switch in ON position)
2	1	0 V (switch in OFF position)
2	2	115 V (switch in ON position)
2	2	115 V (switch in OFF position)

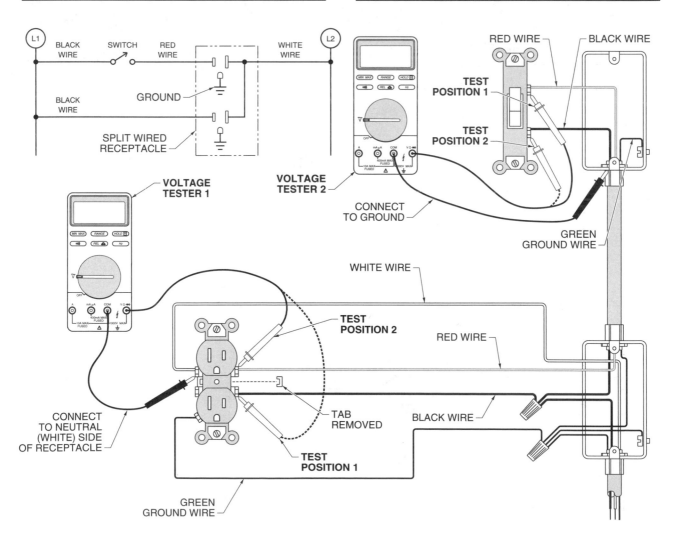

2. A voltage tester is used to test a lamp that is controlled by two 3-way switches.

_____ **A.** The problem is ___.   _____ **B.** The problem is ___.

VOLTAGE TEST 3		
Voltage Tester	Test Position	Voltage Tester Reading
1	1	115 V (at all times)
1	2	0 V (switches in any positon)
2	1	115 V (switch 1 in up position)
2	1	0 V (switch 1 in down position)
2	2	0 V (switch 1 in down position)
3	1	0 V (at all times)
3	2	0 V (switch 2 in up position)
3	2	0 V (switch 2 in down position)

VOLTAGE TEST 4		
Voltage Tester	Test Position	Voltage Tester Reading
1	1	115 V (at all times)
1	2	0 V (switches in any positon)
2	1	115 V (switch 1 in up position)
2	1	115 V (switch 1 in down position)
2	2	115 V (at all times)
3	1	115 V (switch 2 in up position)
3	1	115 V (switch 2 in down position)

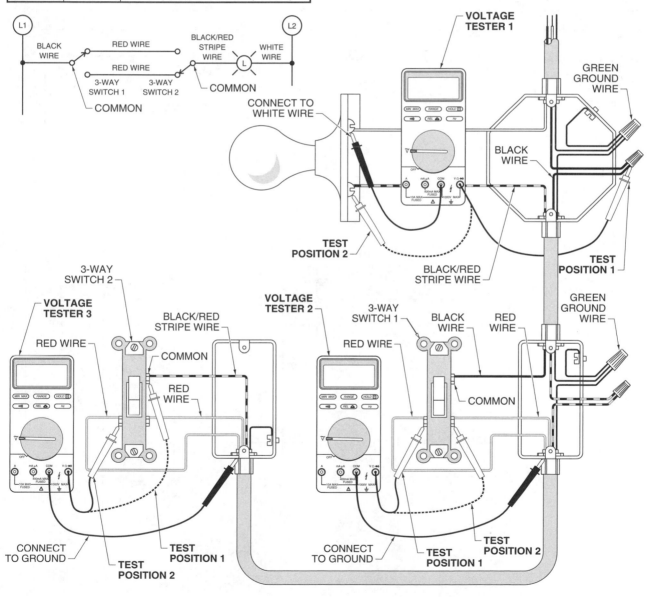

ACTIVITY 5-4 — Voltmeters

1. Set the selector switch(es) and connect each voltmeter to measure the circuit voltage. Draw the position on each meter. Connect Voltmeter 1 to measure the voltage out of the transformer feeding the motor power circuit. Connect Voltmeter 2 to measure the voltage out of the fuse in the motor power circuit. Connect Voltmeter 3 to measure the voltage into the magnetic motor starter. Connect Voltmeter 4 to measure the voltage at the motor's armature when the motor is operating in the forward direction.

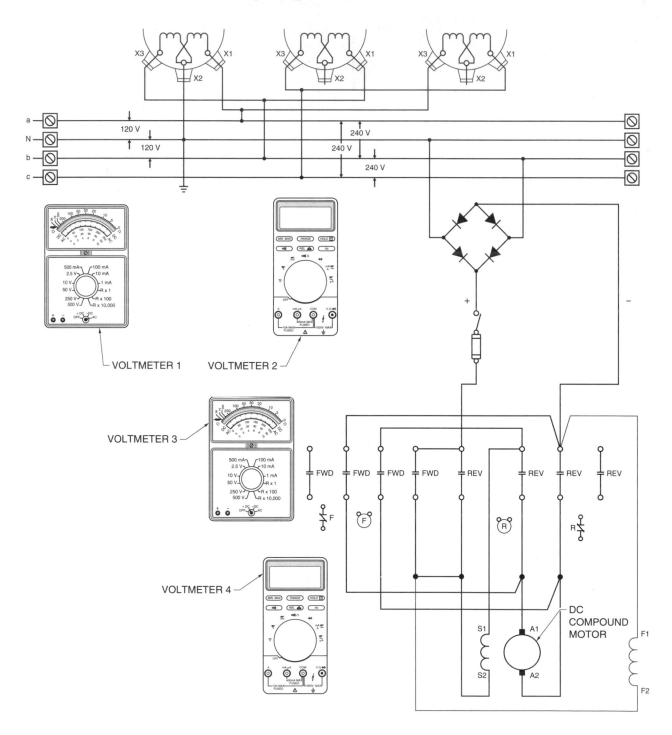

ACTIVITY 5-5—In-Line Ammeters

1. Complete the Power Pack Charging Circuit adding the in-line ammeters. Set the selector switches to measure the circuit current. Draw the position on each meter. Connect Ammeter 1 to measure current through the 120 VAC circuit. Connect Ammeter 2 to measure current in the stepped down AC circuit. Connect Ammeter 3 to measure current out of the positive side of the rectifier circuit.

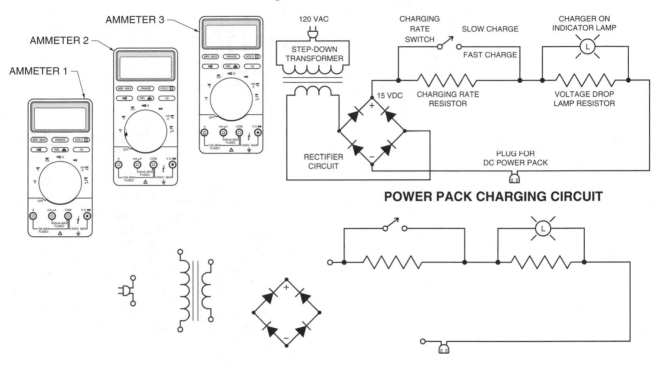

POWER PACK CHARGING CIRCUIT

ACTIVITY 5-6—Clamp-On Ammeters

Current in a circuit is indirectly measured using a clamp-on ammeter by wrapping a current-carrying conductor around the meter jaws. The indirect method is used when the current in the circuit is lower than the lowest limit of the meter. Wrapping the conductor around the meter jaws increases the strength of the magnetic field. The actual current value in the circuit is calculated by dividing the current reading on the meter by the number of times the conductor is wrapped around the jaws. Determine the actual value of current flowing in the circuit. List the current value in the units given.

_____ **1.** The reading equals ___ mA.

_____ **2.** The reading equals ___ mA.

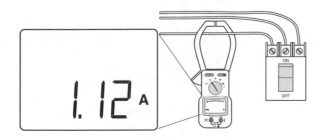

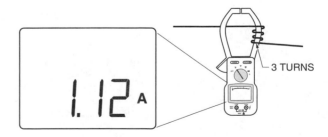

_____ **3.** The reading equals
___ A.

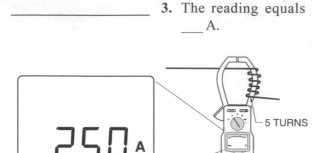
5 TURNS

_____ **4.** The reading equals
___ µA.

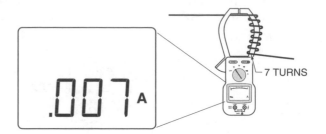
7 TURNS

ACTIVITY 5-7—Ohmmeters

List the correct reading for the function/range settings.

_____ **1.** The reading equals
___ kΩ.

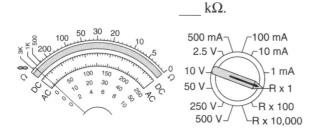

_____ **2.** The reading equals
___ Ω.

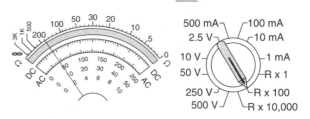

_____ **3.** The reading equals
___ Ω.

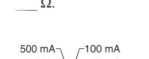
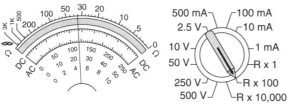

_____ **4.** The reading equals
___ Ω.

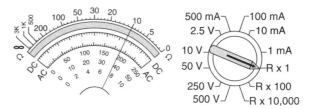

_____ **5.** The reading equals
___ Ω.

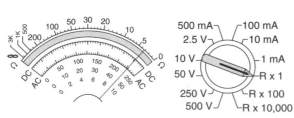
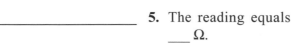

_____ **6.** The reading equals
___ MΩ.

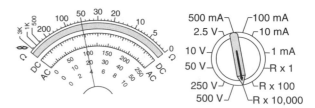
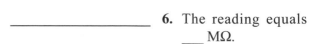

ACTIVITY 5-8—Continuity Checkers

Using the meter connections, determine whether the device under test is good or bad.

_____ **1.** The SPST switch is ___.

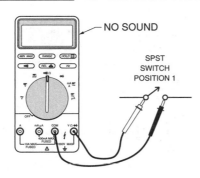

_____ **2.** The three-way switch is ___.

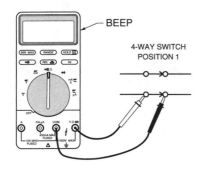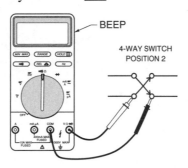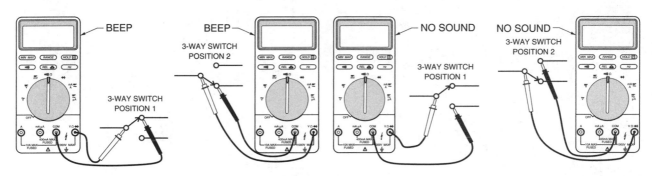

_____ **3.** The four-way switch is ___.

_____ **4.** Fuse 1 is ___. _____ **5.** Fuse 2 is ___.

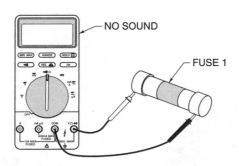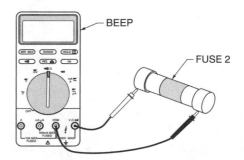

ACTIVITY 5-9—Wattmeters

Determine the circuit apparent power, true power, and efficiency.

_____ 1. Apparent power = ___ W   _____ 4. Apparent power = ___ W

_____ 2. True power = ___ W   _____ 5. True power = ___ W

_____ 3. Efficiency = ___%   _____ 6. Efficiency = ___%

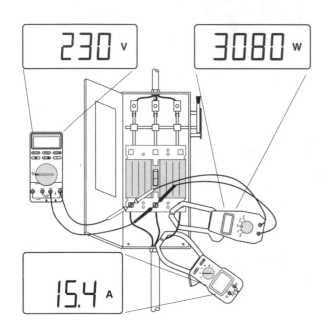

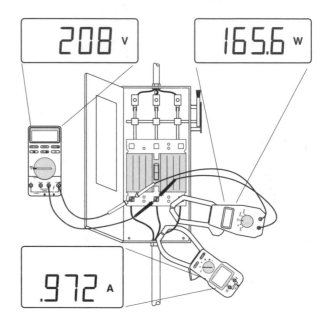

APPLICATION 6-1—Light Meters

A *light meter* is an instrument that measures light levels in footcandles (fc) or lumens (lm). Electricians use light meters to evaluate and document the amount of light present for an application. This ensures that the proper amount of light is available to conform to work, safety, and security standards.

The amount of light illuminating a surface depends on the number of lumens produced by the light source and the overall distance from the light source to a surface. The lower the lumen output or the greater the distance from the light source, the lower the amount of light that falls on a surface. **See Light Meter Measurement Procedure.**

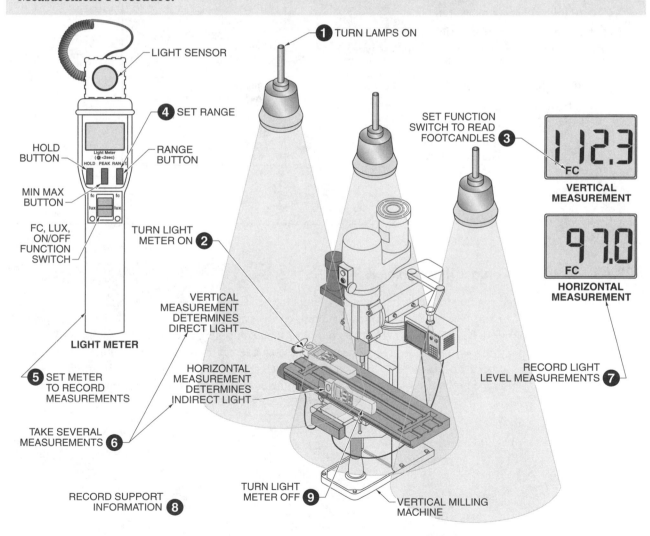

LIGHT METER MEASUREMENT PROCEDURE

To measure the amount of light with a light meter, apply the procedure:

1. Turn the lamps ON in the area where the amount of light is to be measured. If natural light is present (windows, skylights, etc.), readings should be taken at several different times.

2. Turn the light meter ON.

3. Set the function switch to the footcandles or lux mode.

4. Set the selector switch to the expected range, for example 40 fc or 400 fc.

5. Set the light meter to record light measurements with MIN MAX and/or PEAK modes.

6. Take several light-level measurements at various locations and at various angles.

7. Record all light-level measurements.

8. Record any support information, such as when measurements were taken and if shades were open or closed.

9. Turn the light meter OFF.

The light level required on a surface varies widely for each lighting application. For example, an operating table in a hospital requires 2500 fc, while the lobby of a hospital requires only 30 fc. **See Recommended Light Levels.**

RECOMMENDED LIGHT LEVELS	
Area	**Light Level***
Auditorium	
Exhibitions	30
Banks	
Lobby, general	50
Teller station	150
Garages (Auto)	
Parking	10
Repair	50
Hospital/Medical	
Lobby	30
Operating table	2500
Machine Shop	
Rough bench	50
Medium bench	100
Materials handling	
Picking stock	30
Packing, labeling	50
Offices	
Regular office work	100
Accounting	150
Printing	
Proofreading	150
Color inspecting	200
Schools	
Auditoriums	20
Classrooms	60-100
Indoor gyms	30-40
Stores	
Stockroom	30
Service area	100

* in footcandles

 Refer to **Activity 6-1—Light Meters**..

APPLICATION 6-2—Contact Thermometers

A *contact thermometer* is an instrument that measures temperature at a single point. Temperature is measured when troubleshooting because the resistance of most materials changes as the temperature of the materials change. An increase in temperature decreases the performance of electrical equipment and destroys insulation.

Temperature is measured with a contact thermometer by placing a temperature probe in contact with the area to be measured. **See Contact Thermometer.** Contact thermometers measure the temperature of solids, liquids, and gases, depending on the probe used. Most contact thermometer temperature probes are interchangeable.

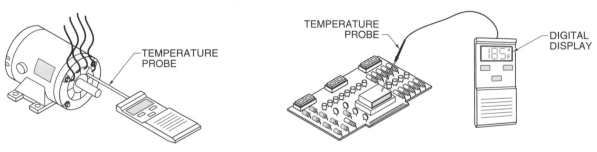

CONTACT THERMOMETER

Contact Thermometer Temperature Measurement

Measuring hot devices presents a safety hazard. Ensure that proper equipment is used and that safety precautions are taken. Use a temperature probe that is rated higher than the highest possible temperature and always wear safety glasses. **See Contact Thermometer Temperature Measurement.**

Warning: Avoid contact with any material that can cause burns.

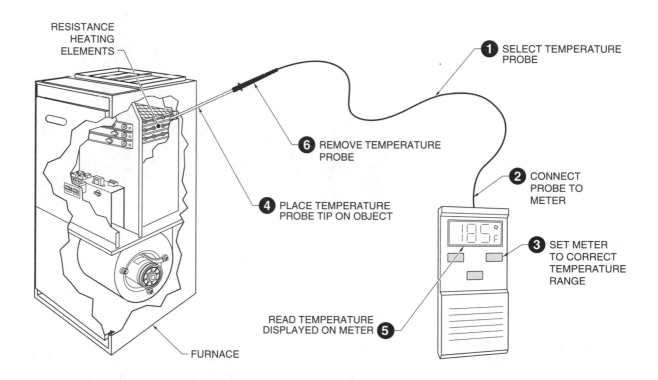

CONTACT THERMOMETER TEMPERATURE MEASUREMENT

To measure temperature using a contact thermometer, apply the procedure:

1. Select a temperature probe. A probe must have a higher temperature rating than the highest temperature it may contact and should have a shape that allows good contact with the device under test.

2. Connect the temperature probe to the meter.

3. Set the meter to the correct temperature range. Select the highest range if the temperature is unknown.

4. Place the temperature probe tip on the object or in the area to be measured.

5. Read the temperature displayed on the meter.

6. Remove the temperature probe from the object or area under test.

 Refer to **Activity 6-2—Contact Thermometers**...

APPLICATION 6-3—Infrared Meters

An *infrared meter* is a meter that measures heat energy by measuring the infrared energy that a material emits. All materials emit infrared energy in proportion to their temperature.

Infrared meters detect small amounts of heat in electrical distribution systems, motors, and switching circuits. A problem is usually caused by a poor electrical connection. The poorer the connection, the higher the heat level in the system.

An infrared meter displays an image on a screen that shows different temperatures indicated by different colors or gives a direct digital temperature readout. **See Infrared Meter.**

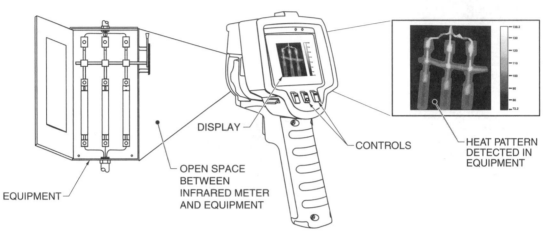

INFRARED METER

Loose, corroded, or dirty electrical connections generate unwanted resistance and heat. The temperature rise at a connection depends on the current flowing through the connection and the resistance of the connection. A temperature rise of 20°F above ambient temperature indicates a fault that requires routine maintenance. Routine maintenance is performed so that the problem is fixed before it causes damage. A 100°F increase requires immediate action. Immediate action involves the immediate shut down of a system for repair of the fault. **See Equipment Condition.**

Infrared Meter Temperature Measurement

Resistance in a normal circuit occurs at the load. Resistance in a circuit that has poor connections, corrosion, or other high resistance paths occurs at points other than the load. Infrared meter temperature measurements prevent problems by locating unwanted heat. **See Infrared Meter Temperature Measurement.**

To measure temperature using an infrared meter, apply the procedure:

1. Aim meter at area to be measured. Focus meter based on distance of the object from meter.
2. Take ambient temperature reading for reference.
3. Take temperature reading of any areas suspected to have temperatures above ambient temperature.

EQUIPMENT CONDITION	
Temperature Difference*	**Indication**
45	Light load on circuit
60	Heavy load on circuit
85	Possible problem. Schedule routine maintenance
100	Dangerous problem. Take immediate corrective action

* in °F

To determine temperature readings in such areas, subtract the ambient temperature from the reading obtained.

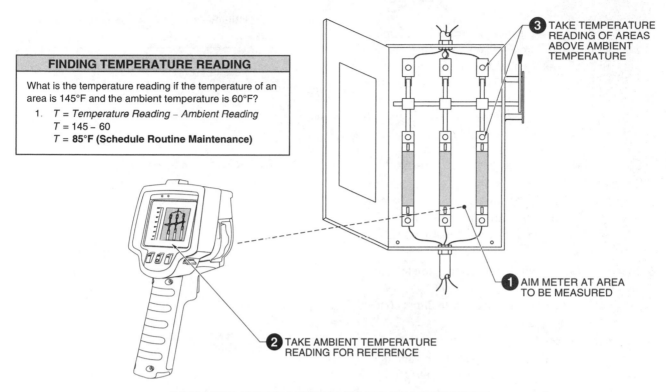

③ TAKE TEMPERATURE READING OF AREAS ABOVE AMBIENT TEMPERATURE

FINDING TEMPERATURE READING

What is the temperature reading if the temperature of an area is 145°F and the ambient temperature is 60°F?

1. *T = Temperature Reading – Ambient Reading*
   T = 145 – 60
   T = **85°F (Schedule Routine Maintenance)**

① AIM METER AT AREA TO BE MEASURED

② TAKE AMBIENT TEMPERATURE READING FOR REFERENCE

INFRARED METER TEMPERATURE MEASUREMENT

 Refer to **Activity 6-3—Infrared Meters**...

APPLICATION 6-4—Contact Tachometers

A *tachometer* is a device that measures the speed of a moving object. Types of tachometers include contact, photo, and strobe tachometers. **See Contact, Photo, and Strobe Tachometers.** A *contact tachometer* is a device that measures the rotational speed of an object through direct contact of the tachometer tip with the object to be measured. A contact tachometer is used to measure the speed of rotating objects, such as motors, gears, belts, and pulleys. It can also be used to measure linear speeds, such as moving conveyors, webs, and printing presses. A contact tachometer measures speeds from .1 rpm to 25,000 rpm.

Electricity drives motors that produce a rotating mechanical force. The amount of force a motor produces depends on the size and speed of the motor. Speed may be rotational or linear.

CONTACT TACHOMETER

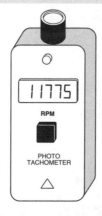

PHOTO TACHOMETER

STROBE TACHOMETER

Rotational speed is distance traveled per unit of time in a circular direction. Rotational speed is measured in revolutions per minute (rpm). *Linear speed* is distance traveled per unit of time in a straight line. Linear speed is measured in miles per hour (mph), feet per second (fps), or inches per second (ips). **See Rotational Speed and Linear Speed.**

Contact Tachometer Speed Measurement

Exercise caution when working around moving objects. Use a photo or strobe tachometer if there is danger of contact with a moving object. **See Contact Tachometer Speed Measurement.**

To measure speed with a contact tachometer, apply the procedure:
1. Place the tip of the tachometer in direct contact with the moving object.
2. Read the speed displayed on the meter.

Rotating speeds are displayed in rpm, and linear speeds are displayed in feet per minute (fpm) or meters per minute (mpm).

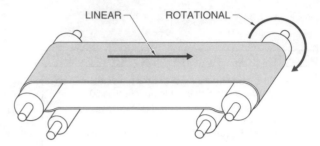

ROTATIONAL SPEED AND LINEAR SPEED

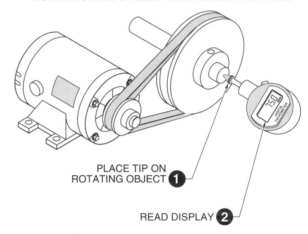

PLACE TIP ON ROTATING OBJECT ❶

READ DISPLAY ❷

CONTACT TACHOMETER SPEED MEASUREMENT

 Refer to **Activity 6-4—Contact Tachometers** ....

APPLICATION 6-5—Photo Tachometers

A *photo tachometer* is a device that measures the speed of an object without direct contact with the object. A photo tachometer measures speed by focusing a light beam on a reflective area or a section of reflective tape.

A photo tachometer is used in applications in which the rotating object cannot be reached or touched. A photo tachometer cannot measure linear speeds. Photo tachometers measure speeds from 1 rpm to 99,999 rpm.

Photo Tachometer Speed Measurement

Exercise caution when working around moving objects. Turn OFF, lockout, and tag disconnected switches of motors and rotating machinery as required. Use a photo or strobe tachometer if there is danger of contact with a moving object. **See Photo Tachometer Speed Measurement.**

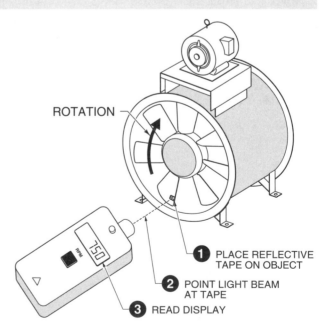

ROTATION

❶ PLACE REFLECTIVE TAPE ON OBJECT

❷ POINT LIGHT BEAM AT TAPE

❸ READ DISPLAY

PHOTO TACHOMETER SPEED MEASUREMENT

To measure speed with a photo tachometer, apply the procedure:
1. Place reflective tape on the object to be measured.
2. Point the light beam of the photo tachometer at the reflective tape.
3. Read the speed displayed on the meter.

Rotating speeds are displayed in rpm.

 Refer to **Activity 6-5—Photo Tachometers** ....

APPLICATION 6-6—Strobe Tachometers

A *strobe tachometer* is a device that uses a flashing light to measure the speed of a moving object. A strobe tachometer measures speed by synchronizing its light's flash rate with the speed of the moving object.

A strobe tachometer allows speed measurements through glass. This eliminates the possibility of direct exposure with hazardous areas. Strobe tachometers are also used for analysis of motion and vibration. Strobe tachometers measure speeds from 20 rpm to 100,000 rpm.

Strobe Tachometer Speed Measurement

Exercise caution when working around moving objects. Use a photo or strobe tachometer if there is danger of contact with a moving object. **See Strobe Tachometer Speed Measurement.**

To measure speed with a strobe tachometer, apply the procedure:
1. Set the meter for the best speed range for the application.
2. Turn the tachometer ON and align the visible light beam with the object to be measured.
3. Read the speed displayed on the meter.

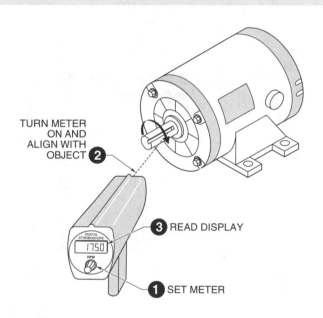

STROBE TACHOMETER SPEED MEASUREMENT

 Refer to **Activity 6-6—Strobe Tachometers** ....

APPLICATION 6-7—Megohmmeters

A *megohmmeter* is a device that detects insulation deterioration by measuring high resistance values under high test voltage conditions. Megohmmeter test voltages range from 50 V to 5000 V. A megohmmeter detects insulation failure or potential failure of insulation caused by excessive moisture, dirt, heat, cold, corrosive vapors or solids, vibration, and aging. **See Megohmmeter.**

Insulation separates conductors from each other and from earth ground. Insulation must have a high resistance in order to prevent current from passing through it. All insulation has a resistance value less than infinity and allows some leakage current. *Leakage current* is current that flows through insulation. Under normal operating conditions, the amount of leakage current is so small (microamperes) that it has no effect on the operation or safety of a circuit.

The total leakage current of insulation is the combination of conductive leakage current, capacitive leakage current, and surface leakage current. All three contribute to the total leakage current in a circuit.

Conductive Leakage Current

Conductive leakage current is the small amount of current that normally flows through the insulation of a conductor. Conductive leakage current flows from conductor to conductor or from a hot conductor to the ground. **See Conductive Leakage Current.** Ohm's law is used to determine conductive leakage current. To calculate conductive leakage current, apply the formula:

$$I_L = \frac{E_A}{R_I}$$

where
I_L = leakage current (in microamperes)
E_A = applied voltage (in volts)
R_I = insulation resistance (in megohms)

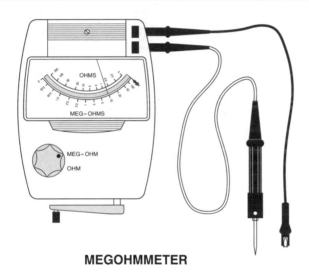

MEGOHMMETER

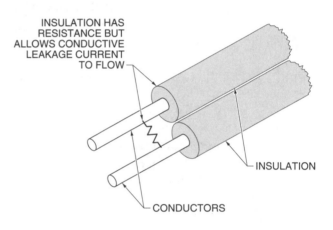

CONDUCTIVE LEAKAGE CURRENT

Example: Calculating Conductive Leakage Current

Calculate the conductive leakage current when the applied voltage is 240 V and the resistance of the insulation is 10 M. **See Leakage Current Calculation.**

$$I_L = \frac{E_A}{R_I}$$

$$I_L = \frac{240}{10,000,000}$$

$$I_L = .000024 \text{ or } 24 \text{ μA}$$

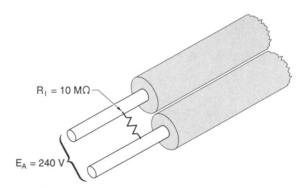

LEAKAGE CURRENT CALCULATION

The resistance of insulation decreases as the insulation ages and is exposed to damaging elements. Conductive leakage current increases as the resistance of the conductor insulation decreases. Additional insulation deterioration results as conductive leakage current increases. Keeping insulation clean and dry minimizes conductive leakage current.

Capacitive Leakage Current

Capacitive leakage current is leakage current that flows through conductor insulation due to a capacitive effect. A *capacitor* is an electronic device used to store an electric charge. A capacitor is made by separating two plates with a dielectric. Two conductors that are run together act as a low-level capacitor. The insulation between the conductors is the dielectric, and the conductors are the plates. **See Capacitive Leakage Current.**

DC voltages produce little capacitive leakage current. Conductors are charged as DC voltage is applied. The capacitive leakage current lasts for a few seconds and then stops. AC voltages produce continuous capacitive leakage current. Conductors are continuously charged and discharged as AC voltage is applied. This is due to the alternating characteristic of AC voltage. Capacitive leakage current flows continuously as AC voltage alternates. Capacitive leakage current is minimized by separating or twisting the conductors.

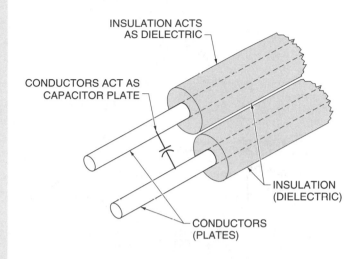

CAPACITIVE LEAKAGE CURRENT

Surface Leakage Current

Surface leakage current is current that flows from areas on conductors where insulation is removed to allow for electrical connections. Conductors are terminated to wire nuts, splices, spade lugs, terminal posts, and other fastening devices at different points along an electrical circuit. The point at which the insulation is removed provides a low resistance path for surface leakage current. Dirt and moisture allow additional surface leakage current to flow. **See Surface Leakage Current.** Surface leakage current results in increased heat at a connection. Increased heat contributes to insulation deterioration and makes a conductor brittle. Surface leakage current is minimized by making good, tight, clean connections.

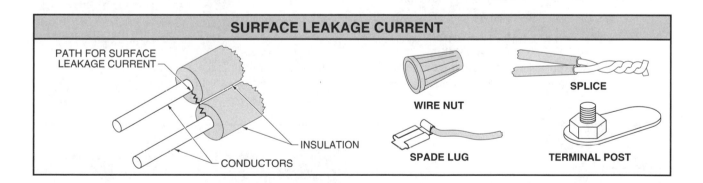

Measuring Resistance with a Megohmmeter

Megohmmeters deliver a high voltage to the circuit under test. **See Megohmmeter Resistance Measurement.**

Warning: Ensure that no voltage is present in a circuit or component under test before taking any resistance measurements. Ensure that no body part contacts the high voltage.

9 CONSULT EQUIPMENT OR METER MANUFACTURER

5 ENSURE THAT METER BATTERIES ARE GOOD

4 PLUG RED LEAD INTO POSITIVE JACK

6 CONNECT LINE TEST LEAD TO CONDUCTOR UNDER TEST

ENSURE THAT ALL POWER IS OFF **1**

BATTERY OPERATED

3 PLUG BLACK LEAD INTO NEGATIVE JACK

2 SET SELECTOR SWITCH TO CORRECT VOLTAGE

7 CONNECT EARTH TEST LEAD TO A SECOND CONDUCTOR OR EARTH GROUND

8 PRESS TEST BUTTON OR TURN CRANK AND READ RESISTANCE DISPLAYED ON METER

MEGOHMMETER RESISTANCE MEASUREMENT

To use a megohmmeter, apply the procedure:

1. Ensure that all power is OFF in the circuit or component under test. Test for voltage using a voltmeter if uncertain.
2. Set the selector switch to the voltage at which the circuit is to be tested. The test voltage should be as high or higher than the highest voltage to which the circuit under test is exposed.
3. Plug the black test lead into the negative (earth) jack.
4. Plug the red test lead into the positive (line) jack.
5. Ensure that the batteries are in good condition. The meter contains no batteries if the meter includes a crank. The meter contains no batteries or crank if the meter plugs into a standard outlet.
6. Connect the line test lead to the conductor under test.
7. Connect the earth test lead to a second conductor in the circuit or earth ground.
8. Press the test button or turn the crank and read the resistance displayed on the meter. Change the meter resistance or voltage range if required.
9. Consult the equipment manufacturer or meter manufacturer for the minimum recommended resistance values. The insulation is good if the meter reading is equal to or higher than the minimum value.

 Refer to **Activity 6-7—Megohmmeters** ....

APPLICATION 6-8—Digital Logic Probes

Digital logic circuits make decisions in control circuits. A digital signal is a signal represented by one of two states. The signal is either high (1) or low (0). A high signal is normally 5 V, but can be from 2.4 V to 5 V. A low signal is normally 0 V, but can be from 0 V to .8 V.

Digital logic gates are used to control electrical circuits. The AND, OR, and NOT logic gates are the three basic logic functions that make up most digital circuit logic. The NOT gate is used to invert the incoming signal to the gate. The NOR gate is a NOT, OR, or inverted OR gate. The NAND gate is a NOT, AND, or inverted AND gate. AND, OR, NOT, NOR, and NAND logic has the same meaning for digital logic, hard-wired electrical logic, and relay logic. **See Basic Logic Functions.**

BASIC LOGIC FUNCTIONS

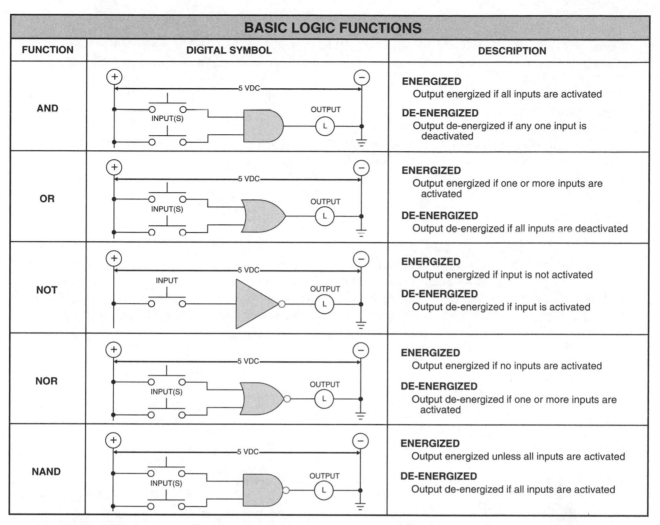

FUNCTION	DIGITAL SYMBOL	DESCRIPTION
AND		**ENERGIZED** Output energized if all inputs are activated **DE-ENERGIZED** Output de-energized if any one input is deactivated
OR		**ENERGIZED** Output energized if one or more inputs are activated **DE-ENERGIZED** Output de-energized if all inputs are deactivated
NOT		**ENERGIZED** Output energized if input is not activated **DE-ENERGIZED** Output de-energized if input is activated
NOR		**ENERGIZED** Output energized if no inputs are activated **DE-ENERGIZED** Output de-energized if one or more inputs are activated
NAND		**ENERGIZED** Output energized unless all inputs are activated **DE-ENERGIZED** Output de-energized if all inputs are activated

Digital Logic Probe

A *digital logic probe* is a special DC voltmeter that detects the presence or absence of a signal. Displays on a digital logic probe include logic high, logic low, pulse light, memory, and TTL/CMOS. **See Digital Logic Probe.**

The high light-emitting diode (LED) lights when the logic probe detects a high logic level (1). The low LED lights when the logic probe detects a low logic level (0).

The pulse LED flashes relatively slowly when the probe detects logic activity present in a circuit. Logic activity indicates that the circuit is changing between logic levels. The pulse light displays the changes between logic levels because the changes are usually too fast for the high and low LEDs to display.

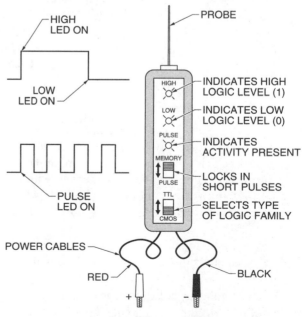

DIGITAL LOGIC PROBE

The memory switch sets the logic probe to detect short pulses, usually lasting a few nanoseconds. Any change from the original logic level causes the memory LED to light and remain ON. The memory LED is the pulse LED switch in the memory position. The memory switch is manually moved to the pulse position and back to the memory position to reset the logic probe.

The TTL/CMOS switch selects the logic family of integrated circuits (ICs) to be tested. *Transistor-transistor logic (TTL) ICs* are a broad family of ICs that employ a two-transistor arrangement. The supply voltage for TTL ICs is 5.0 VDC, ±.25 V.

Complementary metal-oxide semiconductor (CMOS) ICs are a group of ICs that employ MOS transistors. CMOS ICs are designed to operate on a supply voltage ranging from 3 VDC to 18 VDC. Check circuit schematics for CMOS circuit voltages. The supply voltage for CMOS ICs should be greater than −5% of the rated voltage. CMOS ICs are noted for their exceptionally low power consumption.

Pull-Up Resistor

A *floating input* is a digital input that is not high or low at all times. A floating input should not occur in a digital circuit. A pull-up resistor prevents a floating input. A *pull-up resistor* is a resistor that has one side connected to the power supply at all times and the other side connected to one side of the input switch. The connection produces a high signal on one side of the input switch. The other side of the input switch is connected to the ground. The high signal becomes a low signal when the input switch is closed. **See Pull-Up Resistor Use.**

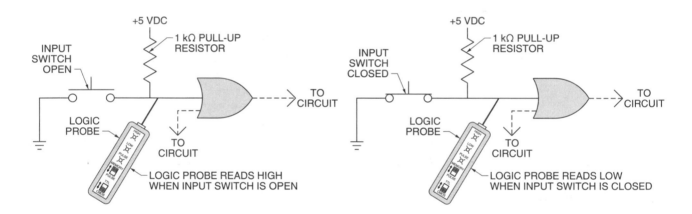

PULL-UP RESISTOR USE

Digital Logic Probe Use

Digital circuits fail because the signal is lost somewhere between the circuit input and output stages. The problem is usually solved by finding the point where the signal is missing and repairing that area. Repair normally involves replacing a component, section, or the entire PC board.

The power supply voltage is checked with a voltmeter if the digital circuit or logic probe has intermittent problems. A logic probe may indicate a high signal, but the supply voltage may be too low for proper circuit operation. **See Digital Logic Probe Use.**

To use a digital logic probe, apply the procedure:
1. Connect the positive (red) power lead to the positive side of the circuit's power supply. The positive power supply is +5 VDC for TTL circuits.
2. Connect the negative (black) power lead to the ground side of the circuit's power supply.
3. Set the selector switch to the logic family (TTL or CMOS) under test.

4. Touch the probe tip to the point being tested. Start at the input side of the circuit and move to the output of the circuit.
5. Note the condition of the LEDs on the logic probe.

Single-shot pulses are stored indefinitely by placing the switch in the memory position.

LOGIC PROBE FUNCTIONS	
LED Condition	**Meaning**
No LEDs lit	Open circuit
High LED lit	Logic 1
Low LED lit	Logic 0
Pulse LED lit	Circuit activity

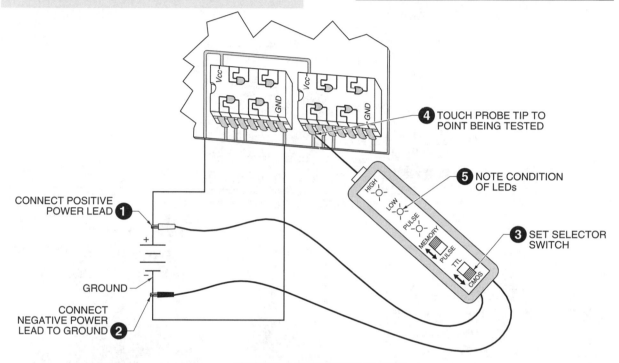

DIGITAL LOGIC PROBE USE

Circuit Board Breaks

Broken traces (conducting paths) result when a PC board is subjected to mechanical stress. The break is repaired by soldering a bridge over the break. A bridge is made by soldering a short piece of wire on both sides of the break. **See PC Board Repair.**

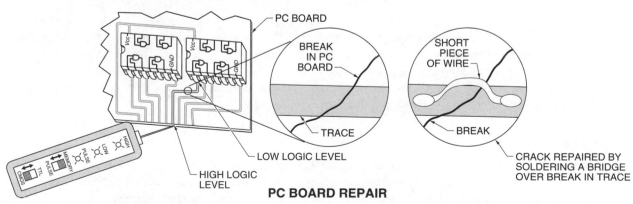

PC BOARD REPAIR

 Refer to **Activity 6-8—Digital Logic Probes** ....

APPLICATION 6-9—Oscilloscopes/Scopemeters

A graphical display of circuit voltage has more meaning than a numerical value in some troubleshooting situations. When circuits include rapidly fluctuating signals, stray signals, and phase shifts, problems are easily detected on a graphical display.

A *scope* is a device that gives a visual display of voltages. A scope shows the shape of a circuit's voltage and allows the voltage level, frequency, and phase to be measured. The two basic types of scopes used in troubleshooting are oscilloscopes and scopemeters. **See Bench Oscilloscope** and **Scopemeters.**

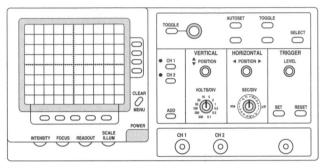

BENCH OSCILLOSCOPE

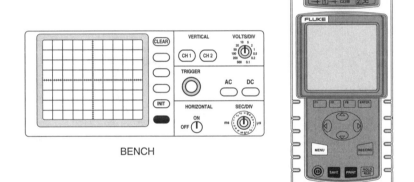

BENCH

PORTABLE SINGLE-PHASE   PORTABLE THREE-PHASE

SCOPEMETERS

Oscilloscopes

An *oscilloscope* is an instrument that displays an instantaneous voltage. An oscilloscope is used to display the shape of a voltage waveform when bench testing electronic circuits. *Bench testing* is testing performed when equipment under test is brought to a designated service area. Oscilloscopes are used to troubleshoot digital circuits, communication circuits, TVs, VCRs, and computers. Oscilloscopes are available in basic and specialized types that can display different waveforms simultaneously.

Scopemeters

A *scopemeter* is a combination oscilloscope and digital multimeter. A scopemeter is used to display the shape of a voltage waveform when troubleshooting circuits in the field. A scopemeter is portable and can be used as a multimeter and a scope. A scopemeter does not have all the features of a specialized oscilloscope.

Scope Displays

A scope displays the voltage under test on the scope screen. The scope screen contains scribed horizontal and vertical axes. The horizontal (x) axis represents time. The vertical (y) axis represents the amplitude of the voltage waveform.

The scribed lines divide the screen into equal divisions. The divisions help to measure the voltage level and frequency of the displayed waveforms. **See Scope Screen.**

Scope Trace

A trace is established on the screen before a circuit under test is connected. A *trace* is a reference point/line that is visually displayed on the face of the scope screen. The trace is normally positioned over the horizontal centerline on the screen.

The starting point of the trace is located near the left side of the screen. *Sweep* is the movement of the displayed trace across the scope screen. The sweep of the scope trace is from left to right. **See Scope Trace.**

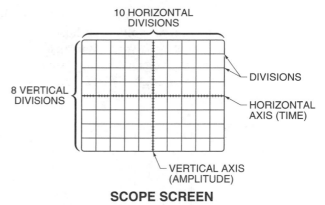

SCOPE SCREEN

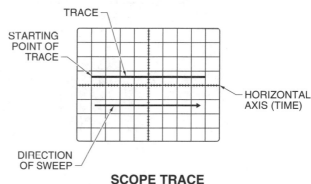

SCOPE TRACE

Manually Operated Controls

Manually operated controls are adjusted to view a waveform. Typical scope adjustment controls include intensity, focus, horizontal positioning, vertical positioning, volts/division, and time/division.

Intensity is the level of brightness. The intensity control sets the level of brightness on the displayed voltage trace. The intensity level is kept as low as possible to keep the trace in focus. The focus control adjusts the sharpness of the displayed voltage trace.

The horizontal control adjusts the left and right positions of the displayed voltage trace. The horizontal control sets the starting point of the trace. The vertical control adjusts the up and down positions of the displayed voltage trace. **See Horizontal Positioning** and **Vertical Positioning.**

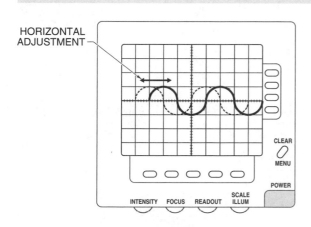

HORIZONTAL POSITIONING

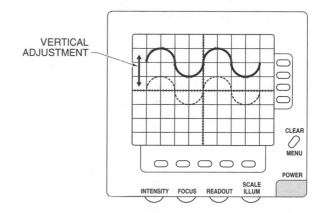

VERTICAL POSITIONING

The *volts/division (volts per division) control* selects the height of the displayed waveform. The setting determines the number of volts each horizontal screen division represents. For example, if a waveform occupies 4 divisions and the volts/division control is set on 20, the peak-to-peak voltage ($V_{p\text{-}p}$) equals 80 V ($4 \times 20 = 80$ V). Eighty volts peak-to-peak equals 40 V peak (V_{max}) ($80 \div 2 = 40$ V). Forty volts peak equals 28.28 V_{rms} ($40 \times .707 = 28.28$ V). **See Volts Per Division Control.**

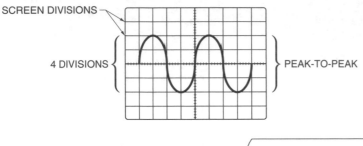

SCREEN DIVISIONS

4 DIVISIONS

PEAK-TO-PEAK

FINDING PEAK-TO-PEAK VALUE

What is the peak-to-peak value when a waveform occupies 4 divisions and the volts/division control is set on 20?

1. $V_{P\text{-}P} = Divisions \times Volts/Division$
   $V_{P\text{-}P} = 4 \times 20$

   $V_{P\text{-}P} = \textbf{80 V}$

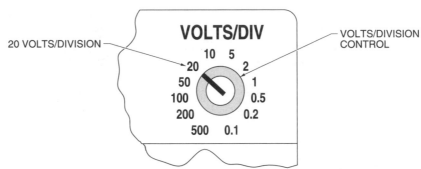

20 VOLTS/DIVISION

VOLTS/DIV

VOLTS/DIVISION CONTROL

10 5
20   2
50   1
100   0.5
200   0.2
500   0.1

VOLTS PER DIVISION CONTROL

The *time/division (time per division) control* selects the width of the displayed waveform. The setting determines the length of time each cycle takes to move across the screen. For example, if the time/division control is set on 10, each vertical screen division equals 10 milliseconds (ms). If one cycle of a waveform equals 4 divisions, the displayed time equals 40 ms ($4 \times 10 = 40$ ms). **See Time Per Division Control.**

To determine the frequency of the displayed waveform, apply the formula:

$$f = \frac{1}{T}$$

where

f = frequency (in hertz)
1 = constant
T = time period (in seconds)

Example: Calculating Frequency

Find the frequency of a waveform if one cycle of the waveform occupies 6 divisions and the time/division control is set on 5 ms (5 ms = .005 s).

1. Calculate time period.

$T = 6 \times 5$

$T = 6 \times .005$

$T = \textbf{.03 s}$

2. Calculate frequency.

$$f = \frac{1}{T}$$

$$f = \frac{1}{.03}$$

$f = \textbf{33.3 Hz}$

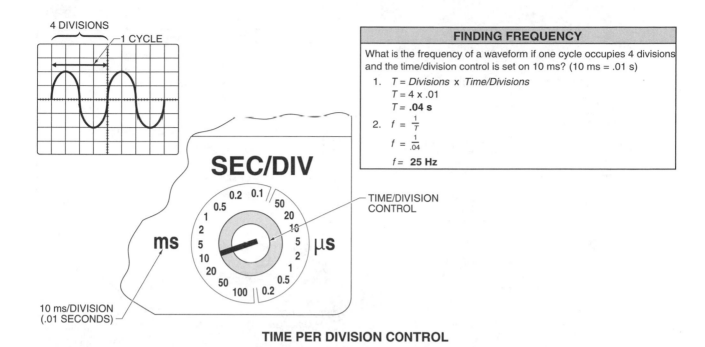

TIME PER DIVISION CONTROL

Scope Measurement—AC Voltage

A scope is connected in parallel with a circuit or component under test. The scope is connected by a probe on the end of a test lead. A 1X probe (1 to 1) is used to connect the input of the scope to the circuit under test when the test voltage is lower than the voltage limit of the scope.

A 10X probe (10 to 1) is used to divide the input voltage by 10. The scope voltage limit equals 10 times the normal rated voltage when a 10X probe is used. When using a 10X probe, the amount of measured voltage displayed on the scope screen must be multiplied by 10 to obtain the actual circuit voltage. For example, if the measured scope voltage is 25 V when using a 10X probe, the actual circuit voltage is 250 V (25 × 10 = 250 V). **See Scope Measurement—AC Voltage.**

To use a scope to measure AC voltage, apply the procedure:

1. Turn the power switch ON and adjust the trace brightness on the screen.

2. Set the AC/DC control switch to AC.

3. Set the volts/division control to display the voltage level under test. Set the control to the highest value if the voltage level is unknown.

4. Connect the scope probe to the AC voltage under test.

5. Adjust the volts/division control to display the full waveform of the voltage under test.

6. Set the time/division control to display several cycles of the voltage under test.

7. Use the vertical control to set the lower edge of the waveform on one of the lower lines.

8. Measure the vertical amplitude of the waveform by counting the number of divisions displayed ($V_{p\text{-}p}$).

To calculate V_{rms}, first calculate $V_{p\text{-}p}$. $V_{p\text{-}p}$ is calculated by multiplying the number of divisions by the volts/division setting. For example, if a waveform occupies 4 divisions and the volts/division setting is 10, $V_{p\text{-}p}$ equals 40 V (4 × 10 = 40 V). V_{max} equals 20 V (40 ÷ 2 = 20 V).

To find V_{rms}, multiply V_{max} by .707. For example, if V_{max} is 20 V, V_{rms} equals 14.14 V (20 × .707 = 14.14 V). V_{rms} is the value of the voltage under test as measured on a voltmeter.

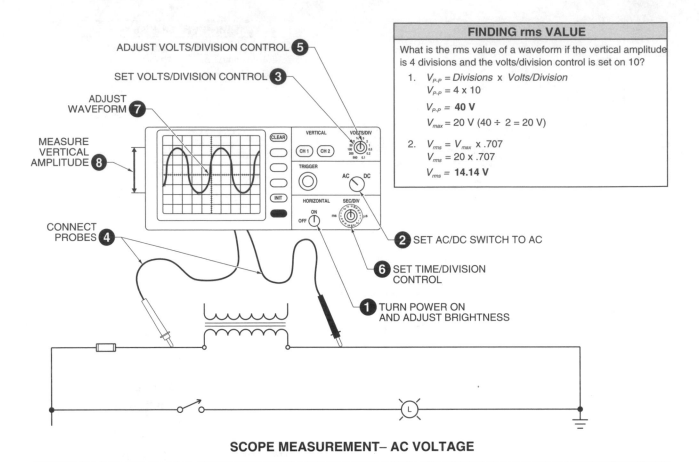

SCOPE MEASUREMENT– AC VOLTAGE

FINDING rms VALUE

What is the rms value of a waveform if the vertical amplitude is 4 divisions and the volts/division control is set on 10?

1. $V_{P\text{-}P} = Divisions \times Volts/Division$
   $V_{P\text{-}P} = 4 \times 10$
   $V_{P\text{-}P} = \mathbf{40\ V}$
   $V_{max} = 20\ V\ (40 \div 2 = 20\ V)$

2. $V_{rms} = V_{max} \times .707$
   $V_{rms} = 20 \times .707$
   $V_{rms} = \mathbf{14.14\ V}$

Scope Measurement—Frequency

In AC applications such as variable frequency motor drives, it may be necessary to measure the frequency in the circuit. A frequency meter should be used in applications where high accuracy is required. A scopemeter gives an adequate reading in most frequency measurement applications. A scopemeter also shows any distortion present in the circuit under test. To measure frequency, the scope probes are connected in parallel with the circuit or component under test. **See Scope Measurement — Frequency.** To use a scope to measure frequency, apply the procedure:

1. Turn the power switch ON and adjust the trace brightness.
2. Set the AC/DC control switch to AC.
3. Set the volts/division control to display the voltage level under test. Set the control to the highest value if the voltage level is unknown.
4. Connect the scope probe to the AC voltage under test.
5. Adjust the volts/division control to display the vertical amplitude of the waveform under test.
6. Set the time/division control to display approximately two cycles of the waveform under test.
7. Set the vertical control so that the center of the waveform is on the centerline of the scope screen.
8. Set the horizontal control so that the start of one cycle of the waveform begins at the vertical centerline on the scope screen.
9. Measure the number of divisions between the start point and end point of one cycle.

To determine frequency, multiply the number of divisions by the time/division setting. This value is the time period for one cycle. To determine the frequency of the waveform, divide the time period by 1.

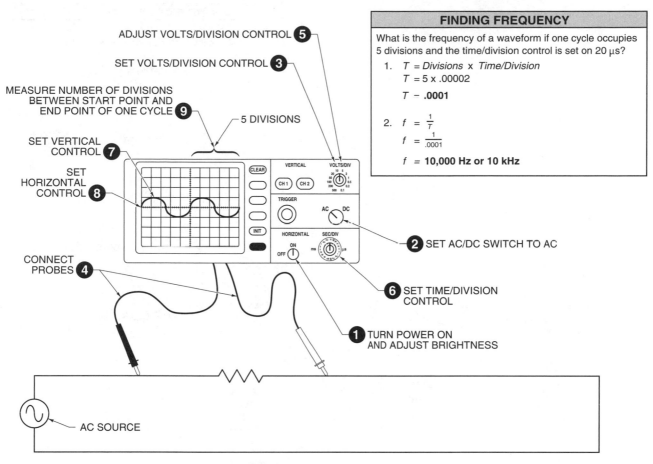

ADJUST VOLTS/DIVISION CONTROL **5**

SET VOLTS/DIVISION CONTROL **3**

MEASURE NUMBER OF DIVISIONS
BETWEEN START POINT AND
END POINT OF ONE CYCLE **9**

5 DIVISIONS

SET VERTICAL CONTROL **7**

SET HORIZONTAL CONTROL **8**

CONNECT PROBES **4**

2 SET AC/DC SWITCH TO AC

6 SET TIME/DIVISION CONTROL

1 TURN POWER ON AND ADJUST BRIGHTNESS

AC SOURCE

FINDING FREQUENCY

What is the frequency of a waveform if one cycle occupies 5 divisions and the time/division control is set on 20 μs?

1. $T = Divisions \times Time/Division$
    $T = 5 \times .00002$
    $T - \textbf{.0001}$

2. $f = \frac{1}{T}$
    $f = \frac{1}{.0001}$
    $f = \textbf{10,000 Hz or 10 kHz}$

SCOPE MEASUREMENT– FREQUENCY

Example: Calculating Frequency

Calculate the frequency if one cycle of a waveform occupies 10 divisions and the time/division setting is 2 μs (microseconds).

1. Calculate time period.
    $T = 10 \times .000002$
    $T = \textbf{.00002 s}$

2. Calculate frequency.
    $f = \frac{1}{T}$
    $f = \frac{1}{.00002}$
    $f = \textbf{50,000 Hz}$

Scope Measurement—DC Voltage

A test probe is connected to the point in the circuit where the DC voltage is to be measured. The ground lead of the scope is connected to the ground of the circuit. The voltage is positive if the trace moves above the centerline. The voltage is negative if the trace moves below the centerline. **See Scope Measurement—DC Voltage.**

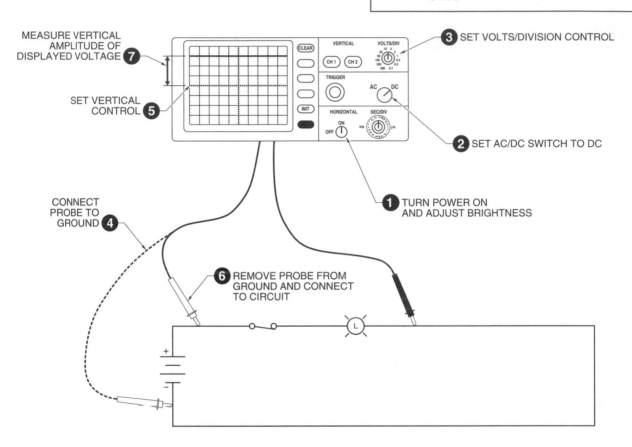

FINDING DC VOLTAGE

What is the circuit voltage if the displayed voltage is 3 divisions above the center line and the volts/division control is set at 5 V?

1. $V = Divisions \times Volts/Division$

    $V = 3 \times 5$

    $V = $ **15 VDC**

SCOPE MEASUREMENT– DC VOLTAGE

To use a scope to measure DC voltage, apply the procedure:

1. Turn the power switch ON and adjust the trace brightness.
2. Set the AC/DC control switch to DC.
3. Set the volts/division control to display the voltage level under test. Set the control to the highest value if the voltage level is unknown.
4. Connect the scope probe to the ground point of the circuit under test.
5. Set the vertical control so that the displayed line is in the center of the screen. The displayed line represents 0 VDC.
6. Remove the scope probe from ground point and connect it to the DC voltage under test. The displayed voltage moves above or below the scope centerline depending on the polarity of the DC voltage under test.
7. Measure the vertical amplitude of the voltage from the centerline by counting the number of divisions from the centerline.

Multiply the number of displayed divisions by the volts/division setting to determine the DC voltage. For example, if a waveform is 3 divisions above the centerline and the volts/division control is set at 5 V, the voltage equals 15 VDC ($3 \times 5 = 15$ V).

Measuring DC voltages helps when troubleshooting DC circuits, such as power supplies that rectify AC to DC. The shape of the waveform can be used to determine when any of the diodes in the rectifier circuit are shorted or open. **See Rectified DC Waveforms.**

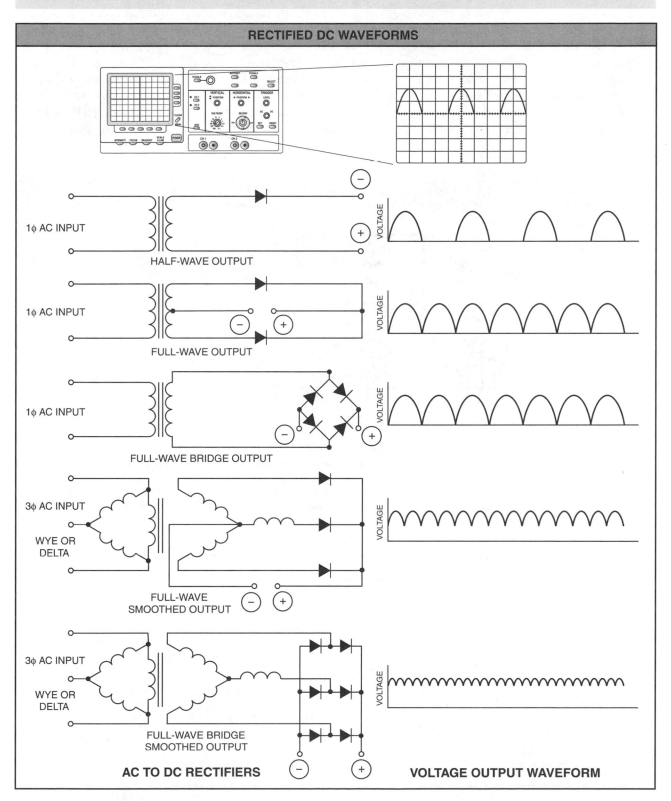

RECTIFIED DC WAVEFORMS

1φ AC INPUT — HALF-WAVE OUTPUT

1φ AC INPUT — FULL-WAVE OUTPUT

1φ AC INPUT — FULL-WAVE BRIDGE OUTPUT

3φ AC INPUT — WYE OR DELTA — FULL-WAVE SMOOTHED OUTPUT

3φ AC INPUT — WYE OR DELTA — FULL-WAVE BRIDGE SMOOTHED OUTPUT

AC TO DC RECTIFIERS　**VOLTAGE OUTPUT WAVEFORM**

Signal (Function) Generators

Much of the work involved when troubleshooting electronics involves signal tracing. Signal tracing follows a signal through a circuit, device, or system to find where the signal disappears or becomes distorted. Signal tracing is accomplished using instruments to either measure an electronic signal already within a system or circuit, or to measure an electronic signal intentionally inserted into a circuit or system.

A test signal, instead of a signal that would naturally be appearing in a circuit or system during normal operation, is used to test electronic circuits and systems. Signal generators are normally used to produce test signals. A *signal (function) generator* is a test instrument that provides a known test signal to a circuit, device, or system for testing purposes. Signal generators normally produce sine waves, square waves, or triangular waves. The produced signal is measured at various points as it travels through a circuit under test to check when the signal is lost, is overly attenuated (reduced in energy), distorted, or clipped. **See Signal (Function) Generators.**

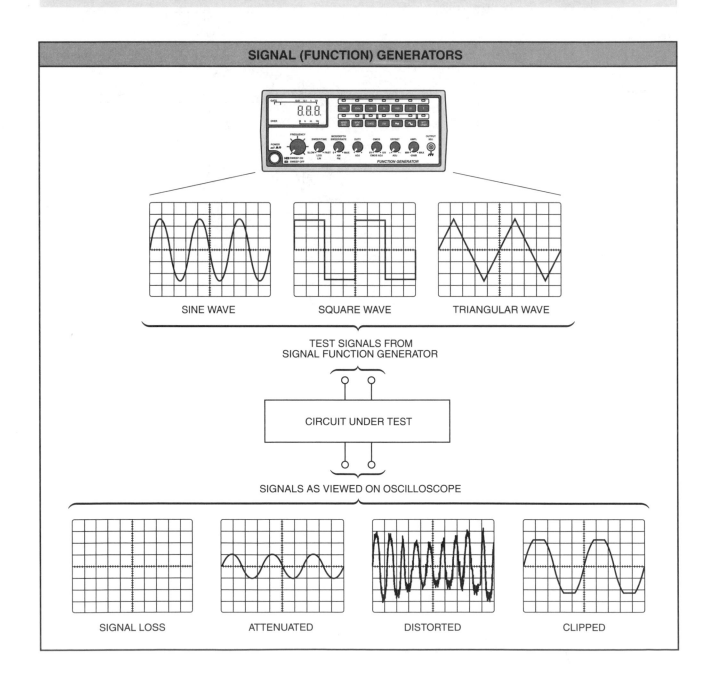

SIGNAL (FUNCTION) GENERATORS

SINE WAVE

SQUARE WAVE

TRIANGULAR WAVE

TEST SIGNALS FROM
SIGNAL FUNCTION GENERATOR

CIRCUIT UNDER TEST

SIGNALS AS VIEWED ON OSCILLOSCOPE

SIGNAL LOSS

ATTENUATED

DISTORTED

CLIPPED

Signal Gain Measurements

Electronic circuits that are used for amplification are tested by inserting a known signal into the input point of the circuit and measuring the resulting output signal. When the output signal is greater than the input signal, the circuit has a gain. When the output signal equals the input signal, there is no gain. When the output signal is less than the input signal, there is a loss of signal. *Gain* is the ratio of the amplitude of the output signal to the amplitude of the input signal.

Signal gain is only one measurement of a signal. In addition to testing for gain, a signal must also be tested to ensure the signal is not distorted or compromised (carrying noise). Oscilloscopes, including handheld oscilloscopes, are used to test the condition of a signal and display any distortions or other problems.

Decibel Gain

Voltage gain can also be expressed in dB. A *decibel* (dB) is an electrical unit used for expressing the ratio of the magnitudes of two electric values such as voltage or current. When an output signal is greater than the input signal, the gain (in dB) is a positive number (e.g., 20 dB or 55 dB). When the output signal equals the input signal, the gain is 0 dB. When the output signal is less than the input signal, the gain (in dB) is a negative number (e.g., –10 dB or –45 dB).

 Refer to **Activity 6-9—Oscilloscopes/Scopemeters** ....

 *Refer to Quick Quiz® on CD-ROM*

 Refer to Chapter 6 in the **Troubleshooting Electrical/Electronic Systems Workbook** *for additional questions.*

Name_____ Date _____

ACTIVITY 6-1—Light Meters

Using recommended light levels and building print, determine the recommended light level for each area.

_____ **1.** Recommended light level at area A = ___ fc.

_____ **2.** Recommended light level at area B = ___ fc.

_____ **3.** Recommended light level at area C = ___ fc.

_____ **4.** Recommended light level at area D = ___ fc.

_____ **5.** Recommended light level at area E = ___ fc.

_____ **6.** Recommended light level at area F = ___ fc.

_____ **7.** Recommended light level at area G = ___ fc.

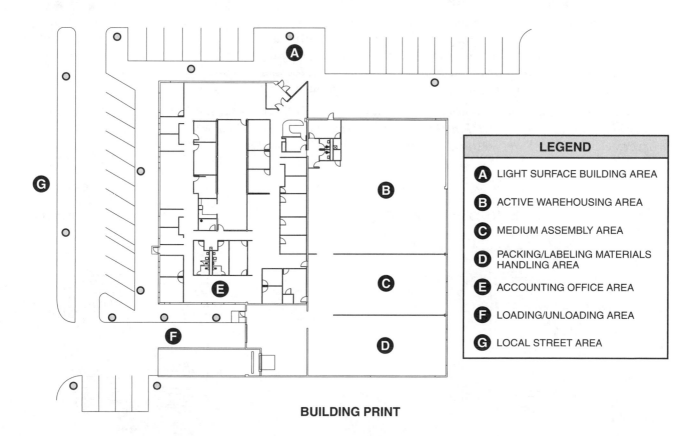

LEGEND

A	LIGHT SURFACE BUILDING AREA
B	ACTIVE WAREHOUSING AREA
C	MEDIUM ASSEMBLY AREA
D	PACKING/LABELING MATERIALS HANDLING AREA
E	ACCOUNTING OFFICE AREA
F	LOADING/UNLOADING AREA
G	LOCAL STREET AREA

BUILDING PRINT

RECOMMENDED LIGHT LEVELS

INTERIOR LIGHTING

Area	Light Level*
Assembly	
Rough, easy seeing	30
Medium	100
Fine	500
Auditorium	
Exhibitions	30
Materials handling	
Picking stock	30
Packing, labeling	50
Offices	
Regular office work	100
Accounting	150
Detailed work	200
Printing	
Proofreading	150
Color inspecting	200
Stores	
Stockroom	30
Service area	100
Warehousing, storage	
Inactive	5
Active	30

* in footcandles

EXTERIOR LIGHTING

Area	Light Level*
Building	
Light surface	15
Dark surface	50
Loading/unloading area	20
Parking areas	
Industrial	2
Shopping	5
Storage yards (Active)	20
Street	
Local	0.9
Expressway	1.4
Car lots	
Front line	100-500
Remaining area	20-75

* in footcandles

FOOTCANDLE - A UNIT OF MEASURE OF THE INTENSITY OF LIGHT FALLING ON A SURFACE, EQUAL TO ONE LUMEN PER SQUARE FOOT

ACTIVITY 6-2—Contact Thermometers

Relative humidity may be measured using a contact thermometer by measuring the dry bulb and wet bulb temperatures of the air. Dry bulb temperature *is the temperature of the air taken with a dry temperature probe.* Wet bulb temperature *is the temperature of the air taken with a wet sock covering the temperature probe. Dry bulb and wet bulb temperatures are plotted on a psychrometric chart to determine relative humidity. A psychrometric chart is a graph that is used to find various properties of air.* **See Psychrometric Charts.**

The point where the dry bulb and wet bulb temperatures intersect is the relative humidity of the air. For example, if the dry bulb temperature is 75.1°F and the wet bulb temperature is 65.2°F, the relative humidity is 60%. Using the test results and the Psychrometric Chart, determine the relative humidity.

1. Relative humidity = ___%   **2.** Relative humidity = ___%

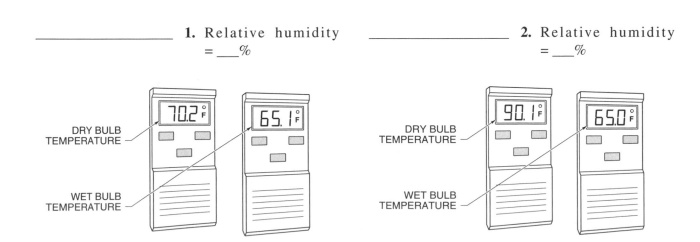

DRY BULB TEMPERATURE   70.2 °F   65.1 °F   WET BULB TEMPERATURE

DRY BULB TEMPERATURE   90.1 °F   65.0 °F   WET BULB TEMPERATURE

3. Relative humidity
= ___%

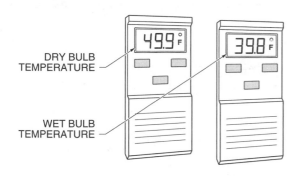

DRY BULB
TEMPERATURE

WET BULB
TEMPERATURE

4. Relative humidity
= ___%

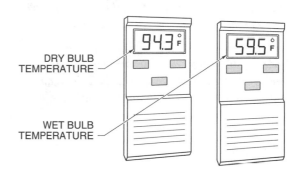

DRY BULB
TEMPERATURE

WET BULB
TEMPERATURE

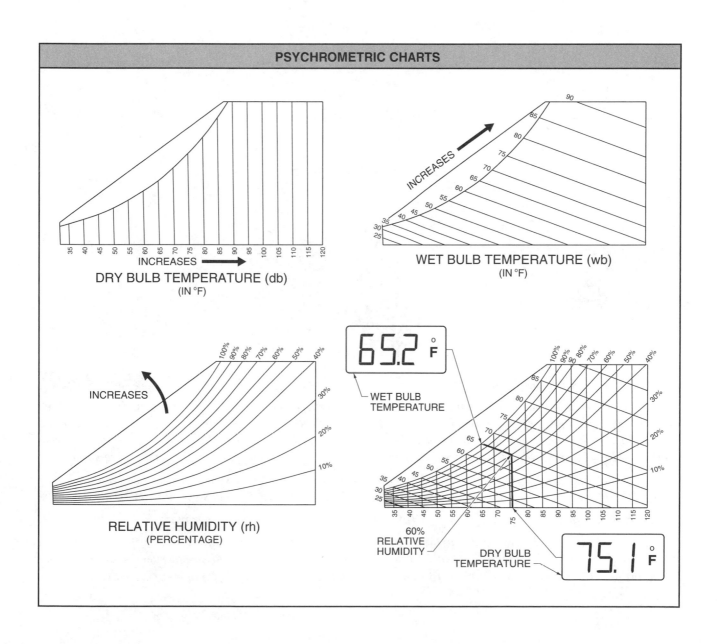

PSYCHROMETRIC CHARTS

DRY BULB TEMPERATURE (db)
(IN °F)

WET BULB TEMPERATURE (wb)
(IN °F)

RELATIVE HUMIDITY (rh)
(PERCENTAGE)

WET BULB
TEMPERATURE

60%
RELATIVE
HUMIDITY

DRY BULB
TEMPERATURE

ACTIVITY 6-3—Infrared Meters

An infrared meter may be used to check motor bearing temperature. When motor bearing temperature exceeds 100°C, the bearing hardness decreases, bearing life is reduced, and the allowable load is reduced. The load on the bearings should be lessened to reduce the temperature of the bearings. For example, the load on a motor with a bearing temperature of 210°C should be reduced to 70%.

Using the Bearing Temperature Derating chart, determine the percent that the load should be reduced to lower bearing temperature to an acceptable level.

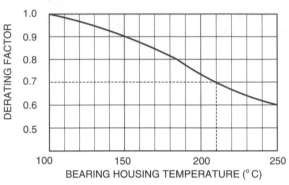

BEARING TEMPERATURE DERATING

_____ **1.** The load should be reduced to ___%.

_____ **2.** The load should be reduced to ___%.

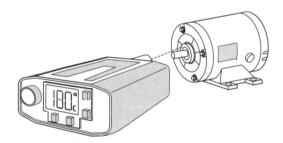

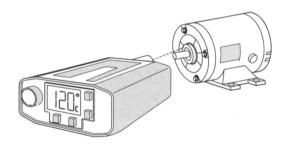

ACTIVITY 6-4—Contact Tachometers

Slip is the difference between the synchronous speed and actual speed of a motor. Normal rated slip is the difference between the synchronous speed and nameplate speed. Normal rated slip is approximately 4%. For example, if the synchronous speed of a motor is 1800 rpm and the nameplate speed is 1725 rpm, the normal rated slip is 75 rpm (1800 – 1725 = 75 rpm). Measured slip is the difference between the synchronous speed and measured speed.

Percent normal rated slip is found by dividing the normal rated slip by the synchronous speed and multiplying by 100. For example, if the normal rated slip equals 75 rpm and synchronous speed equals 1800 rpm, the percent normal rated slip equals 4% ($\frac{75}{1800} \times 100 = 4\%$). Higher slip indicates a problem.

Slip should not exceed 20% more than the normal rated slip in general applications. For example, the normal rated slip of a motor is 75 rpm (4% of synchronous speed). The actual (measured) motor speed should not drop an additional 15 rpm (75 rpm × 20% = 15 rpm). Determine the normal rated slip and measured slip.

1.

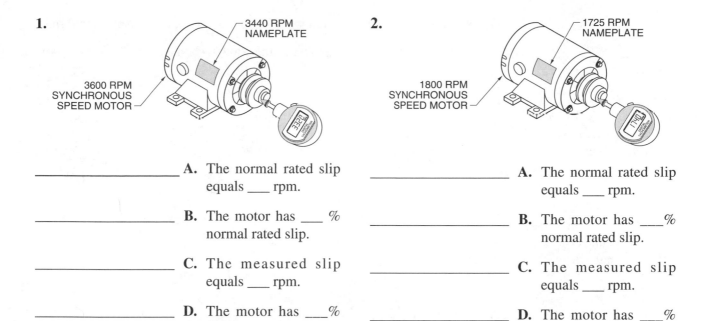

3440 RPM
NAMEPLATE

3600 RPM
SYNCHRONOUS
SPEED MOTOR

_____ **A.** The normal rated slip
equals ___ rpm.

_____ **B.** The motor has ___ %
normal rated slip.

_____ **C.** The measured slip
equals ___ rpm.

_____ **D.** The motor has ___%
measured slip.

_____ **E.** The difference between
normal rated slip and
the measured slip is
___ rpm.

2.

1725 RPM
NAMEPLATE

1800 RPM
SYNCHRONOUS
SPEED MOTOR

_____ **A.** The normal rated slip
equals ___ rpm.

_____ **B.** The motor has ___%
normal rated slip.

_____ **C.** The measured slip
equals ___ rpm.

_____ **D.** The motor has ___%
measured slip.

_____ **E.** The difference between
normal rated slip and
the measured slip is
___ rpm.

ACTIVITY 6-5—Photo Tachometers

A photo tachometer may be used to measure the speed of a pump. Pumps are rated by their maximum operating pressure and output in gallons per minute (gpm) at a given speed. The output of a pump decreases as the pump wears. The speed of a pump and the pump's output are measured when troubleshooting a pump. A pump's output is proportional to its driven speed. Most manufacturers provide a table or graph showing a pump's output at different pressures and speeds.

The pump's rated output may be used to calculate the pump flow rate. The more the pump is worn, the less the output is at higher speeds. To calculate the flow rate for a given drive speed, apply the formula:

$$Q = \frac{N \times V_d}{231}$$

where

Q = flow rate (in gpm)
N = pump drive speed (in rpm)
V = pump displacement (in cu in./rev)
231 = constant

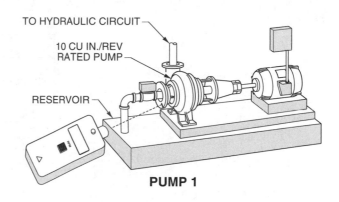

TO HYDRAULIC CIRCUIT

10 CU IN./REV
RATED PUMP

RESERVOIR

PUMP 1

1. Calculate the output of Pump 1.

_____ **A.** The output of Pump 1 is ___ gpm when the photo tachometer displays 1800 rpm.

_____ **B.** The output of Pump 1 is ___ gpm when the photo tachometer displays 1250 rpm.

_____ **C.** The output of Pump 1 is ___ gpm when the photo tachometer displays 900 rpm.

_____ **D.** The output of Pump 1 is ___ gpm when the photo tachometer displays 1640 rpm.

The speed of a pump may be set to deliver the proper flow rate when a given pump output is required. To calculate the required drive speed for a given flow rate, apply the formula:

$$N = \frac{Q \times 231}{V_d}$$

where

N = pump drive speed (in rpm)
Q = pump flow rate (in gpm)
V_d = rated displacement of pump (in cu in./rev)
231 = constant

PUMP 2

2. Calculate the required pump speed of Pump 2.

_____ **A.** The tachometer displays ___ rpm when a 20 gpm output from Pump 2 is required.

_____ **B.** The tachometer displays ___ rpm when a 35 gpm output from Pump 2 is required.

_____ **C.** The tachometer displays ___ rpm when a 28 gpm output from Pump 2 is required.

_____ **D.** The tachometer displays ___ rpm when a 45 gpm output from Pump 2 is required.

ACTIVITY 6-6 — Strobe Tachometers

Proximity switches must be able to activate at a speed faster than the speed of the objects moving down the conveyor to properly detect the objects. A strobe tachometer may be used to measure the speed of moving objects to determine if a given switch may be used. Tachometer readings must be converted from feet per minute (fpm) to operations per second (ops).

To convert fpm to ops, apply the formula:

$$ops = \frac{npf \times V}{60}$$

where

ops = operations per second
npf = objects per foot
V = object speed (in fpm)
60 = constant

PROXIMITY SWITCH RATINGS					
Number	**Type**	**Voltage**	**Sensing Distance (in mm)**	**Operations Per Second (max)**	**Cost ($)**
M01	Mechanical	DC	100	5	25
M02	Mechanical	AC	100	4	25
S01	Solid-state	AC	75	15	20
S02	Solid-state	DC	50	150	25
S03	Solid-state	DC	50	300	27
S04	Solid-state	DC	25	500	27
S05	Solid-state	DC	25	700	29
S06	Solid-state	DC	15	1000	32
S07	Solid-state	DC	10	1500	35
S08	Solid-state	DC	5	2000	38
S09	Solid-state	DC	2	3000	40

Using the Proximity Switch Ratings, determine the least expensive proximity switch that may be used to measure the speed of the products moving down the conveyor.

_____ **1.** Proximity switch number equals ___.

_____ **2.** Proximity switch number equals ___.

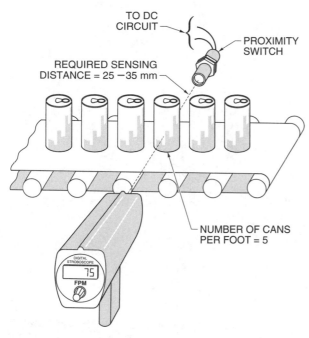

TO DC CIRCUIT

PROXIMITY SWITCH

REQUIRED SENSING DISTANCE = 25 – 35 mm

DIGITAL STROBOSCOPE

75

FPM

NUMBER OF CANS PER FOOT = 5

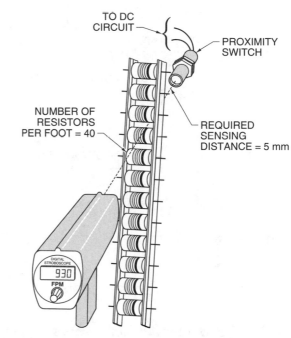

TO DC CIRCUIT

PROXIMITY SWITCH

NUMBER OF RESISTORS PER FOOT = 40

REQUIRED SENSING DISTANCE = 5 mm

DIGITAL STROBOSCOPE

930

FPM

ACTIVITY 6-7—Megohmmeters

Answer the questions using the megohmmeter readings.

_____ **1.** Leakage current equals ___ mA.

_____ **2.** Is the insulation resistance within the manufacturer's recommendations?

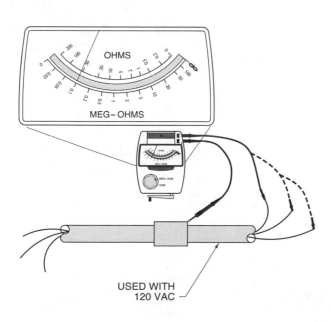

OHMS

MEG-OHMS

USED WITH 120 VAC

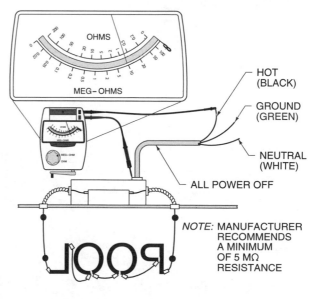

OHMS

MEG-OHMS

HOT (BLACK)

GROUND (GREEN)

NEUTRAL (WHITE)

ALL POWER OFF

NOTE: MANUFACTURER RECOMMENDS A MINIMUM OF 5 MΩ RESISTANCE

_____ **3.** Is the leakage current enough to produce a shock that can be felt?

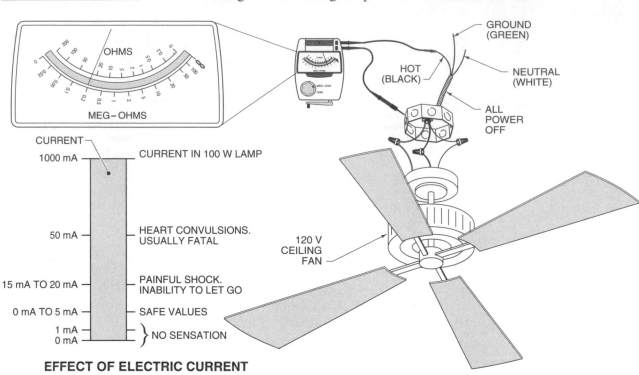

EFFECT OF ELECTRIC CURRENT

ACTIVITY 6-8—Digital Logic Probes

1. Two 115 V loads are controlled by two pushbuttons in a digital circuit. Using Test Circuit 1, complete the Truth Table for the switch positions.

TRUTH TABLE			
A	**B**	**Load 1**	**Load 2**
O	O	ON	OFF
O	X		
X	O		
X	X		

X = switch closed
O = switch open

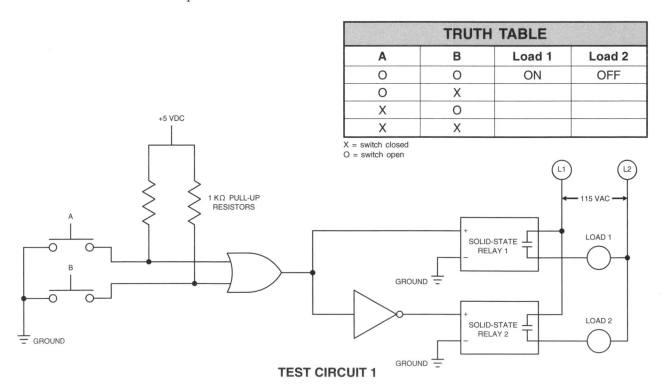

TEST CIRCUIT 1

2. Digital logic probes and voltmeters are added to Test Circuit 1. Using the meter indications, determine the problem. Select the answer from the following list.

1 = No 115 VAC supply voltage in system   6 = Problem with Inverter Gate

2 = Problem with Load 1   7 = Problem with OR Gate

3 = Problem with Load 2   8 = Problem with Switch A

4 = Problcm with Solid-state Relay 1   9 = Problem with Switch B

5 = Problem with Solid-state Relay 2

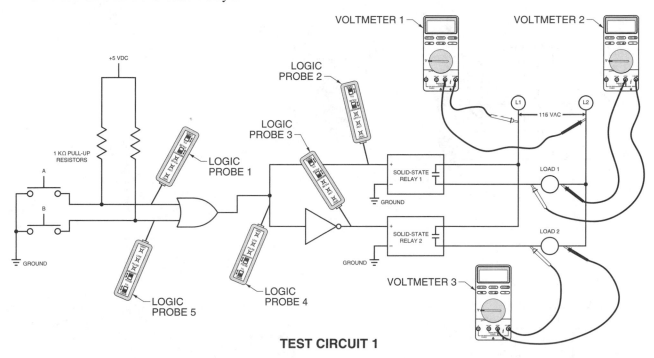

TEST CIRCUIT 1

Switch Positions		Meter Indication								Load	
		Logic Probe					Voltmeter				
A	B	1	2	3	4	5	1	2	3	1	2
A. Open	Open	High	High	High	High	High	115 V	115 V	115 V	ON	ON
B. Closed	Closed	Low	Low	High	Low	Low	115 V	0 V	115 V	OFF	OFF
C. Closed	Closed	Low	High	Low	High	Low	115 V	115 V	0 V	ON	OFF
D. Open	Closed	High	High	Low	High	High	115 V	115 V	0 V	ON	OFF
E. Open	Open	High	High	Low	High	High	115 V	0 V	0 V	OFF	OFF
F. Closed	Closed	Low	Low	High	Low	Low	0 V	0 V	0 V	OFF	OFF
G. Closed	Open	Low	Low	High	Low	High	115 V	0 V	0 V	OFF	OFF
H. Open	Open	High	High	Low	High	High	115 V	115V	0 V	OFF	OFF

_____ **A.** The problem equals ___.   _____ **E.** The problem equals ___.

_____ **B.** The problem equals ___.   _____ **F.** The problem equals ___.

_____ **C.** The problem cquals ___.   _____ **G.** The problem equals ___.

_____ **D.** The problem equals ___.   _____ **H.** The problem equals ___.

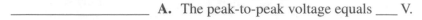

ACTIVITY 6-9—Oscilloscopes/Scopemeters

1. Determine the peak-to-peak voltage, rms voltage, and frequency.

_____ **A.** The peak-to-peak voltage equals ___ V.

_____ **B.** The rms voltage equals ___ V.

_____ **C.** The frequency equals ___ Hz.

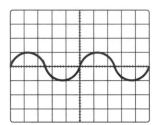
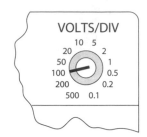
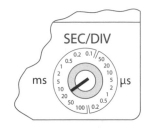

2. Determine the voltage.

_____ **A.** The voltage equals ___ V.

_____ **B.** The voltage equals ___ V.

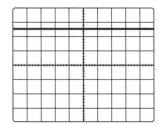
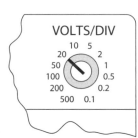
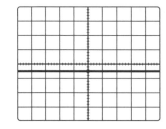
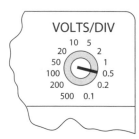

3. Determine the minimum and maximum peak voltage.

_____ **A.** The minimum peak voltage equals ___ V.

_____ **B.** The maximum peak voltage equals ___ V.

_____ **C.** The minimum peak voltage equals ___ V.

_____ **D.** The maximum peak voltage equals ___ V.

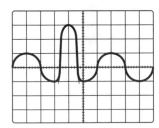
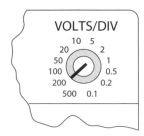
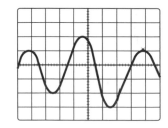
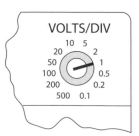

APPLICATION 7-1—Branch Circuit Voltage Drop Measurement

Branch circuit conductors must be properly sized for proper and safe equipment operation. Conductors are sized large enough to prevent no more than a 3% voltage drop at the furthest outlet. Voltage is measured under no-load and full-load conditions to test for excessive voltage drop.

To measure voltage drop in a branch circuit, apply the procedure:
1. Turn all loads connected to the branch circuit OFF.
2. Measure the voltage at the furthest outlet on the branch circuit to obtain the circuit's no-load voltage.
3. Turn all loads connected to the branch circuit ON.
4. Measure the voltage at the furthest outlet on the branch circuit to obtain the circuit's full-load voltage.

The percentage of voltage drop in the branch circuit is calculated by applying the formula:

$$\%V_D = \frac{V_{NL} - V_{FL}}{V_{NL}} \times 100$$

where
$\%V_D$ = percent voltage drop (in volts)
V_{NL} = no-load voltage drop (in volts)
V_{FL} = full-load voltage drop (in volts)
100 = constant

Example: Calculating Percent Voltage Drop— Branch Circuit

A branch circuit has a measured full-load voltage of 117 V and a measured no-load voltage of 120 V. Calculate the percent voltage drop. **See Branch Circuit.**

$$\%V_D = \frac{V_{NL} - V_{FL}}{V_{NL}} \times 100$$

$$\%V_D = \frac{120 - 117}{120} \times 100$$

$$\%V_D = \frac{3}{120} \times 100$$

$$\%V_D = .025 \times 100$$

$$\%V_D = \mathbf{2.5\%}$$

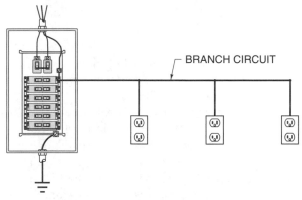

BRANCH CIRCUIT

BRANCH CIRCUIT

 Refer to *Activity 7-1—Branch Circuit Voltage Drop Measurement* ....

APPLICATION 7-2—Recording Minimum and Maximum Readings

Intermittent problems are found by recording a circuit's minimum and maximum readings. A meter must have a minimum/maximum (min/max) function to record readings over time. Most meters that have a min/max function hold any change that lasts more than one second. **See Recording Minimum and Maximum Readings.**

The meter test leads are connected to a circuit such as a voltmeter when measuring voltage or an ammeter when measuring current. The min/max button is pressed after the meter is connected to the circuit. The minimum or maximum value is read by pressing the min/max button anytime after the meter is connected to the circuit.

Min or max is displayed indicating which reading is being displayed. Most meters have a beeper that sounds each time a new minimum or maximum value is recorded. A continuous sound usually indicates the meter overloaded.

Low voltage problems caused by large motor start-ups high voltage surges, or temporary voltage losses are detected when the meter is connected to a circuit such as a voltmeter and the min/max button is pressed.

The maximum current used in a circuit is detected when the meter is connected to a circuit such as an ammeter and the min/max button is pressed. This information is used to determine if a circuit is overloaded or if there is room for additional loads.

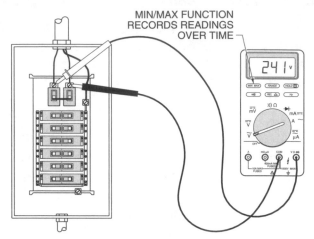

RECORDING MINIMUM AND MAXIMUM READINGS

 Refer to **Activity 7-2—Recording Minimum and Maximum Readings** ....

ACTIVITY 7-3—Meter Loading Effect

A meter connected to an electrical circuit causes the current and voltage values in the circuit to change. This occurs because the meter becomes an additional load in the circuit. *Meter loading effect* is the additional resistance a meter adds to the total resistance in a circuit. The meter is equivalent to adding a high resistance load in parallel with the load under test. A meter should load a circuit under test as little as possible for accurate readings. The better the meter, the less loading effect produced.

Loading effect is a problem when measuring voltages in circuits that include high resistances. For example, in a circuit that has two 200 kΩ resistors connected in series with a 120 V power supply, the voltage drop across each resistor is 60 V. The current through each resistor is .0003 A (300 µA). **See Voltmeter Loading Effect.**

A meter with a 100 kΩ resistance significantly loads the circuit if used to measure the voltage drop across a resistor. The meter loads the circuit to the extent that the voltage drop across the measured resistor changes to 30 V. This change causes a 50% meter error. The current in the circuit also increases. A meter with at least 10 times the input resistance should be used in a high resistance circuit to prevent a large meter error.

The loading effect a meter has on a circuit depends on the input resistance of the meter. The higher the input resistance, the less the loading effect. Analog meters are rated at the amount of ohms per volt (Ω/V) resistance they have. For example, a meter may be rated at 20,000 Ω/V. This means that for every volt of a given range the voltmeter has 20 kΩ of resistance. The total meter loading effect is 200,000 Ω/V voltmeter is set on the 10 V range.

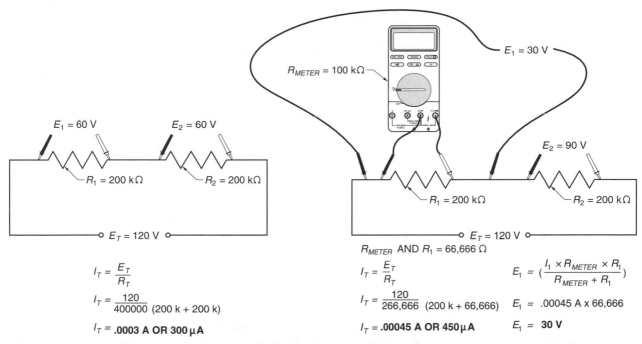

$E_1 = 60$ V $\quad$ $E_2 = 60$ V

$R_1 = 200$ kΩ $\quad$ $R_2 = 200$ kΩ

$\circ E_T = 120$ V $\circ$

$$I_T = \frac{E_T}{R_T}$$

$$I_T = \frac{120}{400000} \text{ (200 k + 200 k)}$$

$I_T = \textbf{.0003 A OR 300 μA}$

R_{METER} AND $R_1 = 66,666$ Ω

$$I_T = \frac{E_T}{R_T} \qquad E_1 = \left(\frac{I_1 \times R_{METER} \times R_1}{R_{METER} + R_1} \right)$$

$$I_T = \frac{120}{266,666} \text{ (200 k + 66,666)} \qquad E_1 = .00045 \text{ A} \times 66,666$$

$I_T = \textbf{.00045 A OR 450 μA} \qquad E_1 = \textbf{30 V}$

VOLTMETER LOADING EFFECT

AC analog meters generally have a higher loading effect than DC meters because AC meters have less input resistance than DC meters. A typical analog meter has 5000 Ω/V for AC and 20,000 Ω/V for DC.

Analog meters generally have a greater loading effect than digital meters because digital meters have a high input resistance (typically 10 MΩ). A 10 MΩ input resistance usually does not cause a significant loading effect. Some analog voltmeters use a high resistance amplifier circuit to develop a high input resistance. This input circuit uses field effect transistors (FETs) for high resistance. Voltmeters that use FETs typically have a 10 MΩ input resistance regardless of the range on which the meter is set.

A meter with a resistance at least ten times greater than the resistance of the component under test does not significantly load the circuit. The higher the resistance of the meter, the more accurate the reading is due to less loading effect.

Refer to **Activity 7-3—Meter Loading Effect** ....

APPLICATION 7-4—*Voltmeter Circuit Connections*

Voltmeter Load Connections

Voltmeters are normally connected across (in parallel) loads such as lights, solenoids, motors, heating elements, and other devices. A voltmeter is connected across a load to determine if voltage is present and at the correct level.

When a voltmeter is connected across a load, the meter reads 0 V when the control switch in the circuit is open. The voltmeter reads 0 V if the load is good or has a fault because no current flows to the load. **See Voltmeter Load Connection.**

When the control switch is closed, the voltmeter reads the full power supply voltage (120 V), provided the supply voltage is present in the circuit. The full supply voltage is read if the load is operating or not. The meter reading indicates that the supply voltage is delivered to the load.

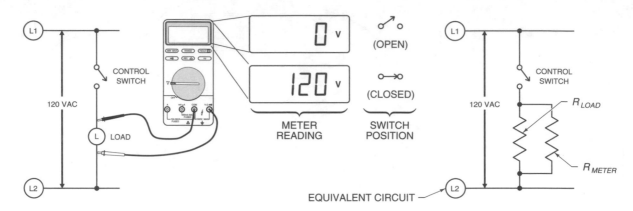

VOLTMETER LOAD CONNECTION

Voltmeter Switch Connections

A voltmeter may be connected across a control switch that controls a load. When a voltmeter is connected across an open control switch, the meter reads a voltage level provided the load does not have an open (or short) circuit. The meter is equivalent to adding a high resistance load in series with the load under test. **See Voltmeter Switch Connection.**

The voltmeter reads a voltage level because the meter completes the path from the power lines. The voltage level the meter reads depends on the resistance of the load and the resistance of the meter.

A meter normally has a large resistance in relation to the resistance of the load. A meter normally reads almost the full supply voltage. The higher the meter resistance in relation to the load resistance, the higher the meter voltage reading. If a load has an open (incomplete circuit) the voltmeter reads 0 V when the switch is open.

When a voltmeter is connected across a closed control switch, the meter reads 0 V if the load is good, faulty, or has an open. The voltmeter reads 0 V because the closed control switch shorts the meter leads. No voltage reading is indicated on the meter.

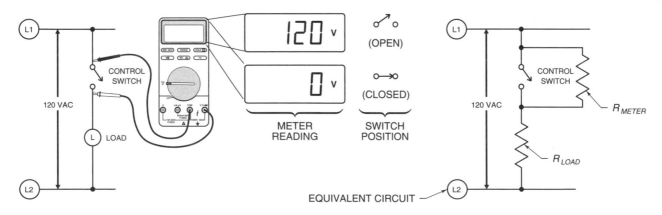

VOLTMETER SWITCH CONNECTION

 Refer to **Activity 7-4—Voltmeter Circuit Connections** ....

APPLICATION 7-5—Harmonic Distortion

Harmonic distortion is the voltage and/or current in a power line that is a multiple of the fundamental line frequency. For example, the fundamental line frequency in the U.S. is 60 Hz. The 60 Hz fundamental line frequency may have harmonic distortion comprised of 120 Hz (second harmonic) and 180 Hz (third harmonic) frequencies. **See Fundamental and Harmonic Waveforms.**

Harmonic distortion is created by electronic circuits such as variable speed motor drives, electronic ballasts (used in lighting circuits), personal computers, printers, and medical test equipment that draw current from power lines in short pulses. Harmonic distortion is created on power lines as a result of the pulsed power draw. Equipment efficiency is improved when electronic circuits draw current in short pulses.

Harmonic distortion is created when diodes are used to rectify AC power and charge capacitors with the rectified AC power. The result is that the capacitors draw current only during the peak of the AC sine wave. A normal power line voltage appears on a scope as a near perfect sine wave. Harmonic distortion distorts the waveform. **See Distorted Waveforms.**

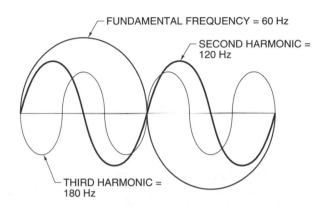

Note: Harmonic frequency does not have to be in-phase with the fundamental frequency

FUNDAMENTAL AND HARMONIC WAVEFORMS

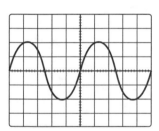

PURE SINE WAVE

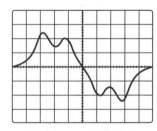

DISTORTED CURRENT
WAVEFORM CAUSED BY
HARMONIC DISTORTION

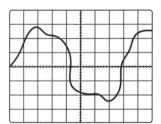

DISTORTED VOLTAGE
WAVEFORM CAUSED BY
HARMONIC DISTORTION

DISTORTED WAVEFORMS

Harmonic distortion overheats transformers and neutral conductors, burns out motors, and trips circuit breakers. Harmonic distortion in a pure 60 Hz AC power line, causes overheating by preventing the normal cancellation between conductors. Overheating may occur when voltage and current readings appear normal.

Harmonic Distortion Measurement

A harmonic analyzer measures harmonic distortion. A *harmonic analyzer* is a special meter that detects and measures harmonic distortion. An oscilloscope or scopemeter may be used to give a visual display of distortion on AC power lines. Two ammeters may be used to measure harmonic distortion when an analyzer or scope is not available. To measure harmonics with two ammeters, a meter that measures average current and one that measures true (rms) current are required. **See Harmonic Distortion Measurement.**

HARMONIC DISTORTION MEASUREMENT

To measure harmonic distortion using two meters, apply the procedure:
1. Measure the current in the power line with the average ammeter.
2. Measure the current in the power line with the true ammeter.

Divide the average meter reading by the true meter reading. A ratio of 1 indicates little or no harmonic distortion. A ratio less than 1 indicates harmonic distortion. A ratio of .5 indicates major harmonic distortion.

Example: Determining Harmonic Distortion

The average current measured in a circuit is 14.4 A. The rms current measured in a circuit is 15 A. Determine the amount of harmonic distortion. **See Current Levels.**

$$D_H = \frac{I_{avg}}{I_{rms}}$$

$$D_H = \frac{14.4}{15}$$

$$D_H = .96$$

The circuit has little harmonic distortion.

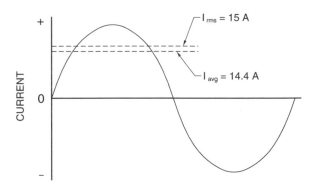

CURRENT LEVELS

The following steps may be taken to help eliminate the effect of harmonic distortion.
- Add harmonic filters to the circuit. The filters are added at the loads.
- Balance the loads in the circuit to help even the current in the lines.
- Oversize neutral conductors that are shared by hot conductors or add a neutral conductor for each hot power line.
- Derate transformers feeding harmonic loads. The amount of derating depends on the amount of harmonic distortion. The greater the harmonic distortion, the more the transformers should be derated.

 Refer to **Activity 7-5—Harmonic Distortion** ....

APPLICATION 7-6—Troubleshooting Fuses

A *fuse* is an overcurrent protection device with a fusible link that melts and opens the circuit on an overcurrent condition. Fuses are connected in series with a circuit to protect a circuit from overcurrents or shorts. Fuses may be one-time or renewable. *One-time fuses* are fuses that cannot be reused after they have opened. One-time fuses are the most common. *Renewable fuses* are designed so that the fusible link can be replaced. A multimeter or voltmeter is used to test fuses. **See Troubleshooting Fuses.**

To troubleshoot fuses, apply the procedure:
1. Turn the handle of the safety switch or combination starter OFF.
2. Open the door of the safety switch or combination starter. The operating handle must be capable of opening the switch. If it is not, replace the switch.
3. Check the enclosure and interior parts for deformation, displacement of parts, and burning. Such damage may indicate a short, fire, or lightning strike. Deformation requires replacement of the part or complete device. Any indication of arcing damage or overheating, such as discoloration or melting of insulation, requires replacement of the damaged part(s).
4. Check the voltage between each pair of power leads. Incoming voltage should be within 10% of the voltage rating of the motor. A secondary problem exists if voltage is not within 10%. This secondary problem may be the reason the fuses have blown.
5. Test the enclosure for grounding if voltage is present and at the correct level. To test for grounding, connect one side of a voltmeter to an unpainted metal part of the enclosure and touch the other side to each of the incoming power leads. A voltage difference is indicated if the enclosure is grounded. The line-to-ground voltage probably does not equal the line-to-line voltage reading taken in Step 4.

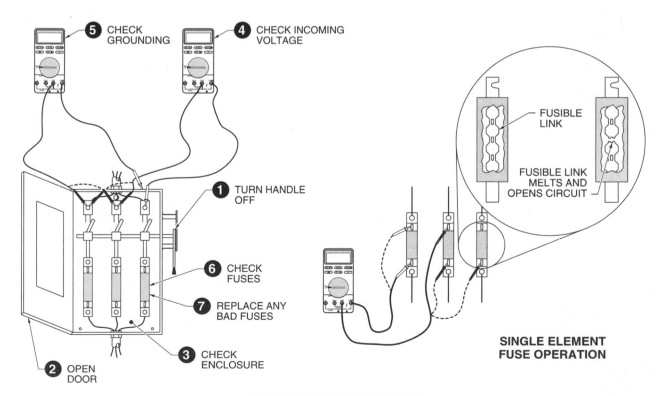

TROUBLESHOOTING FUSES

6. Turn the handle of the safety switch or combination starter ON to test the fuses. One side of a voltmeter is connected to one side of an incoming power line at the top of one fuse. The other side of the voltmeter is connected to the bottom of each of the remaining fuses. A voltage reading indicates the fuse is good. If no voltage reading is obtained, the fuse is open and no voltage passes through. The fuse must be replaced (not at this time). Repeat this procedure for each fuse. When testing the last fuse, the voltmeter is moved to a second incoming power line.

7. Turn the handle of the safety switch or combination starter OFF to replace the fuses. Use a fuse puller to remove bad fuses. Replace all bad fuses with the correct type and size replacement. Close the door on the safety switch or combination starter and turn the circuit ON.

 Refer to **Activity 7-6—Troubleshooting Fuses** ....

APPLICATION 7-7—*Troubleshooting Circuit Breakers*

A *circuit breaker* (CB) is a reusable overcurrent protective device that opens a circuit automatically at a predetermined overcurrent. CBs are connected in series with the circuit. CBs protect a circuit from overcurrents or short circuits. CBs are thermally or magnetically operated and are reset after an overload. A multimeter is used to test CBs. Circuit breakers perform the same function as fuses and are tested the same way. **See Troubleshooting Circuit Breakers.**

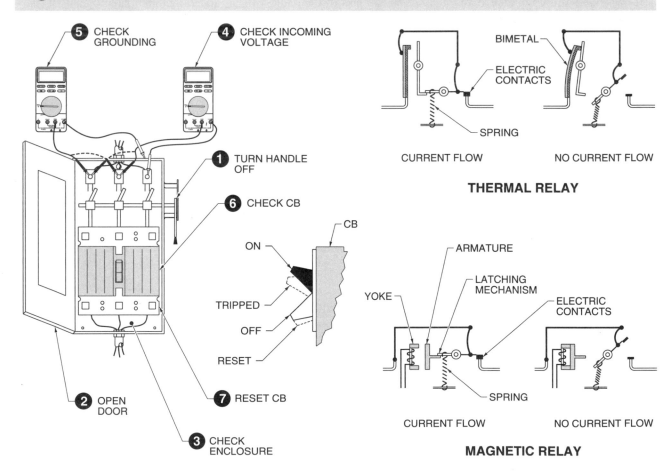

TROUBLESHOOTING CIRCUIT BREAKERS

To troubleshoot CBs, apply the procedure:

1. Turn the handle of the safety switch or combination starter OFF.

2. Open the door of the safety switch or combination starter. The operating handle must be capable of opening the switch. Replace the operating handle if it does not open the switch.

3. Check the enclosure and interior parts for deformation, displacement of parts, and burning.

4. Check the voltage between each pair of power leads. Incoming voltage should be within 10% of the voltage rating of the motor.

5. Test the enclosure for grounding if voltage is present and at the correct level.

6. Examine the CB. It will be in one of three positions, ON, TRIPPED, or OFF.

7. If no evidence of damage is present, reset the CB by moving the handle to OFF and then to ON. CBs must be cooled before they are reset. CBs are designed so they cannot be held in the ON position if an overload or short is present. Check the voltage of the reset CB if resetting the CB does not restore power. Refer to Troubleshooting Fuses, Step 6. Replace all faulty CBs. Never try to service a faulty CB.

 Refer to **Activity 7-7 — Troubleshooting Circuit Breakers**...

 Refer to **Quick Quiz**® *on CD-ROM*

Refer to Chapter 7 in the **Troubleshooting Electrical/Electronic Systems Workbook** *for additional questions.*

Name_____ Date _____

ACTIVITY 7-1—Branch Circuit Voltage Drop Measurement

Determine the percentage of voltage drop.

_____ **1.** The voltage drop equals ___%.

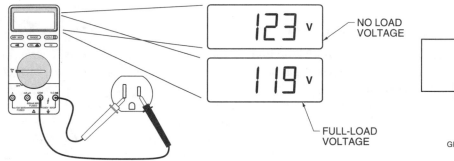

_____ **2.** The voltage drop equals ___%.

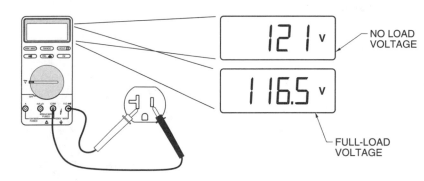
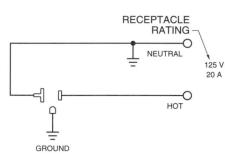

_____ **3.** The voltage drop equals ___%.

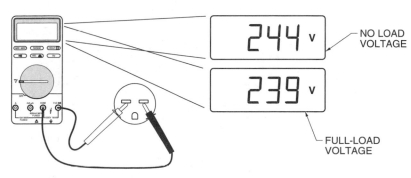
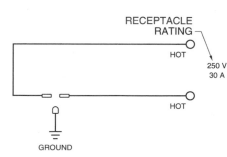

_____ **4.** The voltage drop equals ___%.

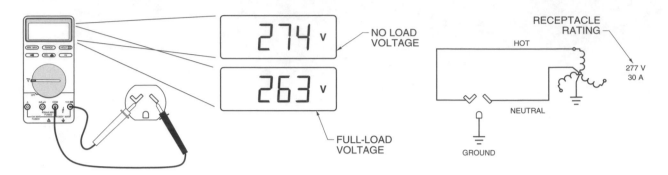

ACTIVITY 7-2—Recording Minimum and Maximum Readings

Minimum and maximum readings are recorded when troubleshooting thermostat differential problems. Thermostat differential problems occur when there is too large or too small of a temperature difference between the turn ON and turn OFF points of a furnace or air conditioner.

Room temperature varies excessively and may be uncomfortable when the temperature differential is too large. The furnace or air conditioner cycles repeatedly when the temperature differential is too small. Overcycling causes motor burnout. The differential should remain between 3°F for comfort and equipment reliability. For example, a thermostat set at 72°F turns ON the furnace when room temperature falls to 72°F, and turns OFF the furnace when the room temperature rises to 75°F. The thermostat differential is equal to 3°F.

Determine the differential temperature and the number of degrees higher or lower the thermostat differential adjustment should be set to maintain a 3°F differential.

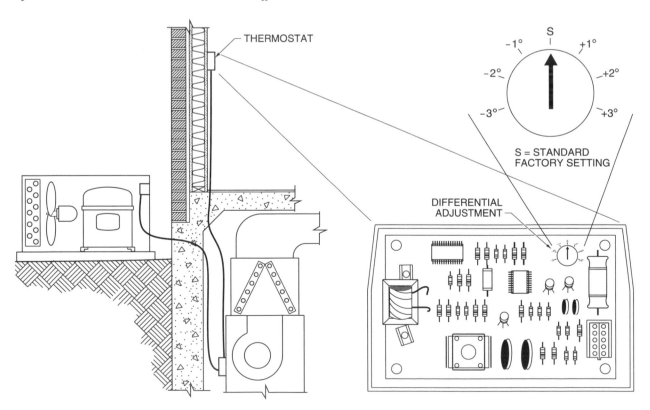

_____ 1. The differential
temperature equals
___°F.

_____ 2. The required adjustment
equals ___°F.

_____ 3. The differential
temperature equals
___°F.

_____ 4. The required adjustment
equals ___°F.

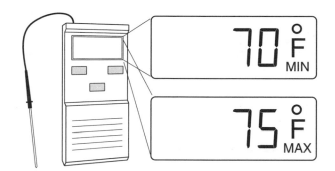

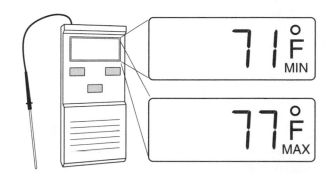

ACTIVITY 7-3—Meter Loading Effect

Determine the meter error caused by the meter loading effect.

_____ 1. The meter reading of Model A equals ___ V.

_____ 2. The meter error equals ___%.

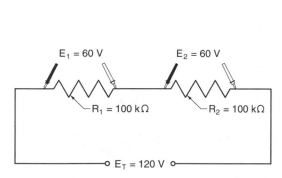

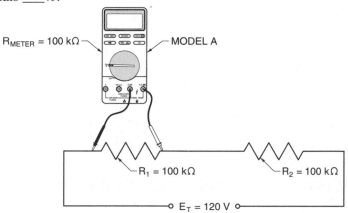

_____ 3. The meter reading of Model B equals ___V.

_____ 4. The meter error equals ___%.

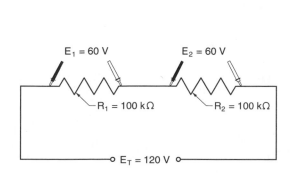

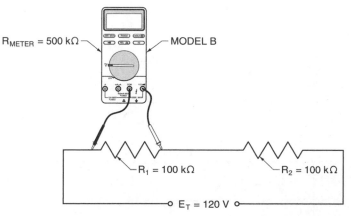

_____ **5.** The meter reading of Model C equals ___ V.

_____ **6.** The meter error equals ___%.

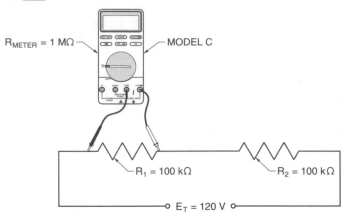

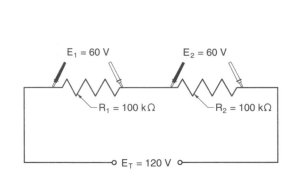

_____ **7.** The meter reading of Model D equals ___ V.

_____ **8.** The meter error equals ___%.

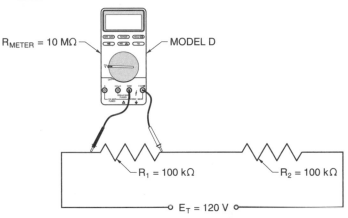

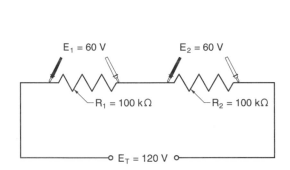

ACTIVITY 7-4—Voltmeter Circuit Connections

Determine the voltmeter reading(s).

_____ **1.** The reading of the voltmeter equals ___ V.

_____ **2.** The reading of Voltmeter 1 equals ___ V.

_____ **3.** The reading of Voltmeter 2 equals ___ V.

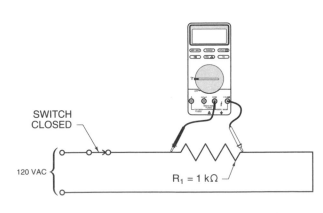

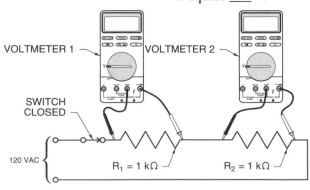

_____ **4.** The reading of Voltmeter 3 equals ___ V.

_____ **5.** The reading of Voltmeter 4 equals ___ V.

_____ **6.** The reading of Voltmeter 5 equals ___ V.

_____ **7.** The reading of Voltmeter 6 equals ___ V.

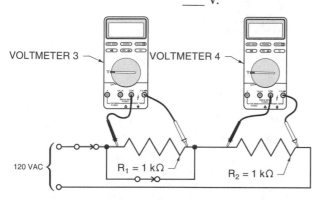

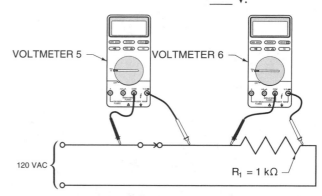

ACTIVITY 7-5—Harmonic Distortion

Using Circuits 1-4, answer the questions.

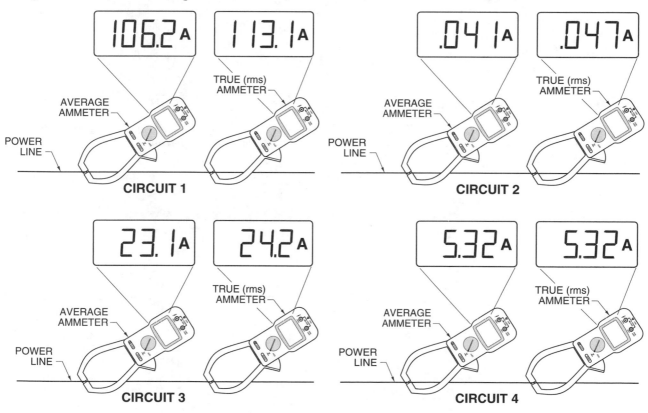

_____ **1.** Circuit ___ has the greatest harmonic distortion.

_____ **2.** Circuit ___ has the least harmonic distortion.

_____ **3.** Circuit ___ has no harmonic distortion.

ACTIVITY 7-6—Troubleshooting Fuses

Using the meter readings, determine whether the fuse is good, bad, or questionable. A fuse is questionable when the meter readings do not provide enough information to determine whether the fuse is definitely good or bad.

_____ **1.** Fuse 1 is ___.

_____ **2.** Fuse 2 is ___.

_____ **3.** Fuse 3 is ___.

_____ **4.** Fuse 1 is ___.

_____ **5.** Fuse 2 is ___.

_____ **6.** Fuse 3 is ___.

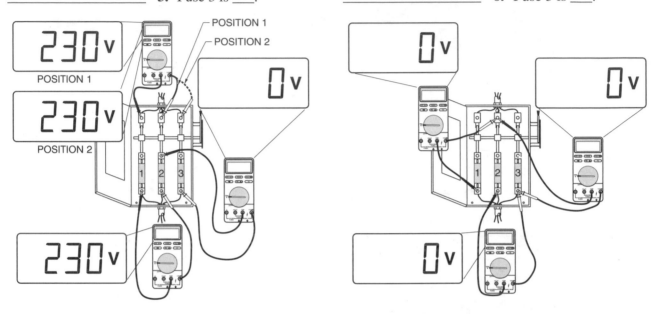

ACTIVITY 7-7—Troubleshooting Circuit Breakers

Using the meter readings, determine whether the circuit breaker is good, bad, or questionable. The circuit breaker is questionable when the meter readings do not provide enough information to determine whether the breaker is definitely good or bad.

_____ **1.** The circuit breaker is ___.

_____ **2.** The circuit breaker is ___.

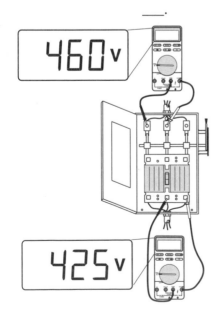

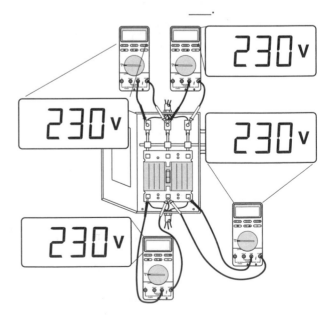

Relays and Motor Starters

APPLICATION 8-1—Relays

A *relay* is an interface that controls one electrical circuit by opening and closing contacts in another circuit. The two types of relays are electromechanical and solid-state relays.

Electromechanical Relays

An *electromechanical relay* is a device that allows the connection of two different components, voltage levels, voltage types (AC/DC), or systems. An electromechanical relay is normally used to increase the number of contacts and allow a low current to switch a high current. By increasing the number of contacts, the current across an individual contact is reduced and allows the control of additional loads. **See Electromechanical Relay.**

An electromechanical relay increases the number of contacts by using a coil to magnetically move several different contacts simultaneously. One coil normally controls three or more contacts.

Contact arrangements are described by their poles, throws, and breaks. *Poles* are the number of completely isolated circuits that can pass through the switch at one time. *Throws* are the number of different closed contact positions per pole that are available on the switch. *Breaks* are the number of separate contacts the switch uses to open or close each individual circuit. Abbreviations are used to simplify relay contact identification. **See Relay Contact Abbreviations.**

An electromechanical relay usually has multipole and multithrow contact arrangements that allow the relay to simultaneously switch several different circuits. The contacts switch to either AC or DC. **See Relay Contact Arrangements.**

An electromechanical relay has a low contact voltage drop; therefore, a heat sink is not required to dissipate heat produced by the switching of the contacts. A *heat sink* is a device that conducts and dissipates heat away from a component.

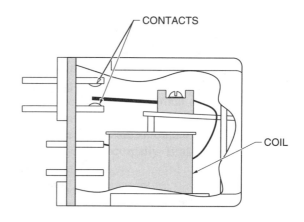

ELECTROMECHANICAL RELAY

RELAY CONTACT ABBREVIATIONS	
Abbreviation	**Meaning**
SP	Single pole
DP	Double pole
ST	Single throw
DT	Double throw
NO	Normally open
NC	Normally closed
SB	Single break
DB	Double break

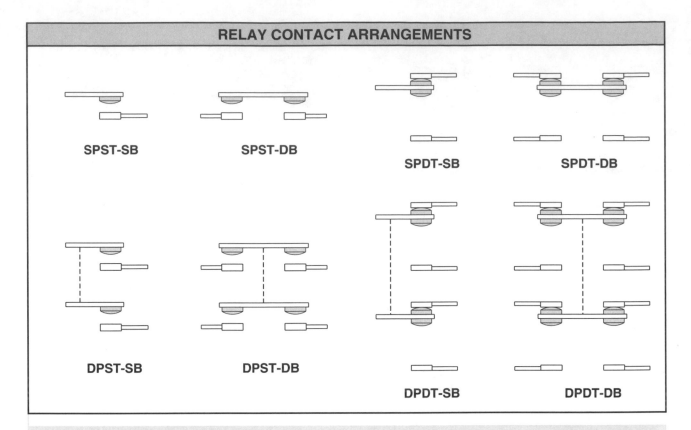

RELAY CONTACT ARRANGEMENTS

SPST-SB   SPST-DB   SPDT-SB   SPDT-DB

DPST-SB   DPST-DB   DPDT-SB   DPDT-DB

An electromechanical relay is relatively inexpensive, resistant to voltage transients, and does not have off-state leakage current through open contacts. *Off-state leakage current* is the amount of current that leaks through a solid-state switch when the switch is turned OFF.

Electromechanical relay contacts have a limited life which is shortened when used for rapid switching applications. An electromechanical relay has poor performance when switching high inrush currents which generate electromagnetic noise and interference on power lines.

Electromechanical Relay Life

Electromechanical relay life expectancy is rated in contact life and mechanical life. *Contact life* is the number of times a relay's contacts switch the load controlled by the relay before malfunctioning. Typical contact life ratings are 100,000 to 500,000 operations. *Mechanical life* is the number of times a relay's mechanical parts operate before malfunctioning. Typical mechanical life ratings are 1,000,000 to 10,000,000 operations.

Relay contact life expectancy is lower than mechanical life expectancy because the life of a contact depends on the application. The contact rating of a relay is based on the contact's full-rated power. Contact life is increased when contacts switch loads less than their full-rated power. Contact life is reduced when contacts switch loads that develop destructive arcs. *Arcing* is the discharge of an electric current across a gap such as when an electric switch is opened. Arcing causes contact burning and temperature rise. **See Arcing.**

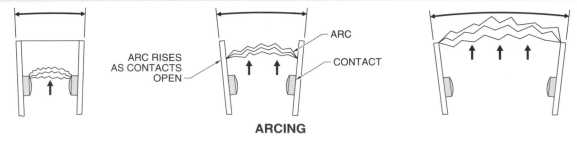

ARC

ARC RISES
AS CONTACTS
OPEN

CONTACT

ARCING

Arcing is minimized by an arc suppressor and by using the correct contact material for the application. An *arc suppressor* is a device that dissipates the energy present across opening contacts. Arc suppression is used in applications that switch arc producing loads such as solenoids, coils, motors, and other inductive loads.

Arc suppression is also accomplished by using a contact protection circuit. A *contact protection circuit* is a circuit that protects contacts by providing a nondestructive path for generated voltage as a switch is opened. A contact protection circuit may contain a diode, a resistance/capacitance (RC) circuit, or a varistor.

A diode is used as contact protection in DC circuits. The diode does not conduct electricity when the load is energized. The diode conducts electricity and shorts the generated voltage when the switch is opened. By shorting the generated voltage, the voltage is dissipated across the diode and not the relay contacts. **See Contact Protection Circuits.**

An RC circuit and varistor are used as contact protection in AC circuits. The capacitor in an RC circuit is a high impedance to the 60 Hz line power and a short circuit to generated high frequencies. The short circuit dissipates generated voltage. A *varistor* is a resistor whose resistance is inversely proportional to the voltage applied to it. The varistor becomes a low impedance circuit when its rated voltage is exceeded. The low impedance circuit dissipates generated voltage when a switch is opened.

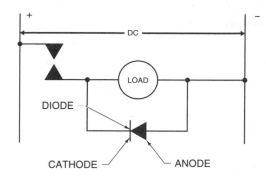

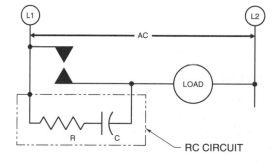

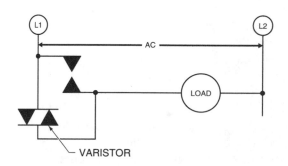

CONTACT PROTECTION CIRCUITS

Contact Material

Relay contacts are available in fine silver, silver-cadmium, gold-flashed silver, and tungsten. Fine silver has the highest electrical and thermal properties of all metals. Fine silver sticks, welds, and is subject to sulfidation when used for many applications. *Sulfidation* is the formation of film on the contact surface. Sulfidation increases the resistance of the contacts. Silver is alloyed with other metals to reduce sulfidation.

Silver is alloyed with cadmium to produce a silver-cadmium contact. Silver-cadmium contacts have good electrical characteristics and low resistance which helps the contact resist arcing but not sulfidation. Silver or silver alloy contacts are used in circuits that switch several amperes at more than 12 V, which burns off the sulfidation.

Sulfidation can damage silver contacts when used in intermittent applications. Gold-flashed silver contacts are used in intermittent applications to minimize sulfidation and provide a good electrical connection. Gold contacts are not used in high-current applications because the gold burns off quickly. Gold-flashed silver contacts are good for switching loads of one ampere or less.

Tungsten contacts are used in high-voltage applications because tungsten has a high melting temperature and is less affected by arcing. Tungsten contacts are used when high repetitive switching is required.

Contact Failure

In most applications a relay fails due to contact failure. In some low-current applications, the relay contacts may look clean but may have a thin film of sulfidation, oxidation, or contaminates on the contact surface. This film increases the resistance to the flow of current through the contact. Normal contact wiping or arcing usually removes the film. In low-power circuits this action may not take place. In most applications, contacts are oversized for maximum life. Low-power circuit contacts are not oversized to the extent that they switch just a small fraction of their rated value.

Contacts are often subject to high-current surges. High-current surges reduce contact life by accelerating sulfidation and contact burning. For example, a 100 W incandescent lamp has a current rating of about 1 A. The life of the contacts is reduced if a relay with 5 A contacts is used to switch the lamp because the lamp's filament has a low resistance when cold. When first turned ON, the lamp draws 12 or more amperes. The 5 A relay switches the lamp, but does not switch it for the rated life of the relay. Contacts are oversized in applications that have high-current surges.

Solid-State Relays

A *solid-state relay* is an electronic switching device that has no moving parts. A solid-state relay is used for applications requiring highly repetitive switching, such as a flashing light.

A solid-state relay provides faster switching, longer life, smaller size, and the ability to handle more complex functions than an electromechanical relay. A solid-state relay differs from an electromechanical relay in its operating characteristics, sensitivity, and ability to operate correctly in different environments. **See Solid-State Relay.**

A solid-state relay has a long life when properly applied. A solid-state relay does not have contacts and does not produce arcing which generates electromagnetic interference. A solid-state relay is resistant to shock and vibration and has fast-switching capability. A solid-state relay has logic compatible to programmable controllers, digital circuits, and computers.

A solid-state relay normally has only one contact per relay. The voltage drop across the contact requires a heat sink. A solid-state relay can switch only AC or DC depending on the type used and has off-state leakage current when the contact is open. A solid-state relay must be the correct size and type for an application.

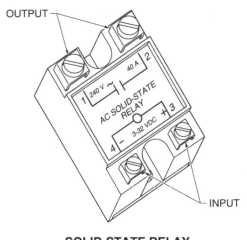

SOLID-STATE RELAY

Solid-State Relay Temperature Problems

Temperature rise is the largest problem in applications using a solid-state relay. As temperature increases, the failure rate of a solid-state relay increases. The operation of a solid-state relay is directly affected by its operating temperature. The higher the heat in a solid-state relay, the more problems occur. **See Temperature Effect On Solid-State Relays.**

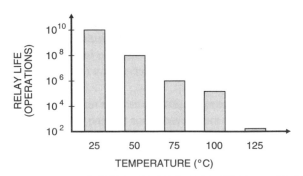

TEMPERATURE EFFECT ON SOLID-STATE RELAYS

The failure rate of most solid-state relays doubles for every 10°C temperature rise above an ambient temperature of 40°C. An ambient temperature of 40°C is considered standard by most manufacturers.

Solid-state relay manufacturers specify the maximum relay temperature permitted. The relay must be properly cooled to ensure that the temperature does not exceed the specified maximum safe value. Proper cooling is accomplished by installing the solid-state relay to the correct heat sink. A heat sink is chosen based on the maximum amount of load current controlled.

Heat Sinks

The performance of a solid-state relay is affected by ambient temperature. The ambient temperature of a relay is a combination of the temperature of the relay location and the type of enclosure used. The temperature inside an enclosure may be much higher than the ambient temperature of an enclosure that does not allow good air flow.

The temperature inside an enclosure increases if the enclosure is located next to a heat source or in the sun. The electronic circuit and solid-state relay also produce heat. Forced cooling is required in some applications.

Selecting Heat Sinks

A low resistance to heat flow is required to remove the heat produced by a solid-state relay. The opposition to heat flow is thermal resistance. *Thermal resistance (R_{TH})* is the ability of a device to impede the flow of heat. Thermal resistance is a function of the surface area of a heat sink and the conduction coefficient of the heat sink material. Thermal resistance is expressed in degrees Celsius per Watt (°C/W). **See Heat Sink Selections.**

Heat sink manufacturers list the thermal resistance of heat sinks. The lower the thermal resistance number, the easier the heat sink dissipates heat. The larger the thermal resistance number, the less effectively the heat sink dissipates heat. The thermal resistance value of a heat sink is used with a solid-state relays Load Current/Ambient Temperature chart to determine the size of the heat sink required.

A relay can control a large amount of current when a heat sink with a low thermal resistance number is used. A relay can control the least amount of current when no heat sink (free air mounting) is used. **See 40 A Relay Load Current/Ambient Temperatures.**

HEAT SINK SELECTIONS		
Type	**H × W × L (mm)**	**R$_{TH}$ (°C/W)**
01	15 × 79 × 100	2.5
02	15 × 100 × 100	2.0
03	25 × 97 × 100	1.5
04	37 × 120 × 100	.9
05	40 × 60 × 150	.5
06	40 × 200 × 150	.4

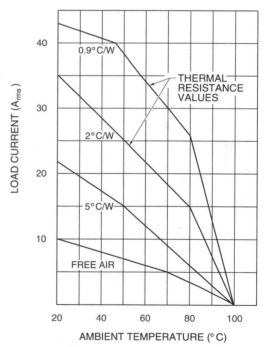

40 A RELAY LOAD CURRENT/AMBIENT TEMPERATURES

To maximize heat conduction through a relay and into a heat sink:
- Use heat sinks made of a material that has a high thermal conductivity. Silver has the highest thermal conductivity rating. Copper has the highest practical thermal conductivity rating. Aluminum has a good thermal conductivity rating and is the most cost effective and widely used heat sink.

- Keep the thermal path as short as possible.
- Use the largest cross-sectional surface area in the smallest space.
- Always use thermal grease or pads between the relay housing and the heat sink to eliminate air gaps and aid in thermal conductivity.

Mounting Heat Sinks

A heat sink must be correctly mounted to assure proper heat transfer.

To properly mount a heat sink:
- Choose a smooth mounting surface. The surfaces between a heat sink and a solid-state device should be as flat and smooth as possible. Ensure mounting bolts and screws are securely tightened.
- Locate heat producing devices so that the temperature is spread over a large area. This helps prevent increased temperature areas.
- Use heat sinks with fins to maintain as large a surface area as possible.
- Ensure that the heat from one heat sink does not add to the other.

Relay Current Problems

The overcurrent passing through a solid-state relay must be kept below the maximum load current rating of the solid-state relay. An overload protection fuse is used to prevent overcurrents from damaging a solid-state relay.

An overload protection fuse opens the circuit when the current is increased to a higher value than the nominal-load current. The fuse should be an ultra-fast fuse used for the protection of semiconductors. **See Overcurrent Protection.**

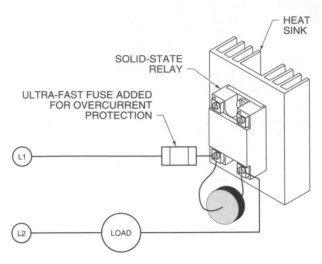

OVERCURRENT PROTECTION

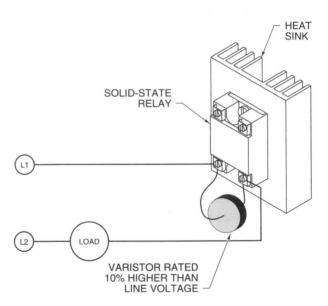

OVERVOLTAGE PROTECTION

Relay Voltage Problems

Most AC power lines contain voltage spikes superimposed on the voltage sine wave. Voltage spikes are produced by switching motors, solenoids, transformers, motor starters, contactors, and other inductive loads. Large spikes are also produced by lightning striking the power distribution system.

The output element of a relay can exceed its breakdown voltage and turn ON for part of a half period if overvoltage protection is not provided. This short turn ON can cause problems in the circuit.

Varistors are added to the relay output terminals to prevent an overvoltage problem. A varistor should be rated for 10% higher than the line voltage of the output circuit. The varistor bypasses the transient current. **See Overvoltage Protection.**

Voltage Drop

A voltage drop in the switching component is unavoidable in a solid-state relay. The voltage drop produces heat. The larger the current passing through the relay, the greater amount of heat produced. The generated heat affects relay operation and can destroy the relay if not removed. **See Relay Voltage Drop.**

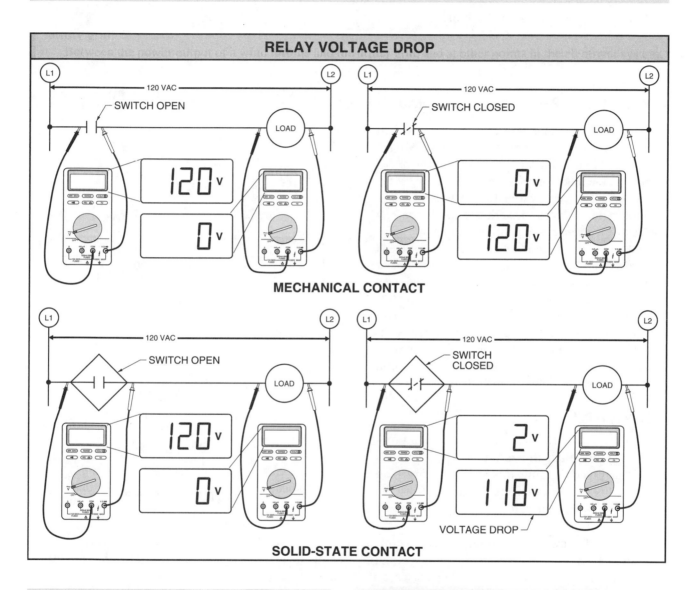

The voltage drop in a solid-state relay is usually 1 V to 1.6 V, depending on the load current. For small loads (less than 1 A), the heat produced is safely dissipated through the relay's case. High-current loads require a heat sink to dissipate the extra heat. **See Solid-State Relay Voltage Drop.**

SOLID-STATE RELAY VOLTAGE DROP		
Load Current (in A)	**Voltage Drop (in V)**	**Power at Switch (in W)**
1	2	2
2	2	4
5	2	10
10	2	20
20	2	40
50	2	100

For example, if the load current in a circuit is 1 A and a solid-state relay switching device has a 2 V drop, the power generated in the device is 2 W. The 2 W of power generates heat that can be dissipated through the relay's case.

If the load current in a circuit is 20 A and the solid-state relay switching device has a 2 V power drop, the power generated in the device is 40 W. The 40 W of power generates heat that requires a heat sink to safely dissipate the heat.

 *Refer to **Activity 8-1 — Relays** ....*

APPLICATION 8-2 — Contactors and Motor Starters

Contactors and motor starters are used to switch high-load currents. A *contactor* is a control device that uses a small control current to energize or de-energize the load connected to it. A *motor starter* is a contactor with overload protection added. A motor starter includes an overload relay that detects excessive current passing through a motor.

Contactors are used to switch non-motor loads such as lights, heating elements, and transformers. Motor starters are used to switch all types and sizes of motors. Contactors and motor starters control the loads connected to them either by using a magnetic coil and contacts, or by solid-state switching. Contactors and motor starters are available in sizes that can switch loads of a few amperes to several hundred amperes. **See Motor Starter.**

Overload Relays

An *overload relay* in a motor starter is a time-delay device that allows temporary overloads without disconnecting the load. The overload relay trips and disconnects the motor from the circuit when an overload occurs and lasts longer than the preset time.

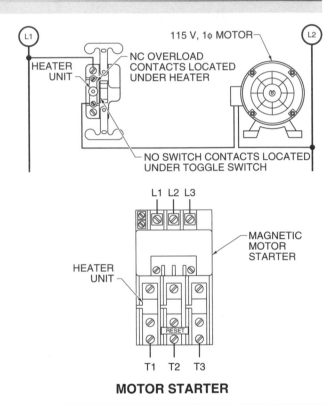

MOTOR STARTER

Overload relays are electromagnetic or thermal. An *electromagnetic overload relay* is an electromechanical relay operated by the current flow in a circuit. When the level of current in the circuit reaches a preset value, the increased magnetic field opens a set of contacts.

A *thermal overload relay* is an electromechanical relay that operates by heat developed in the relay. When the level of current in a circuit reaches a preset value, the increased temperature opens a set of contacts. The increased temperature opens the contacts through a bimetallic strip or by melting an alloy that activates a mechanism that opens the contacts.

Overload relays are connected in series with a motor. The same amount of current passes through the overload relay and the motor. Overload relays are designed to operate with their contacts normally closed (NC). The contacts are connected in series with the starting coil. When the contacts open, the starting coil is de-energized and the motor is disconnected from power.

Overload Tripping

Overloads are installed in motor starters to protect the motor when running. Overloads protect a motor by disconnecting the motor from the power supply when the heat generated in the motor windings approaches a damaging level.

Motors can draw several times their normal running current when starting. Time delay is required to permit a motor to start and allow overloads of short duration. Overloads trip when an overload exists for more than a short time. The time it takes for an overload to trip depends on the length of time the overload exists and the ambient temperature in which the overloads are located. **See Motor Current Draw.**

Overload relays on manual starters are reset after tripping by pressing the stop button and then the start button or by pressing the reset button. The overloads on magnetic starters set for manual reset are reset by pressing the reset button next to the overload contact. The overloads on magnetic starters set for automatic reset are reset automatically after the unit has cooled.

Trip Time

The time in which it takes an overload to trip depends on the length of time the overload current exists. A Heater Trip Characteristics chart shows the relationship between the time an overload takes to trip and the current flowing in the circuit. A Heater Trip Characteristics chart is based on the standard 40°C ambient temperature installation. The larger the overload (horizontal axis), the shorter the time required to trip the overload (vertical axis). **See Heater Trip Characteristics.**

Standard ambient temperature is 40°C unless otherwise stated. Any change from ambient temperature affects the tripping time of an overload. For temperatures higher than 40°C, the overloads trip at a current rating less than the value of the overload. For example, at 50°C the overloads trip at 90% of their rated value. For temperatures lower than 40°C, the overloads trip at a current rating greater than the rated value of the overload. **See Heater Ambient Temperature Correction.**

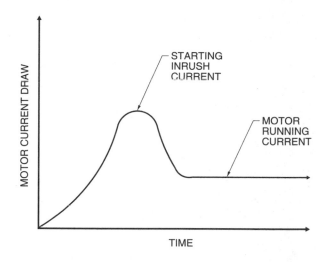

MOTOR CURRENT DRAW

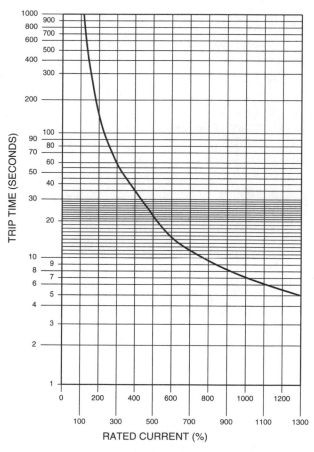

HEATER TRIP CHARACTERISTICS

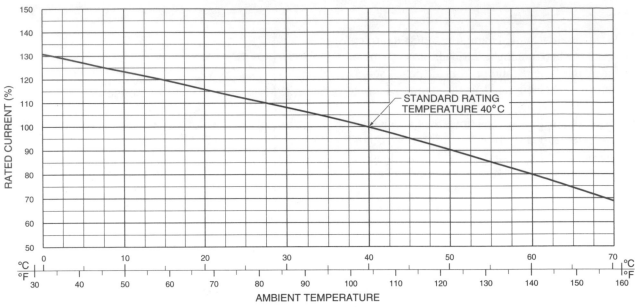

HEATER AMBIENT TEMPERATURE CORRECTION

Selecting Overloads

The ambient temperature in which a starter and motor is located must be considered when selecting overloads because a high ambient temperature reduces overload trip time. Reduced overload trip time can lead to nuisance tripping if a motor is located in a cooler ambient temperature than the starter. Reduced overload trip time can lead to motor burnout when the motor is located in a hotter ambient temperature than the starter.

An overload heater with a full-load current closest to the full-load current value listed on the motor nameplate is selected when the ambient temperature is the same at the motor and the starter. This provides the motor with overload protection between 110% and 120% (115% is standard for most manufacturers) of the motor's full-load current.

A higher overload heater value is selected when the ambient temperature at the starter is higher than the temperature at the motor. A lower overload heater value is selected when the ambient temperature at the starter is lower than the temperature at the motor.

 Refer to **Activity 8-2—Contactors and Motor Starters**..

 Refer to Quick Quiz® on CD-ROM

 *Refer to Chapter 8 in the* **Troubleshooting Electrical/Electronic Systems Workbook** *for additional questions.*

Name_____ **Date** _____

ACTIVITY 8-1—Relays

1. An electromechanical relay is used to control a solenoid-operated valve. The valve opens when the solenoid is energized and closes when the solenoid is de-energized. A new relay must be correctly wired and the contact pin numbers identified when replacing a faulty relay with a new one. Identify each set of relay contact pin numbers.

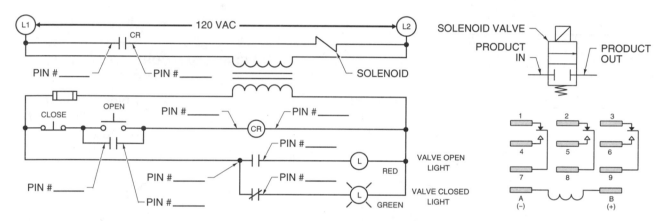

The temperature of a relay coil increases when the relay coil is energized, the relay is placed in a high ambient temperature, or the relay is turned ON and OFF repeatedly. The power draw of the coil increases as the temperature of the relay coil increases. Manufacturers rate coil power according to the coil's temperature. Using the coil power graph, determine the amount of coil power the wattmeter reads if the coil is good.

_____ **2.** The wattmeter reads ___ W.

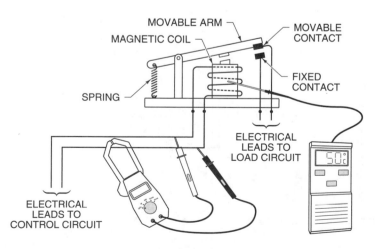

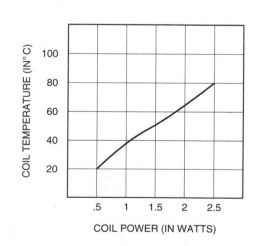

3. A 40 A solid-state relay is used in a control circuit. Using Heat Sink Selections and 40 A Relay Load Current/Ambient Temperatures, determine the minimum heat sink type (01, 02, etc.) required to dissipate the heat from the relay.

HEAT SINK SELECTIONS		
Type	H × W × L (mm)	R_{TH} (°C/W)
01	15 × 97 × 100	2.5
02	15 × 100 × 100	2.0
03	25 × 97 × 100	1.5
04	37 × 120 × 100	.9
05	40 × 160 × 150	.5
06	40 × 200 × 150	.4

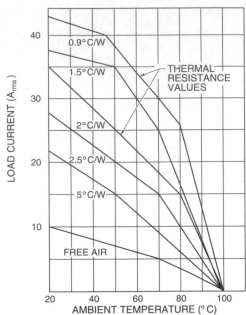

40 A RELAY LOAD CURRENT/AMBIENT TEMPERATURES

_____ **A.** A type ___ heat sink is required to switch a 15 A load.

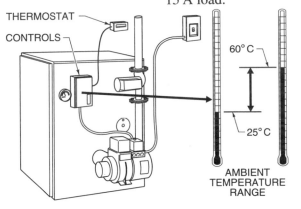

_____ **B.** A type ___ heat sink is required to switch a 30 A load.

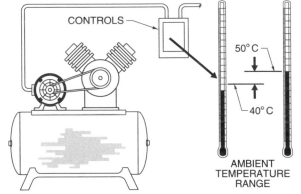

_____ **C.** A type ___ heat sink is required to switch a 27 A load.

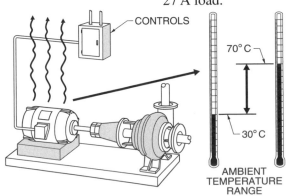

_____ **D.** A type ___ heat sink is required to switch a 20 A load.

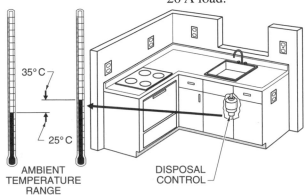

_____ **E.** A type ___ heat sink is required to switch a 15 A load.

_____ **F.** A type ___ heat sink is required to switch a 10 A load.

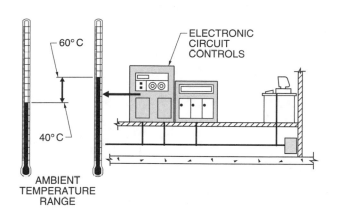

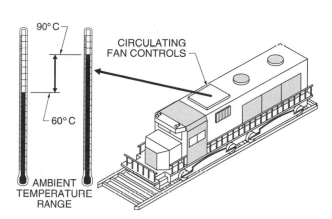

4. Determine the power produced at each solid-state switch.

_____ **A.** ___ W of power is produced at the solid-state switch.

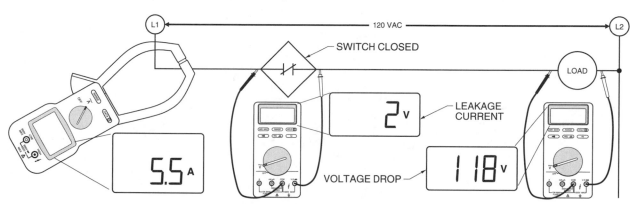

_____ **B.** ___ W of power is produced at the solid-state switch.

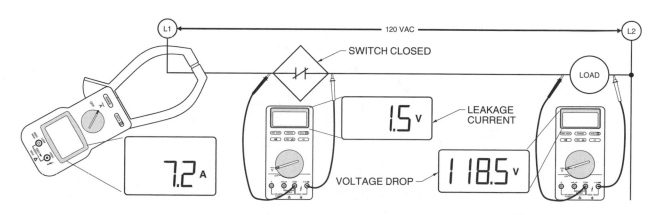

_____ **C.** ___ W of power is produced at the solid-state switch.

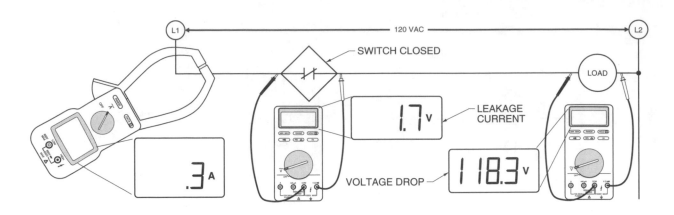

_____ **D.** ___ W of power is produced at the solid-state switch.

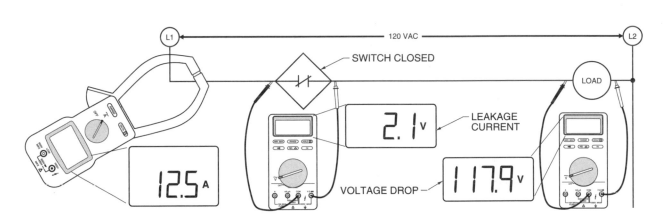

_____ **E.** ___ W of power is produced at the solid-state switch.

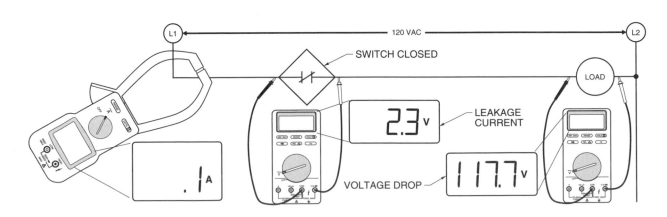

ACTIVITY 8-2—Contactors and Motor Starters

1. Using Heater Trip Characteristics, determine the approximate time (in seconds) that it takes the overloads to trip.

_____ **A.** Overload trip time is ___ seconds.

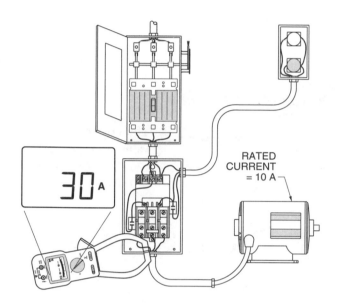

RATED CURRENT = 10 A

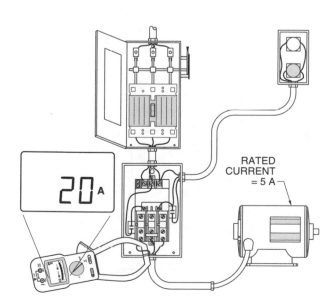

HEATER TRIP CHARACTERISTICS

Y-axis: TRIP TIME (SECONDS)
X-axis: RATED CURRENT (%)

_____ **B.** Overload trip time is ___ seconds.

RATED CURRENT = 5 A

_____ **C.** Overload trip time is ___ seconds.

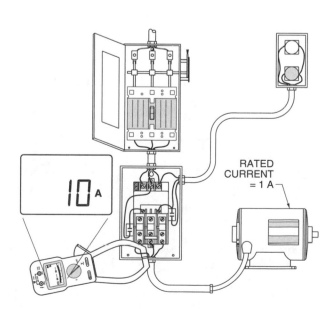

RATED CURRENT = 1 A

2. Using Heater Ambient Temperature Correction, determine the corrected current value used to select an overload for each application.

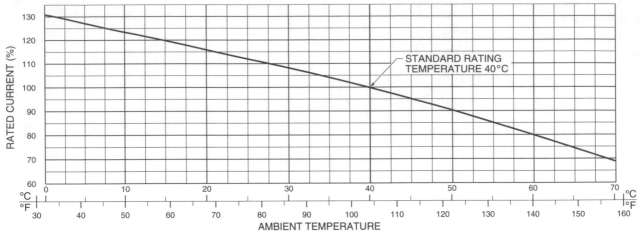

HEATER AMBIENT TEMPERATURE CORRECTION

_____ **A.** The corrected current value is ___ A.

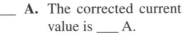

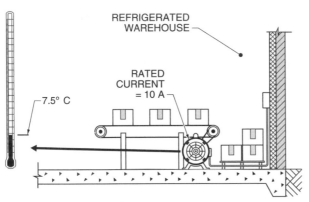

_____ **B.** The corrected current value is ___ A.

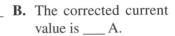

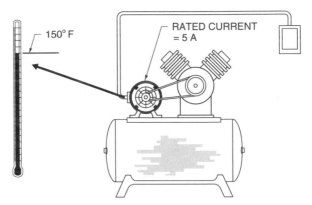

_____ **C.** The corrected current value is ___ A.

_____ **D.** The corrected current value is ___ A.

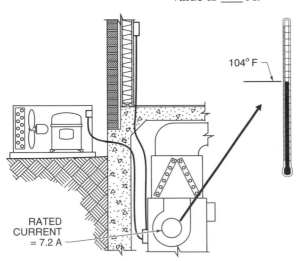

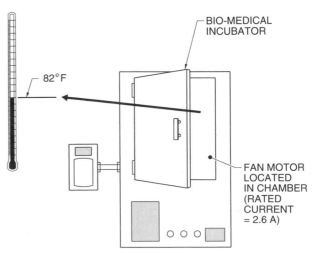

APPLICATION 9-1 — Troubleshooting Electromechanical Relays

When troubleshooting an electromechanical relay, the input and output of the relay are checked to determine if the circuit on the input side of the relay is the problem, if the circuit on the output side of the relay is the problem, or if the relay itself is the problem. The relay coil and contacts are checked to determine if the relay is the problem. The correct voltage must be applied to the relay's coil before it energizes. The relay contacts are checked by energizing and de-energizing the coil. The contacts should have little to no voltage drop across them when closed. The contacts should have nearly full voltage across them when open.

Check for contact sticking or binding if the relay is not functioning properly. Tighten any loose parts. Replace any broken, bent, or badly worn parts. Check all contacts for signs of excessive wear and dirt buildup. Contacts are not harmed by discoloration or slight pitting. Vacuum or wipe contacts with a soft cloth to remove dirt. Never use a contact cleaner on relay contacts. Contacts require replacement when the silver surface has become badly worn. Replace all contacts when severe wear is evident on any contact. Replacing all contacts prevents uneven and unequal contact closing. Never file a contact.

Relay coils should be free of cracks and burn marks. Replace the coil if there is any evidence of overheating, cracking, melting, or burning. Check the coil terminals for the correct voltage level. Overvoltage or undervoltage conditions of more than 10% should be corrected. Use only replacement parts recommended by the manufacturer when replacing parts of a relay. Using non-approved parts can void any manufacturer warranty and may transfer product liability from the manufacturer. Relays are tested by manual operation through the use of a multimeter.

Manual Relay Operation

Most relays can be manually operated. Manually operating a relay determines whether the circuit that the relay is controlling (output side) is working correctly. A relay is manually operated by pressing down on a designated area of the relay. This closes the relay contacts. Electromechanical relays may include a push-to-test button. **See Manual Relay Operation.**

When manually operating relay contacts, the circuit controlling the coil is bypassed. Troubleshoot from the relay through the control circuit when the load controlled by the relay operates manually. Troubleshoot the circuit that the relay is controlling if the load controlled by the relay does not operate when the relay is manually operated.

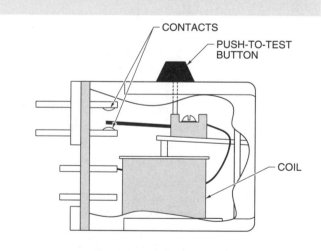

MANUAL RELAY OPERATION

Multimeter Test

A multimeter is also used to test an electromechanical relay. A multimeter is connected across the input and output side of a relay. Troubleshoot from the input of the relay through the control circuit when no voltage is present at the input side of the relay. The relay is the problem if the relay is not delivering the correct voltage.

Troubleshoot from the output of the relay through the power circuit when the relay is delivering the correct voltage. The supply voltage measured across an open contact indicates that the multimeter is completing the circuit across the contact and to the load. The contacts are not closing and the relay is defective if the voltage measured across the contact remains full voltage (when open) when the coil is energized and de-energized. The contacts are welded closed and the relay is defective if the voltage measured across the contacts remains at zero (low) voltage (when closed) when the coil is energized and de-energized. **See Troubleshooting Electromechanical Relays.**

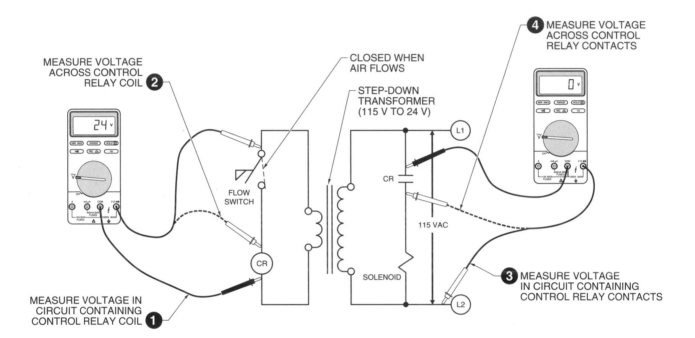

TROUBLESHOOTING ELECTROMECHANICAL RELAYS

To troubleshoot an electromechanical relay, apply the procedure:

1. Measure the voltage in the circuit containing the control relay coil. The voltage should be within 10% of the voltage rating of the coil. The relay coil cannot energize if the voltage is not present. The coil may not energize properly if the voltage is not at the correct level. Troubleshoot the power supply when the voltage level is incorrect.

2. Measure the voltage across the control relay coil. The voltage across the coil should be within 10% of the coil's rating. Troubleshoot the switch controlling power to the coil when the voltage level is incorrect.

3. Measure the voltage in the circuit containing the control relay contacts. The voltage should be within 10% of the rating of the load. Troubleshoot the power supply if the voltage level is incorrect.

4. Measure the voltage across the control relay contacts. The voltage across the contacts should be less than 1 V when the contacts are closed and nearly equal to the supply voltage when open. The contacts have too much resistance and are in need of service if the voltage is more than 1 V when the contacts are closed. Troubleshoot the load when the voltage is correct at the contacts and the circuit does not work.

 *Refer to **Activity 9-1—Troubleshooting Electromechanical Relays**...*

APPLICATION 9-2—*Troubleshooting Solid-State Relays*

Solid-state relays require periodic inspection. Dirt, burning, or cracking should not be present on a solid-state relay. Printed circuit (PC) boards should be properly seated. Ensure that the board locking tabs are in place if used. Consider adding locking tabs if a PC board without locking tabs loosens. Check to ensure that any cooling provisions are working and are free of obstructions.

Troubleshooting a solid-state relay is accomplished by either the exact replacement method or the circuit analysis method. The *exact replacement method* replaces a bad relay with a relay of the same type and size. The exact replacement method involves making a quick check of the relay's input and output voltages. The relay is assumed to be the problem and is replaced when there is only an input voltage (no output voltage) being switched.

The *circuit analysis method* uses a logical sequence to determine the reason for the failure. Steps are taken to prevent the problem from recurring once the reason for a failure is known. The circuit analysis method of troubleshooting is based on three improper relay operations, which are:
- The relay fails to turn OFF the load
- The relay fails to turn ON the load
- Erratic relay operation

Relay Fails to Turn OFF Load

A relay may not turn OFF the load it is connected to when a relay fails. This condition occurs either when the load is drawing more current than the relay can withstand, the relay's heat sink is too small, or transient voltages are causing a breakover of the relay's output.

Transient voltages are temporary, unwanted voltages in an electrical circuit. Overcurrent permanently shorts the relay's switching device if the load draws more current than the rating of the relay. High temperature causes thermal runaway of the relay's switching device if the heat sink does not remove the heat. Replace the relay with one of a higher voltage rating and/or add a transient suppression device to the circuit if the power lines are likely to have transients (usually from inductive loads connected on the same line). **See Relay Fails to Turn OFF Load.**

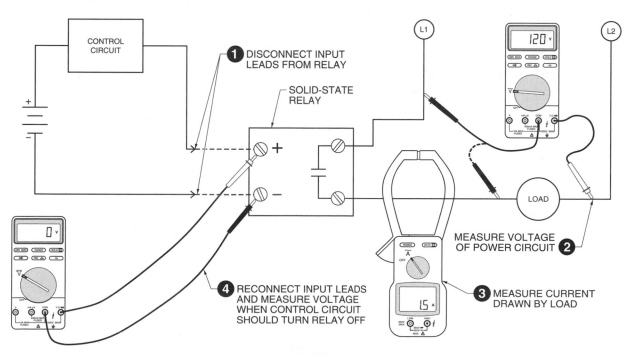

RELAY FAILS TO TURN OFF LOAD

To troubleshoot a solid-state relay that fails to turn OFF a load, apply the procedure:

1. Disconnect the input leads from the solid-state relay. See Step 3 if the relay load turns OFF. The relay is the problem if the load remains ON and the relay is normally open.

2. Measure the voltage of the circuit that the relay is controlling. The line voltage should not be higher than the rated voltage of the relay. Replace the relay with a relay that has a higher voltage rating if the line voltage is higher than the relay's rating. Check to ensure that the relay is rated for the type of line voltage (AC or DC) being used.

3. Measure the current drawn by the load. The current draw must not exceed the relay's rating. For most applications, the current draw should not be more than 75% of the relay's maximum rating.

4. Reconnect the input lead(s) and measure the input voltage to the relay at the time when the control circuit should turn the relay OFF. The control circuit is the problem and needs to be checked if the control voltage is present. The relay is the problem if the control voltage is removed and the load remains ON. Before changing the relay, ensure that the control voltage is not higher than the relay's rated limit when the control circuit delivers the supply voltage. Ensure that the control voltage is not higher than the relay's rated dropout voltage when the control circuit removes the supply voltage. This condition may occur in some control circuits using solid-state switching.

Relay Fails to Turn ON Load

When a relay fails, the relay may fail to turn ON the load connected to it. This condition occurs when the relay's switching device receives a very high voltage spike or the relay's input is connected to a higher-than-rated voltage. A high voltage spike blows open the relay's switching device, preventing the load from turning ON. Excessive voltage on the relay's input side destroys the relay's electronic circuit.

Replace the relay with one that has a higher voltage and current rating and/or add a transient-voltage-suppression device to the circuit if the power lines are likely to have high voltage spikes. **See Relay Fails to Turn ON Load.**

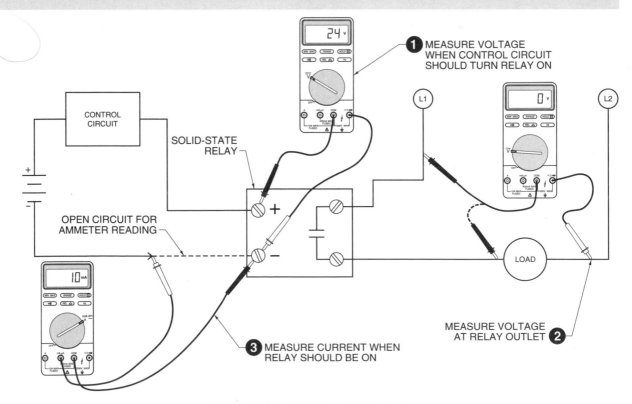

RELAY FAILS TO TURN ON LOAD

To troubleshoot a solid-state relay that fails to turn ON a load, apply the procedure:

1. Measure the input voltage. The relay should be ON. Troubleshoot the circuit ahead of the relay's input if the voltage is less than the relay's rated pick-up voltage. The circuit ahead of the relay is the problem if the voltage is greater than the relay's rated pick-up voltage. The higher voltage may have destroyed the relay. The relay may be a secondary problem caused by the primary problem of excessive applied voltage. Correct the high-voltage problem before replacing the relay. The relay or output circuit is the problem if the input voltage is within the pick-up limits of the relay.

2. Measure the voltage at the output of the relay. The relay is probably the problem if the relay is not switching the voltage. See Step 3. The problem is in the output circuit if the relay is switching the voltage. Check for an open circuit in the load.

3. Insert an ammeter in series with the input leads of the relay. Measure the current. The relay should be ON. The relay input is open if no current is flowing. Replace the relay. The relay is bad if the current flow is within the relay's rating. Replace the relay. The control circuit is the problem if current is flowing but is less than that required to operate the relay.

Erratic Relay Operation

Erratic relay operation is the proper operation of a relay at times, and the improper operation of the relay at other times. Erratic relay operation is caused by mechanical problems (loose connections), electrical problems (incorrect voltage), or environmental problems (high temperature). **See Troubleshooting Erratic Relay Operation.**

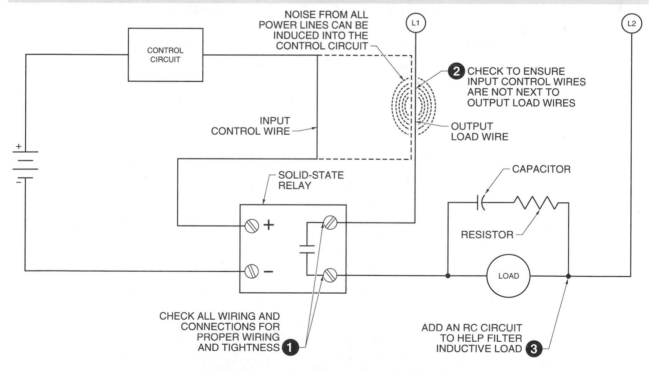

TROUBLESHOOTING ERRATIC RELAY OPERATION

To troubleshoot erratic relay operation, apply the procedure:

1. Check all wiring and connections for proper wiring and tightness. Loose connections cause many erratic problems. No sign of burning should be present at any terminal. Burning at a terminal usually indicates a loose connection.

2. Ensure that the input control wires are not next to the output line or load wires. The noise carried on the output side may cause unwanted input signals.

3. The relay may be half-waving if the load is a chattering AC motor or solenoid. *Half-waving* occurs when a relay fails to turn OFF because the current and voltage in the circuit reach zero at different times. Half-waving is caused by the phase shift inherent in inductive loads. The phase shift makes it difficult for some solid-state relays to turn OFF. Connecting an RC or another snubber circuit across the output load should allow the relay to turn OFF. An *RC circuit* is a circuit in which resistance (R) and capacitance (C) are used to help filter the power in a circuit.

 Refer to **Activity 9-2—Troubleshooting Solid-State Relays**..

APPLICATION 9-3—Troubleshooting Contactors and Motor Starters

Check the tightness of all terminals and busbar connections when troubleshooting control devices. Loose connections in the power circuit of contactors and motor starters cause overheating. Overheating leads to equipment malfunction or failure. Loose connections in the control circuit cause control malfunctions. Loose connections of grounding terminals lead to electrical shock and cause electromagnetic-generated interference.

The contactor or motor starter is the first device checked when troubleshooting a circuit that does not work or has a problem. The contactor or motor starter is checked first because it is the point where the incoming power, load, and control circuit are connected. Basic voltage readings are taken at a contactor or motor starter to determine where the problem lies. Since a motor starter is a contactor with added overload protection, the same basic procedure used to troubleshoot a starter works for contactors as well.

Troubleshoot the power circuit and the control circuit if the control circuit does not correctly operate a motor. The two circuits are dependent on each other but are considered two separate circuits because they are usually at different voltage levels and always at different current levels. **See Troubleshooting Motor Starters.**

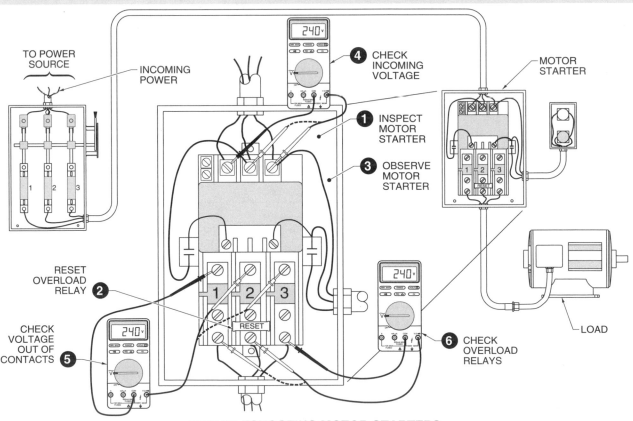

TROUBLESHOOTING MOTOR STARTERS

To troubleshoot a motor starter, apply the procedure:

1. Inspect the motor starter and overload assembly. Service or replace motor starters that show heat damage, arcing, or wear. Replace motor starters that show burning.

2. Reset the overload relay if there is no visual indication of damage. Replace the overload relay if there is visual indication of damage.

3. Observe the motor starter for several minutes if the motor starts after resetting the overload relay. The overload relay continues to open if an overload problem still exists.

4. Check the voltage going into the starter if resetting the overload relay does not start the motor. Check circuit voltage ahead of the starter if the voltage reading is 0 V. The voltage is acceptable if the voltage reading is within 10% of the motor's voltage rating. The voltage is unacceptable if the voltage reading is not within 10% of the voltage rating of the motor.

5. Energize the starter and check the starter contacts if the voltage into the starter is present and at the correct level. The starter contacts are good if the voltage reading is acceptable. Open the starter, turn the power OFF, and replace the contacts if there is no voltage reading.

6. Check the overload relay if voltage is coming out of the starter contacts. Turn the power OFF and replace the overload relay if the voltage reading is 0 V. The problem is downstream from the starter if the voltage reading is acceptable and the motor is not operating.

Troubleshooting Guides

A troubleshooting guide is used when troubleshooting contactors and motor starters. The guide states a problem, its possible cause(s), and corrective action(s) that may be taken. **See Contactor and Motor Starter Troubleshooting Guide.**

CONTACTOR AND MOTOR STARTER TROUBLESHOOTING GUIDE . . .		
Problem	**Possible Cause**	**Corrective Action**
Humming noise	Magnet pole faces misaligned	Realign. Replace magnet assembly if realignment is not possible.
	Too low voltage at coil	Measure voltage at coil. Check voltage rating of coil. Correct any voltage that is 10% less than coil rating.
	Pole face obstructed by foreign object, dirt, or rust	Remove any foreign object and clean as necessary. Never file pole faces.
Loud buzz noise	Shading coil broken	Replace coil assembly.
Controller fails to drop out	Voltage to coil not being removed	Measure voltage at coil. Trace voltage from coil to supply looking for shorted switch or contact if voltage is present.
	Worn or rusted parts causing binding	Clean rusted parts. Replace worn parts.
	Contact poles sticking	Check for burning or sticky substance on contacts. Replace burned contacts. Clean dirty contacts.
	Mechanical interlock binding	Check to ensure interlocking mechanism is free to move when power is OFF. Replace faulty interlock.
Controller fails to pull in	No coil voltage	Measure voltage at coil terminals. Trace voltage loss from coil to supply voltage if voltage is not present.
	Too low voltage	Measure voltage at coil terminals. Correct voltage level if voltage is less than 10% of rated coil voltage. Check for a voltage drop as large loads are energized.
	Coil open	Measure voltage at coil. Remove coil if voltage is present and correct but coil does not pull in. Measure coil resistance for open circuit. Replace if open.

... CONTACTOR AND MOTOR STARTER TROUBLESHOOTING GUIDE		
Problem	**Possible Case**	**Corrective Action**
	Coil shorted	Shorted coil may show signs of burning. The fuse or breakers should trip if coil is shorted. Disconnect one side of coil and reset if tripped. Remove coil and check resistance for short if protection device does not trip. Replace shorted coil. Replace any coil that is burned.
	Mechanical obstruction	Remove any obstructions.
Contacts badly burned or welded	Too high inrush current	Measure inrush current. Check load for problem if higher-than-rated load current. Change to larger controller if load current is correct but excessive for controller.
	Too fast load cycling	Change to larger controller if load cycled ON and OFF repeatedly.
	Too large overcurrent protection device	Size overcurrent protection to load and controller.
	Short circuit	Check fuses or breakers. Clear any short circuit.
	Insufficient contact pressure	Check to ensure contacts are making good connection.
Nuisance tripping	Incorrect overload size	Check size of overload against rated load current. Size up if permissible per NEC®.
	Lack of temperature compensation	Correct setting of overload if controller and load are at different ambient temperatures.
	Loose connections	Check for loose terminal connection.

Refer to **Activity 9-3—Troubleshooting Contactors and Motor Starters** ....

 *Refer to Quick Quiz® on CD-ROM*

 Refer to Chapter 9 in the **Troubleshooting Electrical/Electronic Systems Workbook** *for additional questions.*

Name_____ Date _____

ACTIVITY 9-1—Troubleshooting Electromechanical Relays

Using the Lubricant Circuit Wiring and Line Diagram and the meter readings, answer questions 1-3. A solenoid-operated valve is used to dispense lubricant. Lubricant flows to the brush when the solenoid is energized. Lubricant stops flowing when the solenoid is de-energized.

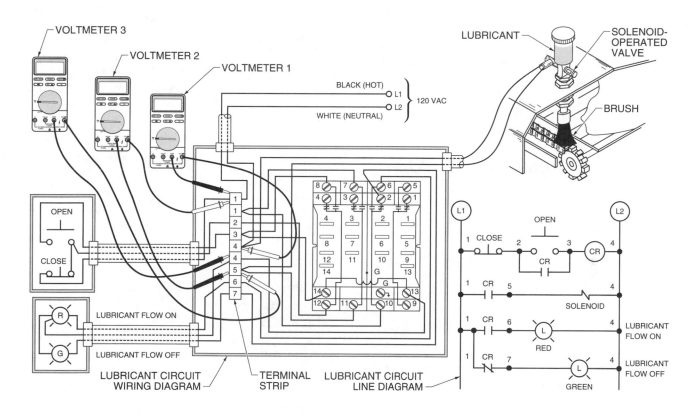

_____ **1.** ___ NO relay contacts are used in the circuit.

_____ **2.** ___ NC relay contacts are used in the circuit.

_____ **2.** ___ and ___ are the pin numbers of the relay coil.

_____ **4.** ___ and ___ are the terminal strip numbers wired to the relay coil.

_____ **5.** ___ and ___ are the relay pin numbers in parallel with the Open Pushbutton.

_____ **6.** The relay contact in parallel with the pushbutton is wired to ___ and ___ terminal strip numbers.

_____ 7. ___ and ___ are the relay pin numbers that control the solenoid valve.

_____ 8. ___ and ___ are the relay pin numbers that control the red light.

_____ 9. ___ and ___ are the relay pin numbers that control the green light.

A machine operator reports that no lubricant is dispensed when the Open Pushbutton is pressed and the red light is ON. The operator also reports that the lubricant tank is full, the green light turns OFF, and the red light turns ON when the circuit is tested by pressing the Open Pushbutton.

_____ **10.** The problem is ___.

The machine operator reports that the red light does not turn ON when the green light turns OFF. The operator also reports that when the lubricant system is working, the green light turns OFF, and the red light does not turn ON when the circuit is tested by pressing the Open Pushbutton.

_____ **11.** The problem is ___.

PROBLEM 1	
Meter	**Reading**
1	120 V (at all times)
2	119.5 V (at all times)
3	.5 V (at all times)

PROBLEM 2	
Meter	**Reading**
1	120 V (at all times)
2	120 V (when Close Pushbutton pressed) 0 V (when Open Pushbutton pressed)
3	120 V (when Open Pushbutton pressed) 0 V (when Close Pushbutton pressed)

ACTIVITY 9-2—Troubleshooting Solid-State Relays

Using the meter readings, determine the problem. Reading 1 is taken when Pushbutton 1 or Pushbutton 2 is pressed. Reading 2 is taken when neither pushbutton is pressed.

_____ **1.** The problem is ___.

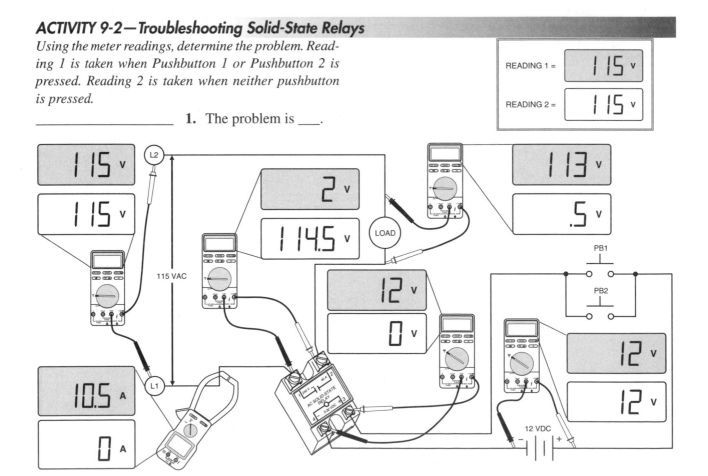

Reading 1 is taken when Pushbutton 1 and Pushbutton 2 are pressed. Reading 2 is taken when neither pushbutton is pressed.

_____ **2.** The problem is ___.

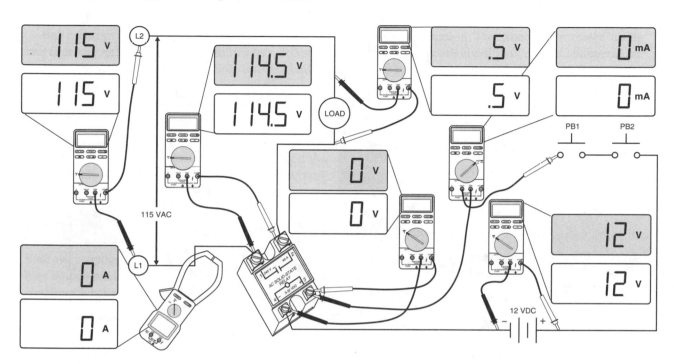

Reading 1 is taken when the pushbutton is not pressed. Reading 2 is taken when the pushbutton is pressed.

_____ **3.** The problem is ___.

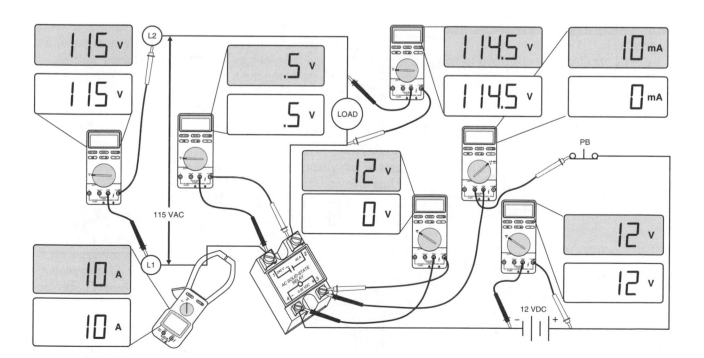

Reading 1 is taken when the pushbutton is pressed. Reading 2 is taken when the pushbutton is not pressed.

_____ **4.** The problem is ___.

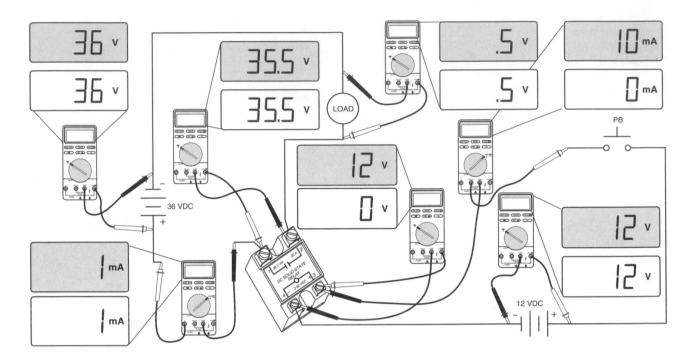

Reading 1 is taken when Pushbutton 1 or Pushbutton 2 is pressed. Reading 2 is taken when neither pushbutton is pressed.

_____ **5.** The problem is ___.

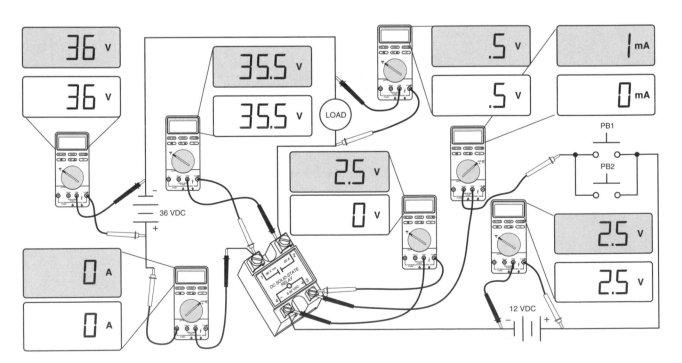

ACTIVITY 9-3—Troubleshooting Contactors and Motor Starters

1. Connect Meter 1 so that it reads Line 1 and Line 2 voltage into the motor starter. Connect Meter 2 so that it reads Line 2 and Line 3 voltage into the motor starter. Connect Meter 3 so that it reads Line 1 voltage out of the starter's power contact. Connect Meter 4 so that it reads Line 2 voltage out of the starter's power contact. Connect Meter 5 so that it reads Line 1 voltage out of the starter's heaters. Connect Meter 6 so that it reads Line 2 and Line 3 voltage out of the starter's power contacts and heaters.

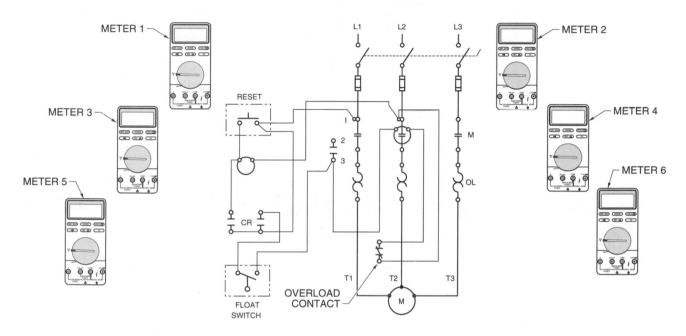

2. Connect Meter 1 so that it reads Line 1 and Line 3 voltage into the motor starter. Connect Meter 2 so that it reads Line 3 voltage out of the starter's power contact. Connect Meter 3 so that it reads Line 3 voltage out of the starter's power contact and the contactor's power contact. Connect Meter 4 so that it reads Line 3 voltage out of the starter's heater.

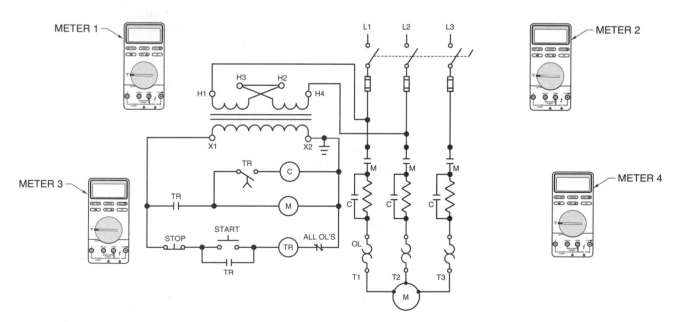

3. Connect Meter 1 so that it reads the line voltage into the motor starter. Connect Meter 2 so that it reads the line voltage at the motor's starting windings when the motor is connected for the forward direction of rotation. Connect Meter 3 so that it reads the line voltage at the motor's running windings when the motor is connected for the forward direction of rotation.

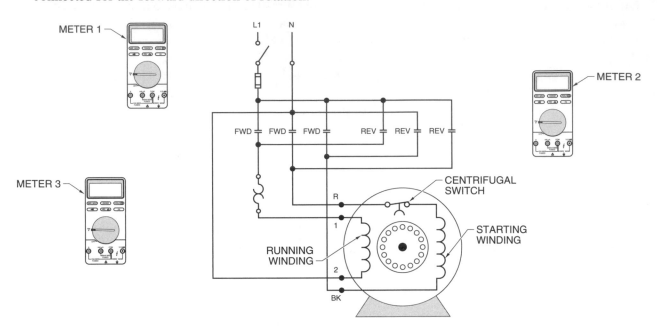

4. Connect Meter 1 so that it reads the line voltage into the motor starter. Connect Meter 2 so that it reads the line voltage at the motor's armature when the motor is connected for the forward direction of rotation. Connect Meter 3 so that it reads the line voltage at the motor's armature when the motor is connected for the reverse direction of rotation. *Note:* The meters must be connected for the correct polarity at the meter test leads.

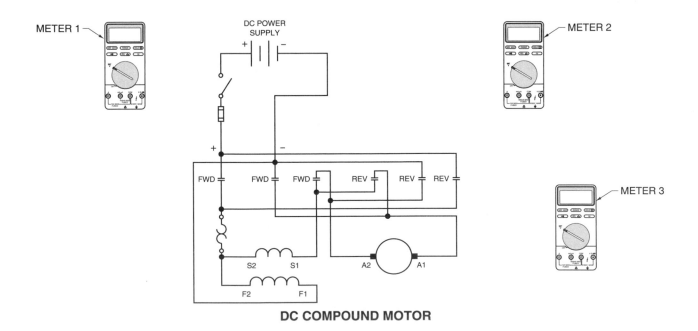

DC COMPOUND MOTOR

Motor Electrical Problems

APPLICATION 10-1—Motor Nameplate Ratings

Electric motors are used to produce work. For a motor to safely produce work for the expected life of the motor, the motor electrical, operating, environmental, and mechanical ratings must be considered. The electrical, operating, environmental, and mechanical ratings of a motor are listed on the motor nameplate. Since the motor nameplate has limited space to convey this information, most information is abbreviated or coded to save space. In addition to the written information, most motor nameplates also include the motor wiring diagram.

Understanding the abbreviated and coded information provided on a motor nameplate is required when selecting, installing, and troubleshooting electric motors. For motors that are already in service, the information provided on the motor nameplate is often the only information available. **See Motor Nameplate.**

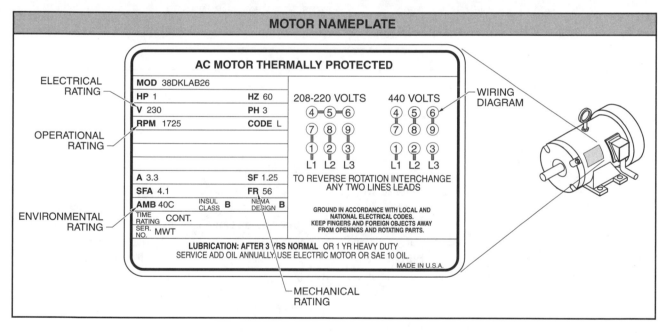

MOTOR NAMEPLATE

ELECTRICAL RATING

OPERATIONAL RATING

ENVIRONMENTAL RATING

AC MOTOR THERMALLY PROTECTED

MOD	38DKLAB26		
HP	1	HZ	60
V	230	PH	3
RPM	1725	CODE	L

A	3.3	SF	1.25
SFA	4.1	FR	56
AMB 40C	INSUL CLASS B	NEMA DESIGN B
TIME RATING	CONT.
SER. NO.	MWT

LUBRICATION: AFTER 3 YRS NORMAL OR 1 YR HEAVY DUTY
SERVICE ADD OIL ANNUALLY USE ELECTRIC MOTOR OR SAE 10 OIL.

208-220 VOLTS   440 VOLTS

④─⑤─⑥   ④ ⑤ ⑥

⑦ ⑧ ⑨   ⑦ ⑧ ⑨

① ② ③   ① ② ③
L1 L2 L3   L1 L2 L3

TO REVERSE ROTATION INTERCHANGE
ANY TWO LINES LEADS

GROUND IN ACCORDANCE WITH LOCAL AND
NATIONAL ELECTRICAL CODES.
KEEP FINGERS AND FOREIGN OBJECTS AWAY
FROM OPENINGS AND ROTATING PARTS.

MADE IN U.S.A.

WIRING DIAGRAM

MECHANICAL RATING

 Refer to **Activity 10-1—Motor Nameplate Ratings** ....

APPLICATION 10-2—Motor Nameplate Electrical Ratings

Electrical ratings, included on most motor nameplates, consist of a power rating, phase rating, voltage rating, current rating, frequency rating, and code letter rating. The electrical ratings of a motor are determined by the manufacturer and vary once the motor is placed in service. These ratings are based on the actual load on the motor shaft and actual applied voltage. **See Motor Nameplate Electrical Ratings.**

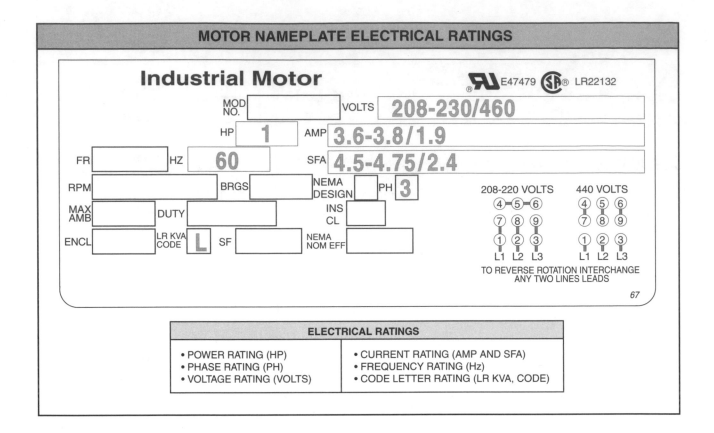

MOTOR NAMEPLATE ELECTRICAL RATINGS

Industrial Motor ®UⓁ E47479 ⓈA® LR22132

MOD NO.	VOLTS 208-230/460
HP 1	AMP 3.6-3.8/1.9
FR HZ 60	SFA 4.5-4.75/2.4
RPM BRGS	NEMA DESIGN PH 3
MAX AMB DUTY	INS CL
ENCL LR KVA CODE L SF	NEMA NOM EFF

208-220 VOLTS
④—⑤—⑥
⑦ ⑧ ⑨
① ② ③
L1 L2 L3

440 VOLTS
④ ⑤ ⑥
⑦ ⑧ ⑨
① ② ③
L1 L2 L3

TO REVERSE ROTATION INTERCHANGE ANY TWO LINES LEADS

67

ELECTRICAL RATINGS

- POWER RATING (HP)
- PHASE RATING (PH)
- VOLTAGE RATING (VOLTS)
- CURRENT RATING (AMP AND SFA)
- FREQUENCY RATING (Hz)
- CODE LETTER RATING (LR KVA, CODE)

Power Rating

All motors convert electrical energy into rotating mechanical energy. The amount of rotating mechanical energy produced by a motor determines the amount of work the motor can perform. Power is defined as the amount of work produced. The two most common units used for measuring motor power are watts (W) and horsepower (HP). For conversion purposes, 746 W equals 1 HP. Motors manufactured in the United States (or designed for the U.S. market) that are ⅙ HP or greater are normally rated in horsepower, and motors that are less than ⅙ HP are normally rated in watts. Motors manufactured in Europe are normally rated in kilowatts (kW), regardless of size.

If a motor is rated in kilowatts, a general comparison to horsepower can be made by multiplying the kilowatt rating of the motor by 1.34 to get the equivalent horsepower rating. For example, a 1.5 kW rated motor is equivalent to 2 HP (1.5 kW × 1.34 ≅ 2 HP). Likewise, a general comparison of kilowatts to horsepower can be made by multiplying the horsepower rating of the motor by 0.746 to get the equivalent kilowatt rating. A motor that is not connected to a load produces less than the nameplate rated power. A motor that is operating the maximum load at which it was designed to safely handle produces the nameplate rated power. An overloaded motor operates the load by trying to produce more power than the motor is rated for. The more power produced by a motor, the higher the current draw and motor temperature.

Phase Rating

AC motors are either single-phase or three-phase. DC motors do not have a phase rating. Three-phase motors are preferred over single-phase motors because they can deliver more power at less operating cost and draw less current from the power lines. Less current draw means smaller conductors (wire) and smaller motor control devices can be used.

Voltage Rating

All motors are designed for optimum performance at a specific voltage level. The voltage rating may be a single voltage rating (e.g., 115 VAC) or multiple voltage ratings (e.g., 208/230 VAC or 460 VAC). Applied voltage to a motor should be within +5% to –10% of the nameplate-rated voltage.

Current Rating

Current changes with a change in motor load. As the load on the motor increases, the current draw of the motor increases. As the load on the motor decreases, the current draw of the motor decreases. The nameplate-rated current of a motor is the amount of current the motor draws when fully loaded (motor nameplate power rating). Motor nameplate-rated current is also referred to as full-load amperes (FLA). The nameplate-rated current is dependent on the applied voltage. For dual-voltage-rated motors, a lower voltage rating corresponds to a higher current rating. Likewise, a higher voltage rating corresponds to a lower current rating.

Frequency Rating

AC motors have either a 50 Hz or 60 Hz frequency rating. Traditionally, motors manufactured in the United States have only a 60 Hz rating and motors manufactured in Europe have only a 50 Hz rating. Modern motors are designed for either rating and have both a 50 Hz and 60 Hz rating on the nameplate.

Motors typically tolerate a ±5% frequency variation without affecting the motor load operation. A higher frequency increases motor speed but reduces motor torque. A lower frequency decreases motor speed and increases motor torque.

Code Letter Rating

When a motor is started at full-line voltage, the motor draws a much higher current than when the motor is running at rated speed. This higher starting current is referred to as inrush current or locked rotor current (LRC). The amount of starting current is a function of the motor size (in horsepower) and the motor design characteristics. The amount of current a motor draws when starting is indicated on the motor nameplate by a code letter.

The list of motor nameplate code letters begins with the letter "A" and ends with "V." The closer the nameplate-listed code letter is to "A," the lower the motor starting current. Likewise, the closer the nameplate-listed code letter is to "V," the higher the motor starting current. For example, a motor with a listed code letter of "G" has a lower starting current than a motor with a listed code letter of "H." Most motors have a code letter in the H to N range.

 Refer to **Activity 10-2—Motor Nameplate Electrical Ratings** ....

APPLICATION 10-3—*Phase Unbalance*

Phase unbalance is the unbalance that occurs when power lines are out-of-phase. Phase unbalance of a 3ϕ power system occurs when 1ϕ loads are applied, causing one or two of the lines to carry more or less of the load. The loads of 3ϕ power systems are balanced by electricians during installation. An unbalance begins to occur as additional 1ϕ loads are added to the system. This unbalance causes the 3ϕ lines to move out-of-phase so the lines are no longer 120 electrical degrees apart. **See Phase Unbalance.**

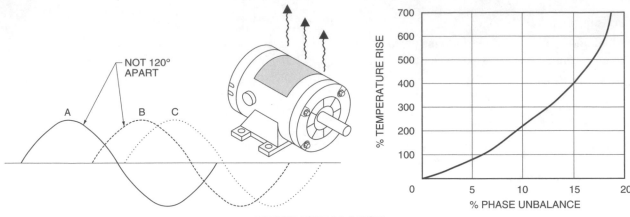

PHASE UNBALANCE

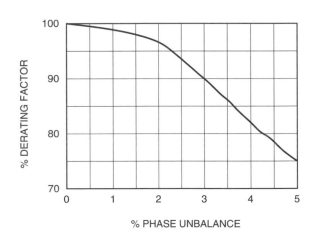

Phase unbalance causes 3ϕ motors to run at temperatures higher than their listed ratings. The greater the phase unbalance, the greater the temperature rise. High temperatures produce insulation breakdown and other related problems.

A 3ϕ motor operating in an unbalanced circuit cannot deliver its rated horsepower. For example, a phase unbalance of 3% causes a motor to work at 90% of its rated power. This requires the motor to be derated. **See Phase Unbalance Derating Factor.**

PHASE UNBALANCE DERATING FACTOR

 Refer to **Activity 10-3—Phase Unbalance** ....

APPLICATION 10-4—Voltage Unbalance

Voltage unbalance is the unbalance that occurs when the voltages at different motor terminals are not equal. One winding overheats, causing thermal deterioration of the winding if the voltage is not balanced. Voltage unbalance results in a current unbalance. Line voltage should be checked for voltage unbalance periodically and during all service calls. Whenever more than 2% voltage unbalance is measured, take the steps:

• Check the surrounding power system for excessive loads connected to one line.

• Adjust the load or motor rating by reducing the load on the motor or oversizing the motor if the voltage unbalance cannot be corrected.

• Notify the power company.

Voltage unbalance is found by applying the procedure:

1. Measure the voltage between each incoming power line. The readings are taken from L1 to L2, L1 to L3, and L2 to L3.

2. Add the voltages.

3. Divide by 3 to find the voltage average.

4. Subtract the voltage average from the voltage with the largest deviation to find the voltage deviation.

5. To find voltage unbalance, apply the formula:

$$V_u = \frac{V_d}{V_a} \times 100$$

where
V_u = voltage unbalance (%)
V_d = voltage deviation (in volts)
V_a = voltage average (in volts)
100 = constant

Example: Calculating Voltage Unbalance

Calculate the voltage unbalance of a feeder system with the following voltage readings. L1 to L2 = 442 V; L1 to L3 = 474 V; L2 to L3 = 456 V. **See Measuring Voltage Unbalance.**

1. Measure incoming voltage. Incoming voltage is 442 V, 474 V, and 456 V.

2. Add voltages. 442 V + 474 V + 456 V = **1372 V**

3. Find voltage average.

$$V_a = \frac{V}{3}$$

$$V_a = \frac{1372}{3}$$

$$V_a = \mathbf{457\ V}$$

4. Find voltage deviation.
$$V_d = V - V_a$$
$$V_d = 474 - 457$$
$$V_d = \mathbf{17\ V}$$

5. Find voltage unbalance.

$$V_u = \frac{V_d}{V_a} \times 100$$

$$V_u = \frac{17}{457} \times 100$$

$$V_u = .0372 \times 100$$

$$V_u = \mathbf{3.72\%}$$

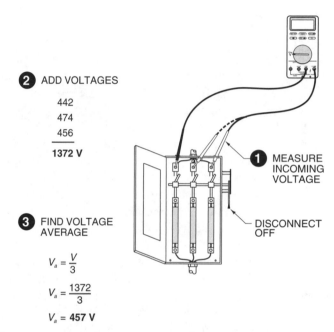

2 ADD VOLTAGES

442
474
456
―――
1372 V

1 MEASURE INCOMING VOLTAGE

DISCONNECT OFF

3 FIND VOLTAGE AVERAGE

$$V_a = \frac{V}{3}$$

$$V_a = \frac{1372}{3}$$

$$V_a = \mathbf{457\ V}$$

4 FIND VOLTAGE DEVIATION

$$V_d = V - V_a$$
$$V_d = 474 - 457$$
$$V_d = \mathbf{17\ V}$$

5 FIND VOLTAGE UNBALANCE

$$V_u = \frac{V_d}{V_a} \times 100$$

$$V_u = \frac{17}{457} \times 100$$

$$V_u = 0.0372 \times 100$$

$$V_u = \mathbf{3.72\%}$$

MEASURING VOLTAGE UNBALANCE

A troubleshooter can observe the blackening of one or two of the stator windings, which occurs when a motor has failed due to voltage unbalance. The winding with the largest voltage unbalance is the darkest. **See Voltage Unbalance Motor Damage.**

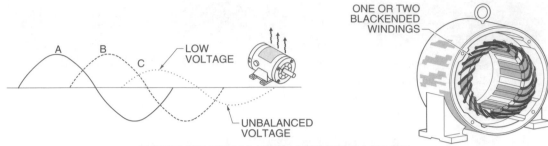

VOLTAGE UNBALANCE MOTOR DAMAGE

 Refer to **Activity 10-4—Voltage Unbalance**...

APPLICATION 10-5—Single Phasing

Single phasing is the operation of a motor designed to operate on three phases, but is operating on only two phases because one phase is lost. Single phasing occurs when one of the 3ϕ lines leading to a 3ϕ motor does not deliver voltage to the motor. Single phasing is the maximum condition of voltage unbalance.

Single phasing occurs when one phase opens on either the primary or secondary power distribution system. This happens when one fuse blows, there is a mechanical failure within the switching equipment, or when lightning takes out one of the lines.

Single phasing can go undetected on most systems because a 3ϕ motor running on 2ϕ continues to run in most applications. The motor usually runs until it burns out. When single phasing, the motor draws all its current from two lines.

Measuring the voltage at a motor does not usually detect a single phasing condition. The open winding in the motor generates a voltage almost equal to the phase voltage that is lost. In this case, the open winding acts as the secondary of a transformer, whereas the two windings connected to power act as the primary.

Single phasing is reduced by using the proper size dual-element fuse and by using the correct heater sizes. In motor circuits, or other types of circuits in which a single phasing condition cannot be allowed to exist for even a short period of time, an electronic phase-loss monitor is used to detect phase loss. When a phase loss is detected, the monitor activates a set of contacts to drop out the starter coil.

A troubleshooter can observe the severe blackening of one of the 3ϕ windings, which occurs when a motor has failed due to single phasing. The coil that experienced the voltage loss indicates obvious and fast damage, which includes the blowing out of the insulation on the one winding. **See Single-Phasing Motor Damage.**

Single phasing is distinguished from voltage unbalance by the severity of the damage. Voltage unbalance causes less blackening (but usually over more coils) and little or no distortion. Single phasing causes severe burning and distortion to one phase coil.

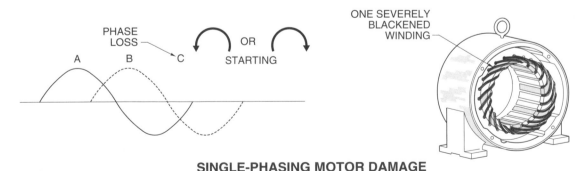

SINGLE-PHASING MOTOR DAMAGE

 Refer to **Activity 10-5—Single Phasing**...

APPLICATION 10-6—*Improper Phase Sequence*

Improper phase sequence is the changing of the sequence of any two phases (phase reversal) in a 3ϕ motor control circuit. Improper phase sequence reverses the motor rotation. Reversing motor rotation can damage driven machinery or injure personnel.

Phase reversal can occur when modifications are made to a power distribution system or when maintenance is performed on electrical conductors or switching equipment. The NEC® requires phase reversal protection on all personnel transportation equipment, such as moving walkways, escalators, and ski lifts. **See Improper Phase Sequence.**

A power monitor can be used to detect improper phase sequence. The monitor operates if the phases are in the correct sequence. The monitor drops out if any two phases are reversed.

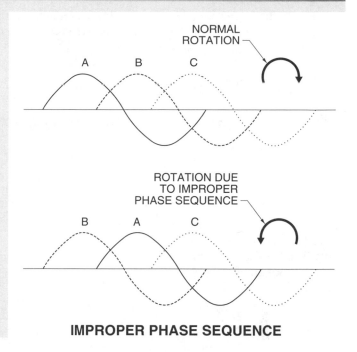

IMPROPER PHASE SEQUENCE

 Refer to **Activity 10-6—*Improper Phase Sequence*** ....

APPLICATION 10-7—*Voltage Surges*

A *voltage surge* is any higher-than-normal voltage that temporarily exists on one or more of the power lines. Lightning is a major cause of large voltage surges. A lightning surge on a power line comes from a direct hit or induced voltage. The lightning energy moves in both directions on the power lines, much like a rapidly moving wave.

This traveling surge causes a large voltage rise in an extremely short period of time. The large voltage is impressed on the first few turns of the motor windings, destroying the insulation and burning out the motor.

A troubleshooter can observe the burning and opening of the first few turns of the windings that occur when a motor has failed due to a voltage surge. The rest of the windings appear normal, with little or no damage. **See Voltage Surge Motor Damage.**

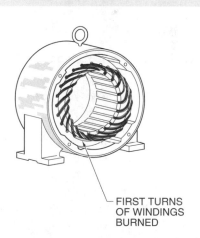

VOLTAGE SURGE MOTOR DAMAGE

Lightning arresters with the proper voltage rating and connection to an excellent ground assure maximum voltage surge protection. Surge protectors are also available. Surge protectors are placed on the equipment or throughout the distribution system.

Voltage surges can also occur from normal switching of high-rated power circuits. Voltage surges occurring from switching high-rated power circuits are of much less magnitude than lightning strikes and normally do not cause any problems in motors. A surge protector should be used on circuits with computer equipment to protect sensitive electronic components.

 *Refer to **Activity 10-7—Voltage Surges** ....*

APPLICATION 10-8—AC Voltage and Frequency Variations

Voltage Variations

Motors are rated for operation at specific voltages. Motor performance is affected when the supply voltage varies from a motor's rated voltage. A motor operates satisfactorily with a voltage variation of ±10% from the voltage rating listed on the motor nameplate. **See Voltage Variation Characteristics.**

VOLTAGE VARIATION CHARACTERISTICS		
Performance Characteristics	**10% above Rated Voltage**	**10% below Rated Voltage**
Starting current	+10% to +12%	−10% to −12%
Full-load current	−7%	+11%
Motor torque	+20% to +25%	−20% to −25%
Motor efficiency	Little change	Little change
Speed	+1%	−1.5%
Temperature rise	−3°C to −4°C	+6°C to +7°C

Frequency Variations

Motors are rated for operation at specific frequencies. Motor performance is affected when the frequency varies from a motor's rated frequency. A motor operates satisfactorily with a frequency variation of ±5% from the frequency rating listed on the motor nameplate. **See Frequency Variation Characteristics**.

FREQUENCY VARIATION CHARACTERISTICS		
Performance Characteristics	**5% above Rated Frequency**	**5% below Rated Frequency**
Starting current	−5% to −6%	+5% to +6%
Full-load current	−1%	+1%
Motor torque	−10%	+11%
Motor efficiency	Slight increase	Slight decrease
Speed	+5%	−5%
Temperature rise	Slight decrease	Slight increase

 *Refer to **Activity 10-8—AC Voltage and Frequency Variations** ....*

APPLICATION 10-9—DC Voltage Variations

DC motors should be operated on pure DC power. *Pure DC power* is power obtained from a battery or DC generator. DC power is also obtained from rectified AC power. Most industrial DC motors obtain power from a rectified AC power supply. DC power obtained from a rectified AC power supply varies from almost-pure DC power to half-wave DC power.

Half-wave rectified power is obtained by placing a diode in one of the AC power lines. Full-wave rectified power is obtained by placing a bridge rectifier (four diodes) in the AC power line. Rectified DC power is filtered by connecting a capacitor in parallel with the output of the rectifier circuit. **See DC Power Types.**

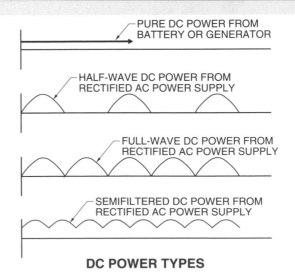

DC POWER TYPES

DC motor operation is affected by a change in voltage. The change may be intentional as in a speed- control application, or the change may be caused by variations in the power supply.

Typically, the power supply voltage should not vary by more than 10% of a motor's rated voltage. Motor speed, current, torque, and temperature are affected if the DC voltage varies from the motor rating. **See DC Motor Performance Characteristics.**

DC MOTOR PERFORMANCE CHARACTERISTICS				
Performance Characteristics	Voltage 10% below Rated Voltage		Voltage 10% above Rated Voltage	
	Shunt	Compound	Shunt	Compound
Starting torque	–15%	–15%	+15%	+15%
Speed	–5%	–6%	+5%	+6%
Current	+12%	+12%	–8%	–8%
Field temperature	Decreases	Decreases	Increases	Increases
Armature temperature	Increases	Increases	Decreases	Decreases
Commutator temperature	Increases	Increases	Decreases	Decreases

 Refer to **Activity 10-9—DC Voltage Variations**...

APPLICATION 10-10—Allowable Motor Starting Time

A motor must accelerate to its rated speed within a limited time period. The longer a motor takes to accelerate, the higher the temperature rise in the motor. The larger the load, the longer the acceleration time. The maximum recommended acceleration time depends on the motor's frame size. Large motor frames dissipate heat faster than small motor frames. **See Maximum Acceleration Time.**

MAXIMUM ACCELERATION TIME	
Frame Number	Maximum Acceleration Time (in seconds)
48 and 56	8
143–286	10
324–326	12
364–505	15

Overcycling

Overcycling is the process of turning a motor ON and OFF repeatedly. Motor starting current is several times the full-load running current of the motor. Most motors are not designed to start more than 10 times per hour. Overcycling occurs when a motor is at its operating temperature and still cycles ON and OFF. This further increases the temperature of the motor, destroying the motor insulation. **See Motor Overcycling.**

Totally enclosed motors better withstand overcycling than open motors because they hold heat longer. When a motor application requires a motor to be cycled often, take the following steps:

- Use a motor with a 50°C rise instead of the standard 40°C.
- Use a motor with a 1.25 or 1.35 service factor instead of a 1.00 or 1.15.
- Provide additional cooling by forcing air over the motor.

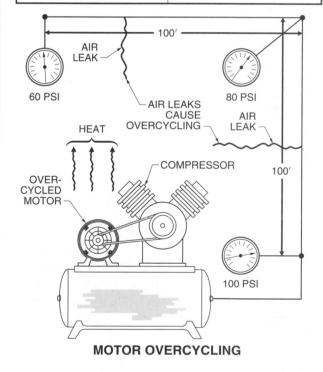

MOTOR OVERCYCLING

 Refer to **Activity 10-10—Allowable Motor Starting Time** ....

APPLICATION 10-11—*Megohmmeter Tests*

A *megohmmeter* is a device that detects insulation deterioration by measuring high resistance values under high test voltage conditions. A megohmmeter detects motor insulation deterioration before a motor fails. A megohmmeter is an ohmmeter capable of measuring very high resistances by using high voltages. Typical megohmmeter test voltages range from 50 V to 5000 V. A megohmmeter is used to perform motor insulation tests to prevent electrical shock and other causes of motor insulation failure, which include excessive moisture, dirt, heat, cold, corrosive vapors or solids, vibration, and aging.

A megohmmeter measures the resistance of different windings or the resistance from a winding to ground. An ohmmeter measures the resistance of common windings and components in a motor circuit. **See Megohmmeter Connections and Ohmmeter Connections.**

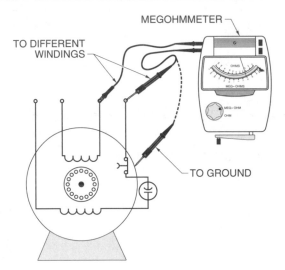

MEGOHMMETER CONNECTIONS

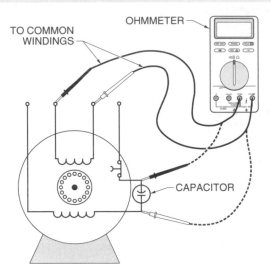

OHMMETER CONNECTIONS

Several megohmmeter readings should be taken over a long period of time because the resistance of good insulation varies greatly. Megohmmeter readings are typically taken when the motor is installed and semiannually thereafter. A motor is in need of service if the megohmmeter reading is below the minimum acceptable resistance. **See Recommended Minimum Resistance.**

Note: A motor with good insulation may have readings of 10 to 100 times the minimum acceptable resistance. Service the motor if the resistance reading is less than the minimum value.

RECOMMENDED MINIMUM RESISTANCE*	
Minimum Acceptable Resistance	Motor Voltage Rating (from nameplate)
100,000 Ω	Less than 208
200,000 Ω	208–240
300,000 Ω	240–600
1 MΩ	600–1000
2 MΩ	1000–2400
3 MΩ	2400–5000

* values for motor windings at 40°C

Caution: A megohmmeter uses very high voltage for testing (up to 5000 V). Avoid touching the meter leads to the motor frame. Always follow the manufacturer's recommended procedures and safety rules. After performing insulation tests with a megohmmeter, connect the motor windings to ground through a 5 kΩ, 5 W resistor. The winding should be connected for 10 times the motor testing time in order to discharge the energy stored in the insulation.

 Refer to **Activity 10-11—*Megohmmeter Tests*** ....

APPLICATION 10-12—*Insulation Spot Test*

An *insulation spot test* is a test that checks motor insulation over the life of the motor. An insulation spot test is taken when the motor is placed in service and every six months thereafter. The test should also be taken after a motor is serviced. **See Insulation Spot Test.**

To perform an insulation spot test, apply the procedure:
1. Connect a megohmmeter to measure the resistance of each winding lead to ground. Record the readings after 60 seconds. Service the motor if a reading does not meet the minimum acceptable resistance. Record the lowest meter reading on an insulation spot test graph if all readings are above the minimum acceptable resistance. The lowest reading is used because a motor is only as good as its weakest point.
2. Discharge the motor windings.
3. Repeat Steps 1 and 2 every six months.

Interpret the results of the test to determine the condition of the insulation. Point A represents the motor insulation condition when the motor was placed in service. Point B represents the effects of aging, contamination, etc., on the motor insulation. Point C represents motor insulation failure. Point D represents motor insulation condition after being rewound.

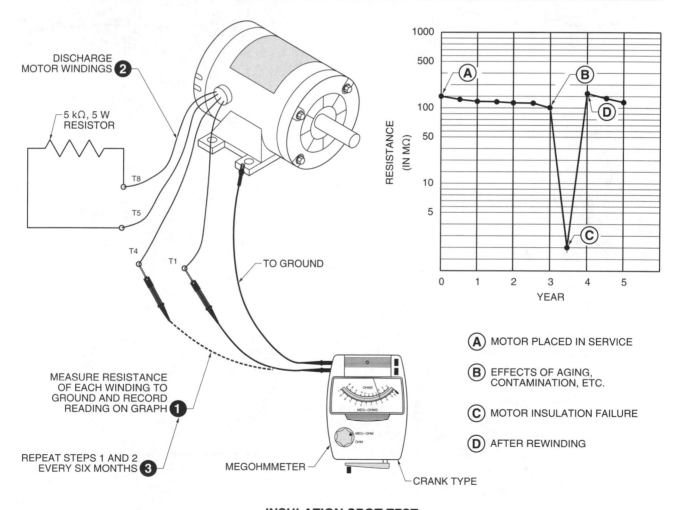

INSULATION SPOT TEST

 Refer to **Activity 10-12—Insulation Spot Test** ....

APPLICATION 10-13—Dielectric Absorption Test

A *dielectric absorption test* is a test that checks the absorption characteristics of humid or contaminated insulation. The test is performed over a 10-minute period. **See Dielectric Absorption Test.** To perform a dielectric absorption test, apply the procedure:

1. Connect a megohmmeter to measure the resistance of each winding lead to ground. Service the motor if a reading does not meet the minimum acceptable resistance. Record the lowest meter reading on a dielectric absorption test graph if all readings are above the minimum acceptable resistance. Record the readings every 10 seconds for the first minute and every minute thereafter for 10 minutes.

2. Discharge the motor windings.

Interpret the results of the test to determine the condition of the insulation. The slope of the curve shows the condition of the insulation. Good insulation (Curve A) shows a continual increase in resistance. Moist or cracked insulation (Curve B) shows a relatively constant resistance.

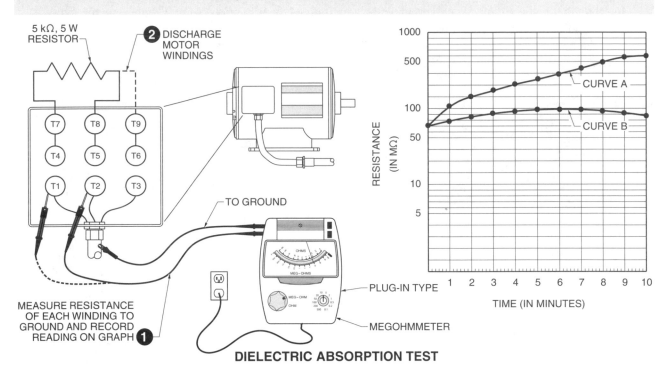

DIELECTRIC ABSORPTION TEST

A polarization index is obtained by dividing the value of the 10-minute reading by the value of the 1-minute reading. The polarization index is an indication of the condition of the insulation. A low polarization index indicates excessive moisture or contamination. **See Minimum Acceptable Polarization Index Values.**

For example, if the 1-minute reading of Class B insulation is 80 MΩ and the 10-minute reading is 90 MΩ, the polarization index is 1.125 (90 MΩ ÷ 80 MΩ = 1.125). The insulation contains excessive moisture or contamination.

MINIMUM ACCEPTABLE POLARIZATION INDEX VALUES	
Insulation	Value
Class A	1.5
Class B	2.0
Class F	2.0

 Refer to *Activity 10-13—Dielectric Absorption Test*..

APPLICATION 10-14—Insulation Step Voltage Test

An *insulation step voltage test is* a test that creates electrical stress on internal insulation cracks to reveal aging or damage not found during other motor insulation tests. The insulation step voltage test is performed only after an insulation spot test. **See Insulation Step Voltage Test.**

To perform an insulation step voltage test, apply the procedure:
1. Set the megohmmeter to 500 V and connect to measure the resistance of each winding lead to ground. Take each resistance reading after 60 seconds. Record the lowest reading.
2. Place the meter leads on the winding that has the lowest reading.
3. Set the megohmmeter on increments of 500 V starting at 1000 V and ending at 5000 V. Record each reading after 60 seconds.
4. Discharge the motor windings.

Interpret the results of the test to determine the condition of the insulation. The resistance of good insulation that is thoroughly dry (Curve A) remains approximately the same at different voltage levels. The resistance of deteriorated insulation (Curve B) decreases substantially at different voltage levels.

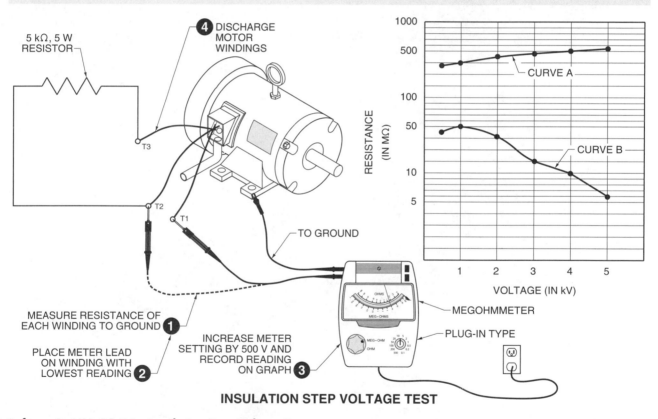

INSULATION STEP VOLTAGE TEST

 Refer to **Activity 10-14—Insulation Step Voltage Test** ....

 Refer to Quick Quiz® on CD-ROM

Refer to Chapter 10 in the **Troubleshooting Electrical/Electronic Systems Workbook** for additional questions.

Name_____ **Date** _____

ACTIVITY 10-1—Motor Nameplate Ratings

Answer the questions using the motor nameplate information and the given meter.

_____ **1.** Can the meter be used to measure the voltage at the motor?

_____ **2.** The meter selector switch should be set to ___ to measure the voltage at the motor.

_____ **3.** Can the meter be used to measure the current at the motor?

_____ **4.** The meter selector switch should be set to ___ to measure the current at the motor.

_____ **5.** Can the meter be used to measure the resistance of the motor windings when all power is OFF?

_____ **6.** The meter selector switch should be set to ___ to measure the resistance of the motor windings when all power is OFF.

AC MOTOR THERMALLY PROTECTED		
MOD 38DKLAB26		
HP 5	**HZ** 60	208-220 VOLTS 440 VOLTS
V 230/460	**PH** 3	④—⑤—⑥ ④ ⑤ ⑥
RPM 1725	**CODE** L	⑦ ⑧ ⑨ ⑦ ⑧ ⑨
		① ② ③ ① ② ③
		L1 L2 L3 L1 L2 L3
A 13.1/6.6	**SF** 1.25	TO REVERSE ROTATION INTERCHANGE
SFA 16.4/8.3	**FR** 56	ANY TWO LINES LEADS
AMB 40C	**INSUL CLASS** B **NEMA DESIGN** B	GROUND IN ACCORDANCE WITH LOCAL AND NATIONAL ELECTRICAL CODES.
TIME RATING CONT.		KEEP FINGERS AND FOREIGN OBJECTS AWAY FROM OPENINGS AND ROTATING PARTS.
SER. NO. MWT		

LUBRICATION: **AFTER 3 YRS NORMAL** OR 1 YR HEAVY DUTY
SERVICE ADD OIL ANNUALLY. USE ELECTRIC MOTOR OR SAE 10 OIL.

MADE IN U.S.A.

ACTIVITY 10-2—Motor Nameplate Electrical Ratings

Answer the questions using the General Purpose Motor Data. If the rating is not given, mark NA (not available).

_____ **1.** The power rating for Stock Item Motor 1 is ___.

_____ **2.** The phase rating for Stock Item Motor 1 is ___.

_____ **3.** The voltage rating for Stock Item Motor 1 is ___.

_____ **4.** The current rating for Stock Item Motor 1 is ___.

_____ **6.** The code letter rating for Stock Item Motor 1 is ___.

_____ **7.** When connected to 230 VAC, the current rating for Stock Item Motor 14 is ___.

_____ **8.** When connected to 460 VAC, the current rating for Stock Item Motor 14 is ___.

GENERAL PURPOSE MOTOR DATA

THREE-PHASE ENERGY-EFFICIENT OPEN DRIPPROOF

- MOUNTING: RIGID BASE
- BEARINGS: BALL
- THERMAL PROTECTION: NONE
- INSULATION CLASS B OR CLASS F (FOR LONGER LIFE)
- ENCLOSURE: ODP
- 60 Hz
- ROTATION: CW/CCW
- MAX. AMBIENT: 40°C
- UL RECOGNIZED AND CSA CERTIFIED
- WARRANTY: 48 TO 56 FRAME, 1 YEAR; 140T FRAME AND ABOVE, 3 YEARS
- CONTINUOUS DUTY RATED

STAINLESS STEEL NAMEPLATE MAINTAINS INFORMATION OVER LONG LIFE. 56HZ FRAMES HAVE ⅞″ × 2¼″ SHAFT AND BASE BOLT-HOLE CONFIGURATION TO MATCH 56, 56H, 143T AND 145T.

USES: GENERAL PURPOSE FOR CLEAN, DRY, NONHAZARDOUS APPLICATIONS WITH PUMPS, VENTILATION EQUIPMENT, MACHINE TOOLS, AND OTHER INDUSTRIAL EQUIPMENT.

HP	STOCK ITEM NUMBER	NAMEPLATE RPM	NEMA FRAME	VOLTS AT 60 Hz	FULL-LOAD AMPERES	NEMA NOMINAL EFFICIENCY	SERVICE FACTOR	FRAME MATERIAL	OVERALL LENGTH*	INSULATION CLASS
¼	1	1725	48	208–230/460	1.2–1.3/0.65	62.0	1.35	Steel	9⁵⁄₁₆	B
	2	1725	56	208–230/460	1.2–1.3/0.65	62.0	1.35	Steel	9¾	B
	3	1140	56	208–230/460	1.3–1.5/0.75	64.0	1.35	Steel	10¼	B
⅓	4	3450	48	208–230/460	1.4–1.4/0.7	68.0	1.35	Steel	8⅝	B
	5	1725	48	208–230/460	1.5–1.6/0.8	67.0	1.35	Steel	9⁹⁄₁₆	B
	6	1725	56	208–230/460	1.5–1.6/0.8	67.0	1.35	Steel	10	B
	7	1140	56	208–230/460	1.9–2.2/1.1	62.0	1.35	Steel	9¹⁵⁄₁₆	B
½	8	3450	56	208–230/460	1.9–2.2/1.1	66.0	1.25	Steel	10⁷⁄₁₆	B
	9	1725	56	208–230/460	2.1–2.2/1.1	75.7	1.25	Steel	10	B
	10	1725	56	208–230/460	2.0–2.2/1.1	70.0	1.25	Steel	10⅜	B
¾	11	1725	56	208–230/460	2.9–2.8/1.4	76.5	1.25	Steel	10⅜	B
1	12	3450	56	208–230/460	3.4–3.7/1.85	74.0	1.25	Steel	10⅜	B
	13	1755	182	208–230/460	3.6–3.3/1.6	82.5	1.15	Steel	13⅛	F
	14	1725	56	208–230/460	3.6–3.8/1.9	78.4	1.25	Steel	10¾	B
	15	1725	56	208–230/460	3.4–3.4/1.7	76.5	1.15	Steel	11½	B
	16	1140	56HZ	208–230/460	3.6–3.2/1.6	80.0	1.15	Steel	12¼	F
	17	1140	145T	208–230/460	3.6–3.2/1.6	80.0	1.15	Steel	12¼	F
1½	18	1725	56H	208–230/460	4.8–4.8/2.4	80.0	1.15	Steel	12	B
	19	1725	56H	208–230/460	4.8–4.8/2.4	80.0	1.15	Steel	10¹⁵⁄₁₆	F
2	20	1170	184T	208–230/460	7.8–7.1/3.5	85.5	1.15	Steel	13⅝	F
	21	1170	213	208–230/460	6.8–6.2/3.1	85.5	1.15	Steel	16¹³⁄₁₆	F
10	22	3540	254U	208–230/460	25.9–23.4/11.7	89.5	1.15	Steel	21¹³⁄₁₆	F
	23	3490	213T	208–230/460	26.6–24.0/12.0	88.5	1.15	Steel	17³⁄₁₆	F
	24	1770	256U	208–230/460	27.4–24.8/12.4	89.5	1.15	Steel	21¹³⁄₁₆	F
	25	1740	215T	208–230/460	26.7–24.2/12.1	89.5	1.15	Steel	17³⁄₁₆	F
15	26	1760	254T	208–230/460	38.8–35.1/17.5	91.0	1.15	Steel	22¹⁄₁₆	F
	27	1175	284T	208–230/460	41.5–37.5/18.8	90.2	1.15	Steel	24¹¹⁄₁₆	F
20	28	3520	254T	208–230/460	50.3–45.5/22.7	90.2	1.15	Cast Iron	20⁷⁄₁₆	F
	29	1755	256T	208–230/460	51.7–46.8/23.4	91.0	1.15	Cast Iron	22³⁄₁₆	F
	30	1175	286T	208–230/460	54.5–49.3/24.6	91.0	1.15	Cast Iron	24⅜	F

* in in.

ACTIVITY 10-3—Phase Unbalance

Balancing a system requires connecting the loads so that each main power line draws about the same amount of current. Determine the ammeter readings.

_____ 1. Ammeter 1 reading equals ___ A.

_____ 2. Ammeter 2 reading equals ___ A.

_____ 3. Ammeter 3 reading equals ___ A.

_____ 4. Ammeter 4 reading equals ___ A.

_____ 5. Ammeter 5 reading equals ___ A.

Circuit Device	Load Condition
Range	ON
Refrigerator	ON
Electric Heating Unit 1	ON
Electric Heating Unit 2	ON
Electric Heating Unit 3	ON
Dryer	ON
Furnace	ON
Air Conditioner	ON
Motor 1	ON
Motor 2	ON
Motor 3	ON
Motor 4	ON

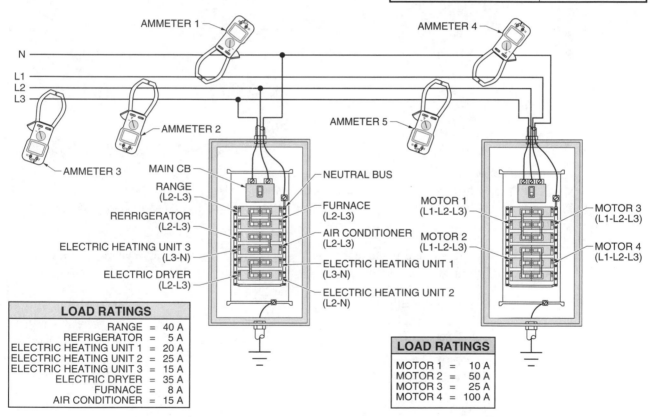

LOAD RATINGS

RANGE	= 40 A
REFRIGERATOR	= 5 A
ELECTRIC HEATING UNIT 1	= 20 A
ELECTRIC HEATING UNIT 2	= 25 A
ELECTRIC HEATING UNIT 3	= 15 A
ELECTRIC DRYER	= 35 A
FURNACE	= 8 A
AIR CONDITIONER	= 15 A

LOAD RATINGS

MOTOR 1	= 10 A
MOTOR 2	= 50 A
MOTOR 3	= 25 A
MOTOR 4	= 100 A

ACTIVITY 10-4—Voltage Unbalance

Determine the average voltage, voltage deviation, and voltage unbalance.

_____ 1. The average voltage equals ___ V.

_____ 2. The voltage deviation equals ___ V.

_____ 3. The voltage unbalance equals ___%.

_____ 4. The average voltage equals ___ V.

_____ 5. The voltage deviation equals ___ V.

_____ 6. The voltage unbalance equals ___%.

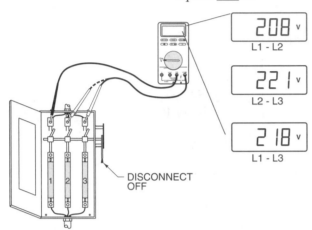

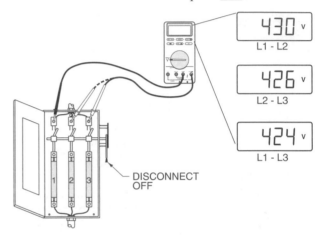

ACTIVITY 10-5—Single Phasing

Single phasing is measured when the load is a 3φ motor and the motor is ON. An uneven voltage is measured at a running motor because the open winding has an induced voltage applied to it from the powered windings. Each motor has lost power in one incoming power line. Determine the voltage unbalance and identify the incoming power line that has lost power.

_____ 1. The voltage unbalance equals ___%.

_____ 2. The incoming power line that has lost power is Line ___.

_____ 3. The voltage unbalance equals ___%.

_____ 4. The incoming power line that has lost power is Line ___.

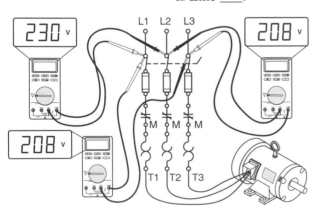

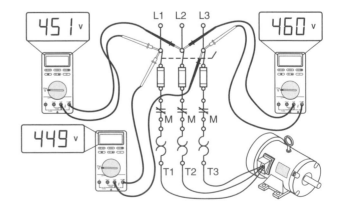

ACTIVITY 10-6—*Improper Phase Sequence*

A phase sequence tester is used to determine which 3ϕ phase power lines are powered and the direction the motor shaft will rotate. The tester is connected to the supply voltage before the motor is connected. Lights are used to indicate the circuit's condition. Three lights indicate the presence of line power and two lights indicate motor shaft rotation. The motor shaft rotation (CW or CCW) is determined when viewing the shaft end of the motor. Determine if the phase sequence is correct or incorrect.

_____ 1. The phase sequence is ___.

_____ 2. The phase sequence is ___.

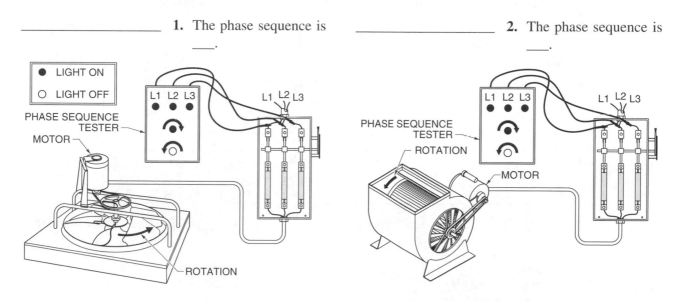

ACTIVITY 10-7—*Voltage Surges*

A surge protector may be added to the power lines to limit any voltage surges to a set limit. Determine the maximum peak voltage that the surge protector allows as measured on the scopemeter.

_____ 1. The maximum peak voltage of Scopemeter 1 equals ___.

_____ 2. The maximum peak voltage of Scopemeter 2 equals ___.

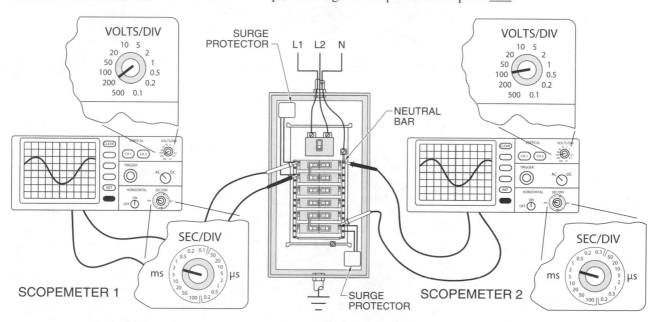

ACTIVITY 10-8—AC Voltage and Frequency Variations

Motor torque and current are reduced when the voltage at the motor terminals is less than the motor rated nameplate voltage. Determine the amount of torque and current based on the measured terminal voltage.

MOTOR REDUCED VOLTAGE STARTING CHARACTERISTICS		
Motor Terminal Voltage (% nameplate voltage)	Torque Reduction (%)	Current Reduction (%)
100	0	0
95	11	6
90	20	11
85	30	16
80	39	22
75	47	27
70	55	32
65	60	37
60	68	42
55	73	48
50	78	54

_____ 1. The derated torque equals ___ lb-ft.

_____ 2. The derated current equals ___ A.

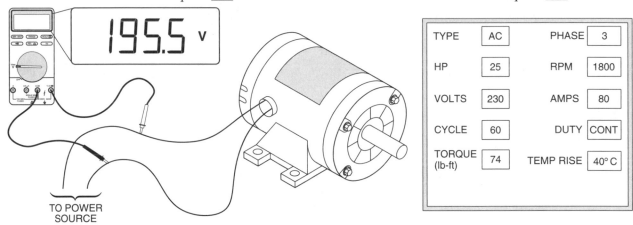

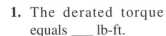

_____ 3. Derated torque equals ___ lb-ft.

_____ 4. Derated current equals ___ A.

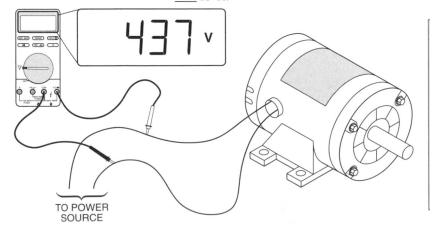

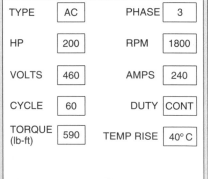

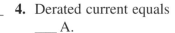

ACTIVITY 10-9—DC Voltage Variations

Match each waveform to the scopemeter on which it should appear.

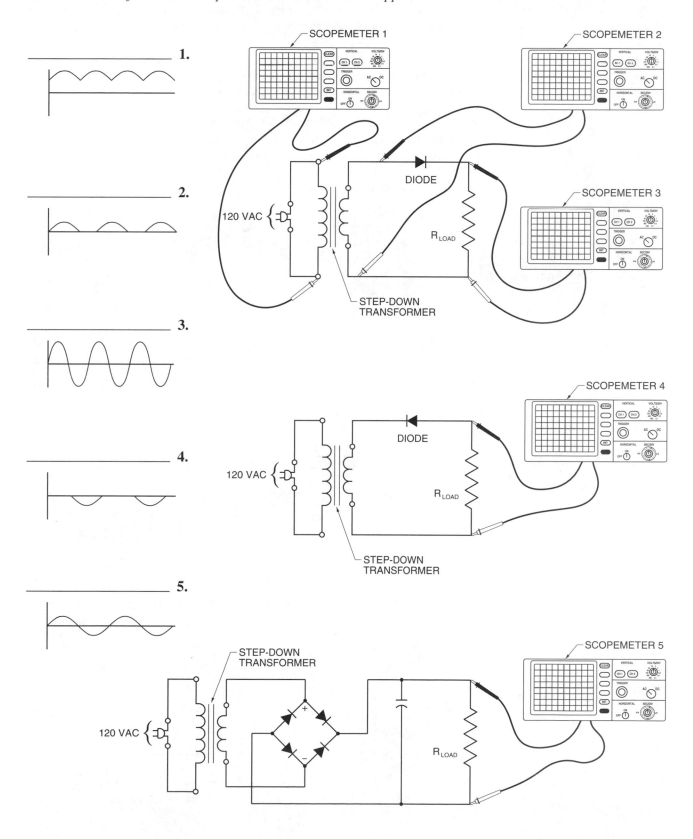

ACTIVITY 10-10—Allowable Motor Starting Time

A tachometer or ammeter may be used to measure the acceleration time of a motor. A tachometer measures the speed of a motor to determine when the motor has reached its rated speed. An ammeter measures the current draw of a motor. All motors draw high inrush current when starting. A motor is at full speed when the motor current draw has leveled.

The true acceleration condition of a motor is measured when both a tachometer and ammeter are used. When a motor is operating properly, a tachometer indicates that the motor has constant acceleration to its rated speed, and an ammeter indicates that the high starting current has leveled to its rated (or less) value.

_____ 1. Using the actual readings, calculate the percent of rated speed and percent of rated current for each second of acceleration. Draw the graph of the acceleration condition of the motor based on its percent rated speed and current.

_____ 2. Does the chart indicate a problem?

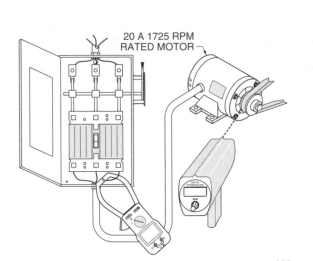

20 A 1725 RPM RATED MOTOR

Time*	Readings		Percent	
	rpm	Amps	rpm	Amps
1	172	50		
2	345	47		
3	517	44		
4	690	41		
5	862	38		
6	1035	35		
7	1207	32		
8	1380	29		
9	1552	26		
10	1725	23		
11	1725	20		
12	1725	20		

* in sec

$$\% \text{ RATED RPM} = \frac{\text{MEASURED RPM}}{\text{RATED RPM}} \times 100$$

$$\% \text{ RATED AMPS} = \frac{\text{MEASURED AMPS}}{\text{RATED AMPS}} \times 100$$

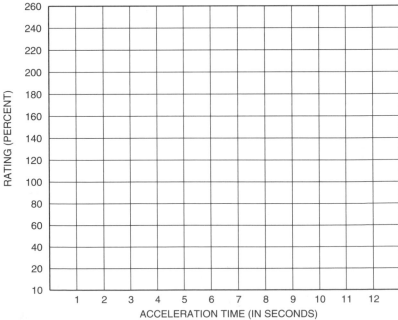

RATING (PERCENT)

ACCELERATION TIME (IN SECONDS)

ACTIVITY 10-11—Megohmmeter Tests

1. Connect Megohmmeter 1 to measure the resistance of the armature to ground circuit. Connect Megohmmeter 2 to measure the resistance of the series winding to ground.

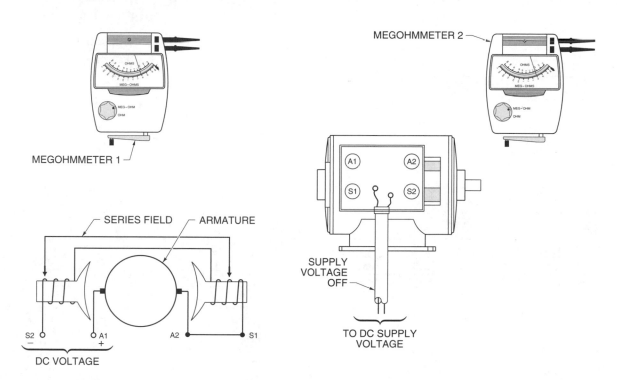

ACTIVITY 10-12—Insulation Spot Test

1. A 120 V split-phase motor was placed in service in January 2005. Semiannual resistance readings are taken. Develop the Insulation Spot Test Graph from the readings. Mark the point at which the motor requires service.

INSULATION SPOT TEST	
Test Taken	**Resistance Reading (in MΩ)**
1-05	300
7-05	300
1-06	250
7-06	225
1-07	100
7-07	100
1-08	90
7-08	10
1-09	.1
7-09	.05
1-10	.03

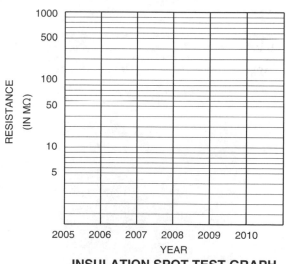

INSULATION SPOT TEST GRAPH

ACTIVITY 10-13—Dielectric Absorption Test

Semiannual resistance readings were taken during a dielectric absorption test at 240 V with Class A insulation.

1. Develop the Dielectric Absorption Test Graph.

2. The polarization index of the insulation is ___.

3. Is the motor insulation good or bad?

DIELECTRIC ABSORPTION TEST	
Test Taken	**Resistance Reading (in MΩ)**
At 10 seconds	200
At 20 seconds	200
At 30 seconds	210
At 40 seconds	220
At 50 seconds	250
At 60 seconds	250
At 2 minutes	225
At 3 minutes	200
At 4 minutes	195
At 5 minutes	190
At 6 minutes	180
At 7 minutes	180
At 8 minutes	175
At 9 minutes	175
At 10 minutes	175

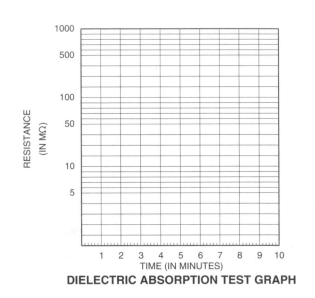

DIELECTRIC ABSORPTION TEST GRAPH

ACTIVITY 10-14—Insulation Step Voltage Test

Resistance readings were taken during an insulation step voltage test.

1. Develop the Insulation Step Voltage Test Graph.

2. Is the motor insulation good or bad?

INSULATION STEP VOLTAGE TEST	
Applied Voltage (in V)	**Resistance Reading (in MΩ)**
500	200
1000	200
1500	225
2000	225
2500	220
3000	220
3500	210
4000	200
4500	195
5000	190

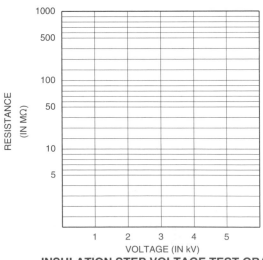

INSULATION STEP VOLTAGE TEST GRAPH

APPLICATION 11-1—Motor Nameplate Operating Ratings

Motors are selected for an application based on their operating ratings. Operating ratings include motor usage, service factor, speed, operating time (duty cycle), and efficiency. These ratings are listed on the motor nameplate and are required when performing installation and troubleshooting procedures. **See Motor Operating Ratings.**

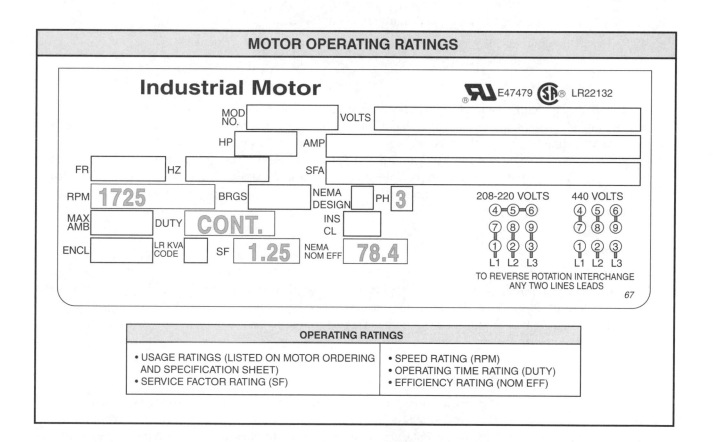

MOTOR OPERATING RATINGS

OPERATING RATINGS	
• USAGE RATINGS (LISTED ON MOTOR ORDERING AND SPECIFICATION SHEET) • SERVICE FACTOR RATING (SF)	• SPEED RATING (RPM) • OPERATING TIME RATING (DUTY) • EFFICIENCY RATING (NOM EFF)

Usage Rating

Motors are rated for general-purpose usage or specific-purpose usage. Motors that are rated for general-purpose usage are used in a wide range of general applications, such as conveyors, machine tools, and belt-driven equipment.

Motors rated for specific-purpose usage are designed for applications in which the motor use is defined by the specific application, such as the following:

- Washdown rated (food, beverage, and chemical plants)

- Submersible pumps (sump pumps, drainage pumps, water wells, and septic systems)

- Other pumps (wastewater treatment, water treatment, and water recirculation systems)

- Hazardous locations and severe-duty (dry cleaning plants, paint factories, and grain elevators)

- Instantly reversible (hoists, gates, cranes, and mechanical doors)

- Extra-high torque (hard-starting loads)

- Farm duty/agricultural (protection against dust, dirt, and chemicals)

- Irrigation (corrosion resistance from high moisture and chemical environments)

- Auger drive (augers and drilling systems)

- HVAC (heating/ventilating/air conditioning systems)

- Inverter duty (variable frequency drives)

- Pools (swimming pools, water parks, and whirlpool hot tubs)

- AC/DC vacuum (commercial vacuum systems, carwash, and sprayer/fogger systems)

Service Factor Rating

A motor will attempt to drive a load, even if the load exceeds the motor power rating. A motor service factor rating indicates whether a motor can safely handle an overloaded condition. A nameplate service rating of 1 (or no listed rating) indicates the motor is not designed to safely handle an overloaded condition above the motor rated power (in horsepower or kilowatts). A nameplate service rating higher than 1 indicates the motor is designed to develop more than its nameplate rated power without causing damage to the motor insulation. For example, a 10 HP rated motor with a service factor of 1.15 can be operated as an 11.5 HP motor ($10 \times 1.15 = 11.5$).

Speed Rating

Motors have two speed ratings: the synchronous speed and the operating speed. The synchronous speed of an AC motor is based on the number of stator poles and the applied frequency. The operating speed is the actual nameplate-listed speed at which the motor develops rated horsepower at rated voltage and frequency. The difference between the motor theoretical speed (synchronous speed) and actual speed (nameplate rating) is referred to as motor slip.

Operating Time Rating (Duty Cycle)

Most motors can be operated for any length of time. However, some motors are designed to operate for short durations only. Motors designed to operate for unlimited time periods are marked "CONT" (continuous) on the motor nameplate, or have no designation. Motors designed to operate for intermittent time periods before being turned off and allowed to cool are marked "INTER" on the nameplate, or have a time rating. Typical intermittent-duty motor time ratings are 5 min, 15 min, 30 min, or 60 min.

Intermittent-duty rated motors are used in applications such as waste disposal systems (garbage disposals), electric hoists, gate openers, and other applications in which the motor is turned on for short time periods to meet the application requirements.

Efficiency Rating

Motor efficiency is a measure of the effectiveness with which a motor converts electrical energy to mechanical energy. Motor efficiency is the ratio of motor power output to supply power input. All motors require more power to operate than they can produce because of power loss within the motor. Power loss occurs due to losses from friction and heat within the motor.

Motor manufacturers produce standard and energy-efficient motors. A standard motor normally operates between 75% to 93% efficiency. An energy-efficient motor normally operates between 83% to 96% efficiency. Energy-efficient motors are more expensive than standard motors but cost less to operate. Using energy-efficient motors saves money and reduces power usage because motors consume over 60% of all electrical power produced.

 Refer to **Activity 11-1—Motor Nameplate Operating Ratings** ....

APPLICATION 11-2—Motor Nameplate Environmental Ratings

Motor environmental ratings include temperature ratings and insulation ratings. Motors must operate under rated temperature conditions. Operating a motor in a condition that is excessively hot or with insulation that is not rated for the application could cause premature failure and damage to the motor and related equipment. **See Motor Environmental Ratings.**

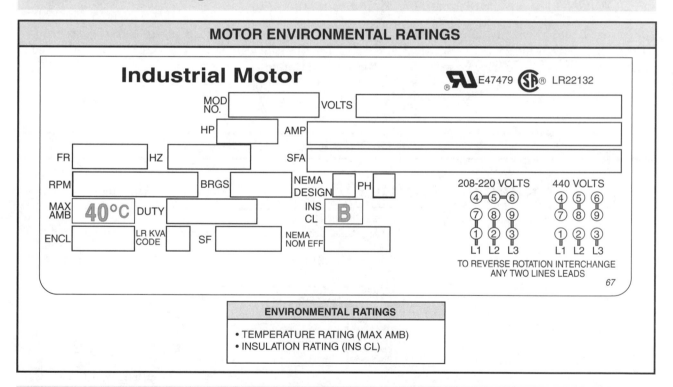

Temperature Rating

Ambient temperature is the temperature of the air surrounding a motor. *Temperature rise* is the difference between the motor winding temperature when running and the ambient temperature. The temperature rise produced at full load does not harm the motor as long as the ambient temperature does not exceed the nameplate listed temperature rating.

Higher temperatures caused by either an increase in ambient temperature or by overloading the motor damage the motor insulation. Normally, for every 10°F above the temperature rating of a motor, the motor life is cut in half. Heat destroys insulation, and the higher the heat, the greater damage occurs at a faster rate. Most motors have a temperature rating of 40°C (104°F). A motor without a temperature rating is normally rated at 40°C. Motors designed to operate in high-ambient-temperature areas should have additional cooling provided or have a higher rating (55°C nameplate rating). Although a motor nameplate lists the maximum ambient temperature at which a motor is designed to operate, a motor also has a low temperature rating that can be found in the motor specifications. This low limit is normally –25°C, unless stated otherwise.

Insulation Rating

Insulation breakdown is the main cause of motor failure. Motor insulation is rated according to its thermal breakdown resistance. The four motor insulation classes are Class A, Class B, Class F, and Class H. Class A is the least common motor insulation. Class F is the most common motor insulation. Class H is the best-rated insulation and should be used in any application in which a motor drive is used to operate the motor.

 *Refer to **Activity 11-2—Motor Nameplate Environmental Ratings** ....*

APPLICATION 11-3—Motor Nameplate Mechanical Ratings

Mechanical ratings on a motor nameplate include frame size rating, design rating, enclosure type, and motor bearings. Following the guidelines on mechanical ratings when installing or servicing a motor results in improved efficiency and lower operating cost. **See Motor Mechanical Ratings.**

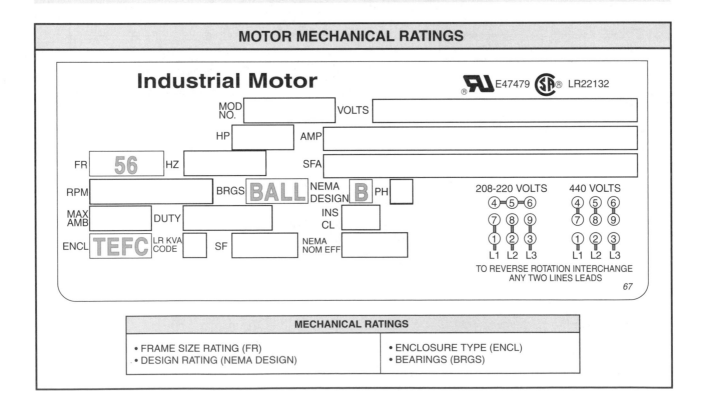

Frame Size Rating

All motors have a frame to protect the working parts of the motor and provide a means of mounting. Motor frames follow standard ratings that are used to designate the physical size and measurements of the motor. Using standardized dimensions aids in mounting the motor and allows for interchangeability among different motor manufacturers.

Frame sizes follow standards established by the National Electrical Manufacturers Association (NEMA) or International Electrotechnical Commission (IEC). Dimensionally, NEMA standards are expressed in English units and IEC standards are expressed in metric units. Frame size is listed on the motor nameplate as FR. For both NEMA and IEC, the larger the frame size number, the larger the motor.

Design Rating

Most motor loads take more torque to get them started than to keep them running. Motor design ratings (letters) represent the torque characteristics of a motor. Motor design ratings may be listed as Design A, B, C, D, or E. The higher the design rating, the higher the motor starting torque. Designs B, C, and D are the most common motor design ratings. Design B has a starting torque of up to 200% of the motor running torque. Design C has a starting torque of up to 250% of the motor running torque. Design D has a starting torque of over 250% of the motor running torque.

Enclosure Type

Motors include an enclosure to protect the motor working parts from the outside environment as well as to protect individuals from the electrical and rotating parts of the motor. Protecting individuals from the electrical and rotating parts of a motor can be accomplished using a basic enclosure. However, protecting a motor from all environmental conditions in which they must operate requires more than a basic enclosure. The following are environmental conditions that motors must occasionally operate in:
- Wet locations (rain, snow, sleet, washdown)
- Dirty locations (dirt, noncombustible dust)
- Oily locations (lubricants, cutting oils, coolants)
- Corrosive locations (salt, chlorine, fertilizers, chemicals)
- Extremely low- and high-temperature locations
- Gas and other hazardous locations
- Combustible dust locations (grain and chemical)

The type of enclosure a motor has is listed on the motor nameplate as the ENCL rating. Typical motor enclosure ratings include the following:
- ODP (open dripproof) for use in clean, dry, nonhazardous locations
- TENV (totally enclosed nonventilated) for use in moist, dirty, nonhazardous locations
- TEFC (totally enclosed fan-cooled) for use in the same locations as TENV but can tolerate higher temperatures
- TEAO (totally enclosed air over) for use in nonhazardous high-temperature locations

Hazardous location motors include a hazardous location listing specifying the hazardous classification/class/group rating (such as a Class I, Group D rating).

Bearings

Motor bearings are either sleeve or ball designs. Both sleeve bearings and ball bearings are used with different-size motors ranging from fractional horsepower to hundreds of horsepower. Sleeve bearings are used where a low noise level is important, such as fan and blower motors. Ball bearings are used where high-load capacity is required or periodic lubrication is not practical. The type of bearing used on a motor is listed on the motor nameplate as BALL or SLEEVE.

 Refer to **Activity 11-3—Motor Nameplate Environmental Ratings** ....

APPLICATION 11-4—Heat Problems

Excessive heat is a major cause of motor failure and a sign of other motor problems. Heat destroys motor insulation. When motor insulation is destroyed, the windings are shorted, and the motor is no longer functional.

As the heat in a motor increases beyond the temperature rating of the insulation, the life of the insulation is shortened. The higher the temperature, the sooner the insulation fails. The temperature rating of motor insulation is listed as the insulation class. **See Motor Insulation Class.**

The insulation class is given in Celsius (°C) and/or Fahrenheit (°F). A motor nameplate normally lists the insulation class of the motor. Heat buildup in a motor can be caused by the following conditions:

- Incorrect motor type or size for the application
- Improper cooling, normally from dirt buildup
- Excessive load, normally from improper use
- Excessive friction, normally from misalignment or vibration
- Electrical problems, normally voltage unbalance, phase loss, or a voltage surge

MOTOR INSULATION CLASS		
Class	°C	°F
A	105	221
B	130	266
F	155	311
*H	180	356

*Motor class H recommended for any motor contacted by a drive.

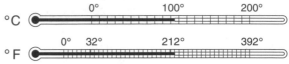

Improper Ventilation

All motors produce heat as they convert electrical energy to mechanical energy. This heat must be removed to prevent destruction of motor insulation. Motors are designed with air passages that permit a free flow of air over and through the motor. This airflow removes the heat from the motor. Anything that restricts airflow through a motor causes the motor to operate at a higher temperature than it is designed for.

Airflow through a motor may be restricted by the accumulation of dirt, dust, lint, grass, pests, rust, etc. If a motor becomes coated with oil from leaking seals or from overlubrication, airflow is restricted much faster. **See Improper Ventilation.**

Overheating can also occur if a motor is placed in an enclosed area. A motor overheats due to the recirculation of heated air when a motor is installed in a location that does not permit the heated air to escape. Vents added at the top and bottom of the enclosed area allow a natural flow of heated air.

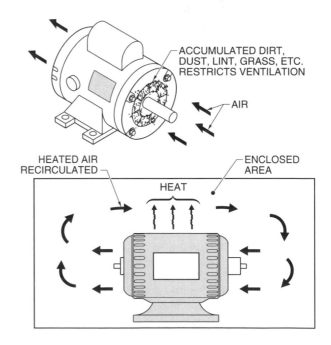

IMPROPER VENTILATION

Overloads

An *overload* is the application of excessive load to a motor. Motors attempt to drive the connected load when the power is ON. The larger the load, the more power is required. All motors have a limit to the load they can drive. For example, a 5 HP, 460 V, 3φ motor should draw no more than 7.6 A. See NEC® Table 430-150.

Overloads should not harm a properly protected motor. Any overload present longer than the built-in time delay of the protection device is detected and removed. Properly sized heaters in the motor starter assure that an overload is removed before any damage is done. **See Motor Overloading.**

A troubleshooter can observe the even blackening of all motor windings, which occurs when a motor has failed due to overloading. The even blackening is caused by the motor's slow destruction over a long period of time. No obvious damage or isolated areas of damage to the insulation are visible.

Current readings are taken at a motor to determine an overload problem. A motor is working to its maximum if it is drawing rated current. A motor is overloaded if it is drawing more than rated current. The motor size may be increased or the load on the motor decreased if overloads are a problem. **See Motor Current Readings.**

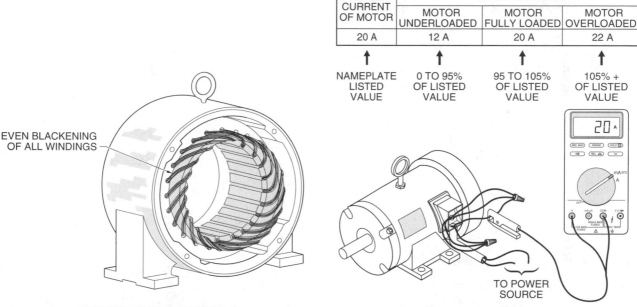

RATED CURRENT OF MOTOR	METER READING		
	MOTOR UNDERLOADED	MOTOR FULLY LOADED	MOTOR OVERLOADED
20 A	12 A	20 A	22 A
NAMEPLATE LISTED VALUE	0 TO 95% OF LISTED VALUE	95 TO 105% OF LISTED VALUE	105% + OF LISTED VALUE

EVEN BLACKENING OF ALL WINDINGS

TO POWER SOURCE

MOTOR OVERLOADING

MOTOR CURRENT READINGS

 Refer to **Activity 11-4—Heat Problems** ....

APPLICATION 11-5—Altitude Correction

Temperature rise of motors is based on motor operation at altitudes of 3300′ or less. A motor with a service factor of 1.0 is derated when it operates at altitudes above 3300′. A motor with a service factor above 1.0 is derated based on the altitude and service factor. **See Motor Altitude Deratings.**

MOTOR ALTITUDE DERATINGS			
Altitude Range (in ft)	Service Factor		
	1.0	1.15	1.25	1.35
3300–9000	93%	100%	100%	100%
9000–9900	91%	98%	100%	100%
9900–13,200	86%	92%	98%	100%
13,200–16,500	79%	85%	91%	94%
Over 16,500	Consult manufacturer			

 Refer to **Activity 11-5—Altitude Correction** ....

APPLICATION 11-6—*Motor Mounting and Positioning*

Motors that are not mounted properly are more likely to fail from mechanical problems. A motor must be mounted on a flat, stable base. A flat, stable base helps reduce vibration and misalignment problems. An adjustable motor base aids in proper mounting and alignment.

To ensure a long life span, a motor should be mounted so that it is kept as clean as possible. To reduce the chance of damaging material reaching a motor, a belt cleaner should be used in any application in which the belts are likely to bring damaging material to the motor.

Adjustable Motor Base

An adjustable motor base makes the installation, tensioning, maintenance, and replacement of belts easier. An *adjustable motor base* is a mounting base that allows a motor to be easily moved over a short distance. **See Adjustable Motor Base.** An adjustable motor base simplifies the installation of the motor and the tightening of belts and chains.

Mounting Direction

The position of the driven machine usually determines whether a motor is installed horizontally or vertically. Standard motors are designed to be mounted with the shaft horizontal. The horizontal position is the best operating position for the motor bearings. A specially designed motor is used for vertical mounting. Motors designed to operate vertically are more expensive and require more preventive maintenance.

Motor Belt Tension and Mounting

Belt drives provide a quiet, compact, and durable form of power transmission and are widely used in industrial applications. A belt must be tight enough not to slip, but not so tight as to overload the motor bearings.

Belt tension is usually checked by placing a straightedge from pulley to pulley and measuring the amount of deflection at the midpoint, or by using a tension tester. Belt deflection should equal $\frac{1}{64}''$ per inch of span. For example, if the span between the center of a drive pulley and the center of a driven pulley is 16", the belt deflection is $\frac{1}{4}''$ ($16 \times \frac{1}{64}'' = \frac{1}{4}''$). Belt tension is usually adjusted by moving the drive component away from or closer to the driven component. **See Belt Tension.**

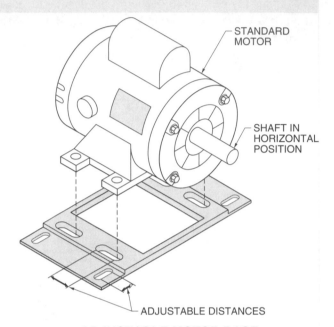

ADJUSTABLE MOTOR BASE

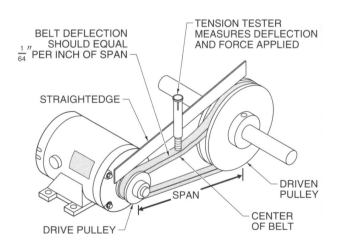

BELT TENSION

When using a belt drive for a horizontal application, belt sag should be on top. Maximum belt contact is made with the drive pulley when the taut part of the belt is on the bottom. Whenever possible, align pulley centers on the same horizontal plane. **See Horizontal Pulley Drive.**

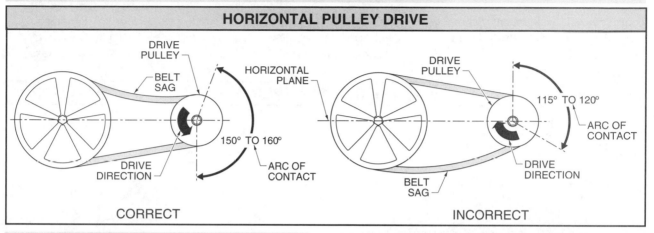

HORIZONTAL PULLEY DRIVE

CORRECT   INCORRECT

Vertical belt drives cause more problems than horizontal belt drives. If an application requires a vertical belt drive, mount the drive pulley on top. Maximum belt contact is made with the drive pulley when the taut part of the belt is pulling up. **See Vertical Pulley Drive.**

Whenever possible, align pulley centers on the same vertical plane. If the pulley centers cannot be aligned on the same vertical plane, an angle of 45° or less between the pulley centers and the vertical plane is permissible.

Belt Cleaner

In many motor conveyor applications, the motor's belts are likely to come in contact with material used in production, such as chemicals, dirt, sand, wood pulp, food, etc. If material adheres to a motor belt, a belt wears faster, misalignment and vibration occur, and the motor is required to deliver additional power.

The motor can be damaged from reduced airflow and corrosion if a foreign material is delivered to the motor. A belt cleaner should be installed to reduce material buildup in any application in which foreign material is likely to come in contact with motor belts. A belt cleaner is a specially designed device that removes foreign material on a belt and deposits the material at one location. **See Belt Cleaner.**

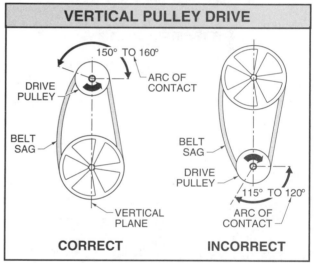

VERTICAL PULLEY DRIVE

CORRECT   INCORRECT

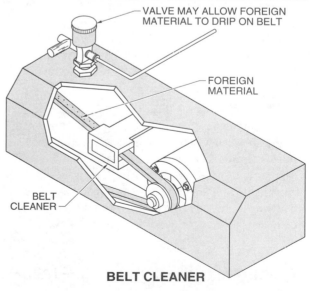

BELT CLEANER

Misalignment and Vibration

Misalignment of a motor and driven load is a major cause of motor failure. If a motor and driven load are misaligned, premature failure of the motor bearings, load, or both may occur.

Equipment shafts should be properly aligned on all new installations and checked during periodic maintenance inspections. Misalignment is normally corrected by placing shims under the feet of the motor or driven equipment. A coupling designed to allow some misalignment is used if misalignment cannot be corrected. Couplings used in misaligned applications include rubber-in-shear, flexible spring, and all-metal flex link types. **See Motor Couplings.**

MOTOR COUPLINGS

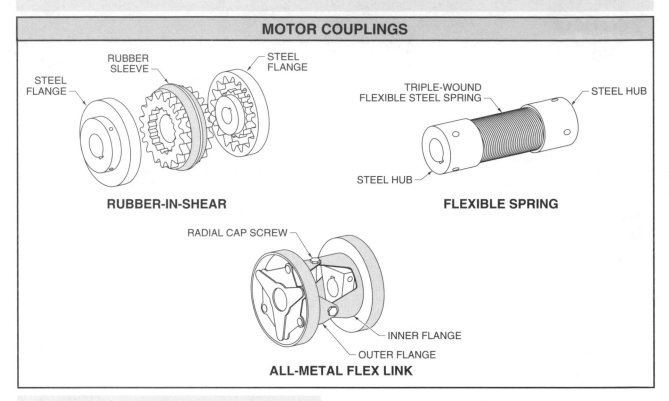

RUBBER-IN-SHEAR

FLEXIBLE SPRING

ALL-METAL FLEX LINK

Misalignment may be angular or parallel. *Angular misalignment* is misalignment when two shafts are not parallel. *Parallel misalignment* is misalignment when two shafts are parallel but not on the same line. **See Angular Misalignment** and **Parallel Misalignment.**

Motor Coupling Rating

Motor couplings are rated according to the amount of torque they can handle. Couplings are rated in pound-inches (lb-in.) or pound-feet (lb-ft). The coupling torque rating must be correct for the application to prevent the coupling from bending or breaking. A bent coupling causes misalignment and vibration. A broken coupling prevents the motor from doing work.

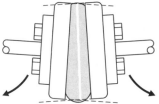

ANGULAR MISALIGNMENT

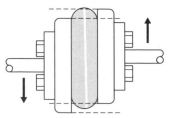

PARALLEL MISALIGNMENT

 Refer to **Activity 11-6—Motor Mounting and Positioning**...

APPLICATION 11-7—Loose Connections

All motors produce vibration as they rotate. The vibration can loosen mechanical and electrical connections. Loose mechanical connections generally cause noise and are easily detected. Loose electrical connections do not cause noise, but do cause a voltage drop to the motor and excess heat. Always check mechanical and electrical connections when troubleshooting a motor.

 Refer to **Activity 11-7—Loose Connections** ....

APPLICATION 11-8—Motor Defect

A motor can fail due to a motor defect or motor damage. A *motor defect* is an imperfection created during the manufacture of a motor that impairs its use. The defect is usually caught by the manufacturer if it impairs initial motor operation. If the defect manifests itself after the motor has been in operation for some time, the troubleshooter determines that the problem is a defect in the motor. Motors with defects should be replaced and the manufacturer should be notified.

A troubleshooter can observe the effect when a motor has failed due to a defect, which is usually confined to a small area of the motor. Typical defects that may occur in a motor include windings grounded in the slot, windings grounded at the edge of the slot, windings shorted phase-to-phase, and shorted connections. **See Motor Defects.**

Motor damage is any damage that occurs to a properly manufactured motor. The damage may occur before or during installation and during operation. A sound maintenance schedule and proper operation of a motor minimize the occurrence of motor damage.

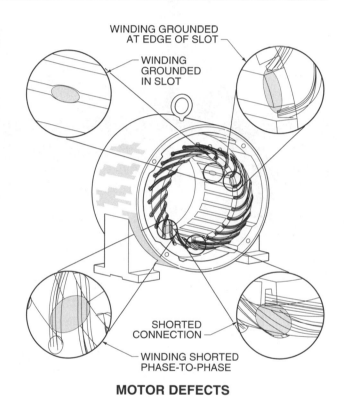

MOTOR DEFECTS

 Refer to **Activity 11-8—Motor Defect**..

 Refer to Quick Quiz® on CD-ROM

*Refer to Chapter 11 in the **Troubleshooting Electrical/Electronic Systems Workbook** for additional questions.*

Motor Mechanical Problems

Name_____ **Date** _____

ACTIVITY 11-1—Motor Nameplate Operating Ratings

Answer the questions using the General-Purpose Motor Data table. If the rating is not given, write "NA" (not available).

GENERAL-PURPOSE MOTOR DATA

THREE-PHASE ENERGY-EFFICIENT OPEN DRIPPROOF

- MOUNTING: RIGID BASE
- BEARINGS: BALL
- THERMAL PROTECTION: NONE
- INSULATION CLASS B OR CLASS F (FOR LONGER LIFE)
- ENCLOSURE: ODP
- 60 Hz
- ROTATION: CW/CCW
- MAX. AMBIENT: 40°C
- UL RECOGNIZED AND CSA CERTIFIED
- WARRANTY: 48 TO 56 FRAME, 1 YEAR; 140T FRAME AND ABOVE, 3 YEARS
- CONTINUOUS DUTY RATED

STAINLESS STEEL NAMEPLATE MAINTAINS INFORMATION OVER LONG LIFE. 56HZ FRAMES HAVE 7/8″ × 2¼″ SHAFT AND BASE BOLT-HOLE CONFIGURATION TO MATCH 56, 56H, 143T AND 145T.

USES: GENERAL PURPOSE FOR CLEAN, DRY, NONHAZARDOUS APPLICATIONS WITH PUMPS, VENTILATION EQUIPMENT, MACHINE TOOLS, AND OTHER INDUSTRIAL EQUIPMENT.

HP	STOCK ITEM NUMBER	NAMEPLATE RPM	NEMA FRAME	VOLTS AT 60 Hz	FULL-LOAD AMPERES	NEMA NOMINAL EFFICIENCY	SERVICE FACTOR	FRAME MATERIAL	OVERALL LENGTH*	INSULATION CLASS
¼	1	1725	48	208–230/460	1.2–1.3/0.65	62.0	1.35	Steel	9⁵⁄₁₆	B
	2	1725	56	208–230/460	1.2–1.3/0.65	62.0	1.35	Steel	9¾	B
	3	1140	56	208–230/460	1.3–1.5/0.75	64.0	1.35	Steel	10¼	B
⅓	4	3450	48	208–230/460	1.4–1.4/0.7	68.0	1.35	Steel	8⅝	B
	5	1725	48	208–230/460	1.5–1.6/0.8	67.0	1.35	Steel	9⁹⁄₁₆	B
	6	1725	56	208–230/460	1.5–1.6/0.8	67.0	1.35	Steel	10	B
	7	1140	56	208–230/460	1.9–2.2/1.1	62.0	1.35	Steel	9¹⁵⁄₁₆	B
½	8	3450	56	208–230/460	1.9–2.2/1.1	66.0	1.25	Steel	10⁷⁄₁₆	B
	9	1725	56	208–230/460	2.1–2.2/1.1	75.7	1.25	Steel	10	B
	10	1725	56	208–230/460	2.0–2.2/1.1	70.0	1.25	Steel	10⅜	B
¾	11	1725	56	208–230/460	2.9–2.8/1.4	76.5	1.25	Steel	10⅜	B
1	12	3450	56	208–230/460	3.4–3.7/1.85	74.0	1.25	Steel	10⅜	B
	13	1755	182	208–230/460	3.6–3.3/1.6	82.5	1.15	Steel	13⅛	F
	14	1725	56	208–230/460	3.6–3.8/1.9	78.4	1.25	Steel	10¾	B
	15	1725	56	208–230/460	3.4–3.4/1.7	76.5	1.15	Steel	11½	B
	16	1140	56HZ	208–230/460	3.6–3.2/1.6	80.0	1.15	Steel	12¼	F
	17	1140	145T	208–230/460	3.6–3.2/1.6	80.0	1.15	Steel	12¼	F
1½	18	1725	56H	208–230/460	4.8–4.8/2.4	80.0	1.15	Steel	12	B
	19	1725	56H	208–230/460	4.8–4.8/2.4	80.0	1.15	Steel	10¹⁵⁄₁₆	F
2	20	1170	184T	208–230/460	7.8–7.1/3.5	85.5	1.15	Steel	13⅝	F
	21	1170	213	208–230/460	6.8–6.2/3.1	85.5	1.15	Steel	16¹³⁄₁₆	F
10	22	3540	254U	208–230/460	25.9–23.4/11.7	89.5	1.15	Steel	21¹³⁄₁₆	F
	23	3490	213T	208–230/460	26.6–24.0/12.0	88.5	1.15	Steel	17³⁄₁₆	F
	24	1770	256U	208–230/460	27.4–24.8/12.4	89.5	1.15	Steel	21¹³⁄₁₆	F
	25	1740	215T	208–230/460	26.7–24.2/12.1	89.5	1.15	Steel	17³⁄₁₆	F
15	26	1760	254T	208–230/460	38.8–35.1/17.5	91.0	1.15	Steel	22¹⁄₁₆	F
	27	1175	284T	208–230/460	41.5–37.5/18.8	90.2	1.15	Steel	24¹¹⁄₁₆	F
20	28	3520	254T	208–230/460	50.3–45.5/22.7	90.2	1.15	Cast Iron	20⁷⁄₁₆	F
	29	1755	256T	208–230/460	51.7–46.8/23.4	91.0	1.15	Cast Iron	22³⁄₁₆	F
	30	1175	286T	208–230/460	54.5–49.3/24.6	91.0	1.15	Cast Iron	24⅜	F

* in in.

_____ **1.** The usage rating for Stock Item Motor 1 is ___.

_____ **2.** The service factor rating for Stock Item Motor 1 is ___.

_____ **3.** The speed rating for Stock Item Motor 1 is ___.

_____ **4.** The operating time rating for Stock Item Motor 1 is ___.

_____ **5.** The efficiency rating for Stock Item Motor 1 is ___.

Determine the amount of power that is saved for each motor type. Use the following formula to determine the amount of power a motor consumes when operating.

$$P = \frac{HP \times 746}{E_{ff}}$$

where

P = power consumed (in W)
HP = motor horsepower rating
746 = constant
E_{ff} = % motor efficiency

_____ **6.** Motors 14 and 15 are both 1 HP, 1725 rpm motors that operate on the same amount of voltage. How much power does motor 14 consume when operating?

_____ **7.** How much power does motor 15 consume when operating?

_____ **8.** How many fewer watts does motor 14 use as compared to motor 15?

_____ **9.** Are any of the ⅓ HP motors designed to be used as sump pumps?

_____ **10.** Can the 20 HP motors be used in an application that requires them to operate between 2 hr to 16 hr at a time?

ACTIVITY 11-2—Motor Nameplate Environmental Ratings

Answer the questions using the General-Purpose Motor Data table. If the rating is not given, write "NA" (not available).

_____ **1.** The temperature rating for Stock Item Motor 1 is ___.

_____ **2.** The insulation rating for Stock Item Motor 1 is ___.

_____ **3.** Does motor 18 or 19 have the better insulation that is designed for longer life?

_____ **4.** Do the fractional-horsepower-rated motors have better insulation than motors rated 10 HP and above?

ACTIVITY 11-3—Motor Nameplate Mechanical Ratings

Answer the questions using the General-Purpose Motor Data table. If the rating is not given, write "NA" (not available).

_____ **1.** The frame size rating for Stock Item Motor 1 is ___.

_____ **2.** The design rating for Stock Item Motor 1 is ___.

_____ **3.** The enclosure type for Stock Item Motor 1 is ___.

_____ **4.** The bearings for Stock Item Motor 1 are ___.

_____ **5.** Does ¼ HP, 1725 rpm Stock Item Motor 1 or Stock Item Motor 2 have the bigger frame?

_____ **6.** Can the 2 HP motors be used in a food processing operation that requires several wash downs each day?

ACTIVITY 11-4 — Heat Problems

Using the Ambient Temperature Correction Chart, calculate the horsepower rating.

_____ **1.** The motor rating adjusted for ambient temperature is ___ HP.

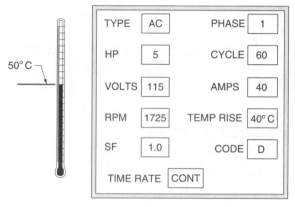

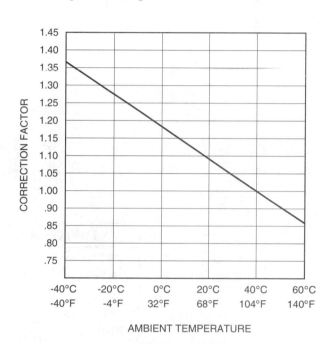

AMBIENT TEMPERATURE CORRECTION CHART

_____ **2.** The motor rating adjusted for ambient temperature is ___ HP.

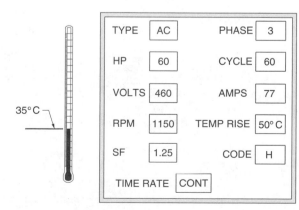

_____ **3.** The motor rating adjusted for ambient temperature is ___ HP.

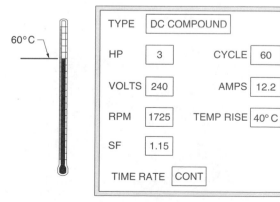

4. The motor rating adjusted for ambient temperature is ___ HP.

5. The motor rating adjusted for ambient temperature is ___ HP.

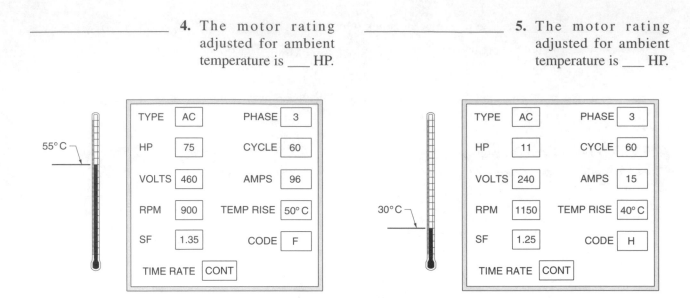

TYPE	AC	PHASE	3
HP	75	CYCLE	60
VOLTS	460	AMPS	96
RPM	900	TEMP RISE	50° C
SF	1.35	CODE	F
TIME RATE	CONT		

55°C

TYPE	AC	PHASE	3
HP	11	CYCLE	60
VOLTS	240	AMPS	15
RPM	1150	TEMP RISE	40° C
SF	1.25	CODE	H
TIME RATE	CONT		

30°C

ACTIVITY 11-5—Altitude Correction

Using Motor Altitude Deratings, calculate the horsepower rating.

MOTOR ALTITUDE DERATINGS				
Altitude Range (in ft)	**Service Factor**			
	1.0	**1.15**	**1.25**	**1.35**
3300–9000	93%	100%	100%	100%
9000–9900	91%	98%	100%	100%
9900–13,200	86%	92%	98%	100%
13,200–16,500	79%	85%	91%	94%
Over 16,500	Consult manufacturer			

1. The altitude adjusted rating equals ___ HP.

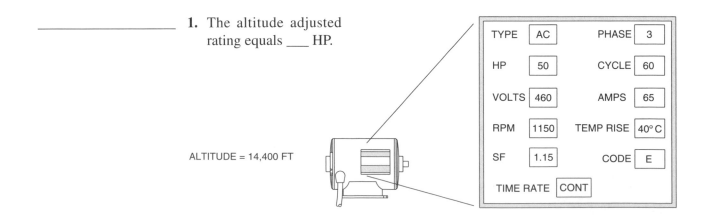

ALTITUDE = 14,400 FT

TYPE	AC	PHASE	3
HP	50	CYCLE	60
VOLTS	460	AMPS	65
RPM	1150	TEMP RISE	40° C
SF	1.15	CODE	E
TIME RATE	CONT		

_____ **2.** The altitude adjusted
rating equals ___ HP.

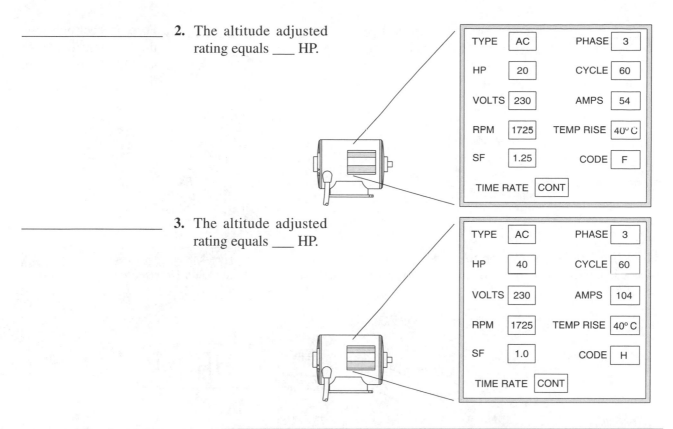

TYPE	AC	PHASE	3
HP	20	CYCLE	60
VOLTS	230	AMPS	54
RPM	1725	TEMP RISE	40° C
SF	1.25	CODE	F
TIME RATE	CONT		

_____ **3.** The altitude adjusted
rating equals ___ HP.

TYPE	AC	PHASE	3
HP	40	CYCLE	60
VOLTS	230	AMPS	104
RPM	1725	TEMP RISE	40° C
SF	1.0	CODE	H
TIME RATE	CONT		

ACTIVITY 11-6—Motor Mounting and Positioning

Determine the proper direction (CW or CCW) of rotation for the drive (small) pulleys for problems 1 through 4.

_____ **1.** The direction of rotation
equals ___.

_____ **2.** The direction of rotation
equals ___.

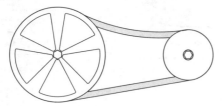

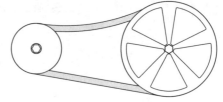

_____ **3.** The direction of rotation
equals ___.

_____ **4.** The direction of rotation
equals ___.

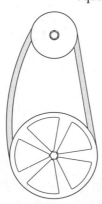

_____ **5.** For proper belt tightness, the belt tension tester should not read more than ___".

_____ **6.** For proper belt tightness, the belt tension tester should not read more than ___".

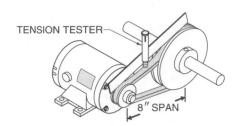

TENSION TESTER

8" SPAN

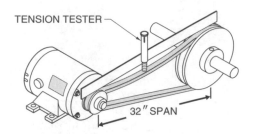

TENSION TESTER

32" SPAN

ACTIVITY 11-7—Loose Connections

A loose connection may cause a circuit to have a low voltage condition at the motor. Determine the point in each circuit that has the problem.

_____ **1.** The problem is located at ___.

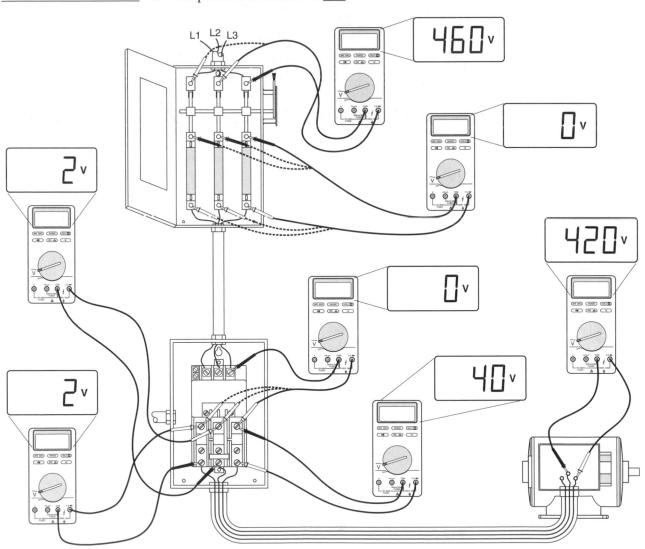

_____ **2.** The problem is located at ___.

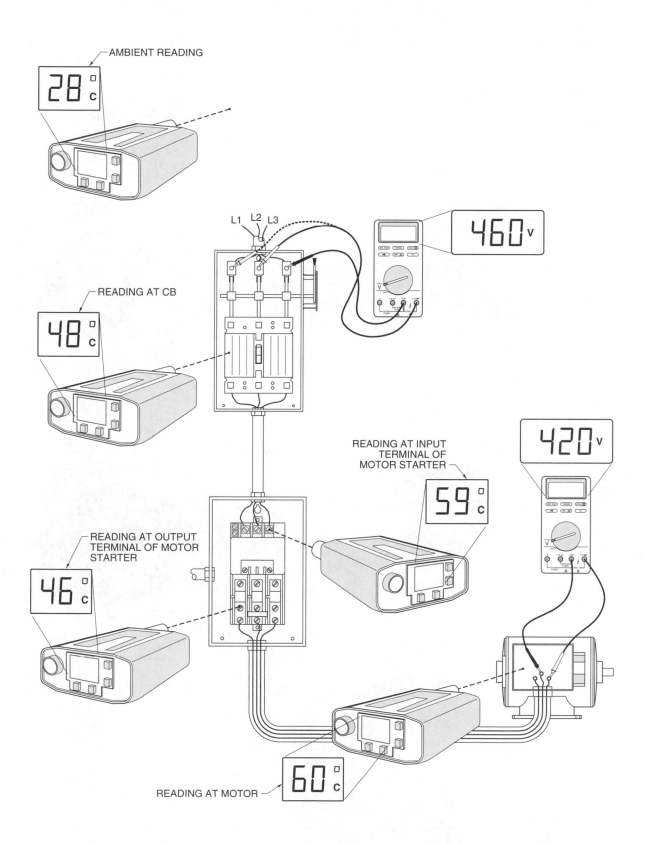

AMBIENT READING

READING AT CB

READING AT OUTPUT TERMINAL OF MOTOR STARTER

READING AT INPUT TERMINAL OF MOTOR STARTER

READING AT MOTOR

ACTIVITY 11-8—Motor Defect

Each motor failed in the first month of service. The motors were replaced and sent to the maintenance shop for inspection. Company policy is to send a motor back to the manufacturer for further inspection if it appears that a motor has failed because of a manufacturing defect. List each motor failure as defect or non-defect motor failure.

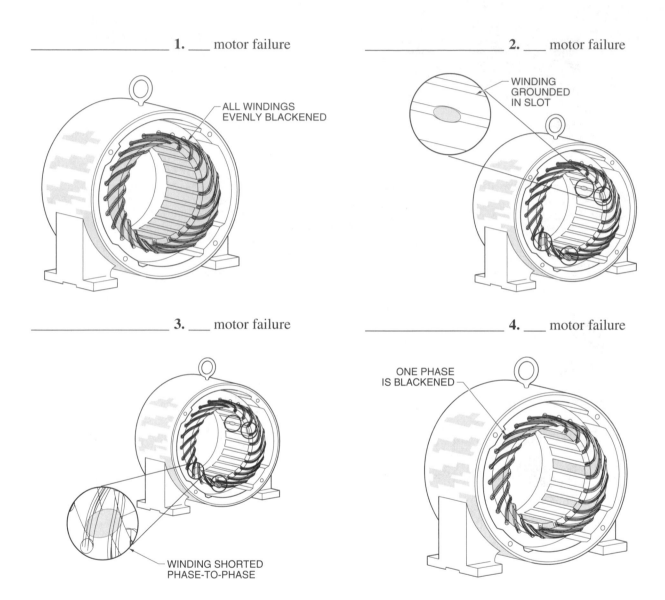

_____ 1. ___ motor failure

ALL WINDINGS EVENLY BLACKENED

_____ 2. ___ motor failure

WINDING GROUNDED IN SLOT

_____ 3. ___ motor failure

WINDING SHORTED PHASE-TO-PHASE

_____ 4. ___ motor failure

ONE PHASE IS BLACKENED

APPLICATION 12-1—DC Motors

A *direct current (DC) motor* is a motor that uses direct current connected to a field and armature to produce rotation. A *field* is the stationary windings, or magnets, of a DC motor. Field windings develop a magnetic field in a DC motor. An *armature* is the rotating part of a DC motor that is mounted on the motor shaft. **See DC Motor.**

When current passes through the armature located in the magnetic field, a mechanical force is exerted on the armature. A *commutator* is a series of copper segments connected to the armature. *Brushes* are the sliding contacts that provide contact between the external power source and the commutator. Brushes are made of carbon or graphite. **See DC Motor Brushes.**

The amount of force exerted on the motor shaft depends on the factors:

- Intensity (strength) of the magnetic field

- Current passing through the armature inside the magnetic field

- Length (or number) of the armature inside the magnetic field

The force on the motor shaft is increased by increasing any of these factors. Normally, the intensity of the field and the current are changed to increase force. The length of the armature is fixed when the motor is manufactured.

The armature windings, commutator, and brushes are arranged so that the flow of current is in one direction in the windings on one side of the armature and is in the opposite direction in the windings on the opposite side of the armature. DC power is delivered to the armature through the brushes, which ride on the commutator sections. The commutator is mounted on the same shaft as the armature and rotates with it.

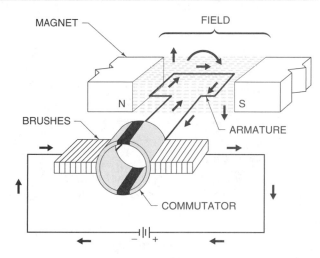

DC MOTOR

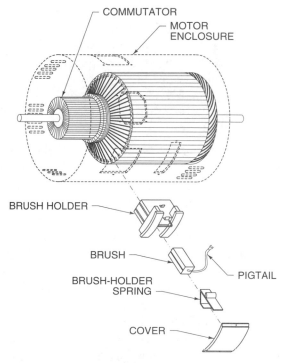

DC MOTOR BRUSHES

To increase the torque of the motor, the armature has a large number of coils connected to a large number of commutator sections. The field coils may be permanent magnets or electromagnets.

DC Motor Types

There are four basic types of DC motors:

- DC series motor
- DC shunt motor
- DC compound motor
- DC permanent-magnet motor

These motors have the same external appearance, but are different in their internal construction and output performance. The selection of the type of DC motor to use is based on the mechanical requirements of the applied load.

DC Series Motor

A *DC series motor* is a motor with the field connected in series with the armature. The field carries the load current passing through the armature. The field has comparatively few turns of heavy-gauge conductor. The conductors extending from the series coil are marked S1 and S2. The conductors extending from the armature are marked A1 and A2. **See DC Series Motor.**

A DC series motor is used as a traction motor because it produces the highest torque of all DC motors. The DC series motor can develop 500% of its full-load torque on starting. Typical applications include traction bridges, hoists, gates, and starting motors in automobiles.

DC Shunt Motor

A *DC shunt motor* is a motor with the field connected in shunt (parallel) with the armature. The field has numerous turns of wire, and the current in the field is independent of the armature. The field windings extending from the shunt field of a DC shunt motor are marked F1 and F2. The armature windings are marked A1 and A2. **See DC Shunt Motor.**

A DC shunt motor has excellent speed control. A DC shunt motor is used where constant or adjustable speed is required and starting conditions are moderate. Typical applications include fans, blowers, centrifugal pumps, conveyors, elevators, woodworking machinery, and metalworking machinery.

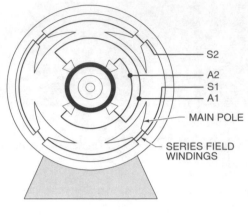

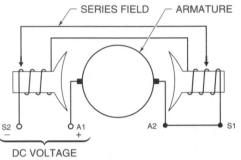

DC SERIES MOTOR

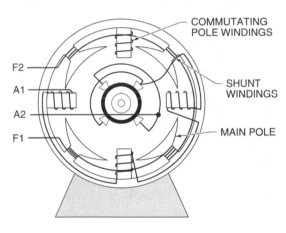

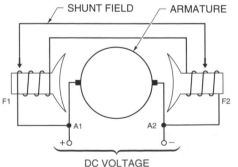

DC SHUNT MOTOR

DC Compound Motor

A *DC compound motor* is a motor with a field connected in both series and shunt with the armature. In a DC compound motor, the field coil is a combination of the series field (S1 and S2) and shunt field (F1 and F2). The series field is connected in series with the armature. The shunt field is connected in parallel with the series field and armature. **See DC Compound Motor.**

A DC compound motor has high torque and constant speed. A DC compound motor is used when starting torque and fairly constant speed are required. Typical applications include punch presses, shears, bending machines, and hoists.

DC Permanent-Magnet Motor

A *DC permanent-magnet motor* is a motor that uses magnets for the field winding. The DC permanent-magnet motor has molded magnets mounted in a steel shell. The permanent magnets are the field coils. DC power is supplied only to the armature. **See DC Permanent-Magnet Motor**

A DC permanent-magnet motor is used in automobiles to control power seats, power windows, and windshield wipers. A DC permanent-magnet motor produces relatively high torque at low speeds and provides some self-braking when removed from power. Not all DC permanent-magnet motors are designed to run continuously because they overheat rapidly. Overheating destroys the permanent magnets.

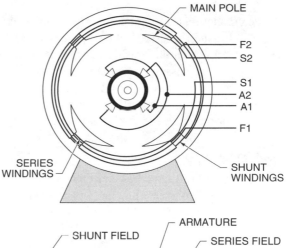

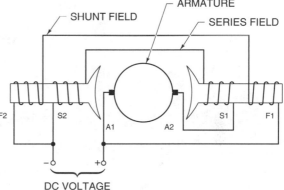

DC COMPOUND MOTOR

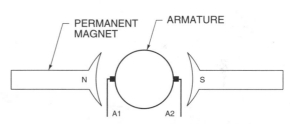

DC PERMANENT-MAGNET MOTOR

 Refer to **Activity 12-1—DC Motors** ....

APPLICATION 12-2—Troubleshooting DC Motors

Direct current motors are used in applications that require high torque. To produce high torque, the outside power supply is connected to both the armature and field. A commutator and brushes are used to supply power to the rotating field. Because of their brushes, DC motors generally require more repair than motors that do not use brushes. The brushes should be checked first when troubleshooting DC motors.

Brushes wear faster than any other component of a DC motor. The brushes ride on the fast-moving commutator. Bearings and lubrication are used to reduce friction when two moving surfaces touch. No lubrication is used between the moving brushes and the commutator because the brushes must carry current from the armature. Sparking occurs as the current passes from the commutator to the brushes. Sparking causes heat, burning, and wear of electric parts.

The brushes and commutator are subject to wear. The brushes are designed to wear as the motor ages. Replacing worn brushes is easier and less expensive than servicing or replacing a worn commutator. Most DC motors are designed so that the brushes and the commutator can be inspected without disassembling the motor. Some motors require disassembly for close inspection of the brushes and commutator.

Observe the brushes as a motor operates. The brushes should be riding smoothly on the commutator with little or no sparking. There should be no brush noise, such as chattering. Brush sparking, chattering, or a rough commutator indicates service is required.

Troubleshooting Brushes

The condition of the brushes and their holders is extremely important for good motor operation. The brushes should be checked every time the motor is serviced. **See Troubleshooting Brushes.**

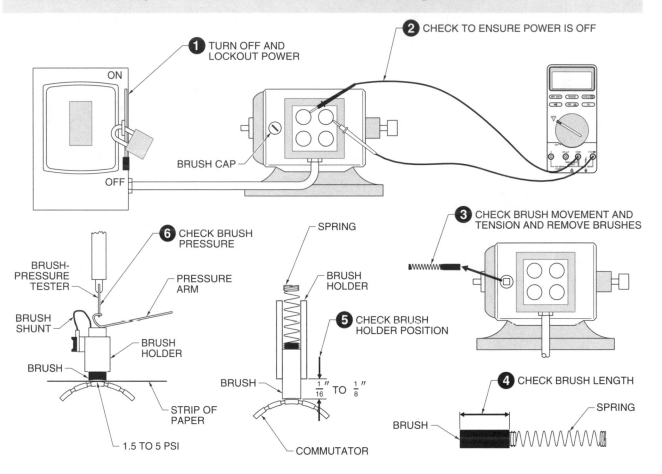

TROUBLESHOOTING BRUSHES

To troubleshoot brushes, apply the procedure:
1. Turn the handle of the safety switch or combination starter OFF. Lockout and tag the starting mechanism per company policy.
2. Measure the voltage at the motor terminals to ensure the power is OFF.
3. Check the brush movement and tension. Remove the brushes. The brushes should move freely in the brush holder. The spring tension should be approximately the same on each brush.

4. Check the length of the brushes. Brushes should be replaced when they have worn down to about half of their original size. Replace all brushes if any brush is less than half its original length. Never replace only one brush.

   Brush compositions include high-grade carbon, electrographite, natural graphite, and carbon graphite. Each composition has its advantages and disadvantages. Always replace brushes with brushes of the same composition. Check manufacturer's recommendations for type of brushes to use.

5. Check the position of the brush holder in relation to the commutator. The brush holder should be $\frac{1}{16}''$ to $\frac{1}{8}''$ from the commutator. The commutator may be damaged if the brush holder is too close. The brush may break if the brush holder is too far away.

6. Check for proper brush pressure. Brush pressure is critical to proper operation. Too little pressure causes the brushes to arc excessively and groove the commutator. Too much pressure causes the brushes to chatter and wear faster than normal.

   Brush pressure varies with the composition of the brush. Brush pressure is usually about 1.5 to 5 psi of surface area. The original spring should provide the proper pressure if it is in good condition. Replace the spring with one of the same type if it is not in good condition. Brush pressure is checked using a brush-pressure tester. Follow the manufacturer's procedures.

   When checking brush pressure, remove the endbell on the side in which the commutator is located. Pull back on the gauge, noting the pressure at which the piece of paper is free to move. Divide this reading by the contact area of the brush to get actual brush pressure in psi. Check the measured pressure to ensure it falls within the manufacturer's listed range.

Troubleshooting Commutators

Brushes wear faster than the commutator. After brushes have been changed once or twice, the commutator usually needs servicing. Any markings on the commutator, such as grooves or ruts, or discolorations other than a polished, brown color where the brushes ride, indicate a problem. **See Troubleshooting Commutators.**

To troubleshoot commutators, apply the procedure:

1. Make a visual check of the commutator. The commutator should be smooth and concentric. A uniform dark copper oxide carbon film should be present on the surface of the commutator. This naturally occurring film acts like a lubricant by prolonging the life of the brushes and reducing wear on the commutator surface.

2. Check the mica insulation between the commutator segments. The mica insulation separates and insulates each commutator segment. The mica insulation should be undercut (lowered below the surface) approximately $\frac{1}{32}''$ to $\frac{1}{16}''$, depending on the size of the motor. The larger the motor, the deeper the undercut. Replace or service the commutator if the mica is raised.

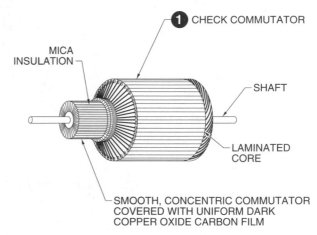

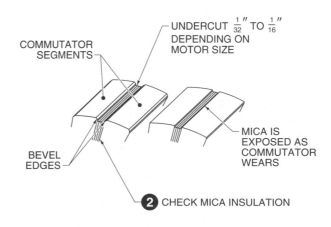

TROUBLESHOOTING COMMUTATORS

Troubleshooting for Grounded, Open, or Short Circuits

A DC motor is tested for a grounded, open, or short circuit by using a test light. A *grounded circuit* is a circuit in which current leaves its normal path and travels to the frame of the motor. A grounded circuit is caused when insulation breaks down or is damaged and touches the metal frame of the motor.

An *open circuit* is a circuit that no longer provides a path for current to flow. An open circuit is caused when a conductor or connection has physically moved apart from another conductor or connection.

A *short circuit* is a circuit in which current takes a shortcut around the normal path of current flow. A short circuit is caused when the insulation of two conductors from different parts of a circuit touch. Short circuits are usually a result of insulation breakdown.

A test light is preferred for a quick check of a motor. A test light gives good results for obvious problems. Additional testing equipment, such as ohmmeters and megohmmeters is required for less obvious problems. For example, a megohmmeter is used to test for insulation breakdown. **See Troubleshooting Circuit Faults.**

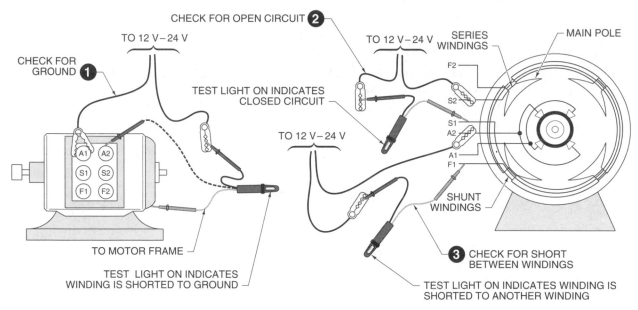

TROUBLESHOOTING CIRCUIT FAULTS

To troubleshoot for a grounded circuit, open circuit, or short circuit, apply the procedure:

1. Check for a grounded circuit. Connect one lead of the test-light to the frame of the motor. Touch the other test-light lead from one motor lead to the other. A grounded circuit is present if the test light turns ON. Service and repair the motor.

2. Check for an open circuit. Connect the two test-light leads to the motor field and armature circuits as follows:

COMPOUND MOTORS	SHUNT MOTORS	SERIES MOTORS
A1 to A2	A1 to A2	A1 to A2
F1 to F2	F1 to F2	S1 to S2
S1 to S2		

The circuits are complete if the test light turns ON. The circuits are open if the test light does not turn ON. Service and repair the motor.

3. Check for a short circuit between windings. Connect the two test-light leads to the motor field and armature circuits as follows:

COMPOUND MOTORS	SHUNT MOTORS	SERIES MOTORS
A1 to F1, A2 to F2	A1 to F1	A1 to S1
A1 to F2, A2 to S1	A1 to F2	A2 to S1
A1 to S1, A2 to S2	A2 to F1	A2 to S2
A1 to S2, F1 to S1	A2 to F2	
A2 to F1, F1 to S2		

The circuit is shorted if the test light turns ON. The circuit is not shorted if the test light does not turn ON. Service and repair the motor.

Troubleshooting for Grounded or Shorted Commutator Windings

A commutator is grounded whenever one or more of its segments (bars) makes contact with the iron core of the commutator. An armature winding is also grounded whenever one or more windings makes contact with the iron core. **See Troubleshooting Commutator Windings.**

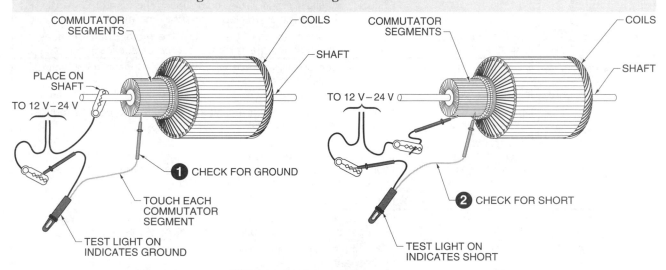

TROUBLESHOOTING COMMUTATOR WINDINGS

To troubleshoot for a grounded commutator or armature winding, apply the procedure:

1. Check for a commutator segment grounded to the shaft. Connect one lead of a test-light to the shaft. Touch the other test lead to each commutator segment. The test light turns ON if voltage passes through the commutator segments. The commutator or the winding is grounded if the test light turns ON. Service and repair the commutator.

2. Check for a short between adjacent commutator segments. Connect one test-light lead to a commutator segment. Touch the other test-light lead to each adjacent commutator segment. The test light turns ON if voltage passes through the commutator segments. The adjacent segments are shorted if the test light turns ON. Service and repair the commutator. There should also be no sparking or arcing between any segments. This indicates a partial short. Service and repair the commutator.

Troubleshooting Guides

Troubleshooting guides for motors state a problem, its possible cause(s), and corrective action(s) that may be taken. These guides may be used to quickly determine potential problems and possible courses of action. **See Direct Current Motor Troubleshooting Guide.**

DIRECT CURRENT MOTOR TROUBLESHOOTING GUIDE		
Problem	**Possible Cause**	**Corrective Action**
Motor will not start.	Blown fuse or open CB	Test the OCPD. If voltage is present at the input, but not the output of the OCPD, the fuse is blown or the CB is open. Check the rating of the OCPD. It should be at least 125% of the motor's FLC.
	Motor overload on starter tripped.	Allow overloads to cool. Reset overloads. If reset overloads do not start the motor, test the starter.
	No brush contact	Check brushes. Replace, if worn.
	Open control circuit between incoming power and motor	Check for cleanliness, tightness, and breaks. Use a voltmeter to test the circuit starting with the incoming power and moving to the motor terminals. Voltage generally stops at the problem area.
Fuse, CB, or overloads retrip after service.	Excessive load	If the motor is loaded to excess or is jammed, the circuit OCPD will open. Disconnect the load from the motor. If the motor now runs properly, check the load. If the motor does not run and the fuse or CB opens, the problem is with the motor or control circuit. Remove the motor from the control circuit and connect it directly to the power source. If the motor runs properly, the problem is in the control circuit. Check the control circuit. If the motor opens the fuse or CB again, the problem is in the motor. Replace or service the motor.
	Motor shaft does not turn.	Disconnect the motor from the load. If the motor shaft still does not turn, the bearings are frozen. Replace or service the motor.
Brushes chip or break.	Brush material is too weak or the wrong type for motor's duty rating.	Replace with better grade or type of brush. Consult manufacturer if problem continues.
	Brush face is overheating and losing brush bonding material.	Check for an overload on the motor. Reduce the load as required. Adjust brush holder arms.
	Brush holder is too far from commutator.	Too much space between the brush holder and the surface of the commutator allows the brush end to chip or break. Set correct space between brush holder and commutator.
	Brush tension is incorrect.	Adjust brush tension so the brush rides freely on the commutator.
Brushes spark.	Worn brushes	Replace worn brushes. Service the motor if rapid brush wear, excessive sparking, chipping, breaking, or chattering is present.
	Commutator is concentric.	Grind commutator and undercut mica. Replace commutator if necessary.
	Excessive vibration	Balance armature. Check brushes. They should be riding freely.
Rapid brush wear	Wrong brush material, type, or grade	Replace with brushes recommended by manufacturer.
	Incorrect brush tension	Adjust brush tension so the brush rides freely on the commutator.
Motor overheats.	Improper ventilation	Clean all ventilation openings. Vacuum or blow dirt out of motor with low-pressure, dry, compressed air.
	Motor is overloaded.	Check the load for binding. Check shaft straightness. Measure motor current under operating conditions. If the current is above the listed current rating, remove the motor. Remeasure the current under no-load conditions. If the current is excessive under load but not when unloaded, check the load. If the motor draws excessive current when disconnected, replace or service the motor.

 Refer to Activity 12-2—Troubleshooting DC Motors...

 *Refer to Quick Quiz® on CD-ROM*

 *Refer to Chapter 12 in the **Troubleshooting Electrical/Electronic Systems Workbook** for additional questions.*

Troubleshooting DC Motors

Name_____ **Date**_____

ACTIVITY 12-1—DC Motors

The voltage and current draw of a motor are measured when testing a motor. Complete the wiring of the motor. Connect the ammeter to measure the current flowing through the positive power line. Connect the voltmeter to measure the voltage applied to the motor.

1. Connect the series motor.

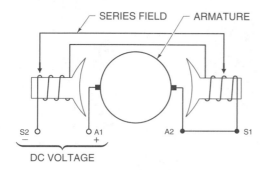

SERIES FIELD — ARMATURE

S2 A1
−   +

DC VOLTAGE

DC CLAMP-ON
AMMETER

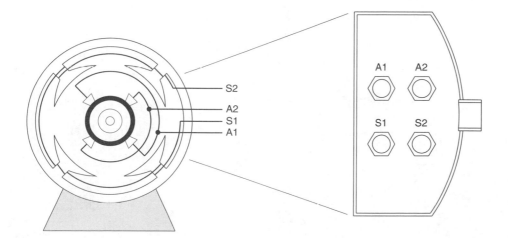

S2
A2
S1
A1

A1 A2

S1 S2

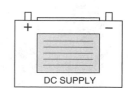

+   −

DC SUPPLY

249

2. Connect the DC-compound motor.

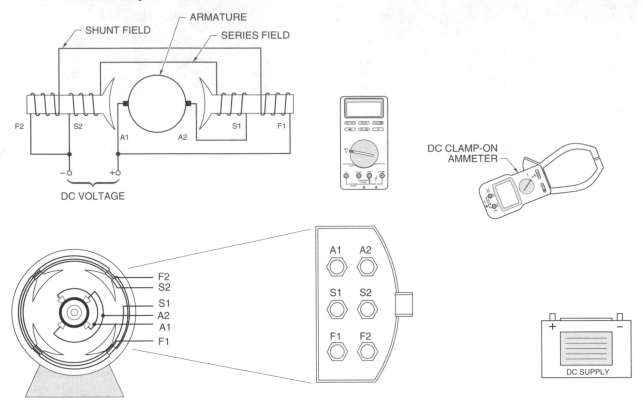

3. Connect the DC-shunt motor.

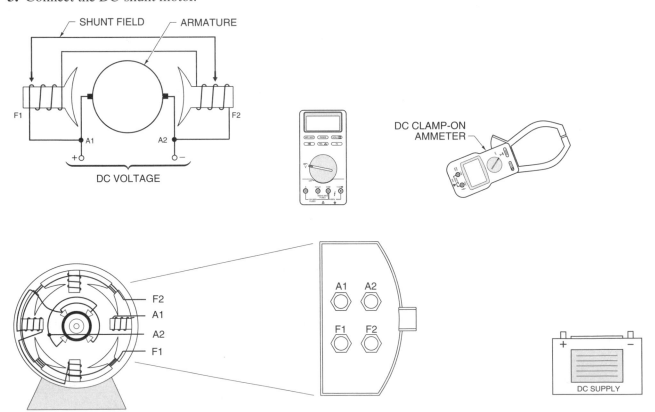

4. Complete the wiring of the motor. Connect Ammeter 1 to measure the current flowing through the shunt field. Connect Ammeter 2 to measure the current flowing through the armature.

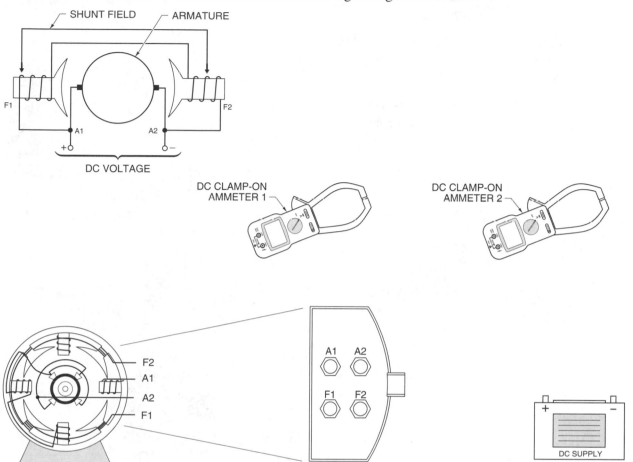

ACTIVITY 12-2—Troubleshooting DC Motors

Determine if the test light should be ON or OFF when the motor is good.

_____

1. Test Light 1 should be ___ when the motor is good.

_____

2. Test Light 2 should be ___ when the motor is good.

_____

3. Test Light 3 should be ___ when the motor is good.

_____ 4. Test Light 4 should be ___ when the motor is good.

_____ 5. Test Light 5 should be ___ when the motor is good.

_____ 6. Test Light 6 should be ___ when the motor is good.

_____ 7. Test Light 7 should be ___ when the motor is good.

_____ 8. Test Light 8 should be ___ when the motor is good.

_____ 9. Test Light 9 should be ___ when the motor is good.

_____ 10. Test Light 10 should be ___ when the motor is good.

APPLICATION 13-1—AC Motors

An *alternating current (AC) motor* is a motor that uses alternating current to produce rotation. AC motors have a rotor and stator. A *rotor* is the rotating part of an AC motor. The rotor rotates the motor shaft and delivers the work. A *stator* is the stationary part of an AC motor to which the power lines are connected. The stator produces a rotating magnetic field. **See AC Motor.**

AC motors are 1ϕ or 3ϕ. Single-phase motors include shaded-pole, split-phase, capacitor-start, capacitor-run, and capacitor start-and-run motors. Single-phase motors require a way of starting the motor. The method used to start a 1ϕ motor determines the motor name.

Shaded-Pole Motors

A *shaded-pole motor* is an AC motor that uses a shaded stator pole for starting. Shading the stator poles is the simplest method used to start a 1ϕ motor. **See Shaded-Pole Motor.**

Shading the poles is accomplished by applying a short-circuited conductor on one side of each stator pole. The shaded pole is normally a solid single turn of copper wire placed around a portion of the main pole laminations.

The shaded pole delays the magnetic flux in the area of the pole that is shaded. Shading causes the magnetic flux at the pole area to be positioned about 90° apart from the magnetic flux of the main stator pole. The offset flux causes the rotor to move from the main pole toward the shaded pole. This movement determines the starting direction of the motor.

Shaded-pole motors are commonly 1/20 HP or less and have low starting torque. Common applications of shaded-pole motors include small appliance cooling fans found in computers and stereos.

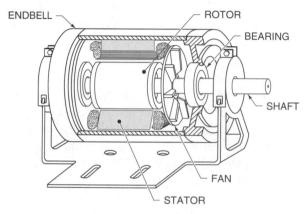

AC MOTOR

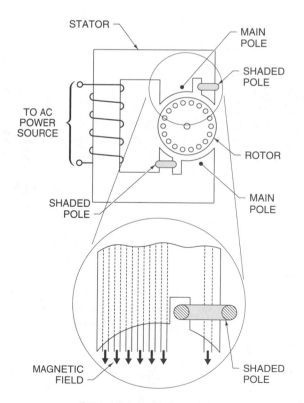

SHADED-POLE MOTOR

Split-Phase Motors

A *split-phase motor* is an AC motor that can run on one or more phases. A split-phase motor has a main winding and an auxiliary winding. The main winding is the running winding. The auxiliary winding is the starting winding. The two windings are placed in the stator slots and positioned 90° apart. **See Split-Phase Motor.**

The running winding is made of larger wire and has a greater number of turns than the starting winding. When the motor is first connected to power, the inductive reactance of the running winding is higher and the resistance is lower than the starting winding. *Inductive reactance* is the opposition to the flow of alternating current in a circuit due to inductance.

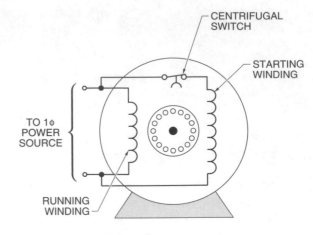

SPLIT-PHASE MOTOR

The starting winding is made of relatively small wire and has a fewer number of turns than the running winding. When the motor is first connected to power, the inductive reactance of the starting winding is lower and the resistance is higher than the running winding.

When power is first applied, both the running winding and the starting winding are energized. The running winding current lags the starting winding current because of its different inductive reactance. This produces a phase difference between the starting and running windings. A 90° phase difference is required to produce maximum starting torque, but the phase difference is commonly much less. A rotating magnetic field is produced because the two windings are out of phase.

The rotating magnetic field starts the rotor turning. With the running and starting windings out of phase, the current changes in magnitude and direction, and the magnetic field moves around the stator. This movement forces the rotor to rotate with the rotating magnetic field.

Centrifugal Switch

To minimize energy loss and prevent heat build-up in the starting winding once the motor is started, a centrifugal switch is used to remove the starting winding when the motor reaches a set speed. A *centrifugal switch* is a switch that opens to disconnect the starting winding when the rotor reaches a certain preset speed and to reconnect the starting winding when the speed falls below a preset value. In most motors the centrifugal switch is located inside the enclosure on the shaft. **See Centrifugal Switch.**

As the motor shaft rotates and the speed of the motor accelerates, the switch is activated by centrifugal force. *Centrifugal force* is the force that moves rotating bodies away from the center of rotation. For some motors the centrifugal switch is often located outside the enclosure for easy repair.

The centrifugal switch is connected in series with the starting winding. The centrifugal switch automatically de-energizes the starting winding at a set speed. The set speed is usually about 60% to 80% of the running speed. After the starting winding is removed, the motor continues to operate on the running winding only.

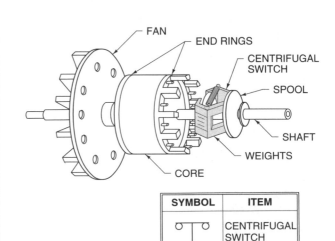

SYMBOL	ITEM
○─┬─○ └	CENTRIFUGAL SWITCH

CENTRIFUGAL SWITCH

The split-phase motor is one of the oldest and most common motor types. A split-phase motor is used in applications such as fans, business machines, machine tools, and centrifugal pumps. A split-phase motor is generally available in sizes ranging from 1/30 HP to 1/20 HP.

Thermal Protection

Many motors contain a thermal switch in addition to a centrifugal switch. A *thermal switch* is a switch that operates its contacts when a preset temperature is reached. A thermal switch is located inside a motor and is used to protect the motor windings from burnout. **See Thermal Protection.**

A thermal switch is activated by high temperatures and automatically removes both the starting and running windings from the power source. A higher-than-normal temperature may be caused by improper ventilation, too high of a motor load, high ambient temperature, or a mechanical problem that prevents rotation.

During normal operation, a thermal switch is in the closed position. As more current than normal is allowed to pass through the motor windings, the switch begins to heat. At a preset temperature, the switch opens. This automatically removes the motor windings from power. When the motor windings are removed from power, they begin to cool. The thermal switch closes as it cools, and the motor is automatically restarted.

The thermal switch constantly recycles the motor ON and OFF if the problem that caused the original overheating is not corrected.

Reversing

The direction of the connections for either the starting or the running windings are reversed to change the direction of rotation of a split-phase motor. Reversing the starting winding is the industrial standard for reversing the direction of rotation of a 1φ motor.

Capacitor Motors

A *capacitor motor* is a 1φ motor with a capacitor connected in series with the stator windings to produce phase displacement in the starting winding. The capacitor is added to provide a higher starting torque at lower starting current than is delivered by a split-phase motor. Applications of capacitor motors include refrigerators, air conditioners, air compressors, and some power tools.

In a capacitor motor, the capacitor causes the starting winding current to lead the applied voltage by about 40°. Since the starting and running

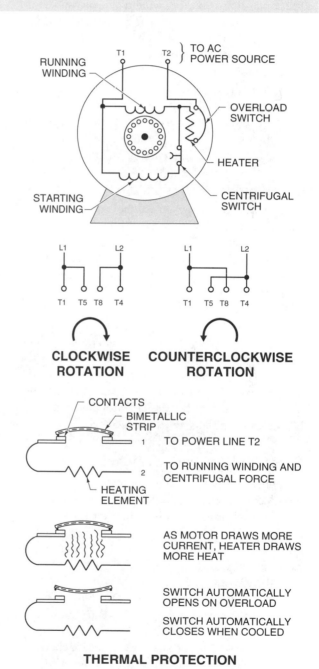

THERMAL PROTECTION

windings are about 90° out of phase, the motor operating characteristics are improved, a higher starting torque is produced, and the motor has a better power factor with a lower current draw. Three types of capacitor motors are the capacitor-start, capacitor-run, and capacitor start-and-run motors.

Capacitor-Start Motors

A *capacitor-start motor* is a motor that has the capacitor connected in series with the starting winding. The capacitor is connected only during starting. The capacitor and starting winding are disconnected from the power by a centrifugal switch at a set speed. **See Capacitor-Start Motor.**

A capacitor-start motor develops considerably more locked rotor torque per ampere than a split-phase motor because of the capacitor in the circuit. *Locked rotor torque* is the torque a motor produces when the rotor is stationary and full power is applied to the motor. Capacitor-start motors are used to drive power tools, pumps, and small machines.

Capacitor-Run Motors

A *capacitor-run motor* leaves the starting winding and capacitor in the circuit at all times. The starting winding is not removed as the motor speed increases because there is no centrifugal switch. A capacitor-run motor has a lower full-load speed than a capacitor-start motor because the capacitor remains in the circuit at all times. **See Capacitor-Run Motor.**

The advantage of leaving the capacitor in the circuit is that the motor has more running torque than a capacitor-start motor or split-phase motor. This allows a capacitor-run motor to be used for loads that require a high running torque. Capacitor-run motors are used to drive shaft-mounted fans and blowers ranging in size from ¹⁄₁₆ HP to ¹⁄₃ HP.

Capacitor Start-and-Run Motors

A *capacitor start-and-run motor* uses two capacitors. A capacitor start-and-run motor starts with one value capacitor in series with the starting winding and runs with a different value capacitor in series with the starting winding. This motor is also known as a dual-capacitor motor. **See Capacitor Start-and-Run Motor.**

A capacitor start-and-run motor has the same starting torque as a capacitor-start motor. A capacitor-start-and-run motor has more running torque than a capacitor-start motor or capacitor-run motor because the capacitance is better matched for starting and running.

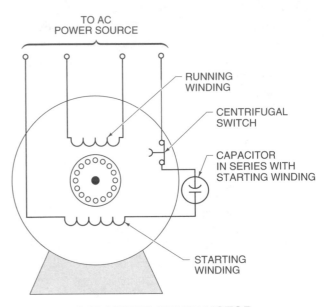

CAPACITOR-START MOTOR

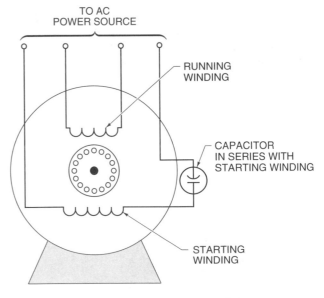

CAPACITOR-RUN MOTOR

In a typical two-capacitor motor, one capacitor is used for starting the motor and the other capacitor remains in the circuit while the motor is running. A larger-value capacitor is used for starting and a smaller-value capacitor is used for running. Capacitor start-and-run motors are used to run refrigerators and air compressors.

Three-Phase Motors

Three-phase motors require a rotating magnetic field. A 3ϕ motor does not require any additional components to produce this rotating magnetic field. The rotating magnetic field is set up automatically in the stator when the motor is connected to 3ϕ power lines. The coils in the stator are connected to form three separate windings (phases). Each phase contains one-third of the total number of individual coils in the motor. These windings (or phases), are the A phase, B phase, and C phase. **See 3ϕ Motor Phases.**

Each phase is positioned 120° from the other phases in the motor. A rotating magnetic field is produced in the stator when each phase reaches its peak value, 120° apart from the other phases.

Electrical degrees and mechanical degrees differ. In electric motors and generators, the distance traveled past one pole represents 180 electrical degrees. One revolution has 360 mechanical degrees.

For example, during one revolution in a 4-pole motor, a rotor passes from a north pole, through a south pole, through a north pole, through a south pole, and back to the original north pole, completing 720 electrical degrees (4 × 180 electrical degrees = 720 electrical degrees). **See Mechanical Versus Electrical Degrees.**

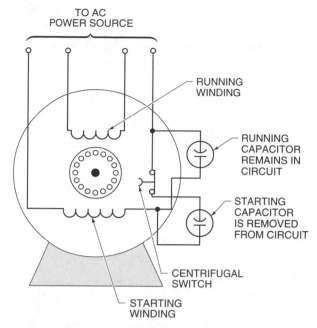

CAPACITOR START-AND-RUN MOTOR

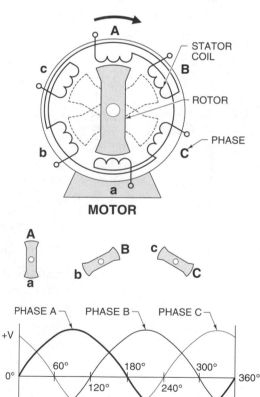

3ϕ MOTOR PHASES

MECHANICAL VERSUS ELECTRICAL DEGREES		
Mechanical Degrees In One Revolution	Motor Poles x 180 =	Electrical Degrees In One Revolution
360		2 x 180 = 360
360		4 x 180 = 720
360		6 x 180 = 1080
360		8 x 180 = 1440

Single-Voltage, 3ϕ Motors

The windings must be connected to the proper voltage to develop a rotating magnetic field in a 3ϕ motor. This voltage level is determined by the manufacturer and stamped on the motor nameplate.

A *single-voltage, 3ϕ motor* is a motor that operates at only one voltage level. Single-voltage, 3ϕ motors are less expensive to manufacture than dual-voltage, 3ϕ motors, but are limited to locations having the same voltage as the motor. Typical single-voltage, 3ϕ motor ratings are 230 V, 460 V, and 575 V. Other single-voltage, 3ϕ motor ratings are 200 V, 208 V, 220 V, and 280 V.

Wye-Connected, 3ϕ Motors

A *wye-connected, 3ϕ motor* has one end of each of the three phases internally connected to the other phases. The remaining end of each phase is brought out externally and connected to the incoming power source. **See Wye-Connected, 3ϕ Motor.**

The leads, which are brought out externally, are labeled terminals one, two, and three (T1, T2, and T3). When connecting a wye-connected, 3ϕ motor to the 3ϕ power lines, the power lines and motor terminals are connected L1 to T1, L2 to T2, and L3 to T3.

Delta-Connected, 3ϕ Motors

A *delta-connected, 3ϕ motor* has each phase wired end-to-end to form a completely closed loop circuit. At each point where the phases are connected, leads are brought out externally to form T1, T2, and T3. **See Delta-Connected, 3ϕ Motor.**

T1, T2, and T3, are connected to the three power lines. L1 is connected to T1, L2 to T2, and L3 to T3. The 3ϕ line supplying power to the motor must have the same voltage and frequency rating as the motor.

Dual-Voltage, 3ϕ Motors

A *dual-voltage, 3ϕ motor* is a motor that operates at more than one voltage level. Most 3ϕ motors are manufactured so they can be connected to either of two voltages. The purpose in making motors for two voltages is to enable the same motor to be used with two different power line voltages.

A typical dual-voltage, 3ϕ motor rating is 230/460 V. Other common dual-voltage, 3ϕ motor ratings are 240/480 V and 208-230/460 V. The dual-voltage rating of a motor is listed on the nameplate of the motor. If both voltages are available, the higher voltage is usually preferred

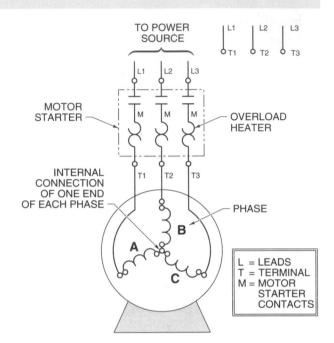

WYE-CONNECTED, 3ϕ MOTOR

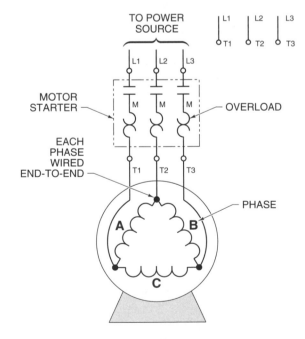

DELTA-CONNECTED, 3ϕ MOTOR

because the motor uses the same amount of power, given the same horsepower output, for either high or low voltage. As the voltage is doubled, the current draw on the power lines is cut in half. With the reduced current, the wire size is reduced and the material cost is decreased.

Dual-Voltage, Wye-Connected, 3ϕ Motors

In a dual-voltage, wye-connected, 3ϕ motor, each phase coil (A, B, and C) is divided into two equal parts. By dividing the phase coils in two, nine terminal leads are available. These motor leads are marked terminals one through nine (T1-T9). The nine terminal leads may be connected for high or low voltage.

To connect a dual-voltage, wye-connected, 3ϕ motor for high voltage, connect L1 to T1, L2 to T2, and L3 to T3 at the motor starter. Using wire nuts, tie T4 to T7, T5 to T8, and T6 to T9. By making these connections, the individual coils in each phase are connected in series. The applied voltage is divided equally among the coils because the coils are connected in series. **See Dual-Voltage, Wye-Connected, 3ϕ Motor.**

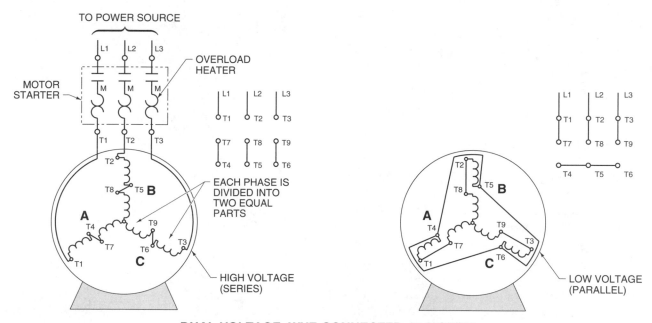

DUAL-VOLTAGE, WYE-CONNECTED, 3ϕ MOTOR

To connect a dual-voltage, wye-connected, 3ϕ motor for low voltage, connect L1 to T1 and T7, L2 to T2 and T8, and L3 to T3 and T9 at the motor starter. Using a wire nut, tie T4, T5, and T6 together. By making these connections, the individual coils in each phase are connected in parallel. The applied voltage is present across each set of coils because the coils are connected in parallel.

Dual-Voltage, Delta-Connected, 3ϕ Motors

In a dual-voltage, delta-connected, 3ϕ motor, each phase coil (A, B, and C) is divided into two equal parts. By dividing the phase coils in two, nine terminal leads are available. These motor leads are marked terminals one through nine (T1-T9). The nine terminal leads can be connected for high or low voltage.

To connect a dual-voltage, delta-connected, 3φ motor for high voltage, connect L1 to T1, L2 to T2, and L3 to T3 at the motor starter. Using wire nuts, tie T4 to T7, T5 to T8, and T6 to T9. By making these connections, the individual coils in each phase are connected in series. Since the coils are connected in series, the applied voltage divides equally among the coils. **See Dual-Voltage, Delta-Connected, 3φ Motor.**

To connect a dual-voltage, delta-connected, 3φ motor for low voltage, connect L1 to T1, L2 to T2, and L3 to T3 at the motor starter. Using wire nuts, tie T1 to T7 and T6, T2 to T8 and T4, and T3 to T9 and T5. By making these connections, the individual coils in each phase are connected in parallel. Since the coils are connected in parallel, the applied voltage is present across each set of coils.

Reversing

The direction of rotation of 3φ motors is reversed by interchanging any two of the 3φ power lines to the motor. The industrial standard is to interchange T1 and T3. This standard holds true for all 3φ motors.

Interchanging T1 and T3 is a standard for safety reasons. When first connecting a motor, the direction of rotation is not usually known until the motor is started. It is common practice to temporarily connect the motor to determine the direction of rotation before making permanent connections. Motor leads of temporary connections are not taped. By always interchanging T1 and T3, T2 may be permanently connected to L2, creating an insulated barrier between T1 and T3.

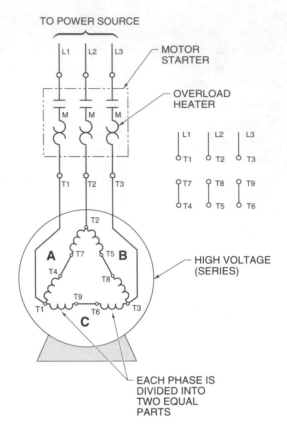

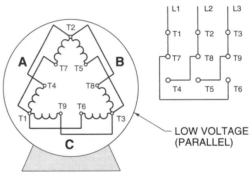

DUAL-VOLTAGE, DELTA-CONNECTED, 3φ MOTOR

 Refer to **Activity 13-1—AC Motors** ....

APPLICATION 13-2—Troubleshooting AC Motors

Most problems with 1φ motors involve the centrifugal switch, thermal switch, or capacitor(s). A motor is usually serviced and repaired if the problem is in the centrifugal switch, thermal switch, or capacitor. A motor is usually replaced if the motor is more than 10 years old and less than 1 HP. A motor is almost always replaced if the motor is less than ⅛, HP. Three-phase motors usually operate for many years without any problems because 3φ motors have fewer components that may malfunction than other motor types.

The motor is serviced or replaced if a 3φ motor is the problem. Servicing usually requires that the motor be sent to a motor repair shop for rewinding. A motor is replaced if the motor is less than 1 HP and more than five years old. A motor may be serviced or replaced if the motor is more than 1 HP but less than 5 HP. A motor is usually serviced if the motor is more than 5 HP.

Troubleshooting Shaded-Pole Motors

Shaded-pole motors that fail are usually replaced. The reason for the motor failure should be discovered. For example, replacing a motor because it failed due to a jammed load, does not solve the problem, etc. **See Troubleshooting Shaded-Pole Motors.**

To troubleshoot a shaded-pole motor, apply the procedure:

1. Turn the power to the motor OFF. Visually inspect the motor. Replace the motor if it is burned, the shaft is jammed, or if there is any sign of damage.

2. Check the stator winding. The stator winding is the only electrical circuit that may be tested without taking the motor apart. Measure the resistance of the stator winding. Set the ohmmeter to the lowest scale for taking the reading. The winding is open if the ohmmeter indicates an infinity reading. Replace the motor. The winding is shorted if the ohmmeter indicates a zero reading. Replace the motor. The winding may still be good if the ohmmeter indicates a low-resistance reading. Check the winding with a megohmmeter before replacing.

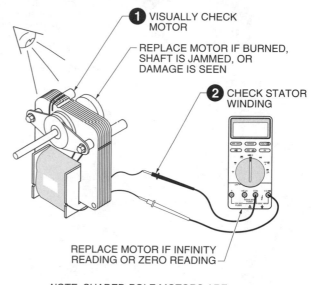

1 VISUALLY CHECK MOTOR

REPLACE MOTOR IF BURNED, SHAFT IS JAMMED, OR DAMAGE IS SEEN

2 CHECK STATOR WINDING

REPLACE MOTOR IF INFINITY READING OR ZERO READING

NOTE: SHADED-POLE MOTORS ARE SMALL AND INEXPENSIVE. FEW TESTS ARE COST EFFICIENT

TROUBLESHOOTING SHADED-POLE MOTORS

Troubleshooting Split-Phase Motors

Some split-phase motors include a thermal switch that automatically turns the motor OFF when it overheats. Thermal switches may have a manual reset or an automatic reset. Caution should be taken with any motor that has an automatic reset should the motor automatically restart at any time. **See Troubleshooting Split-Phase Motors.**

To troubleshoot a split-phase motor, apply the procedure:

1. Turn power to motor OFF. Visually inspect the motor. Replace the motor if it is burned, the shaft is jammed, or if there is any sign of damage.

2. Check to determine if the motor is controlled by a thermal switch. Reset the thermal switch and turn the motor ON if the thermal switch is manual.

3. Use a voltmeter to check for voltage at the motor terminals if the motor does not start. The voltage should be within 10% of the motor's listed voltage. Troubleshoot the circuit leading to the motor if the voltage is not correct. Turn the power to the motor OFF so the motor may be tested if the voltage is correct.

4. Turn the handle of the safety switch or combination starter OFF. Lockout and tag the starting mechanism per company policy.

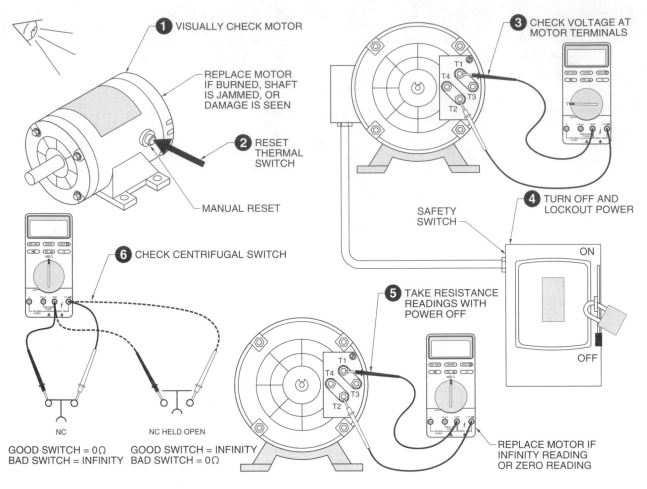

TROUBLESHOOTING SPLIT-PHASE MOTORS

5. With power OFF, connect the ohmmeter to the same motor terminals from which the incoming power leads were disconnected. The ohmmeter reads the resistance of the starting and running windings. Their combined resistance is less than the resistance of either winding alone because the windings are in parallel. A short is present if the meter reads zero. An open circuit is present if the meter reads infinity. Replace the motor. *Note:* The motor size is too small for a repair to be cost efficient.

6. Visually inspect the centrifugal switch for signs of burning or broken springs. Service or replace the switch if any obvious signs of problems are present. Check the switch using an ohmmeter if no obvious signs of problems are present. Manually operate the centrifugal switch. (The endbell on the switch side may have to be removed.) The resistance on the ohmmeter decreases if the motor is good. A problem exists if the resistance does not change. Continue to check the motor to determine the problem.

Thermal Switch Testing

A thermal switch removes the starting and running motor windings from the circuit at a preset temperature. When the windings cool, the contacts close. When a thermal switch is not operating properly, a motor either does not operate or operates without thermal protection. **See Thermal Switch Testing.**

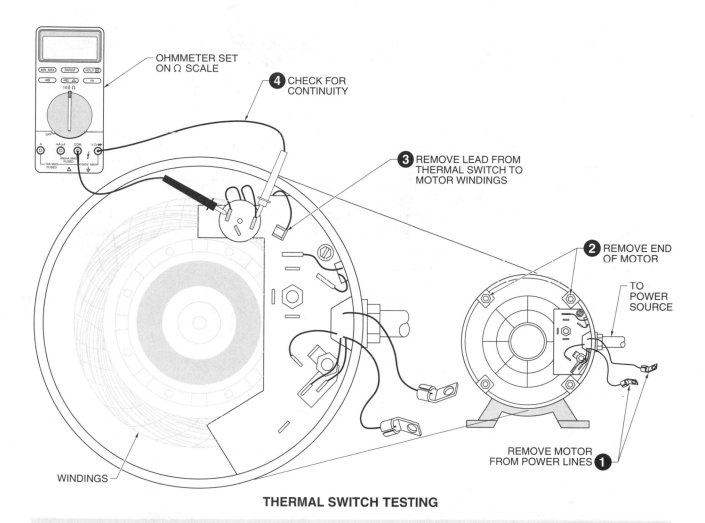

THERMAL SWITCH TESTING

To test a thermal switch, apply the procedure:

1. Remove the motor from the power lines and let cool.

2. Remove the end of the motor that includes the thermal switch.

3. Remove one of the leads running from the thermal switch to the motor windings.

4. In a motor containing a two-terminal thermal switch, check for continuity (very low resistance) using an ohmmeter. Set the ohmmeter on the lowest resistance scale and check across the switch contacts. The contacts are open and the switch is defective if a high-resistance reading is obtained.

   In a motor containing a three-terminal thermal switch, check for continuity across the switch contacts using an ohmmeter set on the lowest scale. The contacts are open and the switch is defective if a high-resistance reading is obtained. Check for continuity across the heater element. The heater element is open and defective if a high-resistance reading is obtained.

   In a motor containing a four-terminal thermal switch, check for continuity across the switch contacts and the heater element using an ohmmeter set on the lowest scale. The contacts or the heater element is open and defective if a high-resistance reading is obtained. Replace the defective component.

Troubleshooting Capacitor Motors

Troubleshooting capacitor motors is similar to troubleshooting split-phase motors. The only additional device to be considered is the capacitor. Capacitors have a limited life and are often the problem in capacitor motors. Capacitors may have a short circuit, an open circuit, or may deteriorate to the point that they must be replaced.

Deterioration may also change the value of a capacitor, which may cause additional problems. When a capacitor short-circuits, the winding in the motor may burn out. When a capacitor deteriorates or opens, the motor has poor starting torque. Poor starting torque may prevent the motor from starting, which usually trips the overloads.

All capacitors are made with two conducting surfaces separated by dielectric material. *Dielectric material* is a medium in which an electric field is maintained with little or no outside energy supply. Dielectric material is used to insulate the conducting surfaces of a capacitor. Capacitors are either oil or electrolytic. Oil capacitors are filled with oil and sealed in a metal container. The oil serves as the dielectric material.

More motors use electrolytic capacitors than oil capacitors. Electrolytic capacitors are formed by winding two sheets of aluminum foil separated by pieces of thin paper impregnated with an electrolyte. An *electrolyte* is a conducting medium in which the current flow occurs by ion migration. The electrolyte is used as the dielectric material. The aluminum foil and electrolyte are encased in a cardboard or aluminum cover. A vent hole is provided to prevent a possible explosion in the event the capacitor is shorted or overheated. AC capacitors are used with capacitor motors. Capacitors that are designed to be connected to AC have no polarity. **See Troubleshooting Capacitor Motors.**

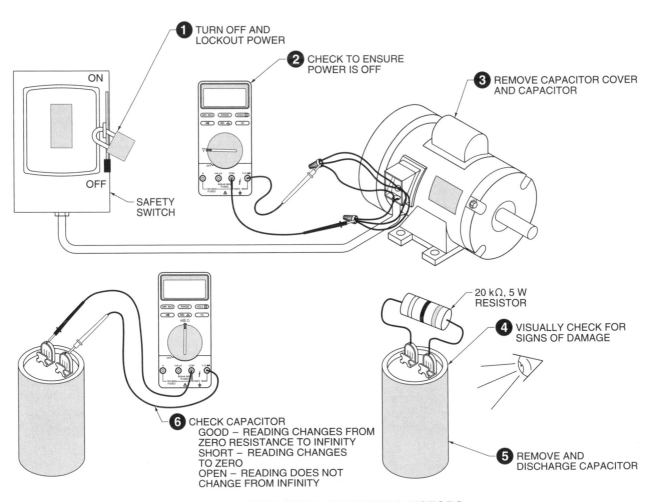

1 TURN OFF AND LOCKOUT POWER

2 CHECK TO ENSURE POWER IS OFF

3 REMOVE CAPACITOR COVER AND CAPACITOR

ON

OFF

SAFETY SWITCH

20 kΩ, 5 W RESISTOR

4 VISUALLY CHECK FOR SIGNS OF DAMAGE

5 REMOVE AND DISCHARGE CAPACITOR

6 CHECK CAPACITOR
GOOD – READING CHANGES FROM ZERO RESISTANCE TO INFINITY
SHORT – READING CHANGES TO ZERO
OPEN – READING DOES NOT CHANGE FROM INFINITY

TROUBLESHOOTING CAPACITOR MOTORS

To troubleshoot a capacitor motor, apply the procedure:

1. Turn the handle of the safety switch or combination starter OFF. Lockout and tag the starting mechanism per company policy.

2. Using a voltmeter, measure the voltage at the motor terminals to ensure the power is OFF.

3. Capacitors are located on the outside frame of the motor. Remove the cover of the capacitor. **Caution:** A good capacitor will hold a charge, even when power is removed.

4. Visually check the capacitor for leakage, cracks, or bulges. Replace the capacitor if present.

5. Remove the capacitor from the circuit and discharge it. To safely discharge a capacitor, place a 20,000 Ω, 5 W resistor across the terminals for five seconds.

6. After the capacitor is discharged, connect the ohmmeter leads to the capacitor terminals. The ohmmeter indicates the general condition of the capacitor. A capacitor is either good, shorted, or open.

   - **Good Capacitor.** The reading changes from zero resistance to infinity. When the reading reaches the halfway point, remove one of the leads and wait 30 seconds. When the lead is reconnected, the reading should change back to the halfway point and continue to infinity. This demonstrates that the capacitor can hold a charge. The capacitor cannot hold a charge and must be replaced when the reading changes back to zero resistance.

   - **Shorted Capacitor.** The reading changes to zero and does not move. The capacitor is bad and must be replaced.

   - **Open Capacitor.** The reading does not change from infinity. The capacitor is bad and must be replaced.

Motor Capacitor Failure

Single-phase motor capacitors fail because of excessive temperature, excessive voltage, excessive duty cycles, internal corrosion, or an open fuse. A substitute capacitor is placed in the circuit to check the motor if a capacitor is suspected to be defective. The original capacitor is defective if motor operation improves. Capacitors normally operate in ambient temperatures up to 80°C (176°F). A capacitor's life is shortened at high temperatures. Operating at low temperatures does not harm a capacitor. The capacitance of a capacitor decreases at temperatures below 0°C.

Excessive voltage causes arcing in the capacitor that leads to permanent damage. Excessive voltage is applied by connecting a motor to a supply voltage higher than the capacitor voltage rating or by a faulty centrifugal switch. The voltage rating of a capacitor is normally listed on the capacitor. The capacitor voltage rating must be equal to or greater than the supply voltage applied to the motor.

Caution: Never replace a capacitor with one that has a lower voltage rating than the original. Never use a capacitor with a voltage rating higher than 10% of the original. Using a capacitor with too high or too low of a voltage rating increases the amperage and wattage draw of the motor. This increase can burn out the motor windings. **See Capacitor Ratings. See Capacitor Ratings in Appendix.**

Faulty centrifugal-switch contacts chatter before opening. The voltage on the capacitor is several times higher than the supply voltage when centrifugal-switch contacts chatter. The high voltage is induced by the collapsing magnetic field of the starting winding.

CAPACITOR RATINGS			
Typical Ratings*	Dimensions**		Model Number***
	Diameter	Length	
88-106	1⁷⁄₁₆	2³⁄₄	EC8815
108-130	1⁷⁄₁₆	2³⁄₄	EC10815
130-156	1⁷⁄₁₆	2³⁄₄	EC13015
145-174	1⁷⁄₁₆	2³⁄₄	EC14515

* in µF
** in inches
*** model numbers vary by manufacturer

A capacitor is damaged by normal voltage if the motor is started many times or if the load applied to the motor requires a long starting time. A starting capacitor is normally in a circuit for less than 3 seconds. Extra acceleration time causes excessive heat buildup in the capacitor, which reduces capacitor life. A capacitor with an open seal allows moisture absorption. Moisture absorption leads to corrosion that destroys the film inside the capacitor.

Many capacitors have an internal fuse that opens when excessive voltage or current is applied to the capacitor. This internal fuse is also opened by improper servicing of the motor. A capacitor with an open internal fuse shows a resistance reading of infinity when checked with an ohmmeter.

Troubleshooting 3ϕ Motors

The extent of troubleshooting a 3ϕ motor depends on the motor's application. Testing is normally limited to checking the voltage at the motor if a motor is used in an application that is critical to an operation or production. The motor is assumed to be the problem if the voltage is present and correct. Unless it is very large, the motor is normally replaced at this time so production may continue. Further tests may be made to determine the exact problem if time is not a critical factor. **See Troubleshooting 3ϕ Motors.**

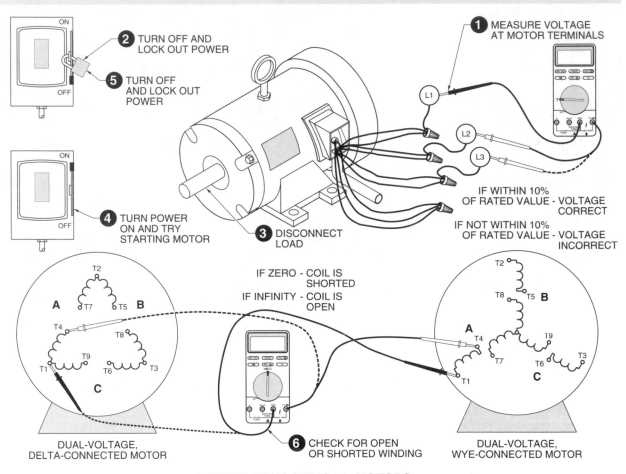

TROUBLESHOOTING 3ϕ MOTORS

To troubleshoot a 3ϕ motor, apply the procedure:
1. Using a voltmeter, measure the voltage at the motor terminals. The motor must be checked if the voltage is present and at the correct level on all three phases. The incoming power supply must be checked if the voltage is not present on all three phases.
2. Turn the handle of the safety switch or combination starter OFF if voltage is present but the motor is not operating. Lockout and tag the starting mechanism per company policy.
3. Disconnect the motor from the load.

4. After the load is disconnected, turn power ON to try restarting the motor. Check the load if the motor starts.

5. Turn the motor OFF and lockout the power if the motor does not start.

6. With an ohmmeter, check the motor windings for any opens or shorts. Take a resistance reading of the T1-T4 coil. This coil must have a resistance reading. The coil is shorted when the reading is zero. The coil is open when the reading is infinity. Since the coil winding is made of wire only, the resistance is low. There is resistance on a good coil winding, however. The larger the motor, the smaller the resistance reading.

After the resistance of one coil has been found, the basic electrical laws of series and parallel circuits are applied. When measuring the resistance of two coils in series, the total resistance is twice the resistance of one coil. When measuring the resistance of two coils in parallel, the total resistance is one half the resistance of one coil.

Troubleshooting Guides

Troubleshooting guides for motors state a problem, its possible cause(s), and corrective action(s) that may be taken. These guides may be used to quickly determine potential problems and possible courses of action. **See Troubleshooting Guides.**

SHADED-POLE MOTOR TROUBLESHOOTING GUIDE		
Problem	**Possible Cause**	**Corrective Action**
Motor will not start.	Blown fuse or open CB	Test the overcurrent protection device (OCPD). If voltage is present at the input, but not the output of the OCPD, the fuse is blown or the CB is open. Check the rating of the OCPD. It should be at least 125% of the motor's FLC.
	Motor overload on starter tripped	Allow overloads to cool. Reset overloads. If reset overloads do not start the motor, test the starter.
	Low or no voltage applied to motor	Check the voltage at the motor terminals. The voltage must be present and within 10% of the motor nameplate voltage. If voltage is present at the motor but the motor is not operating, remove the motor from the load the motor is driving. Reapply power to the motor. If the motor runs, the problem is with the load. If the motor does not run, the problem is with the motor. Replace or service the motor.
	Open control circuit between incoming power and motor	Check for cleanliness, tightness, and breaks. Use a voltmeter to test the circuit starting with the incoming power and moving to the motor terminals. Voltage generally stops at the problem area.
Fuse, CB, or overloads retrip after service.	Excessive load	If the motor is loaded to excess or jammed, the circuit OCPD will open. Disconnect the load from the motor. If the motor now runs properly, check the load. If the motor does not run and the fuse or CB opens, the problem is with the motor or control circuit. Remove the motor from the control circuit and connect it directly to the power source. If the motor runs properly, the problem is in the control circuit. Check the control circuit. If the motor opens the fuse or CB again, the problem is in the motor. Replace or service the motor.
Excessive noise	Unbalanced motor or load	An unbalanced motor or load causes vibration, which causes noise. Realign the motor and load. Check for excessive end play or loose parts. If the shaft is bent, replace the rotor or motor.
	Dry or worn bearings	Dry or worn bearings cause noise. Bearings may be dry due to dirty oil, oil not reaching the shaft, or motor overheating. Oil bearings as recommended. If noise remains, replace the bearings or motor.
	Excessive grease	Ball bearings that have excessive grease may cause bearings to overheat. Overheated bearings cause noise. Remove excess grease.

SPLIT-PHASE MOTOR TROUBLESHOOTING GUIDE

Problem	Possible Cause	Corrective Action
Motor will not start.	Thermal cutout switch is open.	Reset the thermal switch. **Caution:** Resetting the thermal switch may automatically start the motor.
	Blown fuse or open CB	Test the OCPD. If voltage is present at the input, but not the output of the OCPD, the fuse is blown or the CB is open. Check the rating of the OCPD. It should be at least 125% of the motor's FLC.
	Motor overload on starter tripped.	Allow overloads to cool. Reset overloads. If reset overloads do not start the motor, test the starter.
	Low or no voltage applied to motor	Check the voltage at the motor terminals. The voltage must be present and within 10% of the motor nameplate voltage. If voltage is present at the motor but the motor is not operating, remove the motor from the load the motor is driving. Reapply power to the motor. If the motor runs, the problem is with the load. If the motor does not run, the problem is with the motor. Replace or service the motor.
	Open control circuit between incoming power and motor	Check for cleanliness, tightness, and breaks. Use a voltmeter to test the circuit starting with the incoming power and moving to the motor terminals. Voltage generally stops at the problem area.
	Starting winding not receiving power	Check the centrifugal switch to make sure it connects the starting winding when the motor is OFF.
Fuse, CB, or overloads retrip after service.	Blown fuse or open CB	Test the OCPD. If voltage is present at the input, but not the output of the OCPD, the fuse is blown or the CB is open. Check the rating of the OCPD. It should be at least 125% of the motor's FLC.
	Motor overload on starter tripped.	Allow overloads to cool. Reset overloads. If reset overloads do not start the motor, test the starter.
	Low or no voltage applied to motor	Check the voltage at the motor terminals. The voltage must be present and within 10% of the motor nameplate voltage. If voltage is present at the motor but the motor is not operating, remove the motor from the load the motor is driving. Reapply power to the motor. If the motor runs, the problem is with the load. If the motor does not run, the problem is with the motor. Replace or service the motor.
	Open control circuit between incoming power and motor	Check for cleanliness, tightness, and breaks. Use a voltmeter to test the circuit starting with the incoming power and moving to the motor terminals. Voltage generally stops at the problem area.
	Motor shaft does not turn.	Disconnect the motor from the load. If the motor shaft still does not turn, the bearings are frozen. Replace or service the motor.
Motor produces electric shock.	Broken or disconnected ground strap	Connect or replace ground strap. Test for proper ground.
	Hot power lead at motor connecting terminals is touching motor frame.	Disconnect the motor. Open the motor terminal box and check for poor connections, damaged insulation, or leads touching the frame. Service and test motor for ground.
	Motor winding shorted to frame	Remove, service, and test motor.
Motor overheats.	Starting windings are not being removed from circuit as motor accelerates.	When the motor is turned OFF, a distinct click should be heard as the centrifugal switch closes.
	Improper ventilation	Clean all ventilation openings. Vacuum or blow dirt out of motor with low-pressure, dry, compressed air.
	Motor is overloaded.	Check the load for binding. Check shaft straightness. Measure motor current under operating conditions. If the current is above the listed current rating, remove the motor. Remeasure the current under no-load conditions. If the current is excessive under load but not when unloaded, check the load. If the motor draws excessive current when disconnected, replace or service the motor.

	Dry or worn bearings	Dry or worn bearings cause noise. The bearings may be dry due to dirty oil, oil not reaching the shaft, or motor overheating. Oil the bearings as recommended. If noise remains, replace the bearings or the motor.
	Dirty bearings	Clean or replace bearings.
Excessive noise	Excessive end play	Check end play by trying to move the motor shaft in and out. Add end-play washers as required.
	Unbalanced motor or load	An unbalanced motor or load causes vibration, which causes noise. Realign the motor and load. Check for excessive end play or loose parts. If the shaft is bent, replace the rotor or motor.
	Dry or worn bearings	Dry or worn bearings cause noise. The bearings may be dry due to dirty oil, oil not reaching the shaft, or motor overheating. Oil the bearings as recommended. If noise remains, replace the bearings or the motor.
	Excessive grease	Ball bearings that have excessive grease may cause the bearings to overheat. Overheated bearings cause noise. Remove any excess grease.

THREE-PHASE MOTOR TROUBLESHOOTING GUIDE		
Problem	**Possible Cause**	**Corrective Action**
Motor will not start.	Wrong motor connections	Most 3ϕ motors are dual-voltage. Check for proper motor connections.
	Blown fuse or open CB	Test the OCPD. If voltage is present at the input, but not the output of the OCPD, the fuse is blown or the CB is open. Check the rating of the OCPD. It should be at least 125% of the motor's FLC.
	Motor overload on starter tripped.	Allow overloads to cool. Reset overloads. If reset overloads do not start the motor, test the starter.
	Low or no voltage applied to motor	Check the voltage at the motor terminals. The voltage must be present and within 10% of the motor nameplate voltage. If voltage is present at the motor but the motor is not operating, remove the motor from the load the motor is driving. Reapply power to the motor. If the motor runs, the problem is with the load. If the motor does not run, the problem is with the motor. Replace or service the motor.
	Open control circuit between incoming power and motor	Check for cleanliness, tightness, and breaks. Use a voltmeter to test the circuit starting with the incoming power and moving to the motor terminals. Voltage generally stops at the problem area.
Fuse, CB, or overloads retrip after service.	Power not applied to all three lines	Measure voltage at each power line. Correct any power supply problems.
	Blown fuse or open CB	Test the OCPD. If voltage is present at the input, but not the output of the OCPD, the fuse is blown or the CB is open. Check the rating of the OCPD. It should be at least 125% of the motor's FLC.
	Motor overload on starter tripped.	Allow overloads to cool. Reset overloads. If reset overloads do not start the motor, test the starter.
	Low or no voltage applied to motor	Check the voltage at the motor terminals. The voltage must be present and within 10% of the motor nameplate voltage. If voltage is present at the motor but the motor is not operating, remove the motor from the load the motor is driving. Reapply power to the motor. If the motor runs, the problem is with the load. If the motor does not run, the problem is with the motor. Replace or service the motor.

	Open control circuit between incoming power and motor	Check for cleanliness, tightness, and breaks. Use a voltmeter to test the circuit starting with the incoming power and moving to the motor terminals. Voltage generally stops at the problem area.
	Motor shaft does not turn.	Disconnect the motor from the load. If the motor shaft still does not turn, the bearings are frozen. Replace or service the motor.
Motor overheats.	Motor is single phasing.	Check each of the 3ϕ power lines for correct voltage.
	Improper ventilation	Clean all ventilation openings. Vacuum or blow dirt out of motor with low-pressure, dry, compressed air.
	Motor is overloaded.	Check the load for binding. Check shaft straightness. Measure motor current under operating conditions. If the current is above the listed current rating, remove the motor. Remeasure the current under no-load conditions. If the current is excessive under load but not when unloaded, check the load. If the motor draws excessive current when disconnected, replace or service the motor.

 Refer to **Activity 13-2—Troubleshooting AC Motors** ....

 Refer to Quick Quiz® on CD-ROM

Refer to Chapter 13 in the **Troubleshooting Electrical/Electronic Systems Workbook** *for additional questions.*

Name_____ **Date** _____

ACTIVITY 13-1—AC Motors

The voltage and current draw of a motor are measured when testing a motor. Complete the wiring of the motor. Connect the clamp-on ammeter to measure the current flowing through L1 and the voltmeter to measure the voltage applied to the motor.

1. Connect the split-phase motor.

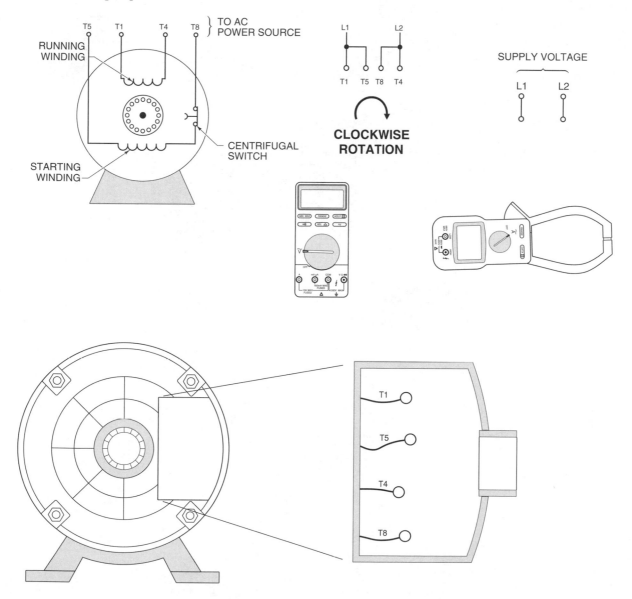

2. Connect the dual-voltage, split-phase motor.

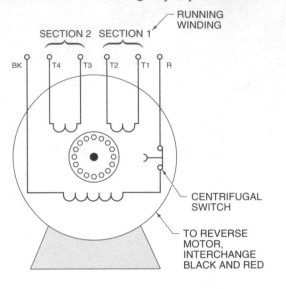

RUNNING WINDING

SECTION 2   SECTION 1

CENTRIFUGAL SWITCH

TO REVERSE MOTOR, INTERCHANGE BLACK AND RED

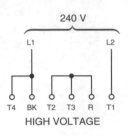

240 V

HIGH VOLTAGE

CLOCKWISE ROTATION

SUPPLY VOLTAGE

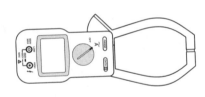

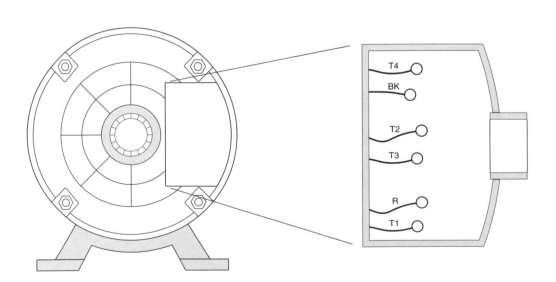

3. Connect the dual-voltage, wye-connected, 3φ motor.

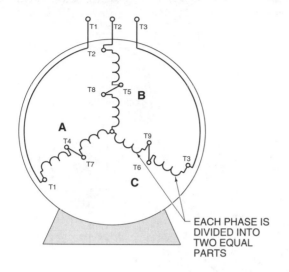

A

B

C

T1 T2 T3

T2

T8 T5

T4

T7

T9

T6

T3

T1

EACH PHASE IS
DIVIDED INTO
TWO EQUAL
PARTS

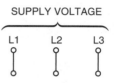

L1 L2 L3
T1 T2 T3

T7 T8 T9
T4 T5 T6

HIGH VOLTAGE (SERIES)

SUPPLY VOLTAGE

L1   L2   L3

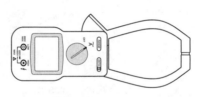

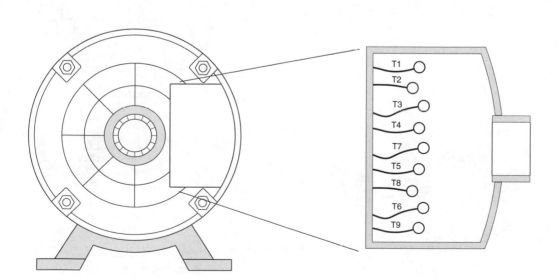

T1

T2

T3

T4

T7

T5

T8

T6

T9

ACTIVITY 13-2—Troubleshooting AC Motors

Determine whether the meter reading indicates an open winding, shorted winding, faulty centrifugal switch, faulty capacitor, faulty (or open) thermal switch, or a normal reading.

_____ **1.** Meter 1 reading indicates ___.

_____ **2.** Meter 2 reading indicates ___.

_____ **3.** Meter 3 reading indicates ___.

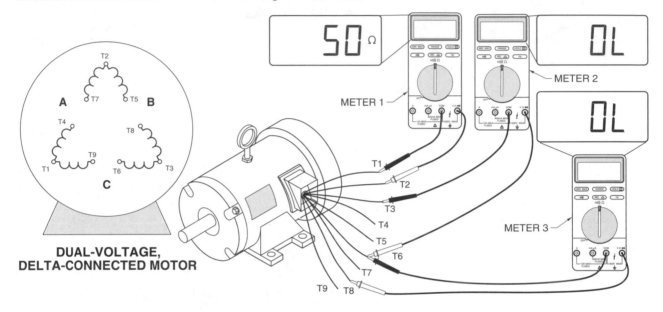

_____ **4.** Meter 4 reading indicates ___.

_____ **5.** Meter 5 reading indicates ___.

_____ **6.** Meter 6 reading indicates ___.

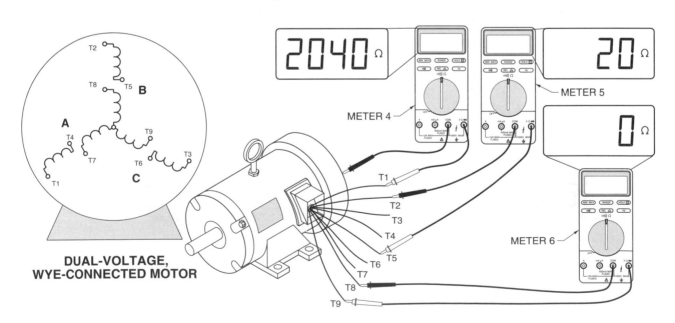

_____   **7.** Meter 7 reading indicates ___.

_____   **8.** Meter 8 reading indicates ___.

_____   **9.** Meter 9 reading indicates ___.

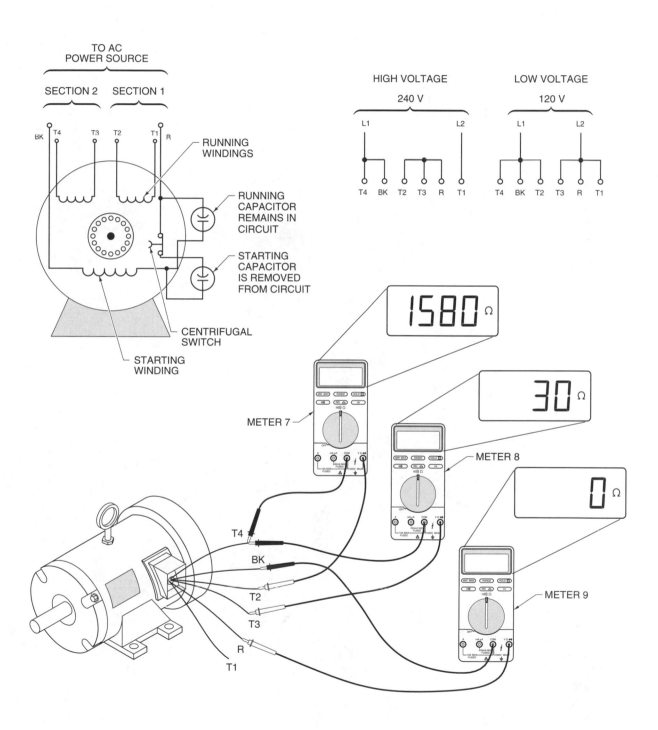

_____ **10.** Meter 10 reading
indicates ___.

_____ **11.** Meter 11 reading
indicates ___.

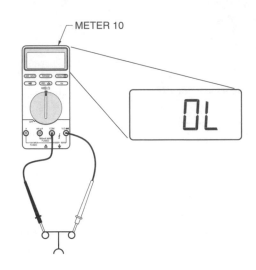

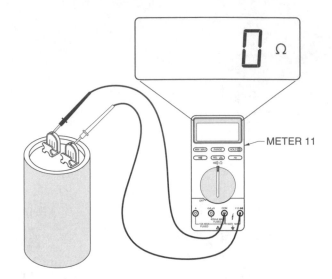

_____ **12.** Meter 12 reading
indicates ___.

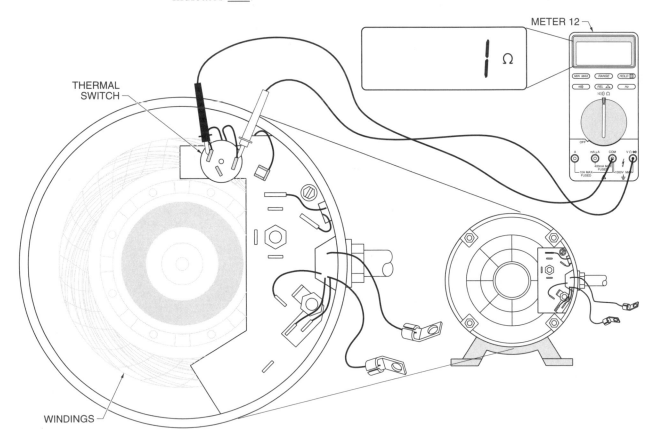

Power Distribution

14

APPLICATION 14-1—Distribution Systems

Electrical power companies generate and distribute high-voltage 3ϕ power. Transformers are used to step down higher voltage transmitted to the voltage level required by the consumer. Power companies primarily use 1ϕ transformers connected in configurations such as wye and/or delta to deliver the correct power to customers.

Different size power distribution systems are used depending on the power required and the type of load(s) to be controlled. For example, power produced for a large commercial application such as a college, small business, or hotel is delivered to the building as high voltage from the utility company. High voltage from the utility company is reduced for usage in building circuits at a substation.

A *substation* is an assemblage of equipment installed for switching, changing, or regulating the voltage of electricity. A substation can be a large outdoor utility distribution center or an in-plant distribution center. For a building application, a substation contains transformers, switchboards, switchgears, transfer switches, and secondary switches that distribute low-voltage levels (208 V – 480 V) to feeder panels, other secondary transformers, busway systems, and branch-circuit panels.

Troubleshooting building power distribution problems vary with the size and type of distribution system. In order to troubleshoot a building power distribution system, an understanding of each type of system (1ϕ, 3ϕ, 3-wire service, 4-wire service, etc.) is required. **See Building Power Distribution.**

Building power distribution systems are used to deliver the required type (DC, 1ϕ, or 3ϕ) and level (120 V, 230 V, 460 V, etc.) to the loads connected on the system.

A *branch circuit* is the portion of a building power distribution system between the final overcurrent protection device and the outlet (receptacle) or load connected to it. Motors are supplied with power from electric motor drives.

The power lines supplying power to the electric motor drive are marked L1 (and/or R), L2 (and/or S), and L3 (and/or T). The power lines delivering power from the electric motor drive to the motor are marked T1 (and/or U), T2 (and/or V), and T3 (and/or W).

In addition to using letters and numbers to identify conductors and terminals, conductors are also color coded. Conductor color coding makes it easier to keep track of the conductors, determine type of service, balance loads among the different phases, and also aids when troubleshooting. Some colors have a definite meaning. For example, the color green always indicates a conductor used for grounding. A *grounding conductor* is a conductor that does not normally carry current, except during a fault (short circuit). Other colors may have more than one meaning depending on the circuit. Some colors are required on conductors to meet NEC® requirements. For example, NEC® Section 110.15 states in a 4-wire delta-connected secondary system, the higher voltage phase should be colored orange (or clearly marked) because it is too high for low-voltage 1ϕ power and too low for 1ϕ high-voltage power.

120/240 V, 1ϕ, 3-Wire Service

A 120/240 V, 1ϕ, 3-wire service is used to supply power to customers that require 120 V and 240 V, 1ϕ power. This service provides 120 V, 1ϕ ; 240 V, 1ϕ; and 120/240 V, 1ϕ circuits. The neutral wire is grounded; therefore, it should not be fused or switched at any point.

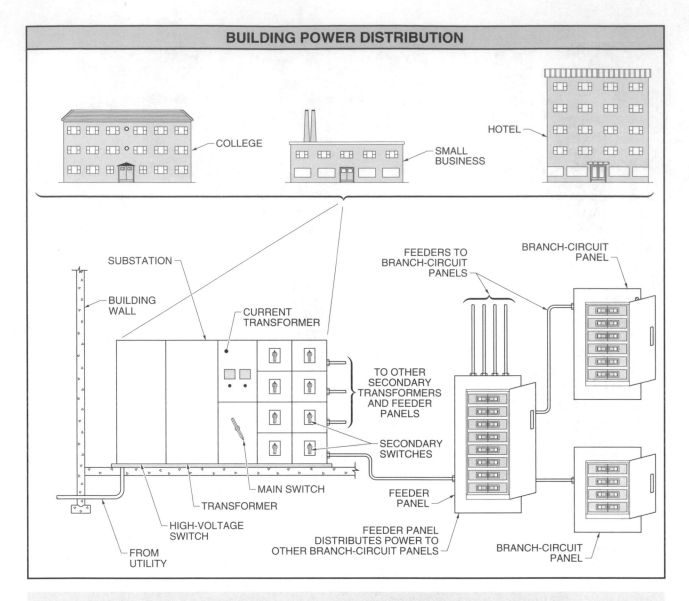

BUILDING POWER DISTRIBUTION

A 120/240 V, 1ϕ, 3-wire service is commonly used for interior wiring for lighting and small appliance use. For this reason, 120/240 V, 1ϕ, 3-wire service is the primary service used to supply most residential buildings. This service is also used for small commercial applications, although a large power panel is used (or there are additional panels).

A 120/240 V, 1ϕ, 3-wire service can be used to supply electric motor drives that operate HVAC systems. Electric motor drives are available that can be connected to either 120 V or 240 V. The electric motor drive then converts the 1ϕ power into 3ϕ for driving 3ϕ motors. **See 120/240 V, 1ϕ, 3-Wire Service.**

NEC® Phase Arrangement

Three-phase circuits include three individual ungrounded (hot) power lines. These power lines are referred to as phases A (L1 or R), B (L2 or S), and C (L3 or T). Phases A, B, and C are connected to a switchboard or panelboard according to NEC® requirements that state the phases must be arranged A, B, C from front to back, top to bottom, and left to right, as viewed from the front of the switchboard or panelboard. **See NEC® Phase Arrangement.**

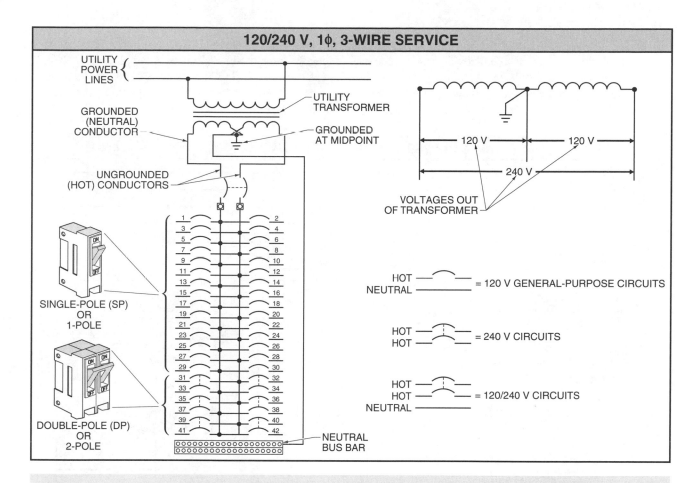

120/208 V, 3φ, 4-Wire Service

A 120/208 V, 3φ, 4-wire service is used to supply customers that require a large amount of 120 V, 1φ power; 208 V, 1φ power; and low-voltage 3φ power. This service includes three ungrounded (hot) lines and one grounded (neutral) line. Each hot line has 120 V to ground when connected to the neutral line.

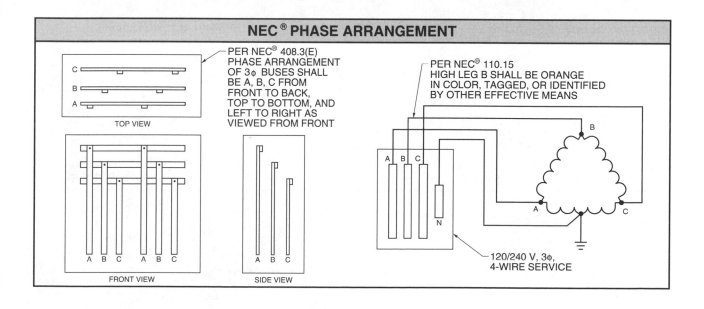

A 120/208 V, 3φ, 4-wire service is used to provide large amounts of low voltage (120 V, 1 φ) power. The 120 V circuits should be balanced to equally distribute the power among the three hot lines. This is accomplished by alternately connecting the 120 V circuits to the power panel so each phase (A to N, B to N, C to N) is divided among the loads (lamps, receptacles, etc.). Likewise, 208 V, 1φ loads such as heating elements and drives should be balanced between phases (A to B, B to C, C to A). **See 120/208 V, 3 φ, 4-Wire Service.**

120/208 V, 3φ, 4-WIRE SERVICE

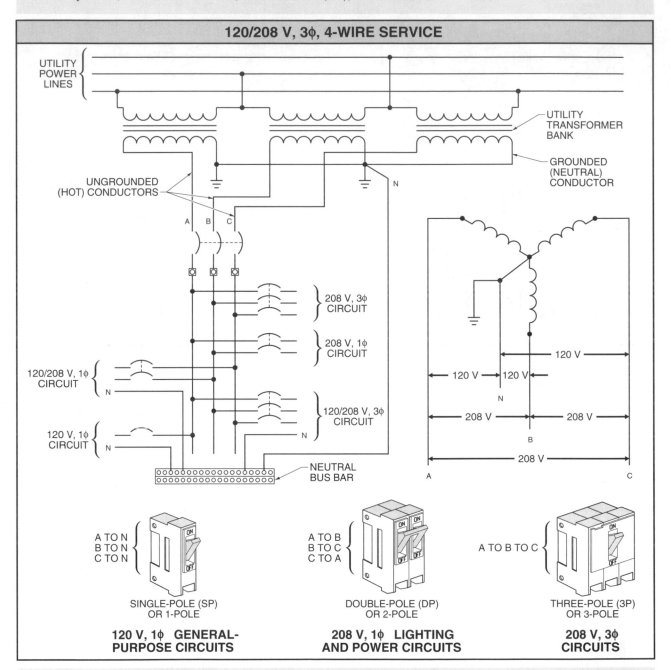

120/240 V, 3φ, 4-Wire Service

A 120/240 V, 3φ, 4-wire service is used to supply customers that require large amounts of 3φ power with some 120 V and 240 V, 1φ power. This service supplies 1φ power delivered by one of the three transformers and 3φ power delivered by using all three transformers. The 1φ power is provided by center tapping one of the transformers.

Because only one transformer delivers all of the 1φ power, this service is used in applications that require mostly 3φ power or 240 V, 1φ power and some 120 V, 1φ power. This service works because in many commercial applications, the total amount of 1φ power used is small when compared to the total amount of 3φ power used. Each transformer may be center tapped if large amounts of 1φ power is required. **See 120/240 V, 3φ, 4-Wire Service.**

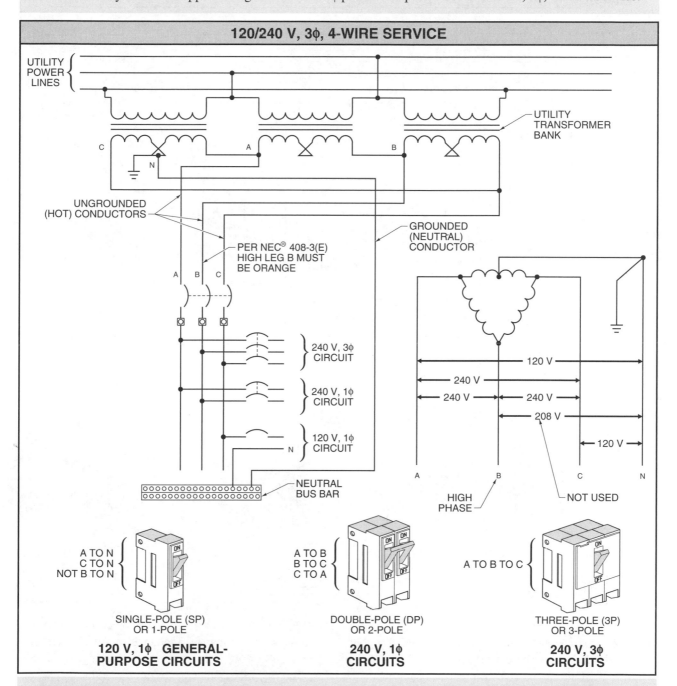

120/240 V, 3φ, 4-WIRE SERVICE

277/480 V, 3φ, 4-Wire Service

The 277/480 V, 3φ, 4-wire service is the same as the 120/240 V, 3φ, 4-wire service except the voltage levels are higher. This service includes three ungrounded (hot) lines and one grounded (neutral) line. Each hot line has 277 V to ground when connected to the neutral or 480 V when connected between any two hot (A to B, B to C, or C to A) lines. **See 277/480 V, 3φ, 4-Wire Service.**

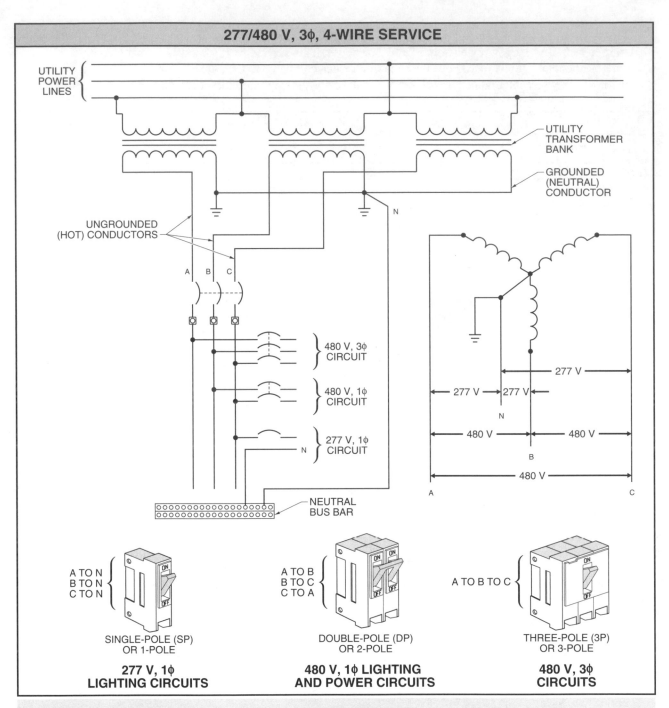

277/480 V, 3ϕ, 4-WIRE SERVICE

UTILITY POWER LINES

UTILITY TRANSFORMER BANK

GROUNDED (NEUTRAL) CONDUCTOR

N

UNGROUNDED (HOT) CONDUCTORS

A B C

480 V, 3ϕ CIRCUIT

480 V, 1ϕ CIRCUIT

277 V, 1ϕ CIRCUIT

N

NEUTRAL BUS BAR

277 V

277 V

277 V

277 V

N

480 V

480 V

480 V

A

B

C

A TO N
B TO N
C TO N

ON

OFF

SINGLE-POLE (SP) OR 1-POLE

277 V, 1ϕ LIGHTING CIRCUITS

A TO B
B TO C
C TO A

ON   ON

OFF

DOUBLE-POLE (DP) OR 2-POLE

480 V, 1ϕ LIGHTING AND POWER CIRCUITS

A TO B TO C

ON

OFF

THREE-POLE (3P) OR 3-POLE

480 V, 3ϕ CIRCUITS

The 277/480 V, 3ϕ, 4-wire service provides 277 V, 1ϕ or 480 V 1ϕ power, but not 120 V, 1ϕ power. For this reason, 277/480 V, 3ϕ, 4-wire service is not used to supply 120 V, 1ϕ general lighting and appliance circuits. However, the 277/480 V, 3ϕ, 4-wire service can be used to supply 277 V and 480 V, 1ϕ lighting circuits. Such high-voltage lighting circuits are used in commercial fluorescent and HID (high-intensity discharge) lighting circuits.

A system that cannot deliver 120 V, 1ϕ power appears to have limited use. However, in many commercial applications (sport complexes, schools, offices, parking lots, etc.), lighting is a major part of the electrical system. Because large commercial applications include several sets of transformer banks, 120 V, 1ϕ power is available through other transformers. Additional transformers can also be connected to the 277/480 V, 3ϕ, 4-wire service to reduce the voltage to 120 V, 1ϕ. **See 277/480 V, 3ϕ, 4-Wire Service Step-Down Transformer Addition.**

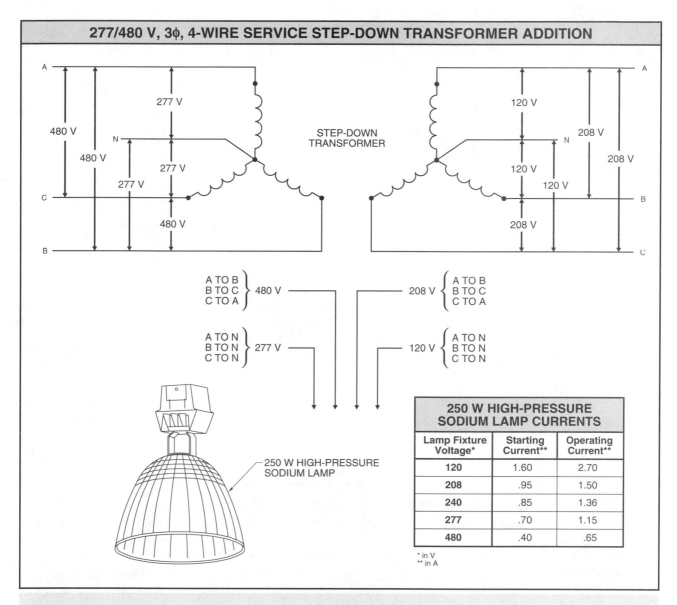

277/480 V, 3ϕ, 4-WIRE SERVICE STEP-DOWN TRANSFORMER ADDITION

STEP-DOWN TRANSFORMER

250 W HIGH-PRESSURE SODIUM LAMP

250 W HIGH-PRESSURE SODIUM LAMP CURRENTS		
Lamp Fixture Voltage*	Starting Current**	Operating Current**
120	1.60	2.70
208	.95	1.50
240	.85	1.36
277	.70	1.15
480	.40	.65

* in V
** in A

When drives are used for larger commercial HVAC systems, the drive is usually connected to the 480 V, 3ϕ power lines. It is always better to connect a drive to a 3ϕ power supply, even if the drive can be connected to a 1ϕ supply. This assures that the internal parts of the drive are evenly balanced to carry only ⅓ of the total power per line.

 Refer to **Activity 14-1—Distribution Systems**...

APPLICATION 14-2—Isolated Ground Receptacles

When a standard receptacle is installed in a metal outlet box, the receptacle ground is connected to the common grounding system. This common grounding system typically includes all metal wiring boxes, conduits, gas pipes, water pipes, and the noncurrent-carrying metal parts of most electrical equipment. When a piece of electrical equipment is plugged into the receptacle, its ground becomes part of the larger grounding system. **See Receptacle Grounding.**

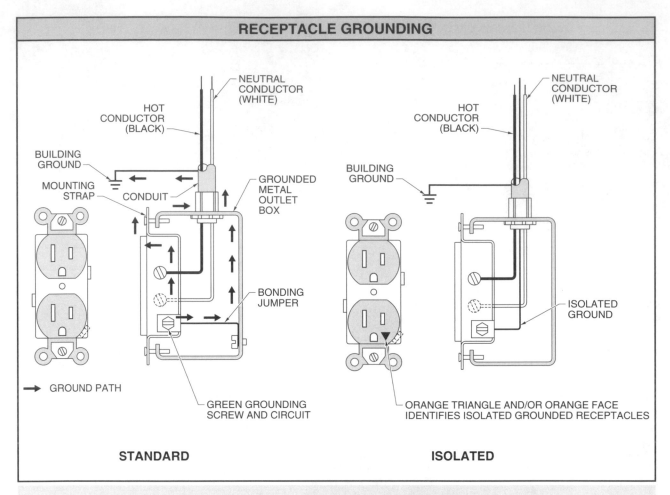

RECEPTACLE GROUNDING

STANDARD

ISOLATED

The common grounding system may act as a large antenna and conduct electrical noise. This electrical noise may cause interference in sensitive electrical equipment such as computer, medical, security, and communication equipment.

An isolated grounded receptacle may be used to minimize problems in sensitive applications or in areas of high electrical noise. An *isolated ground receptacle* is a receptacle that minimizes electrical noise by providing a separate grounding path. The isolation ensures a clean equipment ground for electronic equipment, which may be adversely affected by noise in the equipment-grounding path. Isolated ground receptacles shall be used only with isolated equipment-grounding conductors. Isolated grounded receptacles are identified by an orange triangle on the face of the receptacle and/or an orange colored faceplate.

In the isolated grounding system, a separate ground conductor is run with the circuit conductors. The grounding system must be installed per the NEC®.

 Refer to **Activity 14-2—Isolated Ground Receptacle** ....

APPLICATION 14-3—*Transformer Load Cycle*

Transformers are used to deliver power to a set number of loads. For example, a transformer may be used to deliver power to a school. As loads in the school are switched ON and OFF, the power delivered by the transformer changes. At certain times (night), the power output required from the transformer may be low. At other times (during school hours), the power output required from the transformer may be high. Peak load is the maximum output requirement of a transformer. **See Transformer Load Cycle.**

A transformer is overloaded when it is required to deliver more power than it is rated to carry. A transformer is not damaged when overloaded for a short time period. A transformer is not damaged because the heat storage capacity of a transformer ensures a relatively slow increase in internal transformer temperature.

To help dissipate heat in a transformer, most power transformers are self-cooled or oil immersed. Transformer manufacturers list the length of time a transformer may safely be overloaded at a given peak level. For example, a transformer that is overloaded 15 times its rated current has a permissible overload time of 5.5 sec. **See Temporary Transformer Overloading.**

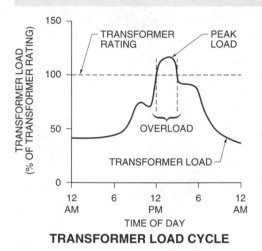

TRANSFORMER LOAD CYCLE

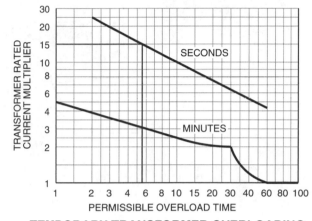

TEMPORARY TRANSFORMER OVERLOADING

 Refer to **Activity 14-3—Transformer Load Cycle**..

APPLICATION 14-4—Wye and Delta Transformer Connections

Transformers are connected in wye and delta configurations. A *wye configuration* is a transformer connection that has one end of each transformer coil connected together. The remaining end of each coil is connected to the incoming power lines (primary side) or used to supply power to the load(s) (secondary side). A *delta configuration* is a transformer connection that has each transformer coil connected end-to-end to form a closed loop. Each connecting point is connected to the incoming power lines or used to supply power to the load(s). The voltage output and type available for the load(s) is determined by whether the transformer is connected in a wye or delta configuration. **See Wye and Delta Configurations.**

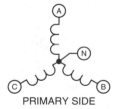 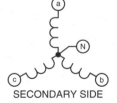 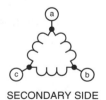

PRIMARY SIDE   SECONDARY SIDE   PRIMARY SIDE   SECONDARY SIDE

WYE CONFIGURATION   **DELTA CONFIGURATION**

 Refer to **Activity 14-4—Wye and Delta Transformer Connections**..

APPLICATION 14-5—Transformer Tap Connections

A transformer with taps is used to compensate for voltage differences in a control circuit. *Taps* are connection points provided along the transformer coil. Taps are usually provided at 2.2% increments along one end of the transformer coil. **See Determining Transformer Tap Connections.**

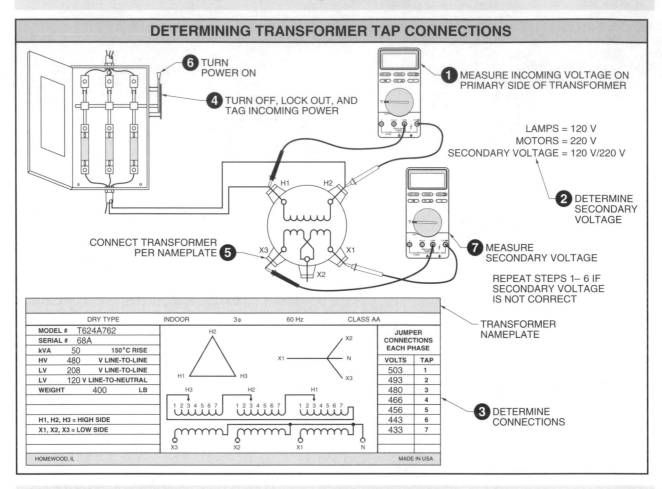

DETERMINING TRANSFORMER TAP CONNECTIONS

To determine the proper transformer tap connections, apply the procedure:

1. Measure the incoming voltage on the primary side of the transformer.
2. Determine the secondary voltage. The secondary voltage is the voltage rating of the loads that are connected to the transformer.
3. Determine the connections using the manufacturer's specifications or transformer nameplate.
4. Turn the power OFF, lock out, and tag the incoming power to the transformer. Ensure the power is OFF by testing the circuit with a multimeter.
5. Connect the transformer per manufacturer's specifications or transformer nameplate. Check each connection twice.
6. Turn the power ON.
7. Measure the secondary voltage of the transformer.

Repeat steps 1–6 if the secondary voltage is not correct.

 Refer to **Activity 14-5—Transformer Tap Connections**...

APPLICATION 14-6—Transformer Sizing and Balancing

A correctly-sized transformer provides years of trouble-free service. Heat destroys a transformer by damaging the insulation on the windings. Heat in a transformer is produced by current flow through the windings. The greater the current flow, the greater the heat. Overcurrent in a transformer is usually the result of overloading a transformer. Overloading a transformer occurs when a transformer is undersized or loads are unbalanced.

Sizing Single-Phase Transformers

When servicing or replacing a burned-out 1φ transformer, the electrical loads and available supply voltage are considered. A transformer is selected based on the electrical load and supply voltage requirements. **See Sizing 1φ Transformers.**

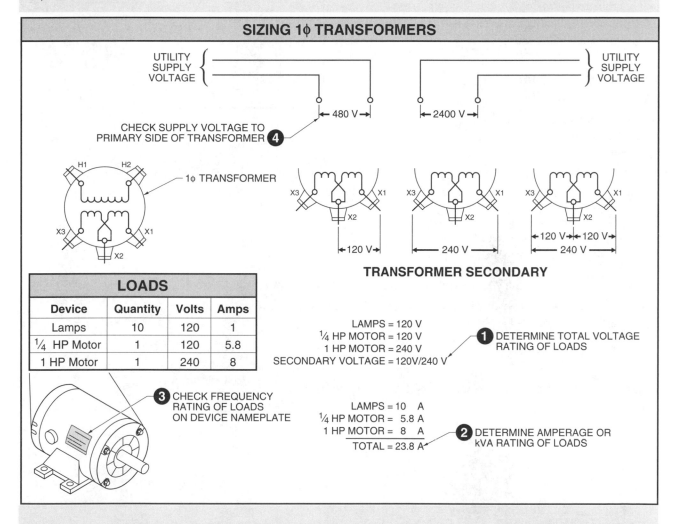

SIZING 1φ TRANSFORMERS

To size a 1φ transformer, apply the procedure:

1. Determine the total voltage required by the loads if more than one load is connected. The secondary side of the transformer must have a rating equal to the voltage of the loads.
2. Determine the amperage rating or kVA capacity required by the load(s). Add all loads that are (or may be) ON concurrently.
3. Check load(s) frequency on the nameplate. The frequency of the supply voltage and the electrical load(s) must be the same. Transformers change voltage levels but not frequency.

4. Check the supply voltage to the primary side of the transformer. The primary side of a transformer must have a rating equal to the supply voltage. Consider each voltage if there is more than one source voltage available. Use a transformer that has primary taps if there is a variation in the supply voltage. The transformer must have a kVA capacity of at least 10% greater than that required by the loads. A conversion table may be used to determine proper kVA capacity when the load rating is given in amperes. **See Single-Phase Full-Load Currents.**

SINGLE-PHASE FULL-LOAD CURRENTS*						
kVA	120 V	208 V	240 V	277 V	380V	480 V
.050	.4	.2	.2	.2	.1	.1
.100	.8	.5	.4	.3	.2	.2
.150	1.2	.7	.6	.5	.4	.3
.250	2.0	1.2	1	.9	.6	.5
.500	4.2	2.4	2.1	1.8	1.3	1
.750	6.3	3.6	3.1	2.7	2	1.6
1	8.3	4.8	4.2	3.6	2.6	2.1
1.5	12.5	7.2	6.2	5.4	3.9	3.1
2	16.7	9.6	8.3	7.2	5.2	4.2
3	25	14.4	12.5	10.8	7.9	62
5	41	24	20.8	18	13.1	10.4
7.5	62	36	31	27	19.7	15.6
10	83	48	41	36	26	20.8
15	125	72	62	54	39	31

* full-load current in amperes

Example: Determining 1ϕ Transformer kVA — Conversion Table

A 240 V circuit has a total load rating of 41 A. Determine the kVA capacity.

A 240 V circuit with a total load rating of 41 A has a kVA capacity of 10 kVA (from conversion table). To calculate kVA capacity when voltage and current are known, apply the formula:

$$kVA_{CAP} = \frac{E \times I}{1000}$$

where

kVA_{CAP} = transformer capacity (in kVA)

E = voltage (in volts)
I = current (in amperes)
1000 = constant

Note: When computing kVA, multiply amperage by voltage. Never use the wattage rating of the loads.

To calculate amperage when kVA capacity and voltage are known, apply the formula:

$$I = \frac{kVA_{CAP} \times 1000}{E}$$

Example: Determining 1ϕ Transformer kVA — Calculation

A 120 V circuit has loads of 5 A and 15 A. Determine the capacity (in kVA) required by the transformer secondary.

$$kVA_{CAP} = \frac{E \times I}{1000}$$

$$kVA_{CAP} = \frac{120 \times 20}{1000}$$

$$kVA_{CAP} = \frac{2400}{1000}$$

$$kVA_{CAP} = \textbf{2.4 kVA}$$

Sizing Three-Phase Transformers

When servicing or replacing a burned-out 3ϕ transformer, the electrical loads and available supply voltage are considered. A transformer is selected based on its electrical load and supply voltage requirements. **See Sizing 3ϕ Transformers.**

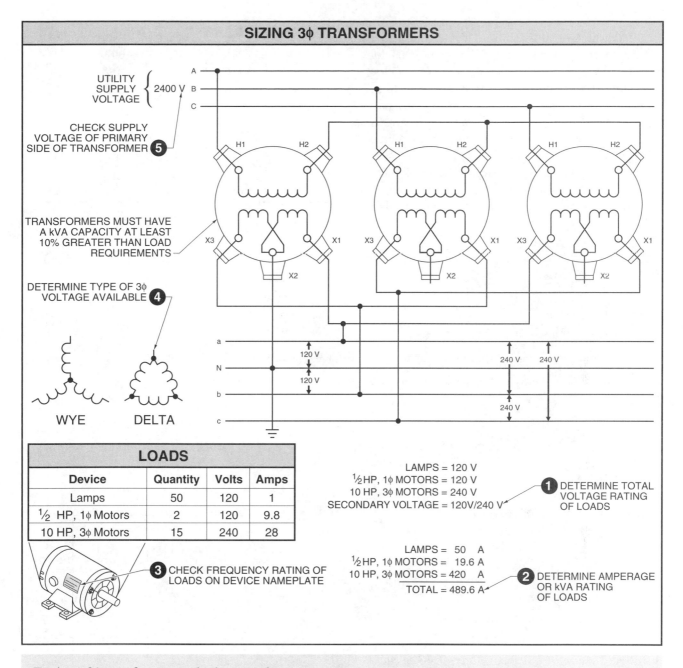

SIZING 3ϕ TRANSFORMERS

UTILITY SUPPLY VOLTAGE { 2400 V

CHECK SUPPLY VOLTAGE OF PRIMARY SIDE OF TRANSFORMER **5**

TRANSFORMERS MUST HAVE A kVA CAPACITY AT LEAST 10% GREATER THAN LOAD REQUIREMENTS

DETERMINE TYPE OF 3ϕ VOLTAGE AVAILABLE **4**

WYE   DELTA

120 V
120 V
240 V   240 V
240 V

LOADS

Device	Quantity	Volts	Amps
Lamps	50	120	1
½ HP, 1ϕ Motors	2	120	9.8
10 HP, 3ϕ Motors	15	240	28

3 CHECK FREQUENCY RATING OF LOADS ON DEVICE NAMEPLATE

LAMPS = 120 V
½ HP, 1ϕ MOTORS = 120 V
10 HP, 3ϕ MOTORS = 240 V
SECONDARY VOLTAGE = 120V/240 V

1 DETERMINE TOTAL VOLTAGE RATING OF LOADS

LAMPS = 50 A
½ HP, 1ϕ MOTORS = 19.6 A
10 HP, 3ϕ MOTORS = 420 A
TOTAL = 489.6 A

2 DETERMINE AMPERAGE OR kVA RATING OF LOADS

To size a 3ϕ transformer, apply the procedure:

1. Determine the total voltage required by the loads if more than one load is connected. The secondary side of the transformer must have a rating equal to the voltage of the loads.

2. Determine the amperage rating or kVA capacity required by the load(s). Add all loads that are (or may be) ON concurrently.

3. Check load(s) frequency on the nameplate. The frequency of the supply voltage and the electrical load(s) must be the same. Transformers change voltage levels, not frequency.

4. Determine the type of 3ϕ voltage available. This includes three-wire no ground or three-wire with ground (four-wire).

5. Check the supply voltage to the primary side of the transformer.

The primary side of a transformer must have a rating equal to the supply voltage. Consider each voltage when there is more than one source of voltage available. Use a transformer that has primary taps when there is a variation in the supply voltage. The transformer must have a kVA capacity of at least 10% greater than that required by the loads.

A conversion table is used to determine the kVA capacity of 3ϕ circuits when the load rating is given in amperes. **See Three-Phase Full-Load Currents.**

THREE-PHASE FULL-LOAD CURRENTS*				
kVA	**208 V**	**240 V**	**480 V**	**600 V**
3	8.3	7.2	3.6	2.9
4	12.5	10.8	5.4	4.3
6	16.6	14.4	7.2	5.8
9	25	21.6	10.8	8.6
15	41	36	18	14.4
22	62	54	27	21.6
30	83	72	36	28
45	124	108	54	43

* full-load current in amperes

To calculate kVA capacity when voltage and current are known, apply the formula:

$$kVA_{CAP} = \frac{E \times 1.732 \times I}{1000}$$

where

kVA_{CAP} = transformer capacity (in kVA)

E = voltage (in volts)
1.732 = constant (for 3ϕ power)
I = current
1000 = constant

Note: When calculating kVA, multiply amperage by voltage. Never use the wattage rating of the loads.

To calculate 3ϕ amperage when kVA capacity and voltage are known, apply the formula:

$$I = \frac{kVA_{CAP} \times 1000}{E \times 1.732}$$

Note: When 3ϕ problems are calculated, the following values* may be substituted to eliminate one mathematical step:

For 208 V × 1.732, use 360
For 230 V × 1.732, use 398
For 240 V × 1.732, use 416
For 440 V × 1.732, use 762
For 460 V × 1.732, use 797
For 480 V × 1.732, use 831
* $\sqrt{3}$ = 1.732

Example: Determining Three-Phase Transformer kVA—Calculation

A 480 V, 3ϕ circuit has two loads of 18 A each. Determine the capacity (in kVA) required by the secondary of the transformer. **See Transformer Capacity Calculation.**

$$kVA_{CAP} = \frac{E \times 1.732 \times I}{1000}$$

$$kVA_{CAP} = \frac{831 \times 36}{1000}$$

$$kVA_{CAP} = \frac{29,916}{1000}$$

$$kVA_{CAP} = \textbf{29.916 kVA}$$

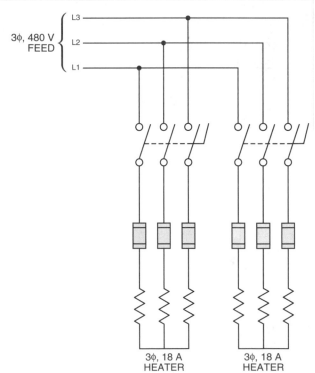

TRANSFORMER CAPACITY CALCULATION

Standard Ambient Compensation Temperature

Temperature rise in a transformer is the temperature of the windings above the existing ambient temperature. Transformers list their maximum temperature rise. Normal ambient temperature is 40° C.

A transformer must be derated if the ambient temperature exceeds 40° C. Transformer derating charts are used to derate transformers in high ambient temperatures. **See Transformer Deratings.**

TRANSFORMER DERATINGS	
Maximum Ambient Temperature (°C)	Maximum Transformer Loading (%)
40	100
45	96
50	92
55	88
60	81
65	80
70	76

Example: Standard Ambient Compensation Temperature

Calculate the rating of a 2.5 kVA transformer installed in an ambient temperature of 50°C.

$kVA = Rated\ kVA \times Maximum\ Load$

$kVA = 2.5 \times .92$ (See Transformer Deratings.)

$kVA = \textbf{2.3 kVA}$

Special Ambient Temperature Compensation

Standard ambient temperature is the average temperature of the air that cools a transformer over a 24-hour period. Standard ambient temperature assumes that maximum temperature does not exceed 10° C above average ambient temperature.

A transformer is derated above the standard values when the maximum temperature exceeds the average temperature by more than 10°C. A transformer is derated by 1½% for each 1°C above 40°C when the maximum ambient temperature exceeds 10°C above the average temperature. This provides additional protection for hot spots (Temperature rise over 10°C). **See Standard Compensation Rating and Special Compensation Rating.**

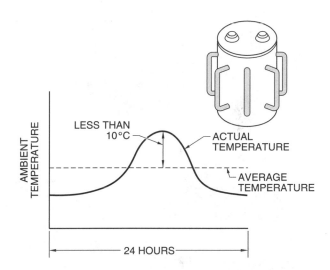

STANDARD COMPENSATION RATING

Example: Special Ambient Temperature Compensation

Calculate the rating of a 2.5 kVA transformer installed in an average ambient temperature of 50°C. The maximum temperature exceeds 10°C above average temperature.

The transformer is derated by 1½% for each degree above 40°C. A 50°C average temperature is 10°C above 40°C, so the transformer must be derated 15% (10°C × 1½% = 15%).

$kVA = Rated\ kVA \times Maximum\ Load$

$kVA = 2.5 \times .85$ (15% derating)

$kVA = \textbf{2.125 kVA}$

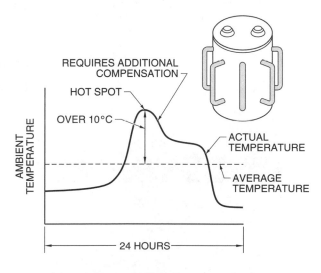

SPECIAL COMPENSATION RATING

Transformer Load Balancing

When connecting loads to a transformer, the loads should be connected so that the transformer is as electrically balanced as possible. *Electrical balance* occurs when loads on a transformer are placed so that each coil of the transformer carries the same amount of current. **See Transformer Load Balancing.**

A clamp-on ammeter is used to check for an unbalanced transformer. The ammeter shows the amount of current draw on each line. The measured current draw on each hot (ungrounded) power line should be nearly equal. The loads should be placed in balance if any current reading is higher than the others and the transformer is at or near its capacity.

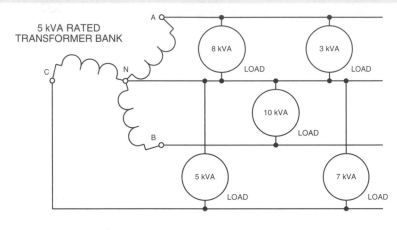

THREE-PHASE

CORRECT LOADING

A TO N = 11 kVA
B TO N = 10 kVA
C TO N = 12 kVA

EACH TRANSFORMER
WINDING LOADED
LESS THAN 15 kVA/φ LIMIT

$$\frac{45\ kVA}{3\phi} = kVA/\phi\ LIMIT$$

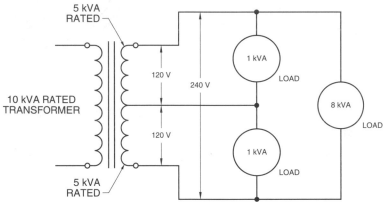

SINGLE-PHASE

CORRECT LOADING

ELECTRICALLY BALANCE
EACH SECONDARY WINDING
RATED AND LOADED TO 5 kVA

TRANSFORMER LOAD BALANCING

 Refer to **Activity 14-6—Transformer Sizing and Balancing** ....

 *Refer to Quick Quiz® on CD-ROM*

 *Refer to Chapter 14 in the **Troubleshooting Electrical/Electronic Systems Workbook** for additional questions.*

Power Distribution

Name_____ **Date** _____

ACTIVITY 14-1—Distribution Systems

A 120/240 V, 1φ, 3-wire service is used to supply power to 120 V loads, 240 V loads, and 120/240 V loads.

1. Connect each load to the proper circuit breaker, breakers, and/or neutral bus bar to supply the required power to the load.

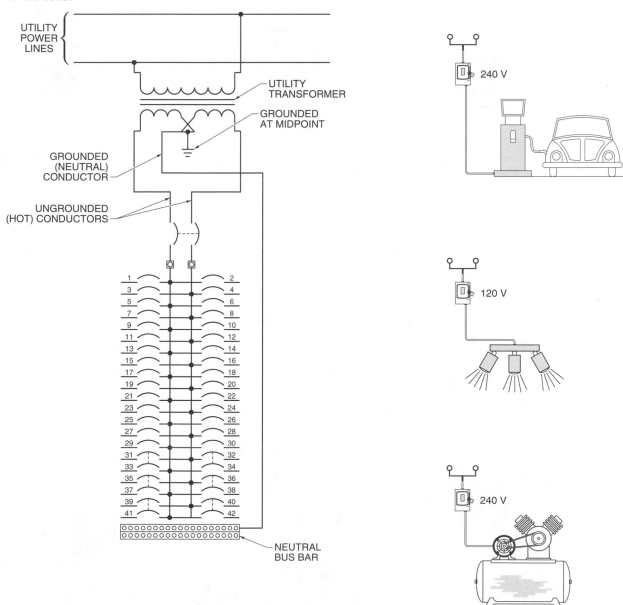

A 120/208 V, 3φ, 4-wire service is used to supply power to 120 V loads, 208 V, 1φ loads, and 208 V, 3φ loads. The 208 V (1φ or 3φ) voltage is on the lower end of the 208 V to 240 V range. Since it is the only voltage supplied within this range, it is still used to supply loads rated at 208 V, 220 V, 230 V, or 240 V. The 208 V usually satisfactorily operates most higher-rated voltage loads, but the loads should be checked and monitored for proper operation. The current draw of the load and the load operation should be checked when first powered and routinely done after that. When a load is supplied with lower than rated voltage, less power is produced. This means that heating elements produce less heat, lamps produce less light, and motors produce less power.

2. Connect each load to the proper circuit breaker, breakers, and/or neutral bus bar to supply the required power to the load.

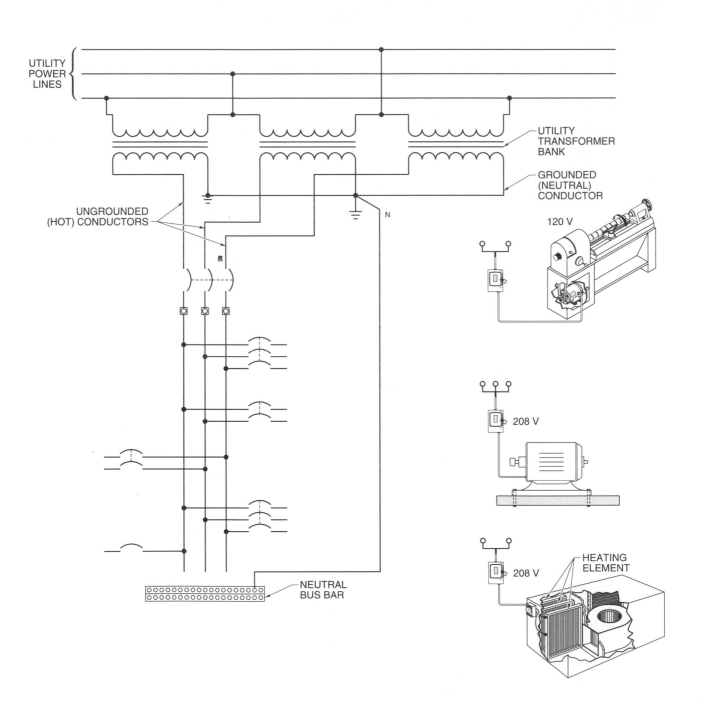

A 120/240 V, 3ϕ, 4-wire service is used to supply power to 120 V loads, 240 V, 1ϕ loads, and 240 V, 3ϕ loads. The 240 V (1ϕ or 3ϕ) voltage is on the higher end of the 208 V to 240 V range. But since it is the only voltage supplied within this range, it is still used to supply loads rated at 208 V, 220 V, 230 V, or 240 V. The 240 V still satisfactorily operates most lower-rated voltage loads, but the loads should be checked and monitored for proper operation. The current draw of the load and the load operation should be done when first powered and routinely done after that. When a load is supplied with higher then rated voltage, more power will be produced. This means that heating elements produce more heat and lamps produce more light.

3. Connect each load to the proper circuit breaker, and/or neutral bus bar to supply the required power to the load.

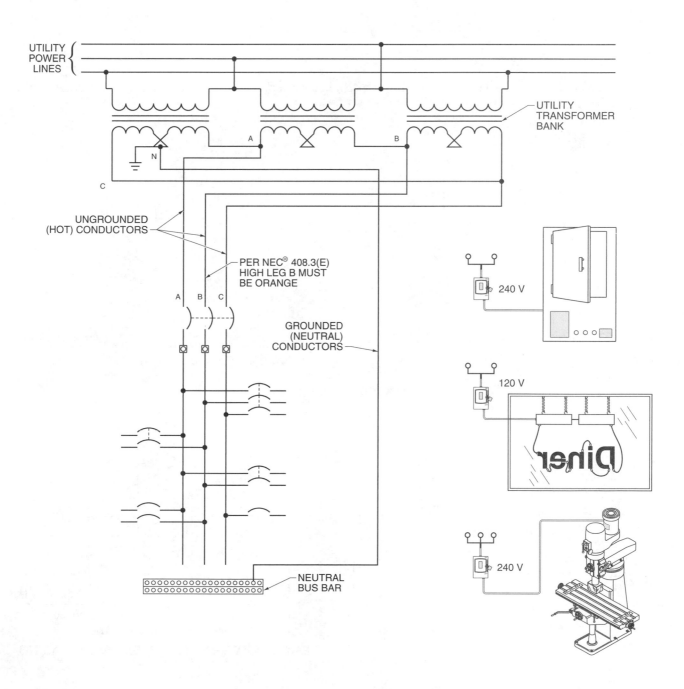

ACTIVITY 14-2—Isolated Ground Receptacles

Draw arrows to indicate the normal path of current flow through the receptacle's conductors when an electrical device is plugged into the receptacle.

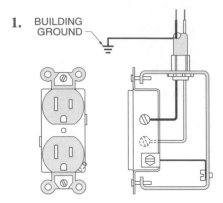

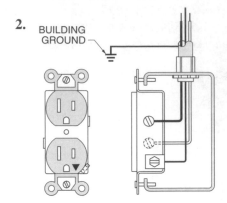

Draw arrows to indicate the path to ground for electrical noise entering the grounded frame of an electrical device plugged into the receptacle.

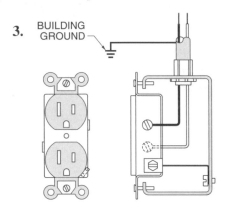

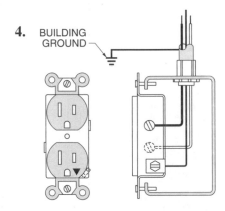

ACTIVITY 14-3—Transformer Load Cycle

Using Temporary Transformer Overloading, determine the permissible overload time for each transformer.

_____ **1.** Permissible overload time equals ___.

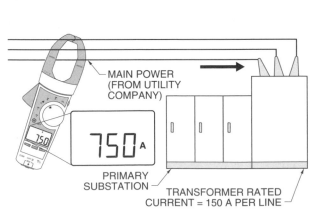

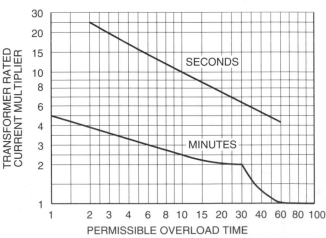

TEMPORARY TRANSFORMER OVERLOADING

2. Permissible overload time equals ___.

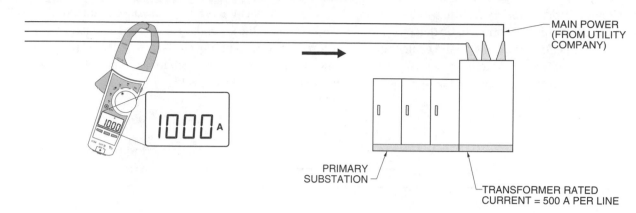

MAIN POWER
(FROM UTILITY
COMPANY)

1000 A

PRIMARY
SUBSTATION

TRANSFORMER RATED
CURRENT = 500 A PER LINE

3. Permissible overload time equals ___.

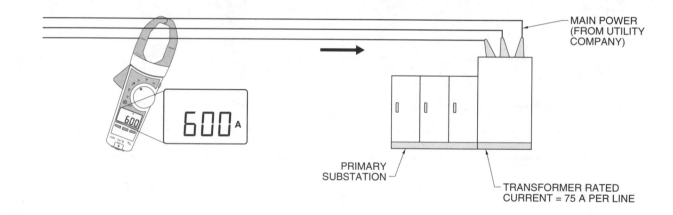

MAIN POWER
(FROM UTILITY
COMPANY)

600 A

PRIMARY
SUBSTATION

TRANSFORMER RATED
CURRENT = 75 A PER LINE

4. Permissible overload time equals ___.

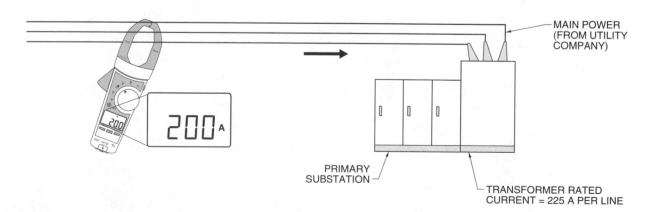

MAIN POWER
(FROM UTILITY
COMPANY)

200 A

PRIMARY
SUBSTATION

TRANSFORMER RATED
CURRENT = 225 A PER LINE

ACTIVITY 14-4—Wye and Delta Transformer Connections

1. Connect Transformer Bank 1 in a delta-delta configuration. Connect the primary side of each transformer to the main power lines to form a basic delta transformer bank. Connect the secondary side of each transformer to form a delta transformer bank with the center tap of Transformer 1 used to produce a 1ϕ low-voltage (120 VAC) neutral. Connect each load to the secondary side of the transformer bank so that each load is connected to the proper voltage. Connect the loads so that the transformer bank is as balanced as possible.

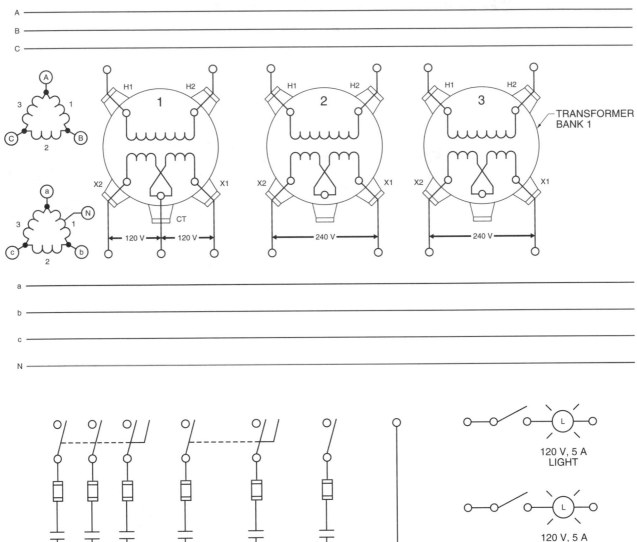

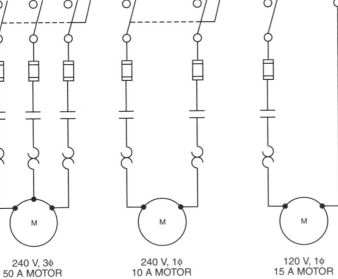

240 V, 3ϕ
50 A MOTOR

240 V, 1ϕ
10 A MOTOR

120 V, 1ϕ
15 A MOTOR

120 V, 5 A
LIGHT

120 V, 5 A
LIGHT

120 V, 5 A
LIGHT

2. Connect Transformer Bank 2 in a delta-wye configuration. Connect the primary side of each transformer to the main power lines to form a basic delta transformer bank. Connect the secondary side of each transformer to form a wye transformer bank with a common neutral. Connect each load to the secondary side of the transformer bank so that each load is connected to the proper voltage. Connect the loads so that the transformer bank is as balanced as possible.

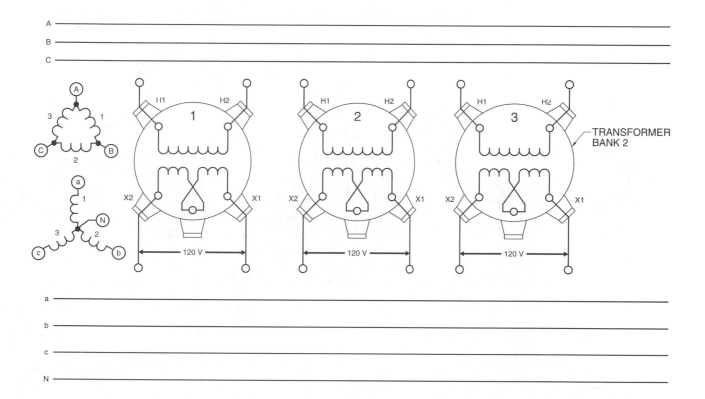

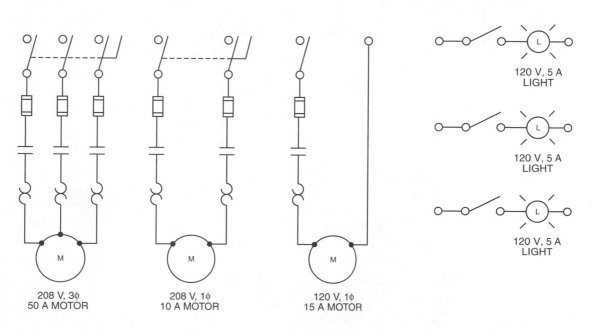

3. Connect Transformer Bank 3 in a wye-wye configuration. Connect the primary side of each transformer to the main power lines to form a basic wye transformer bank with a common neutral. Connect the secondary side of each transformer to form a wye transformer bank with a common neutral. Connect each load to the secondary side of the transformer bank so that each load is connected to the proper voltage. Connect the loads so that the transformer bank is as balanced as possible.

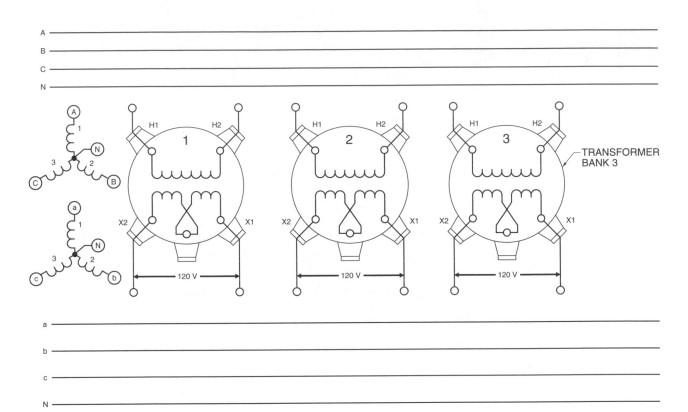

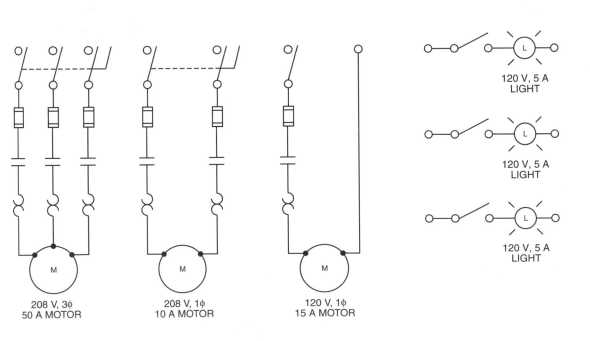

4. Connect Transformer Bank 4 in a wye-delta configuration. Connect the primary side of each transformer to the main power lines to form a basic wye transformer bank with a common neutral. Connect the secondary side of each transformer to form a delta transformer bank. Connect each load to the secondary side of the transformer bank so that each load is connected to the proper voltage. Connect the loads so that the transformer bank is as balanced as possible.

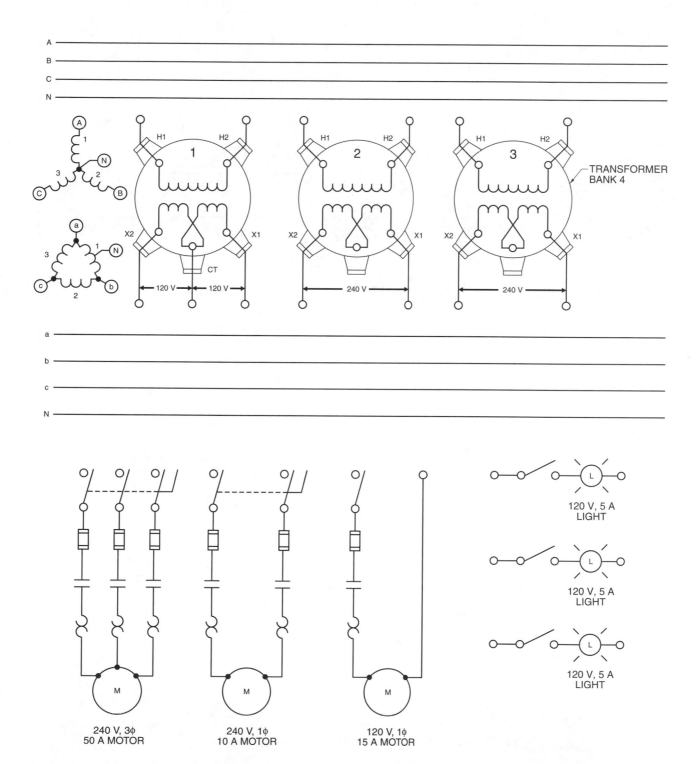

240 V, 3φ
50 A MOTOR

240 V, 1φ
10 A MOTOR

120 V, 1φ
15 A MOTOR

120 V, 5 A
LIGHT

120 V, 5 A
LIGHT

120 V, 5 A
LIGHT

ACTIVITY 14-5—Transformer Tap Connections

Each tap on the primary side of a transformer changes the secondary voltage by about 2½%. Determine the new position of each tap to adjust the output voltage.

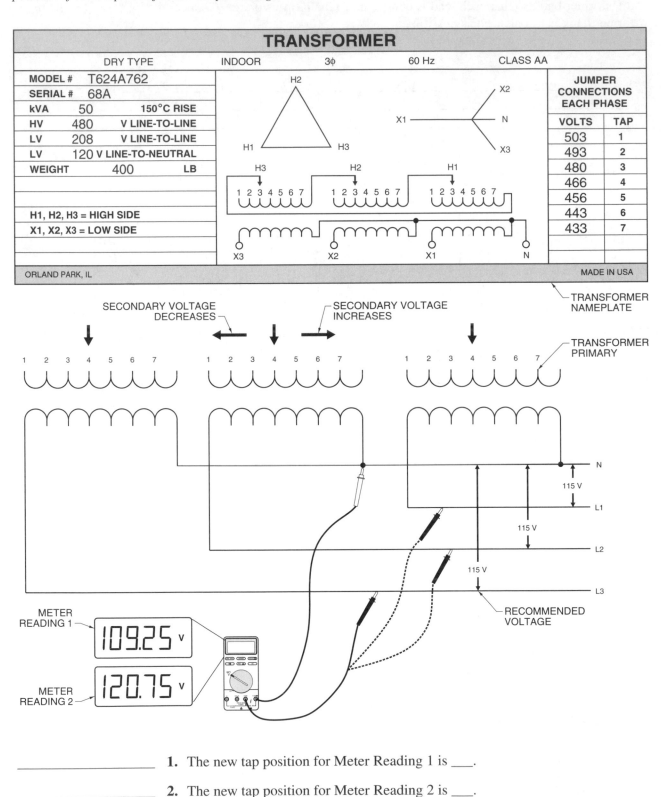

1. The new tap position for Meter Reading 1 is ___.

2. The new tap position for Meter Reading 2 is ___.

ACTIVITY 14-6—Transformer Sizing and Balancing

Determine the kVA capacity of the loads (assume 100% PF). Use Transformer Selections to determine the catalog number of the smallest transformer that could be used for each application. Note: Select a transformer that has at least 10% greater capacity than that required by the loads.

1.

SINGLE-PHASE LOADS				
Device	Quantity	Volts	Amps	Usage
Lamps	50	120	1	50%
¼ HP motor	2	120	5.8	100%
1 HP motor	1	240	8	100%
2 HP motor	2	240	16	50%
Heater	1	120	15	100%
Heater	1	240	30	100%
Heater	1	240	10	100%

TRANSFORMER SELECTIONS				
Catalog Number	Phase	Primary Voltage	Secondary Voltage	kVA Rating
01	1	240/480	120/240	37.5
02	1	240/480	120/240	50
03	1	240/480	120/240	75
04A	1	208	120/240	50
04B	1	208	120/240	75
05A	1	277	120/240	50
05B	1	277	120/240	75
06	3	480 delta	208/120 wye	75
07	3	480 delta	208/120 wye	112.5
08	3	480 delta	208/120 wye	150
09	3	480 delta	208/120 wye	225
10	3	480 delta	240 delta	75
11	3	480 delta	240 delta	150
12	3	600 delta	208/120 wye	75
13	3	600 delta	208/120 wye	112.5
14	3	600 delta	208/120 wye	150
15	3	600 delta	208/120 wye	225

Note: Base all calculations on stated usage.

_____ **A.** The kVA capacity for 120 V loads equals ___ kVA.

_____ **B.** The kVA capacity for 240 V loads equals ___ kVA.

_____ **C.** The kVA capacity for 120 V and 240 V loads equals ___ kVA.

_____ **D.** The KVA capacity of 120V and 240V loads (plus 10%) equals ___ kVA.

2.

THREE-PHASE LOADS				
Device	Quantity	Volts	Amps	Usage
Heater	2	208	25	50%
25 HP motor	2	208	75	50%
10 HP motor	1	208	31	100%
2 HP motor	1	208	7.5	100%

SINGLE-PHASE LOADS				
Device	Quantity	Volts	Amps	Usage
Lamps	100	120	1	75%
Heater	10	120	5	100%
½ HP motor	5	120	9.8	100%

Note: Base all calculations on stated usage.

_____ **A.** The kVA capacity for 3φ loads equals ___ kVA.

_____ **B.** The kVA capacity for 1φ loads equals ___ kVA.

_____ **C.** Total kVA capacity for 1φ and 3φ loads equals ___ kVA.

_____ **D.** Transformer catalog number with 480 primary equals ___.

_____ **E.** Transformer catalog number with 600 primary equals ___.

APPLICATION 15-1 — Power Quality Problems

Electrical equipment, circuits, and systems commonly have power quality problems. The source of power quality problems can be the utility company, the in-house power distribution system, and/or nonlinear loads connected to the system. A *linear load* is any load in which current increases proportionately as the voltage increases and current decreases proportionately as voltage decreases. Pure resistance, inductance, and capacitance loads are linear, such as incandescent lamps, heating elements, motors, alarms, solenoids, and relay coils. A *nonlinear load* is any load in which the instantaneous load current is not proportional to the instantaneous voltage. Nonlinear loads include personal computers, copiers, printers, electronic lighting ballasts, programmable logic controllers (PLCs), and variable frequency drives. A severe power quality problem can cause a malfunction in the system or equipment. Replacing equipment that is malfunctioning can be a short-term solution to the problem. However, the problem can occur again if the root source of the problem has not been identified and corrected. Basic troubleshooting steps include a sequence of tasks in which the system is inspected, measurements are made, and corrective action is taken:

1. Identify the problem by interviewing the personnel affected, observing operation, and taking measurements. Electrical test equipment such as digital multimeters, power quality meters, and power monitoring equipment can be used to determine the source and magnitude of the problem. Inspection provides evidence regarding the problem and its causes, such as overheating, loose connections, improper wiring/grounding, etc. Measurements provide data such as voltage level, current draw, and presence of harmonics.
2. Identify a solution, such as proper grounding, adding transient suppressors, or an uninterruptible power system (UPS).
3. Take corrective action by implementing the solutions.
4. Make observations and take measurements after the solution is implemented to confirm that the problem is corrected.
5. Implement a preventive maintenance program that tracks the operation of the system to verify that the solution works over time and to identify additional problems that could cause damage or downtime.

Troubleshooting Power Quality Problems

Some power quality problems are easier to identify than others. For example, nonlinear loads always produce harmonic distortion when operating, and problems caused by improper grounding and wiring continue until corrected. Other problems, such as sags, swells, undervoltage, and overvoltage, may occur at certain times of the day, week, or month. Some problems, like transients and noise, can occur any time and vary in magnitude. A solution to a power quality problem must solve the current problem and prevent potential future problems. **See Troubleshooting Power Quality Problems.**

Voltage Changes

All electrical and electronic equipment is rated for operation at a specific voltage. The rated voltage is actually a voltage range. Typically, this range used to be ±10%. However, with many components derated to save energy and operating cost, the range now is typically +5% to −10%. This range is used because an overvoltage is generally more damaging than an undervoltage. Equipment manufacturers, utility companies, and regulating agencies must routinely deal with changes in system voltage. Voltage change amount and duration can be broadly classified into standard industrial terms. **See Voltage Changes.**

TROUBLESHOOTING POWER QUALITY PROBLEMS

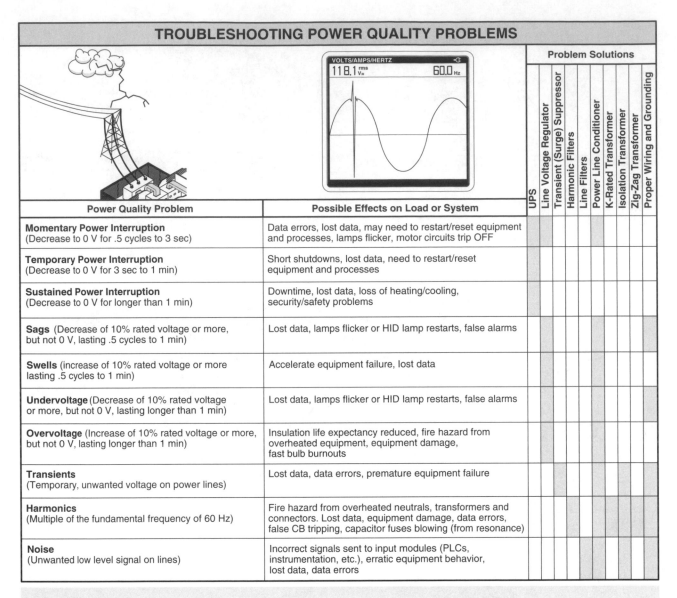

Power Quality Problem	Possible Effects on Load or System	UPS	Line Voltage Regulator	Transient (Surge) Suppressor	Harmonic Filters	Line Filters	Power Line Conditioner	K-Rated Transformer	Isolation Transformer	Zig-Zag Transformer	Proper Wiring and Grounding
Momentary Power Interruption (Decrease to 0 V for .5 cycles to 3 sec)	Data errors, lost data, may need to restart/reset equipment and processes, lamps flicker, motor circuits trip OFF										
Temporary Power Interruption (Decrease to 0 V for 3 sec to 1 min)	Short shutdowns, lost data, need to restart/reset equipment and processes										
Sustained Power Interruption (Decrease to 0 V for longer than 1 min)	Downtime, lost data, loss of heating/cooling, security/safety problems										
Sags (Decrease of 10% rated voltage or more, but not 0 V, lasting .5 cycles to 1 min)	Lost data, lamps flicker or HID lamp restarts, false alarms										
Swells (increase of 10% rated voltage or more lasting .5 cycles to 1 min)	Accelerate equipment failure, lost data										
Undervoltage (Decrease of 10% rated voltage or more, but not 0 V, lasting longer than 1 min)	Lost data, lamps flicker or HID lamp restarts, false alarms										
Overvoltage (Increase of 10% rated voltage or more, but not 0 V, lasting longer than 1 min)	Insulation life expectancy reduced, fire hazard from overheated equipment, equipment damage, fast bulb burnouts										
Transients (Temporary, unwanted voltage on power lines)	Lost data, data errors, premature equipment failure										
Harmonics (Multiple of the fundamental frequency of 60 Hz)	Fire hazard from overheated neutrals, transformers and connectors. Lost data, equipment damage, data errors, false CB tripping, capacitor fuses blowing (from resonance)										
Noise (Unwanted low level signal on lines)	Incorrect signals sent to input modules (PLCs, instrumentation, etc.), erratic equipment behavior, lost data, data errors										

Momentary Power Interruption

A *momentary power interruption* is a decrease to 0 V on one or more power lines lasting from .5 cycles up to 3 sec. All power distribution systems have momentary power interruptions during normal operations. Momentary power interruptions can be caused when lightning strikes nearby, by utility grid switching during a problem (short on one line), or during open circuit transition switching. *Open circuit transition switching* is a process in which power is momentarily disconnected when switching a circuit from one voltage supply (or level) to another.

Temporary Power Interruption

A *temporary power interruption* is a decrease to 0 V on one or more power lines lasting for more than 3 sec up to 1 min. Automatic circuit breakers and other circuit protection equipment protect all power distribution systems. Circuit protection equipment is designed to remove faults and restore power. An automatic circuit breaker typically takes from 20 cycles to about 5 sec to close. If the power is restored, the power interruption is only temporary. If power is not restored, a temporary power interruption becomes a sustained power interruption. A temporary power interruption can also be caused by a time gap between power interruptions and when a back-up power supply (generator) takes over, or if someone accidentally opens the circuit by switching the wrong circuit breaker switch.

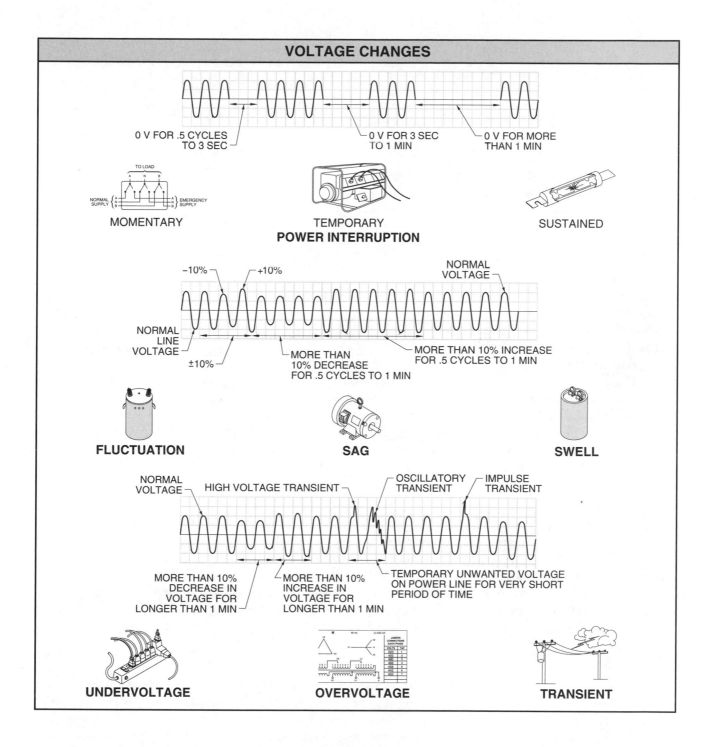

VOLTAGE CHANGES

0 V FOR .5 CYCLES TO 3 SEC

0 V FOR 3 SEC TO 1 MIN

0 V FOR MORE THAN 1 MIN

TO LOAD

NORMAL SUPPLY

EMERGENCY SUPPLY

MOMENTARY

TEMPORARY

POWER INTERRUPTION

SUSTAINED

−10%   +10%

NORMAL VOLTAGE

NORMAL LINE VOLTAGE

±10%

MORE THAN 10% DECREASE FOR .5 CYCLES TO 1 MIN

MORE THAN 10% INCREASE FOR .5 CYCLES TO 1 MIN

FLUCTUATION

SAG

SWELL

NORMAL VOLTAGE

HIGH VOLTAGE TRANSIENT

OSCILLATORY TRANSIENT

IMPULSE TRANSIENT

MORE THAN 10% DECREASE IN VOLTAGE FOR LONGER THAN 1 MIN

MORE THAN 10% INCREASE IN VOLTAGE FOR LONGER THAN 1 MIN

TEMPORARY UNWANTED VOLTAGE ON POWER LINE FOR VERY SHORT PERIOD OF TIME

UNDERVOLTAGE

OVERVOLTAGE

TRANSIENT

Sustained Power Interruption

A *sustained power interruption* is a decrease to 0 V on all power lines for a period of more than 1 min. Even the best power distribution systems have a complete loss of power at some time. Sustained power interruptions (outages) are commonly the result of storms, circuit breakers tripping, fuses blowing, and/or damaged equipment.

The effect of a power interruption on the load depends on the load and the application. If a power interruption could cause equipment, production, and/or security problems that are not acceptable, and uninterruptible power system can be used. An *uninterruptible power system* (UPS) is a power supply that provides constant on-line power when the primary power supply is interrupted. For long-term power interruption protection, a generator UPS is used. For short-term power interruption, a static UPS is used. **See Uninterruptible Power System (UPS).**

A generator UPS is powered by a diesel, gasoline, natural gas, or propane engine connected to an electrical generator. If there is any power interruption in the time period between the loss of main utility power and when the generator starts providing power, the generator is usually classified as a standby (emergency) power supply.

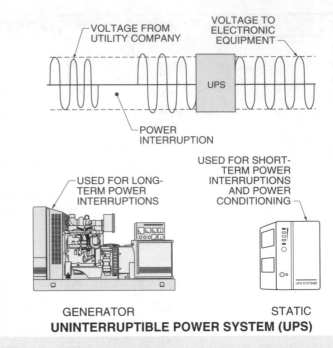

UNINTERRUPTIBLE POWER SYSTEM (UPS)

A static UPS changes AC power supplied from the utility company to DC with a converter, and changes it back into AC with an inverter. The DC power is used to maintain a charge on a bank of batteries/capacitors. The batteries are used to supply power when AC power from the utility is interrupted. A static UPS supplies power for a specified time from minutes to hours depending on the UPS design and load(s) connected to the system. In some applications, a static UPS is used to supply power during the time period when utility power is lost and the generator(s) come(s) on-line.

Voltage Fluctuations

A *voltage fluctuation* is an increase or decease in the normal line voltage within the range of ±10%. In any power distribution system, frequency is generally constant, current constantly changes as loads are added and removed, and voltage is typically held within a normal range of approximately ±5%. Voltage fluctuations are commonly caused by overloaded transformers, imbalanced transformer loading, and/or high impedance caused by long circuit runs, undersized conductors, poor electrical connections, and/or loose connections on the system. Voltage fluctuations usually do not affect equipment performance. Equipment designed as "sensitive loads" has power supplies designed to tolerate normal voltage fluctuations.

Voltage Sags

A *voltage sag (voltage dip)* is more than a 10% decrease (but not to 0 V) below the normal rated line voltage lasting from .5 cycles up to 1 min. Voltage sags commonly occur when a high-current load is turned ON and voltage on the power line drops below the normal voltage fluctuation (–10%) for a short period of time. Voltage sags commonly occur with the switching ON of large motors or temporary short circuits in utility power lines. If the short circuit opens a breaker, a power interruption occurs. Voltage sags are often followed by voltage swells as regulators overcompensate for the voltage sag.

Every electrical or electronic load has a low voltage limit and a high voltage limit in which the load is designed to operate properly. If the voltage dips below the lower voltage limit of a device, damage or problems may result in the loss of memory, data loss, and/or equipment malfunction. If the voltage increases above the voltage limit of a device, permanent equipment or component damage may result. It typically takes much less time to damage hardware with a higher voltage than with a lower voltage.

A digital multimeter (DMM) with MIN MAX Recording mode can be used to measure voltage sags. Voltage should be measured over time. The MIN MAX Recording mode is able to capture and display a low-voltage (minimum voltage) condition. However, the measurements displayed will not indicate when or how long the low-voltage condition was present. It is better to use a power quality meter for the best indication of voltage problems. **See Voltage Sag Measurement.**

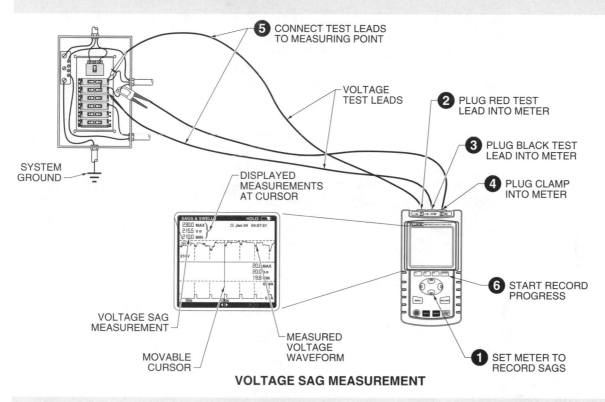

VOLTAGE SAG MEASUREMENT

To measure voltage sags, apply the procedure:
1. Set the meter to record voltage and current over time. If using a power quality meter, the meter will be set to measure SAGs and SWELLs. If using a digital multimeter (DMM), the DMM selector switch should be set to AC voltage.

2. Plug the red test lead into the voltage jack.

3. Plug the black test lead into the common jack.

4. If using a power quality meter, also plug in the current clamp. If two DMMs are available, one can be used for the voltage measurement and the second DMM with a clamp-on current probe accessory is used to measure current. Measurements can be separately made if one DMM is available.

5. Connect the voltage test leads in the circuit at the point measurements are to be taken. Also, connect the current clamp into the circuit, if using a power quality meter.

6. Start the recording process. The record process is started on a power quality meter by pressing RECORD or START. The record process is started on a DMM by pressing the MIN MAX button.

7. Let the meter record over time.

8. Stop the record process and read the displayed information. Look at the minimum (sag) voltage measurement. This is done by scrolling through the MIN MAX mode to the MIN measurement on a DMM, or by reading the displayed graph on a power quality meter.

9. Remove the meter leads, if additional measurements are not going to be made.

For measurement accuracy, voltage should be measured and recorded at different times of the day and at different times of production and/or system operation. Measurements should also be taken as part of a preventive maintenance program. If the voltage fluctuates more than ±8%, a voltage regulator (stabilizer) should be added. A *voltage regulator (stabilizer)* is a device that provides precise voltage regulation to protect equipment from voltage sags (voltage dips) and swells (voltage surges).

Voltage Swells

A *voltage swell (voltage surge)* is more than a 10% increase above the normal rated line voltage lasting from .5 cycles up to 1 min. Voltage swells commonly occur when a large load is turned OFF and voltage on the power line increases above the normal voltage fluctuation (+10%) for a short period of time. For example, a voltage swell commonly occurs in office areas of a plant when production lines with large loads are suddenly shut down.

Voltage swells are not as common as voltage sags. However, voltage swells can be more destructive than voltage sags because it generally takes less time to damage hardware with higher voltage than with a lower voltage. Even a very short high-voltage condition can cause permanent equipment or component damage.

As with voltage sags, a DMM with MIN MAX Recording mode can be used to measure voltage swells. For best results, voltage should be measured over time using a power quality meter. **See Voltage Swell Measurement.**

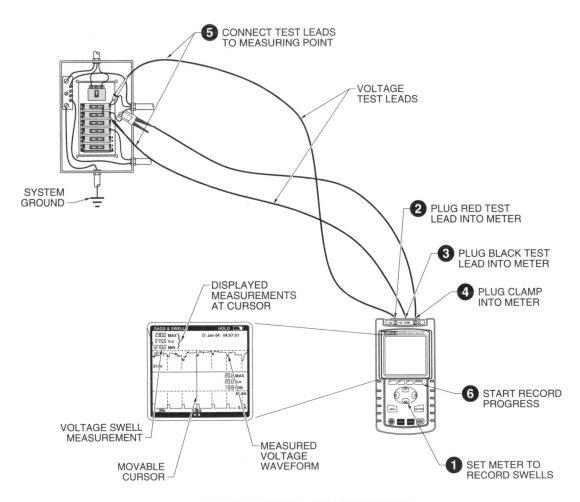

VOLTAGE SWELL MEASUREMENT

To measure voltage swells, apply the procedure:

1. Set the meter to record voltage and current over time. If using a power quality meter, the meter will be set to measure SAGs and SWELLs. If using a (DMM), the DMM selector switch should be set to AC voltage.

2. Plug the red test lead into the voltage jack.

3. Plug the black test lead into the common jack.

4. If using a power quality meter, also plug in the current clamp. If two DMMs are available, one can be used for the voltage measurement and the second DMM with a clamp-on current probe accessory is used to measure current. Measurements can be separately made if one DMM is available.

5. Connect the voltage test leads in the circuit at the point measurements are to be taken. Also, connect the current clamp into the circuit, if using a power quality meter.

6. Start the recording process. The record process is started on a power quality meter by pressing RECORD or START. The record process is started on a DMM by pressing the MIN MAX button.

7. Let the meter record over time.

8. Stop the record process and read the displayed information. Look at the maximum (swell) voltage measurement. This is done by scrolling through the MIN MAX mode to the MAX measurement on a DMM, or by reading the displayed graph on a power quality meter.

9. Remove the meter leads, if additional measurements are not going to be made.

Undervoltage

Undervoltage is more than a 10% decrease (but not to 0 V) below the normal rated line voltage for a period of longer than 1 min. At times, voltage on a power line drops below the normal voltage fluctuation (–10%) for a longer period of time. Undervoltage (low voltage) is more common than overvoltage (high voltage). Undervoltages are commonly caused from overloaded transformers, undersized conductors, excessively long conductor runs, too many loads placed on a circuit, peak power usage periods, and/or brownouts. A *brownout* is the deliberate reduction of the voltage level by a power company to conserve power during peak usage times.

Overvoltage

Overvoltage is more than a 10% increase above the normal rated line voltage for a period of longer than 1 min. Depending on the cause of the overvoltage, voltage increases above the normal voltage fluctuation (+10%) may occur for a long period of time. Overvoltages are sometimes caused when loads are near the beginning of a power distribution system, or if taps on a transformer are set improperly. *Taps* are connection points provided along the transformer coil. Taps are commonly provided at 2.2% increments along one end of the transformer coil.

Line Voltage Regulators

A *line voltage regulator* is a power conditioner that maintains a specified output voltage for a given voltage input fluctuation. AC voltage generated and distributed by the utility company is constantly fluctuating. As long as the voltage fluctuations stay within a normal range of ±10% (+5% to –10% for some equipment), the loads connected to the system should operate properly. If the voltage fluctuation is outside the normal range, a problem may be caused by the sag, swell, undervoltage, or overvoltage. Using a line voltage regulator can prevent short duration voltage fluctuations outside the normal range. Line voltage regulators are available for 1ϕ and 3ϕ protection. Line voltage regulators are rated for standard AC line voltages such as 120 V, 208 V, 240 V, 277 V, and 480 V. A line voltage regulator is designed so large voltage variations on the input side (typically rated for +10% to –25%) do not significantly affect the voltage output. The input variation is typically rated for approximately +10% to –25%, with a ±1% output voltage. Line voltage regulators often include noise suppression and surge protection in addition to voltage regulation. **See Line Voltage Regulator.**

Transient Voltage

A *transient voltage (transient)* is a temporary, unwanted voltage in an electrical circuit. Transient voltages may range from a few volts to several thousand volts and last from a few microseconds up to a few milliseconds. *Oscillatory transient voltages* are transient voltages commonly caused by turning OFF high inductive loads and by switching large utility power factor correction capacitors. *Impulse transient voltages* are transient voltages commonly caused by lightning strikes and result in a short, unwanted voltage placed on the power distribution system. Power factor correction capacitors are used by utility companies to correct power factor on the power lines.

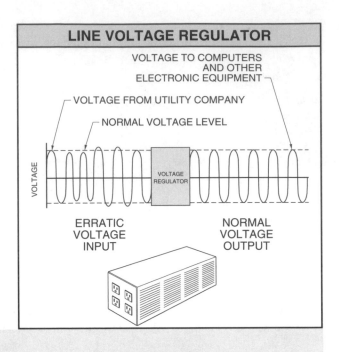

Transient voltage is caused by the sudden release of stored energy due to lightning strikes, unfiltered electrical equipment, contact bounce, arcing, and electricity being switched ON and OFF. The problem with high voltage transients is that it only takes one transient to damage a circuit or component. **See Transient Voltage Measurement.**

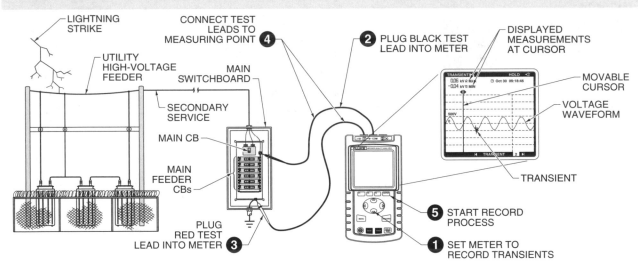

TRANSIENT VOLTAGE MEASUREMENT

To measure transient voltages, apply the procedure:
1. Set the meter to record transients. If using a DMM, the selector switch is set to AC voltage. If using a power quality meter, the meter will be set to the transient voltage measurement mode.
2. Plug the black test lead into the common jack.
3. Plug the red test lead into the voltage jack.
4. Connect the voltage test leads in the circuit at the point measurements are to be taken.
5. Start the recording process. The record process is started on a power quality meter by pressing RECORD or START. The record process is started on a DMM by first pressing the MIN MAX button and then the PEAK button. A DMM must have a "peak" record function in order to capture transients.

6. Let the meter record over a period of time.

7. Stop the record process and read the displayed information. Look at the maximum peak transient. This is done by scrolling through the MIN MAX mode to the MAX display on a DMM, or by scrolling through each captured transient on a power quality meter.

8. Remove the meter leads if additional measurements are not going to be made.

Transient voltages differ from voltage swells and voltage sags. Transient voltages are typically large in magnitude, have a short duration, have a steep (short) rise time, and are very erratic. As with voltage sags and voltage swells, a power quality meter should be used to monitor for transients over a period of time. The size, duration, and time of transients can be displayed at a later time. If transients are identified as the problem, a voltage surge suppressor (surge protection device) should be used. A *surge protection device* is a device that limits voltage surges that may be present on the power lines.

Computers, electronic circuits, and specialized equipment require protection against transient voltage spikes. Protection methods commonly include proper wiring, grounding, shielding of the power lines, and surge suppressors. A *surge suppressor* is an electrical device that provides protection from high-level transients by limiting the level of voltage allowed downstream from the surge suppressor. Surge suppressors should be installed at service entrance panels, distribution panels, and individual loads. **See Surge Suppressor.**

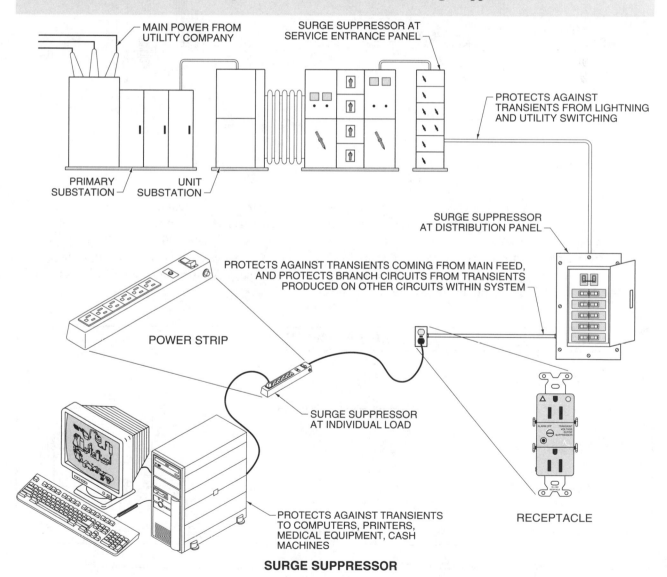

SURGE SUPPRESSOR

Transients can cause significant damage to electronic equipment, and protection should be installed as required in the system. Protection is installed at individual loads and distribution panels to protect from internally generated transient voltage. For example, a surge protector should be installed on 1ϕ and 3ϕ panels in which transient protecting devices are connected along with sensitive electronic equipment. For example, panels that feed computer rooms, hospitals, laboratories, banks, and other areas that include electronic equipment should have surge protectors. **See Power Protection at Panel.**

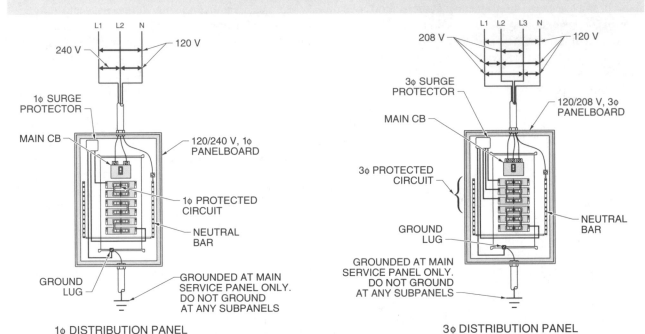

POWER PROTECTION AT PANEL

Many transients can be stopped from entering the system at the source. A transient is produced when inductive loads (motors, solenoids, coils) are turned OFF. These transients can cause problems if allowed to enter the distribution system. A snubber circuit can be used to suppress a voltage transient. Typical snubber circuits use an RC (resistor/capacitor), MOV (metal oxide varistor), or a diode depending on the load type. For example, a snubber circuit for small AC loads uses an RC. A snubber circuit for large AC loads uses an MOV. A snubber circuit for DC loads uses a diode. **See Snubber Circuits.**

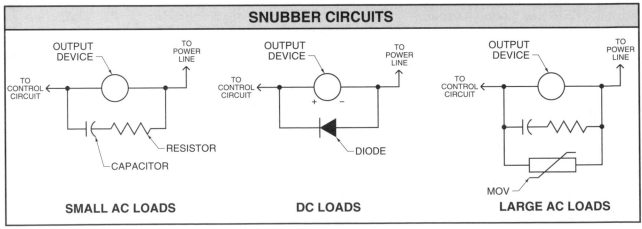

 Refer to Activity 15-1—Power Quality Problems..

APPLICATION 15-2—Harmonics

A *harmonic* is a frequency that is an integer (whole number) multiple (2nd, 3rd, 4th, 5th, etc.) of the fundamental frequency. Harmonics can be either voltage harmonics and/or current harmonics. However, it is current harmonics that produce most problems in the distribution system. Each harmonic has a name (number) or order (5th order harmonic), frequency, and sequence. The 1st harmonic (60 Hz) is the fundamental harmonic. Other multiples of the fundamental harmonic are the 2nd harmonic (120 Hz), 3rd harmonic (180 Hz), 4th harmonic (240 Hz), etc. Some waveforms (ideal sine waves) contain only the fundamental frequency. Waveforms may include odd harmonics, even harmonics, or both odd and even harmonics, in addition to the fundamental frequency. **See Harmonics Classification.**

Harmonic sequence is the phasor rotation with respect to the fundamental (60 Hz) frequency. *Phasor rotation* is the order in which waveforms from each phase (phase A, phase B, and phase C) cross zero. Phasor rotation is simplified by using lines and arrows instead of waveforms to show phase relationships. Phase sequence of a harmonic is important because it determines the effect the harmonic has on the operation of the loads and components such as conductors within a power distribution system.

Positive sequence harmonics (1st, 4th, 7th, etc.) have the same phase sequence as the fundamental (1st) harmonic, and cause additional heat in conductors, circuit breakers, and panels in a power distribution system.

HARMONICS CLASSIFICATION

Harmonics	Frequency*	Sequence
Fundamental (1st)	60	Positive (+)
2nd	120	Negative (−)
3rd	180	Zero (0)
4th	240	(+)
5th	300	(−)
6th	360	(0)
7th	420	(+)
8th	480	(−)
9th	540	(0)
10th	600	(+)

* in Hz

POSITIVE SEQUENCE HARMONICS SEQUENCE (1st, 4th, 7th, ETC.)

NEGATIVE SEQUENCE HARMONICS SEQUENCE (2nd, 5th, 8th, ETC.)

→ A
----→ B
ooooooo→ C

ZERO SEQUENCE HARMONICS (NO ROTATING FIELD) SEQUENCE (3rd, 6th, 9th, ETC.)

PHASOR ROTATION

Negative sequence harmonics (2nd, 5th, 8th, etc.) have a phase sequence opposite the phase sequence of the fundamental (1st) harmonic, which causes a rotating field in the opposite direction. Like positive sequence harmonics, negative sequence harmonics cause additional heat in power distribution system components such as conductors, circuit breakers, and panels. Negative sequence harmonics also cause problems in induction motors because negative sequence harmonics rotate in the reverse direction. The reverse rotation is not enough to cause the motor to reverse direction, but it does reduce forward torque of the motor to cause more motor heat than normal.

Zero sequence harmonics (3rd, 6th, 9th, etc.) do not produce a rotating field in either direction. However, they can result in component and system heating. Zero sequence harmonics do not cancel, but add together in the neutral conductor of 3ϕ, 4-wire systems. This is a major system problem because there is no fuse or circuit breaker in the neutral conductor to limit current flow. The danger of higher than normal current in the neutral is fire. **See Zero Sequence Harmonics in Neutral Conductors.**

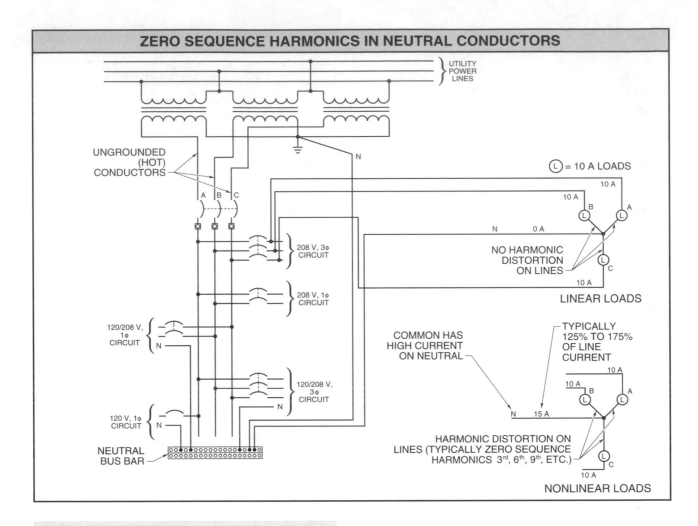

ZERO SEQUENCE HARMONICS IN NEUTRAL CONDUCTORS

UTILITY POWER LINES

UNGROUNDED (HOT) CONDUCTORS

N

(L) = 10 A LOADS

10 A

10 A

B   A

N   0 A

NO HARMONIC DISTORTION ON LINES

208 V, 3φ CIRCUIT

208 V, 1φ CIRCUIT

120/208 V, 1φ CIRCUIT

120/208 V, 3φ CIRCUIT

120 V, 1φ CIRCUIT

NEUTRAL BUS BAR

C

10 A

LINEAR LOADS

COMMON HAS HIGH CURRENT ON NEUTRAL

TYPICALLY 125% TO 175% OF LINE CURRENT

10 A

10 A

B   A

N   15 A

HARMONIC DISTORTION ON LINES (TYPICALLY ZERO SEQUENCE HARMONICS 3rd, 6th, 9th, ETC.)

C

10 A

NONLINEAR LOADS

Knowledge of harmonics present on a power line is important for any personnel working on a power distribution system. A power quality meter can be used to measure the amount of voltage harmonics and current harmonics on a line. The amount of each harmonic (2nd, 3rd, etc.) present on the line and related information is indicated by data and the frequency spectrum on the graphic display. **See Harmonic Identification.**

In general, even number harmonics (2nd, 4th, 6th, 8th, etc.) tend to disappear or occur at levels that do not cause major problems. Likewise, higher harmonics have smaller and smaller amplitudes, and are less important in affecting the overall power system. However, odd number harmonics are more likely to be present and do cause problems. For example, the third harmonic and odd multiples (3rd, 9th, 15th, 21st, etc.) of the third harmonic (triplen harmonics or triplens) cause such problems as overloading of neutral conductors, telephone interference, and transformer heating.

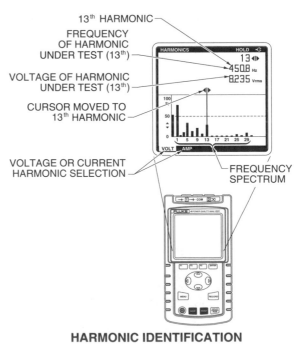

13th HARMONIC

FREQUENCY OF HARMONIC UNDER TEST (13th)

VOLTAGE OF HARMONIC UNDER TEST (13th)

CURSOR MOVED TO 13th HARMONIC

VOLTAGE OR CURRENT HARMONIC SELECTION

FREQUENCY SPECTRUM

HARMONIC IDENTIFICATION

Harmonic Filters

A *harmonic filter* is a device used to reduce harmonic frequencies and total harmonic distortion. Harmonic filters may include different types of circuits or components designed to reduce harmonic currents. A 1ϕ harmonic filter is used to reduce the harmonics from nonlinear 1ϕ loads by minimizing the 3rd and other triplen harmonics. Three-phase harmonic filters, also called trap filters, are used to reduce harmonics produced by 1ϕ nonlinear loads connected to a 3ϕ system, or 3ϕ nonlinear loads such as variable speed (frequency) drives connected to the loads. Harmonic filters should be installed as close as possible to the nonlinear load. With 3ϕ drives, they are typically installed at the service equipment. **See Harmonic Filter.**

Noise

Noise can enter the power distribution system directly on the wires or grounds, or through magnetic coupling of adjacent wires. Noise is produced on power lines from two different points. *Common mode noise* is noise produced between the ground and hot lines, or the ground and neutral lines. *Transverse mode noise* is noise produced between the hot and neutral lines. **See Noise.**

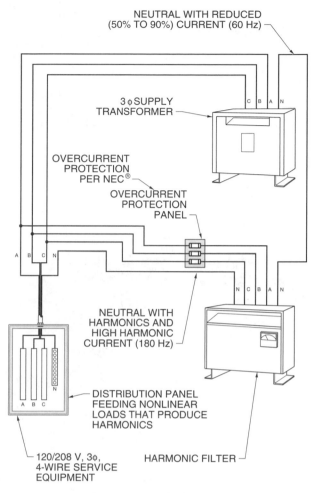

HARMONIC FILTER

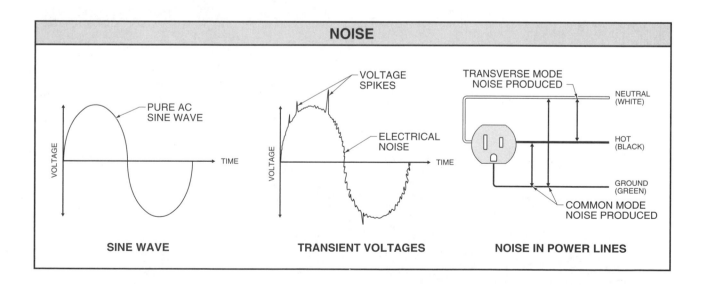

NOISE

SINE WAVE

TRANSIENT VOLTAGES

NOISE IN POWER LINES

Arcing at motor brushes, ground faults, poor grounds, radio transmitters, ignition systems, and the opening of electrical contacts also causes common mode noise. The opening of electrical contacts produces noise because an arc is produced as they are pulled apart. The higher the current in the circuit being opened, the larger and longer the arc. Transverse mode noise is also caused by welders, switched power supplies, and the firing of silicon-controlled rectifiers (SCRs) in electrical equipment.

Noise in a system can produce false signals in electronic circuits leading to processing errors, incorrect data transfer, and printer errors. Using line filters, power line conditioners, noise suppressors, and proper grounding can reduce noise problems.

 Refer to **Activity 15-2—Harmonics** ....

 *Refer to Quick Quiz® on CD-ROM*

Refer to Chapter 15 in the **Troubleshooting Electrical/Electronic Systems Workbook** *for additional questions.*

Troubleshooting Power Quality

15

Name_____ Date _____

ACTIVITY 15-1—Power Quality Problems

A digital multimeter (DMM) with a recording mode can be used to measure voltage changes over time. The MIN MAX recording mode is a meter mode that captures and stores the lowest and highest measurement for later display. Pressing the MIN MAX button after the DMM is connected to the circuit under test indicates minimum or maximum values. The abbreviation MIN MAX is displayed with the value to identify the measured value. Using the meter measurements and given information, answer the questions.

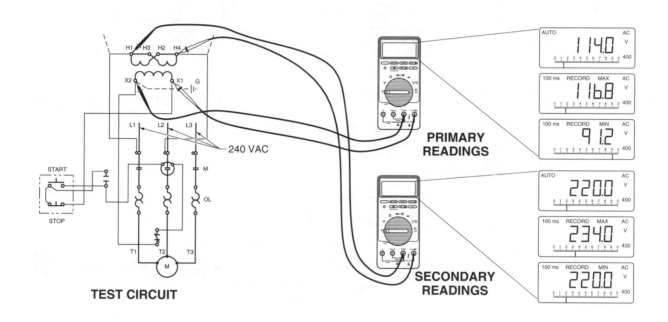

_____ 1. Do the measurements indicate a power interruption problem?

_____ 2. Do the readings indicate a voltage sag or undervoltage condition on the primary side of the transformer?

_____ 3. Do the readings indicate a voltage sag or undervoltage condition on the secondary side of the transformer?

_____ 4. Can the readings on the secondary side of the transformer indicate if the problem was a voltage sag or an undervoltage condition?

_____ 5. Based on the readings, is the transformer undersized?

A recording meter can be used to obtain more valuable troubleshooting data than a DMM. A DMM measurement captures a value only at a particular moment. The DMM MIN MAX Recording mode can be used to capture high and low values. However, the measurements displayed do not indicate when the high or low value was captured or the number of high and low values that were captured. When using a recording meter, the measurement can be captured and recorded over a long time period such as minutes, hours, or days. Using the recorded meter measurements and given information, answer the questions.

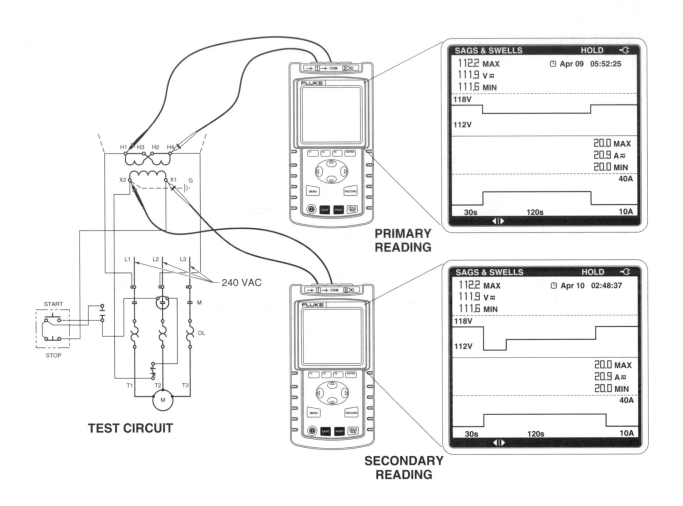

TEST CIRCUIT

PRIMARY READING

SECONDARY READING

_____ **6.** Do the measurements indicate a power interruption problem?

_____ **7.** Do the readings indicate a voltage sag or undervoltage condition on the primary side of the transformer?

_____ **8.** Do the readings indicate a voltage sag or undervoltage condition on the secondary side of the transformer?

_____ **9.** Can the readings on the secondary side of the transformer indicate if the problem was a voltage sag or an undervoltage condition?

_____ **10.** Is the problem either a voltage sag problem or undervoltage problem?

Troubleshooting power quality problems should start at the load(s) and work back through the power distribution system. A troubleshooting checklist provides a sequence for identifying the problem source and documents steps taken to solve the problem. Using the information provided in the load problems checklist, answer the questions.

POWER QUALITY TROUBLESHOOTING CHECKLIST
LOAD PROBLEMS

Problem Observed or Reported:
❑ Overloads Tripping   ☑ Computer Hardware Problems   ❑ Erratic Operation   ❑ Shortened Life
❑ Other _____

Load Type:
☑ Computer   ❑ Printer/Copier   ❑ PLC   ❑ Drive   ❑ Motor   ❑ Other _____

Problem Pattern:
Day(s) of week:
❑ Continuous ☑ Random ❑ Monday ❑ Tuesday ❑ Wednesday ❑ Thursday ❑ Friday ❑ Saturday ❑ Sunday
Time(s):
❑ Continuous ☑ Random ❑ Always Same Time _____ ❑ Morning ❑ Afternoon ❑ Evening ❑ Night

Problem or Load History:
Has problem been observed or reported before? ❑ No ☑ Yes *Second computer problem in week*
Was any corrective action taken? ☑ No ❑ Yes _____
Are other loads affected? ☑ No ❑ Yes _____
Are nonlinear loads in area, or on same circuit? ❑ No ☑ Yes *Printer*
Has there been any recent work or changes made to system lately? ❑ No ☑ Yes *New printer added*

Possible Problem(s):
❑ Operator Error   ❑ Power Interruptions   ❑ Sags or Undervoltage   ❑ Swells or Overvoltage   ❑ Harmonics
☑ Transients or Noise   ❑ Improper Wiring/Grounding   ❑ Undersized Load   ❑ Undersized System
❑ Improperly Sized Protection Devices   ❑ Other _____

Measurements Taken at Load:
Line Voltage _112_ V, Neutral-to-Ground Voltage _2_ V, Current _8_ A, Power _950_ W, _1250_ VA, _320_ VAR
Power Factor _.7_ PF, ____ DPF, Voltage THD ____, Current THD ____, K Factor ____ Other _____

Waveform Shape:
Voltage Waveform:
☑ Sinusoidal   ❑ Non-Sinusoidal   ❑ Flat-Topped   ❑ Other _____
Current Waveform:
☑ Sinusoidal   ❑ Non-Sinusoidal   ❑ Pulsed   ❑ Other _____

Measurements Taken at Load (Over Time):
Normal Voltage _114_ V, Lowest Sag _109_ V at ____ (Time), Highest Swell _119_ V at ____ (Time)
Highest Inrush Current _14_ A at ____ (Time)
Number of transients recorded _16_ over a time period of _30 min_ at a level of _*425_ % above normal

Possible Problem Solution(s):
❑ UPS   ❑ K-Rated Transformer   ❑ Isolation Transformer   ❑ Zig-Zag Transformer   ❑ Line Voltage Regulator
☑ Surge Suppressor   ☑ Power Conditioner   ❑ Proper Wiring and Grounding   ❑ Harmonic Filter   ❑ Derate Load
❑ Proper Fuses/CBs/Monitors   ☑ Other *Move printer to separate circuit*

Note: High transients recorded when the printer on circuit operates.

_____ **11.** Was there evidence of a power interruption problem?

_____ **12.** Was there evidence of a voltage sag or voltage swell problem?

_____ **13.** Was there evidence of an undervoltage or overvoltage problem?

_____ **14.** Based on the findings, should other circuits be checked to determine that computers and printers are not added to the same circuit, unless the computers are protected with a surge suppressor?

Using the information provided in the building distribution problems checklist, answer the questions.

POWER QUALITY TROUBLESHOOTING CHECKLIST
FACILITY TRANSFORMER AND MAIN SERVICE EQUIPMENT PROBLEMS

Problem Observed or Reported:
❏ CBs Tripping/Fuses Blowing ❏ Conduit Overheating ☑ Overheated Neutrals ❏ Electrical Shocks
❏ Damaged Equipment ❏ Other _____

Distribution Type:
❏ 1φ ☑ 3φY ❏ 3φΔ ❏ Fuses ☑ CBs ❏ Voltage(s) _____ V ❏ Amperage Rating _____ A
❏ Other _____

Problem Pattern:
Day(s) of week:
☑ Continuous ❏ Random ❏ Monday ❏ Tuesday ❏ Wednesday ❏ Thursday ❏ Friday ❏ Saturday ❏ Sunday
Time(s):
☑ Continuous ❏ Random ❏ Always Same Time _____ ❏ Morning ❏ Afternoon ❏ Evening ❏ Night

Problem or Distribution History:
Has problem been observed or reported before? ☑ No ❏ Yes _____
Was any corrective action taken? ☑ No ❏ Yes _____
Are other parts of system affected? ☑ No ❏ Yes _____
Have additional loads been added to system? ❏ No ☑ Yes _____
Has there been any recent work or changes made to system lately? ❏ No ☑ Yes *Computer added* ___
Are large power loads being switched ON/OFF? ☑ No ❏ Yes _____
Is main service panel properly grounded? ❏ No ☑ Yes _____
Are any subpanels grounded? ❏ No ☑ Yes _____
Has there been a recent lightning storm? ☑ No ❏ Yes _____
Has there been a recent utility feeder outage? ☑ No ❏ Yes_____

Possible Problem(s):
❏ Conductors Undersized (Hot) ☑ Neutral Conductors Shared/Undersized ☑ High Number of Nonlinear Loads
❏ Voltage/Current Unbalance ☑ Harmonics ❏ System Undersized ❏ Improper Wiring/Grounding
❏ Other *high level of 3rd harmonic line* _____

Measurements Taken:
Taken at Panel *12* _____ Located at *Office area* _____
Voltage *12* V, Current *13* A, Power ____ W, ____ VA, ____ VAR, Power Factor ____ PF, ____ DPF
Voltage THD ____, Current THD ____, K Factor ____, Other _____

Waveform Shape:
Voltage Waveform:
❏ Sinusoidal ❏ Non-Sinusoidal ☑ Flat-Topped ❏ Other _____
Current Waveform:
❏ Sinusoidal ☑ Non-Sinusoidal ❏ Pulsed ❏ Other _____

Measurements Taken at Load (Over Time):
Normal Voltage *12* V, Lowest Sag *108* V, Highest Swell *117* V
Number of transients recorded *0* over a time period of *1 hr.* at a level of _____ % above normal

Possible Problem Solution(s):
❏ Oversize Neutrals ☑ Run Separate Neutrals ❏ Additional Transformer ☑ Harmonic Filter ❏ Change to K-Rated Transformer
❏ Add Subpanel ❏ Separate Loads ❏ Proper Wiring and Grounding ❏ Proper Fuses/CBs/Monitors
❏ Power Factor Correction Capacitors ❏ Change Transformer Size ❏ Surge Suppressor
❏ Other _____

_____ **15.** Was the documented problem life threatening?

_____ **16.** Did the troubleshooter take the correct action by fixing the problem on the spot and not waiting to show a supervisor?

_____ **17.** Should the troubleshooter's recommended solution be acted on immediately or can it wait?

ACTIVITY 15-2—Harmonics

Using the information provided in the facility transformer and main service equipment problems checklist, answer the questions.

POWER QUALITY TROUBLESHOOTING CHECKLIST
BUILDING DISTRIBUTION PROBLEMS

Problem Observed or Reported:
❏ CBs Tripping/Fuses Blowing ❏ Conduit Overheating ❏ Overheated Neutrals ☑ Electrical Shocks
❏ Damaged Equipment ❏ Humming/Buzzing Noise ❏ Other _____

Distribution Type:
❏ 1φ ☑ 3φY ❏ 3φΔ ❏ Fuses ☑ CBs ❏ Voltage(s) _____ V ❏ Amperage Rating _____ A
❏ Other _____

Problem Pattern:
Day(s) of week:
❏ Continuous ☑ Random ❏ Monday ❏ Tuesday ❏ Wednesday ❏ Thursday ❏ Friday ❏ Saturday ❏ Sunday
Time(s):
❏ Continuous ☑ Random ❏ Always Same Time _____ ❏ Morning ❏ Afternoon ❏ Evening ❏ Night

Problem or Distribution History:
Has problem been observed or reported before? ❏ No ☑ Yes _____
Was any corrective action taken? ☑ No ❏ Yes _____
Are other parts of system affected? ☑ No ❏ Yes _____
Are nonlinear loads in area or on same circuit? ❏ No ☑ Yes _____
Has there been any recent work or changes made to system lately? ☑ No ❏ Yes _____
Are large power loads being switched ON/OFF? ☑ No ❏ Yes _____
Is panel properly grounded? ☑ No ❏ Yes *Loose ground found. Tightened as required* ____

Possible Problem(s):
❏ Conductors Undersized (Hot) ❏ Neutral Conductors Shared/Undersized ❏ High Number of Nonlinear Loads
❏ Voltage/Current Unbalance ❏ Harmonics ❏ System Undersized ❏ Improper Wiring/Grounding
❏ Other _____

Measurements Taken:
Taken at Panel _____ Located at _____
Voltage _____ V, Current _____ A, Power _____ W, _____ VA, _____ VAR, Power Factor _____ PF, _____ DPF
Voltage THD _____, Current THD _____, K Factor _____, Other _____

Waveform Shape:
Voltage Waveform:
❏ Sinusoidal ❏ Non-Sinusoidal ❏ Flat-Topped ❏ Other _____
Current Waveform:
❏ Sinusoidal ❏ Non-Sinusoidal ❏ Pulsed ❏ Other _____

Measurements Taken at Load (Over Time):
Normal Voltage _____ V, Lowest Sag _____ V at _____ (Time), Highest Swell _____ V at _____ (Time)
Number of transients recorded _____ over a time period of _____ at a level of _____% above normal

Possible Problem Solution(s):
❏ Oversize Neutrals ❏ Run Separate Neutrals ❏ Additional Transformer ❏ Separate Loads ❏ Harmonic Filter
❏ Proper Wiring and Grounding ❏ Proper Fuses/CBs/Monitors ❏ Additional Subpanel ❏ Surge Suppressor
❏ Power Factor Correction Capacitors ☑ Other *Check system connections for loose or broken connections*

_____ **1.** Are there circuit breakers or fuses added to the neutral line that could trip if the neutral gets too hot?

_____ **2.** Based on the findings, should all the neutrals in the office be checked for overloading?

_____ **3.** Was there evidence of a power interruption problem?

_____ **4.** Was there evidence of a voltage sag or voltage swell problem?

_____ **5.** Was there evidence of an undervoltage or overvoltage problem?

APPLICATION 16-1—Power Loss

The first step in troubleshooting an electrical circuit is to ensure that the circuit has power. Power in a circuit is used by lights, motors, solenoids, etc. All electrical circuits are protected from short circuits, overcurrents, and overloads. Circuit overcurrent protection is provided by fuses, CBs, and overloads. These protection devices are located throughout the electrical system from the point at which power is delivered to a building to the final point at which the electricity is used. **See Power Circuit Distribution System.**

When power is lost in a circuit, the problem is normally caused by a fuse, CB, or an overload contact. When a protection device removes power from a circuit, a troubleshooter must reestablish power and find the reason the protection device removed power.

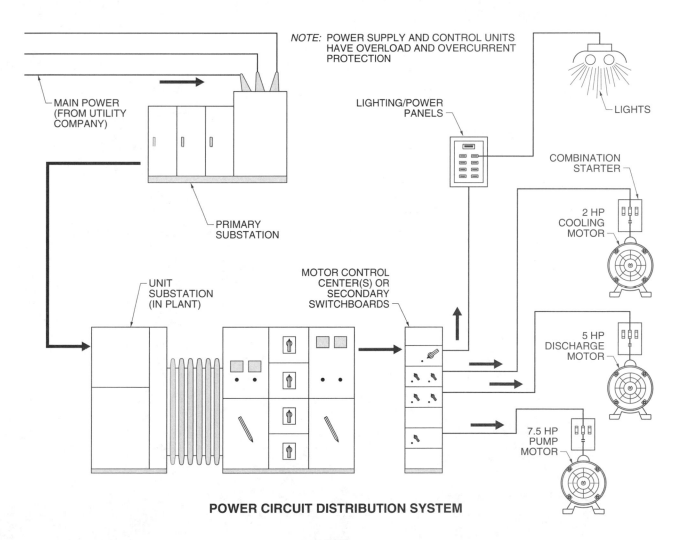

NOTE: POWER SUPPLY AND CONTROL UNITS HAVE OVERLOAD AND OVERCURRENT PROTECTION

MAIN POWER (FROM UTILITY COMPANY)

LIGHTING/POWER PANELS

LIGHTS

COMBINATION STARTER

2 HP COOLING MOTOR

PRIMARY SUBSTATION

UNIT SUBSTATION (IN PLANT)

MOTOR CONTROL CENTER(S) OR SECONDARY SWITCHBOARDS

5 HP DISCHARGE MOTOR

7.5 HP PUMP MOTOR

POWER CIRCUIT DISTRIBUTION SYSTEM

When the power in a circuit is lost, an overall check of the electrical system is required since the power may be lost at any one or more of the protection points in the system. **See System Protection Points.**

Overload Contacts

Overload contacts have a built-in time delay to allow a motor to start. When overload contacts trip, the problem normally occurs because the motor is overloaded when running. Power is reestablished by resetting the overload contacts. The reason that the overload contacts tripped should be determined, especially if the overload contacts trip again. Check for the correct size overload heaters and a motor problem.

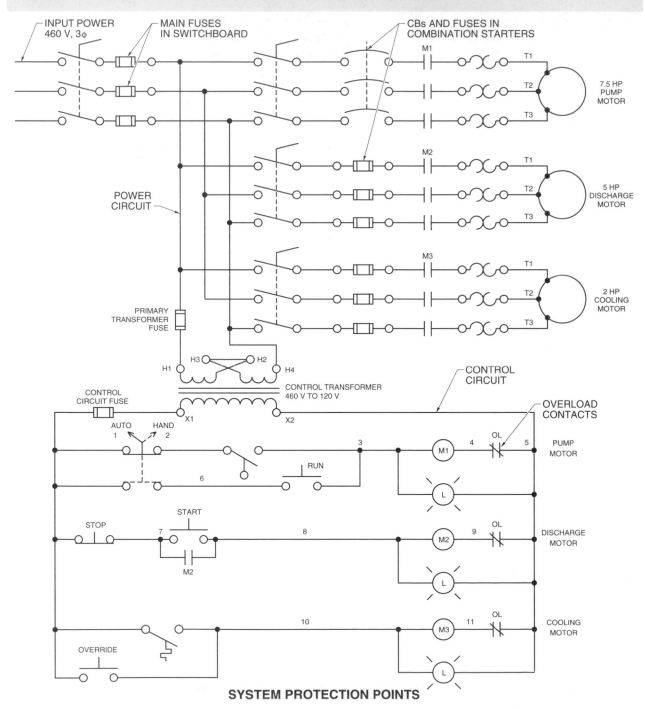

SYSTEM PROTECTION POINTS

Fuses and CBs

Fuses open and CBs trip in the power circuit when there is a short circuit or overcurrent condition. A *short circuit* occurs when the current in a circuit takes a shortcut around the normal path of current flow. An *overcurrent* is any current above the normal current level. **See Normal Current Flow** and **Short Circuit.**

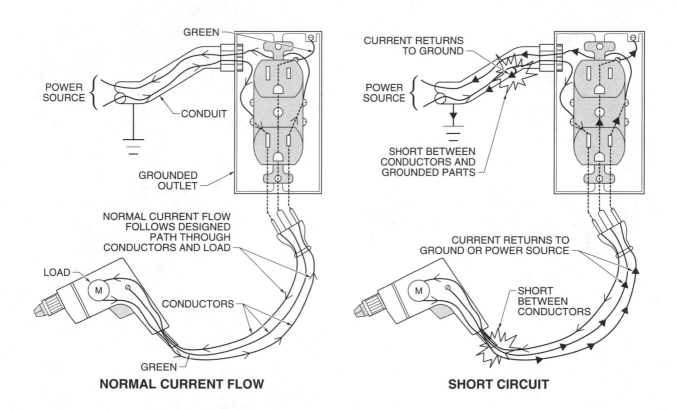

NORMAL CURRENT FLOW   **SHORT CIRCUIT**

When fuses open or CBs trip in the power circuit, the problem is normally a motor starting (not running) problem. Power is reestablished by changing the fuses or resetting the CBs. The reason the fuses opened or the CBs tripped should be determined, especially if the fuses open or the CBs trip again. Check for the correct size fuses or CBs and a short circuit. Fuses open and CBs trip the moment power is turned ON when there is a short circuit. Fuses open and CBs trip a few seconds after power is turned ON when there is an overload.

Fuses open and CBs trip in the control circuit when a short circuit or overcurrent condition exists in the control circuit. A short circuit is caused when power lines (L1 and L2) are connected or one of the ungrounded (hot) power lines is connected to ground. An overcurrent is caused by an excessive current draw in the control circuit. Excessive current draw normally occurs when there is a problem with one of the loads in the control circuit or additional loads are added to the circuit. Power is reestablished by changing the fuses or resetting the CBs. The reason the fuses opened or CBs tripped should be determined, especially if the fuses open or CBs trip again. Check for the correct size control transformer and a short circuit.

 Refer to ***Activity 16-1—Power Loss*** ....

APPLICATION 16-2—Troubleshooting Power Circuits

A *power circuit* is the part of an electrical circuit that connects the loads to the main power lines. *Loads* are devices that convert electrical energy to mechanical energy, heat, light, or sound. The starting point for troubleshooting the power circuit depends on the problem(s) observed. Troubleshooting the power circuit normally involves determining the point in the system where power is lost. This point may be at the load, the primary substation, or any point between the two.

Building Main Problems

A fault in the building main substation, transformers, or utility feed may interrupt power to a building. Problems in one of these areas cause a loss of power to all or most of the building. Common problems in a building main include lightning strikes, short circuits, accidental grounding, and brownouts. The building main and feeder circuit breakers are designed to open the circuit during a fault to prevent fatal injuries, damage to equipment, and electrical fires. When building main power is lost, a troubleshooter must find the location of the open, determine the problem, and reestablish power. **See Troubleshooting Building Mains.**

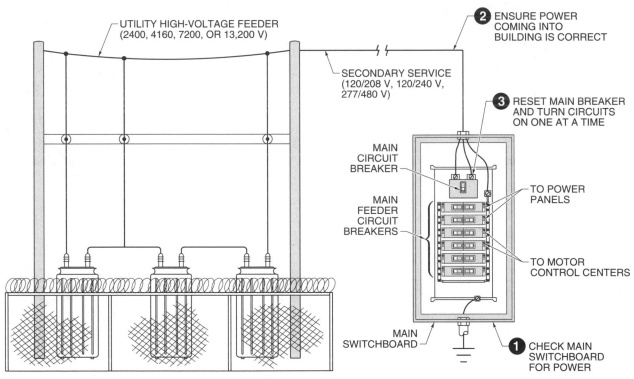

TROUBLESHOOTING BUILDING MAINS

To troubleshoot a building main, apply the procedure:
1. Check the unit substation and main switchboard. Check the power coming into the building when the entire building is without power. Contact the utility company when there is no power coming into the building.

2. Ensure that the power coming into the building is at the correct voltage level and present on each power line. Care must be taken when the power is correct. There is probably a short circuit in the system if a main CB trips. Turn OFF all circuits connected to the main circuit before resetting the main CB.

3. Reset the main CB and turn the circuits ON one at a time. Fuses or CBs trip the moment power is turned ON when there is a short circuit in one of the feeder lines.

Individual Load Problems

Individual load problems occur when one of the loads in a system is not operating. Individual load problems normally indicate a problem with one section (usually the load) of a system. A system problem may exist when the problem occurs at certain times or after repairs are made. The system problem may include low voltage, transient voltages, operator error, or a material or production problem. **See Troubleshooting Individual Load Problems.**

PROBLEM: PUMP MOTOR DOES NOT START

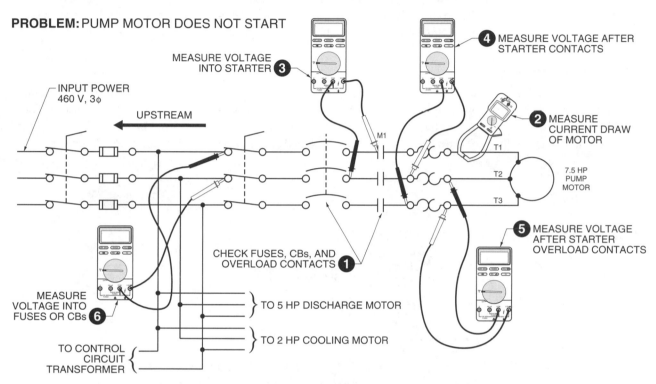

TROUBLESHOOTING INDIVIDUAL LOAD PROBLEMS

To troubleshoot a power circuit containing individual load problems, apply the procedure:

1. Check the fuses, CBs, and overload contacts closest to the load. Reset the devices and try to restart the motor when the overcurrent protection devices are tripped.

2. Measure the current draw of the motor to ensure there is no excessive current draw when the motor starts. Disconnect the load from the motor and measure the current when there is an excessive current draw. Check the motor if the current draw is still excessive. Check for an excessive load if the current returns to normal.

3. Measure the incoming supply voltage on the line side of the starter if the overloads are not tripped. Measure the voltage between L1 and L2, L1 and L3, and L2 and L3. The problem lies upstream from the starter if the voltage is not the same between L1, L2, and L3 or at the correct level. Check the fuses and CBs to which the starter is connected. See Step 6.

4. Measure the voltage after the starter power contacts and before the overload contacts if the voltage is correct. Manually close the overload contacts if they cannot be electrically engaged. Most manufacturers include a manual test lever or point on the starter. The problem lies downstream from the overload contacts when the measured voltage between L1, L2, and L3 is the same and at the correct level. The problem lies with the starter power contacts when the voltage between L1, L2, and L3 is not the same or at the correct level.

5. Measure the voltage after the starter overload contacts. This is the point at which the load is connected to the starter. The problem lies downstream from the overload contacts when the measured voltage between L1, L2, and L3 is the same and at the correct level. Check the motor starter by measuring the voltage at the motor terminals. The problem lies in the overload contacts when the measured voltage between L1, L2, and L3 is not the same and at the correct level. Replace any overload that is open, burned, or has any other problem.

6. Measure the voltage coming into the fuses or CBs when the correct voltage is at the motor starter. The problem lies in the fuses or CBs when the voltage between L1, L2, and L3 is the same and at the correct level. Test for an open fuse or CB. The problem lies upstream from the starter when the measured voltage between L1, L2, and L3 is not the same or at the correct level. Check the next main fuse or CB protecting the unit.

Multiple Load Problems

Multiple load problems occur when more than one load in a system does not operate. Multiple load problems normally indicate malfunctions that have opened the fuses or tripped the CBs which protect the system. The problem is normally a short circuit in the system, a load that has a major problem (drawing excessive current), or all of the loads together are drawing excessive current. **See Troubleshooting Multiple Load Problems.**

PROBLEM: MORE THAN ONE MOTOR DOES NOT START

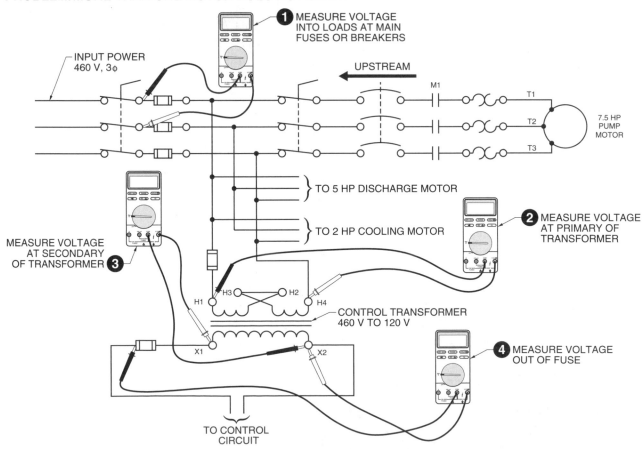

TROUBLESHOOTING MULTIPLE LOAD PROBLEMS

To troubleshoot a power circuit with multiple load problems, apply the procedure:

1. Measure the voltage into the main fuses or CB protecting the loads. Check the motor control center, secondary switchboard, or power panel that feeds the loads. The problem lies upstream from the main fuses or CB when the voltage is not the same and at the correct level. Check the next main fuse or CB feeding the unit. Check the voltage out of the fuses or CB when the voltage is correct. The problem lies in the fuses or CB when the voltage is not correct. Test for an open fuse or CB. Troubleshoot from the fuses or CB to the load when the main fuses or CB has power. Troubleshoot from the fuses or CB back to the main power into the building when the main fuses or CB does not have power.

2. Measure the voltage at the primary side of the control transformer. The problem lies upstream from the control transformer when the voltage is not present or is incorrect. Check the fuses or CB that protects the transformer.

3. Measure the voltage at the secondary of the transformer. The problem lies in the control transformer when the voltage is not present or is incorrect. Replace the transformer.

4. Measure the voltage exiting the fuse which protects the secondary of the transformer. The problem lies in the fuse when the voltage is not present or is incorrect. Replace the fuse.

 Refer to **Activity 16-2—Troubleshooting Power Circuits** ....

APPLICATION 16-3—Troubleshooting Control Circuits

A line diagram illustrates the logic of the control circuit in simplest form. A line diagram is used when troubleshooting the control circuit. **See Troubleshooting Control Circuits.**

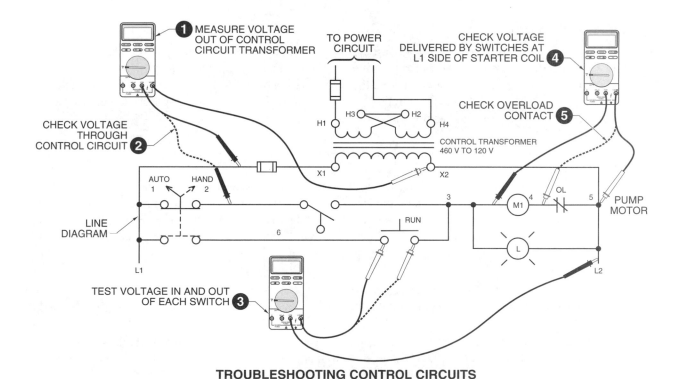

TROUBLESHOOTING CONTROL CIRCUITS

To troubleshoot the control circuit, apply the procedure:

1. Measure the voltage out of the control circuit transformer. Connect a voltmeter across the secondary side of the transformer to measure the power in the control circuit. Troubleshoot the power circuit feeding the transformer when the voltage at the secondary side of the transformer is incorrect. Check the fuses or CBs protecting the power to the transformer when there is no power entering the transformer. The transformer is the problem when there is power entering but no power exiting the transformer. Replace the transformer.

2. Check the voltage through the control circuit when there is power exiting the transformer. Leave one side of the voltmeter lead connected to the X2 side of the control transformer when the secondary side of the transformer has the correct voltage present. Move the other meter lead from the X1 side of the transformer through the control circuit connecting it to the input and output side of each component. Voltage should be present at each point to which the test lead is moved. The point where voltage is lost is the point where the circuit is open.

3. Test the voltage in and out of each switch. Press each button to test the natural open circuit of each selector switch and pushbutton. No voltage should exit a NO switch when the switch is open. Voltage should only exit a NO switch when the switch is closed.

4. Check the voltage delivered by the switches at the L1 side of the starter coil. Move one test lead through the control circuit until the test lead is at the L1 side of the starter coil. Voltage at this point indicates the control switches are delivering the voltage to the starter coil.

5. Check the overload contact. Move one test lead from the X2 side of the transformer to each side of the overload contacts. Leave the other with the other test lead at the L1 side of the starter coil. The problem is an open overload contact when voltage is present at the L2 side of the overload contact but not on the coil side. Check to ensure the reset function on the overload is working. The coil is the problem when the meter measures a voltage when connected directly across the starting coil but the starter is not energized. Replace the starting coil.

Control Panel Testing

A line diagram is used to illustrate the flow of current through the various devices in a control circuit. The line diagram is intended to demonstrate the basic operation of the circuit and not the physical wiring of each device. The actual wiring of the control circuit is found in the circuit wiring diagram. In troubleshooting, the line diagram is used to show circuit logic, and the wiring diagram is used to find the actual test points at which a meter is connected. **See Troubleshooting Inside Control Panels.**

Note: To simplify the control panel wiring, only a portion of the circuit is shown wired.

A voltmeter is used to measure the voltage out of the control transformer. In the line diagram, the meter is shown connected across the secondary side of the transformer and the fuse. In the control panel, the voltmeter is actually connected to the terminal strip. To measure the voltage out of the control transformer, the meter is connected to terminal posts 1 and 5. Using a terminal strip simplifies the troubleshooting process. Start at the terminal strip when troubleshooting inside a control panel.

To troubleshoot inside the control panel, apply the procedure:

1. Use the line diagram as a reference point for determining where to connect the meter and what the expected meter readings should be.

2. Measure the voltage at the terminal strip. After finding the point on the terminal strip at which power is lost, troubleshoot the individual component(s) connected to that point.

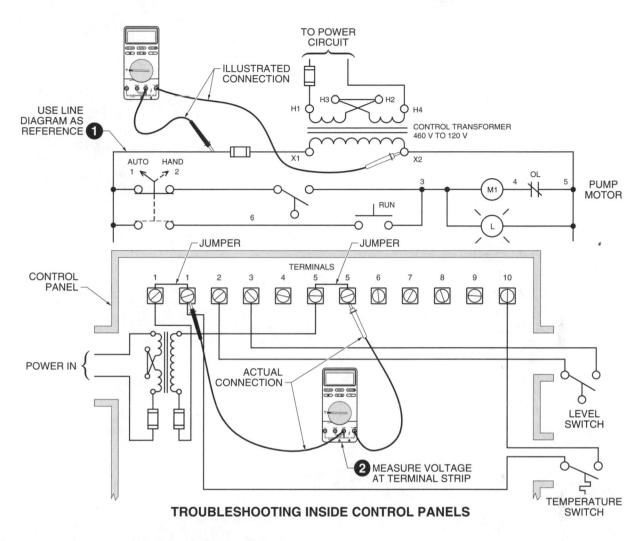

TROUBLESHOOTING INSIDE CONTROL PANELS

 Refer to **Activity 16-3—Troubleshooting Control Circuits**...

Refer to Quick Quiz® on CD-ROM

Refer to Chapter 16 in the **Troubleshooting Electrical/Electronic Systems Workbook** *for additional questions.*

Name_____ **Date**_____

ACTIVITY 16-1—Power Loss

Using Motor Control Circuit 1, determine the starting point for troubleshooting each problem.

1. The operator reports that the pump motor in Motor Control Circuit 1 does not operate. The pump light turns ON when the selector switch is placed in the hand position and the Run Pushbutton is pressed. The pump motor does not turn ON. The discharge motor operates correctly when the Start Pushbutton and Stop Pushbutton are pressed. The cooling motor operates when the Override Pushbutton is pressed.

_____ **A.** Troubleshooting should start at the ___.

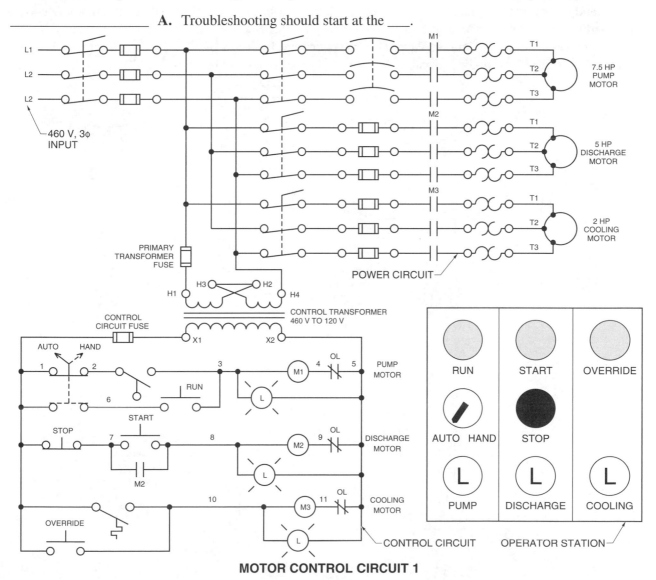

MOTOR CONTROL CIRCUIT 1

2. The operator reports that none of the motors in Motor Control Circuit 1 are operating. None of the motors operate manually when tested by pressing the pushbuttons at the operator station. None of the indicator lights operate. The starter overload contacts and circuit breakers are not tripped. Manually operating the starting contacts by pressing down on the contact assembly starts the pump motor.

_____ **A.** Troubleshooting should start at the ___.

3. The operator reports that none of the motors in Motor Control Circuit 1 operate. None of the motors operate manually when tested by pressing the pushbuttons at the operator station. None of the indicator lights operate. The starter overload contacts and circuit breakers are not tripped. Manually operating the starting contacts by pressing down on the contact assembly does not start the pump motor.

_____ **A.** Troubleshooting should start at the ___.

ACTIVITY 16-2—Troubleshooting Power Circuits

1. Connect each meter to measure the required voltage. Connect Meter A so the meter checks the incoming voltage on L1 and L3 before the main disconnect. Connect Meter B so the meter checks the voltage at motor terminals two and three of Motor 2. Connect Meter C so the meter checks the voltage out of the heaters delivering L1 and L3 power to Motor 1. Connect Meter D so the meter checks Fuse 1 of Motor 2. Connect Meter E so the meter checks Fuse 2 of Motor 1. Connect Meter F so the meter checks Main Fuse 3.

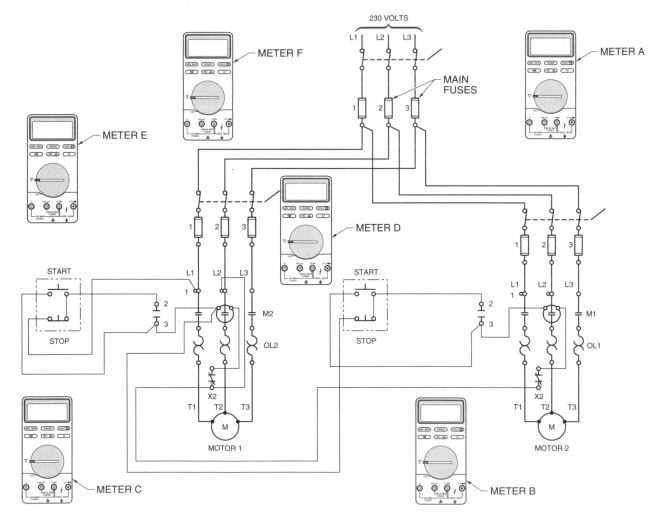

ACTIVITY 16-3—Troubleshooting Control Circuits

1. Using the Operator Station Wiring Diagram, answer the questions.

_____ **A.** Meter A reads ___ V when the selector switch is in the auto position.

_____ **B.** Meter B reads ___ V when the selector switch is in the auto position.

_____ **C.** Meter B reads ___ V when the selector switch is in the hand position and the Run Pushbutton is pressed.

_____ **D.** Meter C reads ___ V when the Override Pushbutton is pressed.

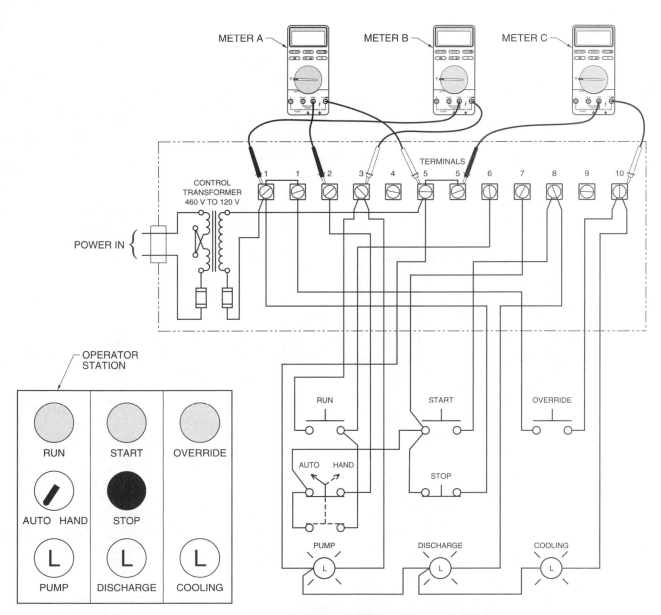

OPERATOR STATION WIRING DIAGRAM

2. Connect each meter to measure the required voltage. Connect Meter A at the terminal strip to measure the voltage directly into the control circuit. Connect Meter B at the terminal strip to measure the voltage out of the Run Pushbutton when the selector switch is in the hand position. Connect Meter C to measure the voltage directly across the light that indicates the cooling motor is ON. Connect Meter D at the terminal strip to measure the voltage out of the Stop Pushbutton. Connect Meter E at the terminal strip to measure the voltage out of the Start Pushbutton.

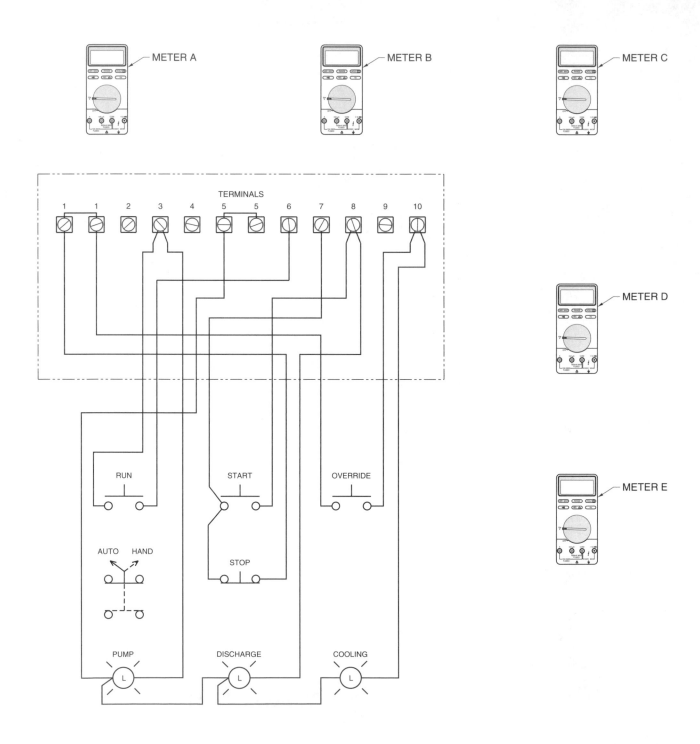

3. Connect each meter to measure the required voltage. Using the Overhead Door Control Circuit, connect Meter A at the terminal strip so that the meter tests the supply voltage into the control circuit. Connect Meter B at the terminal strip so that the meter tests the voltage out of the Stop Pushbutton. Connect Meter C at the terminal strip so that the meter tests the voltage across the starter coil that raises the door. Connect Meter D at the terminal strip so that the meter tests the voltage across the starter coil that lowers the door.

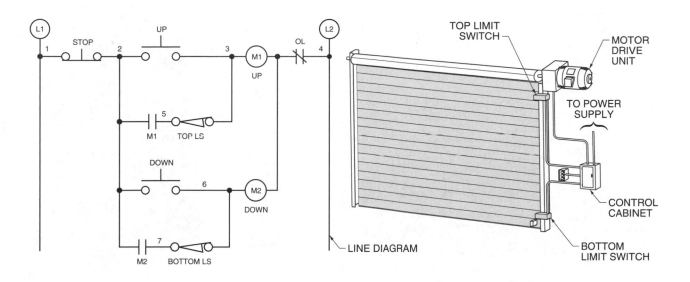

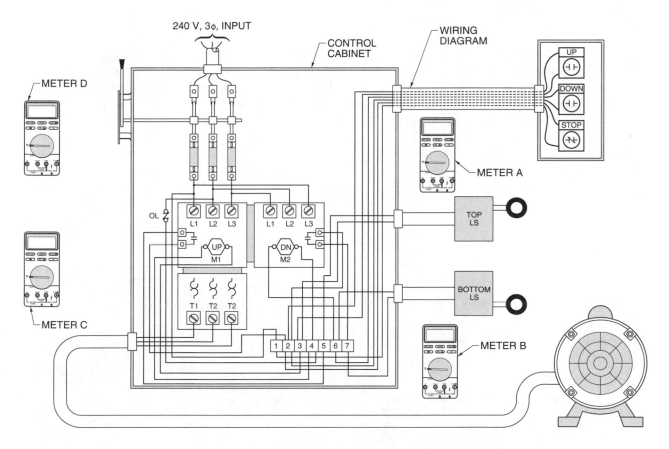

OVERHEAD DOOR CONTROL CIRCUIT

4. Using the Overhead Door Control Circuit, determine the problem. *Note:* The operator reports that the door does not rise when the Up Pushbutton is pressed. Meter A reads 240 V at all times. Meter B reads 240 V when the Up Pushbutton is pressed, but there is no movement of the door. Meter C reads 240 V when the Down Pushbutton is pressed, and the door attempts to move further down.

_____ **A.** The problem is ___.

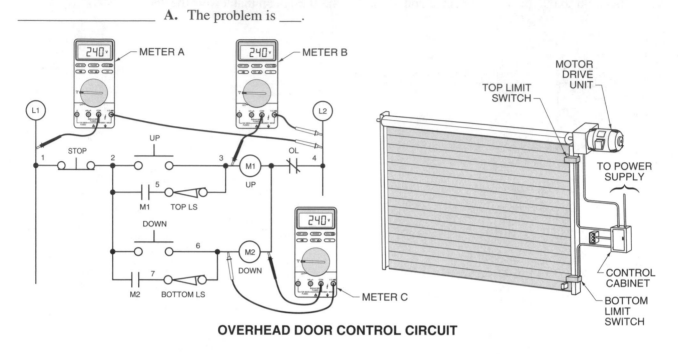

OVERHEAD DOOR CONTROL CIRCUIT

5. Using the Overhead Door Control Circuit, determine the problem. *Note:* The operator reports that the door opens only when the Up Pushbutton is held closed. The door automatically closes after the Down Pushbutton is pressed and released. Meter A reads 240 V only when the Up Pushbutton is held closed. Meter B reads 240 V only when the Up Pushbutton is held closed.

_____ **A.** The problem is ___.

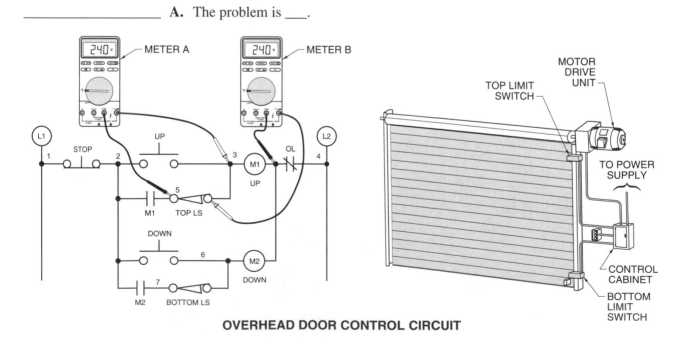

OVERHEAD DOOR CONTROL CIRCUIT

APPLICATION 17-1—*Electric Motor Drive Components*

An *electric motor drive* is an electrical device that controls motor speed between 0 rpm and maximum rpm, and controls motor torque between 0 in-lb and maximum in-lb. An electric motor drive can be classified by the type of drive, the type of motor to be controlled, or the type of incoming supply voltage to the drive. Electric motor drives are manufactured as variable frequency drives, adjustable frequency drives, inverter drives, vector drives, direct torque control drives, closed loop drives, and regenerative drives. The general types of motors are AC motors and DC motors. Supply voltage to an electric motor drive is either AC voltage or DC voltage. An *AC drive* controls AC motor speed by varying frequency. A *DC drive* controls DC motor speed by varying the voltage. **See Motor Speed Control.**

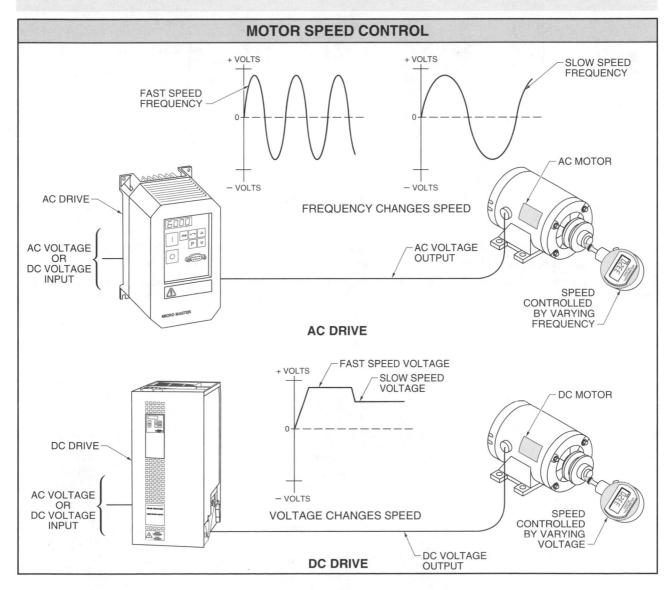

MOTOR SPEED CONTROL

The primary function of a drive is to convert the incoming supply power to an altered voltage level and frequency that can safely control the motor connected to the drive and the load connected to the motor. The three main components of electric motor drives are the converter section, DC bus section, and inverter section.

Converter Section

The converter (rectifier) section receives the incoming AC voltage and changes the voltage to DC. AC input voltage that is different than the AC output voltage sent to a motor requires the converter section first step up or step down the AC voltage to the proper level. For example, an electric motor drive supplied with 115 VAC that must deliver 230 VAC to a motor requires a transformer to step up the input voltage. **See Converter Section.**

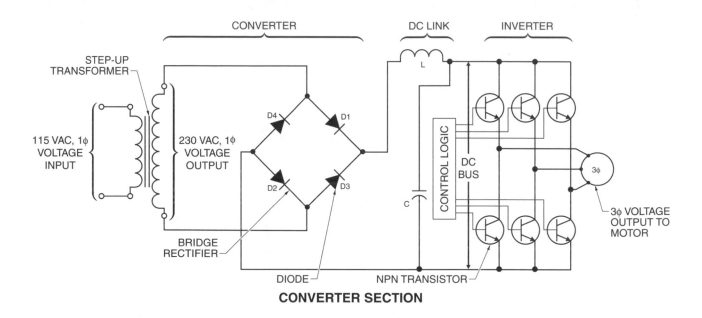

CONVERTER SECTION

Converter sections of electric motor drives are 1φ full-wave rectifiers, 1φ bridge rectifiers, or 3φ full-wave rectifiers. Small electric motor drives supplied with 1φ power use 1φ full-wave or bridge rectifiers. Most electric motor drives are supplied with 3φ power requiring 3φ full-wave rectifiers. **See Three-Phase Full-Wave Rectifier.**

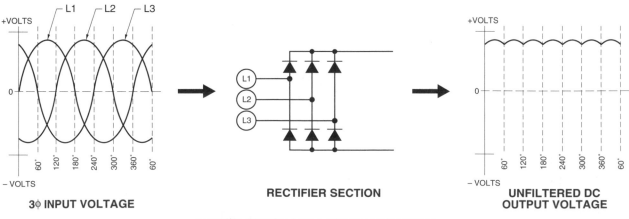

THREE-PHASE FULL-WAVE RECTIFIER

Electric Motor Drive Supply Requirements

In order for a converter section to deliver the proper DC voltage to the DC bus section of an electric motor drive, the rectifier section must be connected to the proper power supply. The proper power supply must not only be at the correct voltage level and frequency, but also provide enough current to operate an electric motor drive at full power. When a power supply cannot deliver enough current, the available voltage to an electric motor drive drops when the drive is required to deliver full power. Current to an electric motor drive is limited by the size of the conductors feeding the drive, fuse and circuit breaker sizes, and the transformer(s) delivering power to the system.

Supply voltage at an electric motor drive must be checked when installing a drive, servicing a drive, or adding additional loads or drives to a system. To determine if an electric motor drive is underpowered, measure the voltage into the drive under no-load and then full-load operating conditions. **See Full-Load Voltage Drop.**

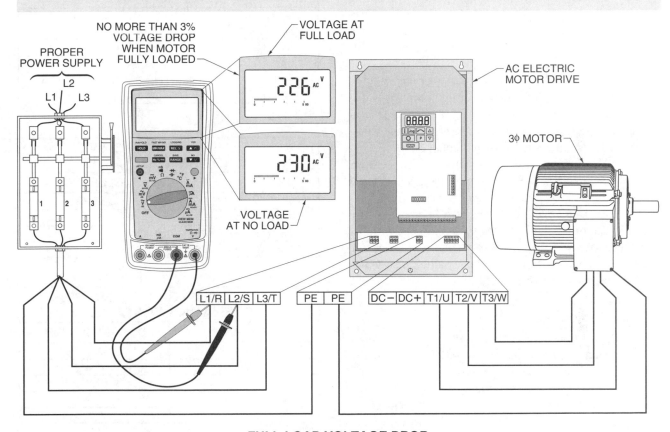

FULL-LOAD VOLTAGE DROP

A voltage difference percentage greater than 3% between no-load and full-load conditions indicates that the electric motor drive is underpowered and/or overloaded. Voltage difference percentage is found by applying the formula:

$$V_\% = \frac{V_D}{V_{NL}} \times 100$$

where

$V_\%$ = percentage of voltage drop (in %)
V_D = volts dropped (in V)
V_{NL} = voltage no load (in V)
100 = constant

For example, what is the voltage difference percentage when an electric motor drive has a 4 V voltage drop with a 230 V no-load measurement?

$$V_\% = \frac{V_D}{V_{NL}} \times 100$$

$$V_\% = \frac{4}{230} \times 100$$

$$V_\% = .01739 \times 100$$

$$V_\% = \mathbf{1.793\%}$$

When measuring the supply voltage to an electric motor drive, it is recommended to check the measured voltage against the drives rated input voltage. **See Electric Motor Drive Supply Voltages.** Larger horsepower electric motor drives are connected to higher voltages to reduce the amount of current required.

AC voltages vary within the power distribution system. AC loads, including electric motor drives and motors, are designed to operate within a specified voltage range. Operating outside the specified voltage range can cause electric motor drive malfunction and/or occur damage over time. Electrical loads operating at lower rated voltages are less likely to be damaged than loads operating at higher rated voltages.

AC loads are rated for proper operation at a voltage that is ±10 of the devices rated voltage. Because higher voltages are more damaging, some higher voltage rated devices have a +5% to –10% voltage rating. **See Electric Motor Drive Operating Voltages.**

Circuit Protection

A bridge rectifier receives incoming AC supply power and converts the AC voltage to a fixed DC voltage. The fixed DC voltage powers the DC bus of the electric motor drive. To prevent damage to the diodes in the converter section and to the electronic circuits, protection against transient voltages must be included in the drive.

ELECTRIC MOTOR DRIVE SUPPLY VOLTAGES	
PHASE & FREQUENCY	VOLTAGES*
1φ, 60 Hz	115, 208, OR 230
1φ, 60 Hz	110, 220, OR 240
3φ, 60 Hz	208, 230, 460, 575, 2300, 4160, OR 4600
3φ, 60 Hz	190, 220, 380, 415, 440, OR 4000

* in VAC

ELECTRIC MOTOR DRIVE OPERATING VOLTAGES	
DRIVE VOLTAGE*	TOLERANCE
ALL 1φ AC DRIVES	± 10%
200 TO 240 3φ DRIVES	± 10%
400 TO 480 3φ DRIVES	± 10%
500 TO 600 3φ DRIVES	+ 5% TO –10%

* in VAC

ELECTRIC MOTOR DRIVE DC BUS OPERATING VOLTAGES	
AC SUPPLY VOLTAGE*	DC BUS VOLTAGE†
208	291
220	308
230	322
460	644
480	672

* in VAC
† in VDC

DC Bus Section

The DC bus section filters the voltage and maintains the proper DC voltage level. The DC bus section delivers the DC voltage to the inverter section for conversion back to AC voltage. The inverter section determines the speed of a motor by controlling frequency and controls motor torque by controlling the voltage sent to a motor. **See Electric Motor Drive DC Bus Operating Voltages.**

A *capacitor* is a device used to store an electrical charge. Capacitors oppose a change in voltage by holding a voltage charge that is discharged back into the circuit any time the circuit voltage decreases. The capacitors in a DC bus section are charged from the rectified DC voltage produced by the converter section. When DC bus voltage starts to drop, capacitors discharge a voltage back into the system to stop the drop in voltage. The main function of capacitors is to maintain proper DC bus voltage levels even when voltage would otherwise fluctuate. **See DC Bus Filter Capacitors.**

Capacitors and inductors (coils) are used together in DC filter circuits. Working together, capacitors and coils maintain a smoother wave form because capacitors oppose a change in voltage and coils oppose a change in current. As current flows through a coil, a magnetic field is produced. The magnetic field remains at maximum potential until the current in the circuit is reduced. As the circuit current is reduced, the collapsing magnetic field around the coil induces current back into the circuit. The coil smoothes or filters the power by building a magnetic field as current is applied and adding current back into the circuit as the magnetic field collapses.

Resistors are used in the DC bus section to limit the charging current to capacitors, discharge capacitors, and absorb unwanted voltages. Current-limiting resistors prevent capacitors from drawing too much current during charging. Braking resistors absorb voltage when a motor becomes a generator after a stop button is pushed.

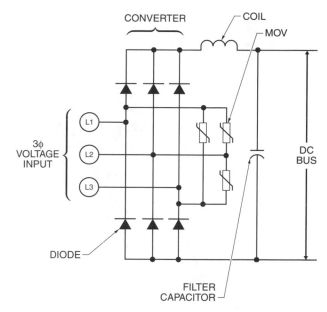

DC BUS FILTER CAPACITORS

Inverter Section

The inverter section of an electric motor drive establishes the voltage level, voltage frequency, and amount of current that a motor receives. Inverter sections have undergone changes in recent years while the rectifier section and DC bus have not changed. Electric motor drive manufacturers are continuously developing inverter sections that can control motor speed and torque with the fewest problems. The main problem for manufacturers is to find a high-current, fast-acting solid-state switch that has the least amount of power loss (voltage drop).

Both DC and AC motors produce work (deliver force) to drive a load by a rotating shaft. The amount of work produced is a function of the amount of torque produced by the motor shaft and the speed of the shaft. The primary function of all electric motor drives is to control the speed and torque of a motor.

Voltage supplied to a DC motor controls the speed of the motor. The higher the applied voltage, the faster a DC motor rotates. DC drives normally control the voltage applied to a motor over the range 0 VDC to the maximum nameplate voltage rating of the motor.

Controlling the amount of current in the armature of a DC motor controls motor torque. Motor torque is proportional to the current in the armature. DC drives are designed to control the amount of voltage and current applied to the armature of DC motors to produce wanted torque and prevent motor damage. The ideal operating condition is to deliver current to a motor to produce enough torque to operate the load without overloading the motor, electric motor drive, or electrical distribution system.

Controlling the amount of frequency (Hz) supplied to an AC motor determines the speed of the motor. AC drives control the frequency applied to a motor over the range 0 Hz to several hundred hertz. AC drives are programmed for a minimum operating speed and a maximum operating speed to prevent damage to the motor or driven load. AC motors should not be operated at a speed higher than 10% above the motor's rated nameplate speed. The upper limits of a motor are based on the motor's voltage and mechanical balancing limits. In general, a motor can be operated at a speed that does not exceed 10% more than the motor rated (nameplate) speed.

Controlling the volts-per-hertz ratio (V/Hz) applied to an AC motor controls motor torque. An AC motor develops rated torque when the V/Hz ratio is maintained. During acceleration (any speed between 1 Hz and maximum hertz), the motor shaft delivers constant torque because the voltage is increased at the same rate as the increase in frequency. Once an electric motor drive reaches the point of delivering full motor voltage, increasing the frequency does not increase torque on the motor shaft because voltage cannot be increased any more to maintain the volts-per-hertz ratio. **See Torque Volts-Per-Hertz Ratio.** The standard units of torque are pound-inches (lb-in) or pound-feet (lb-ft). The exact amount of torque a motor can produce depends on the motors design, applied voltage, and motor current.

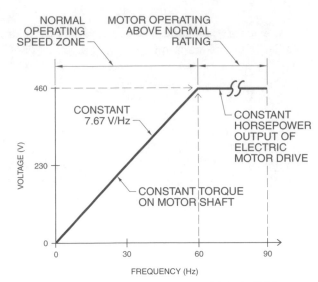

TORQUE VOLTS-PER-HERTZ RATIO

 Refer to **Activity 17-1—Electric Motor Drive Components** ....

APPLICATION 17-2—Pulse Width Modulation

Pulse width modulation (PWM) is the varying of the voltage pulse width to control the amount of voltage supplied to a motor. DC drives and AC drives must control the amount of voltage produced in order to control the speed and torque of a motor. Over the years, different methods of controlling the amount of voltage produced have been used in electric motor drives. Some methods have been replaced by newer technologies, but several older methods are still in use. Silicon-controlled rectifiers (SCRs) were first used in DC drives to control the amount of voltage applied to a motor. An SCR controls the amount of DC voltage output by controlling the amount of AC voltage that is rectified into DC voltage. The amount of DC voltage output is determined by when the gate of an SCR allows current to flow.

Pulse width modulation control offers better performance than SCR control and is used in newer DC drives. PWM controls the amount of voltage output by converting the DC voltage into fixed values of individual DC pulses. The fixed-value pulses are produced by the high-speed switching of transistors (typically insulated gate bipolar transistors) turning ON and OFF. An *insulated gate bipolar transistor (IGBT)* is a transistor with fast switching capabilities. By varying the width of each pulse (time ON) and/or frequency, the amount of voltage can be increased or decreased. The wider the individual pulses, the higher the DC voltage output. The higher the DC output, the faster a DC motor operates. **See Pulse Width Modulation.**

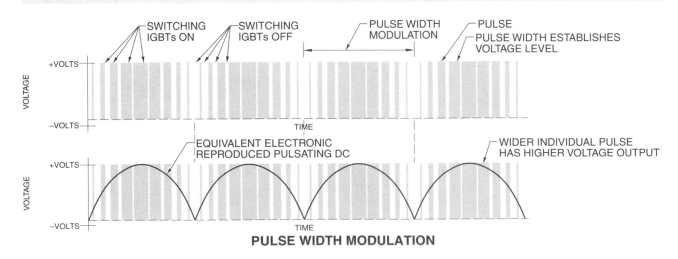

PULSE WIDTH MODULATION

A single-pole single-throw (SPST) mechanical switch can be used to convert pure DC voltage into a pulsating DC at varying voltage levels. If the switch remains closed, the DC voltage output would equal the applied DC voltage input. As the switch is opened and closed, the DC voltage output is equal to a voltage level less than the applied input voltage and greater than 0 V. The longer a switch is left open, the lower the average DC voltage, and the longer a switch is left closed, the higher the average DC voltage. By adding a capacitor or capacitors into the circuit, the waveform is smoothed as the capacitor discharges back into the circuit every time the voltage tries to return to zero (switch opened).

PWM of a DC voltage is also used to reproduce AC sine waves. When using PWM with AC voltage, two IGBTs are used for each phase. One IGBT is used to produce the positive pulses and another IGBT is used to produce the negative pulses of the sine wave. Because AC drives are typically used to control 3ϕ motors, six IGBTs (two per phase) are used. **See IGBT Produced Sine Wave.** The higher the switching frequency of the IGBTs, the closer the simulated AC sine wave is to a real sine wave. The closer the simulated sine wave is to a pure sine wave, the lower the amount of heat produced by the motor.

IGBT PRODUCED SINE WAVE

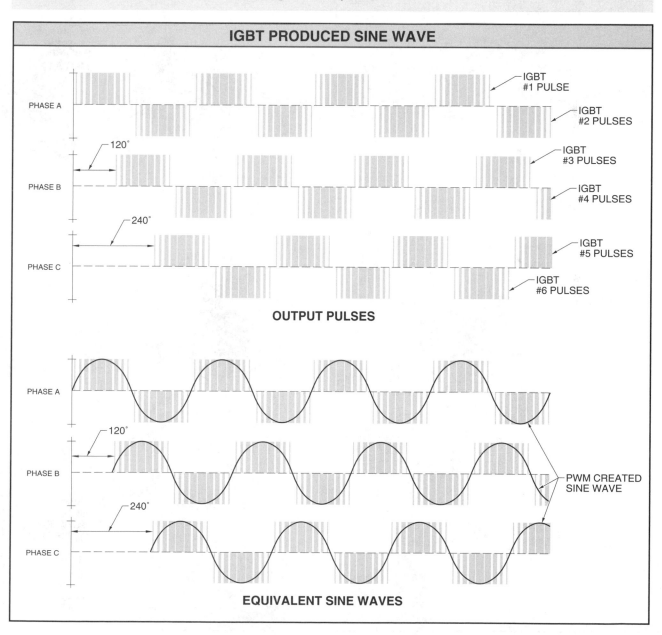

OUTPUT PULSES

EQUIVALENT SINE WAVES

Electric motor drives replace single-pole double-throw (SPDT) switches with fast-acting electronic switches (IGBTs). When two IGBTs are used, an electronically reproduced AC voltage is produced. The fast-acting transistors are performing the same function as mechanical switches. **See IGBTs Controlling Motor Rotation.**

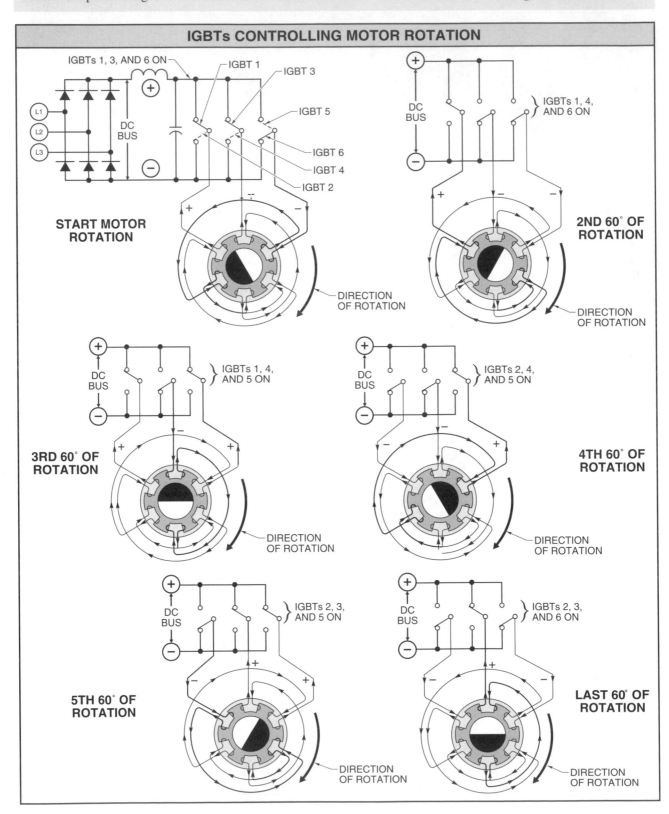

IGBTs CONTROLLING MOTOR ROTATION

Switches (two IGBTs or two SCRs) are used to place either a positive (+) or negative (–) voltage on the motor leads. Three switches (one for each phase) are always closed to apply either a positive or negative voltage on each of the three motor leads (T1, T2, and T3). The positive and negative polarities produce current flow in a fixed direction through the stator windings of a motor. The switches keep changing the direction of current flow to produce a rotating magnetic field around the motor. The rotating magnetic field forces the rotor to rotate.

Carrier Frequencies

Carrier frequency is the frequency that controls the number of times the solid state switches in the inverter section of a PWM electric motor drive turn ON and turn OFF. The higher the carrier frequency, the more individual pulses there are to reproduce the fundamental frequency. *Fundamental frequency* is the frequency of the voltage used to control motor speed. Carrier frequency pulses per fundamental frequency are found by applying the following formula:

$$P = \frac{F_{CARR}}{F_{FUND}}$$

where
P = pulses
F_{CARR} = carrier frequency
F_{FUND} = fundamental frequency

For example, what is the number of pulses per fundamental frequency when a carrier frequency of 1 kHz is used to produce a 60 Hz fundamental frequency?

$$P = \frac{F_{CARR}}{F_{FUND}}$$

$$P = \frac{1000}{60}$$

$$P = \textbf{16.66 pulses}$$

A carrier frequency of 6 kHz used to produce a 60 Hz fundamental frequency would have 100 individual pulses per fundamental cycle. **See Electric Motor Drive Frequencies.**

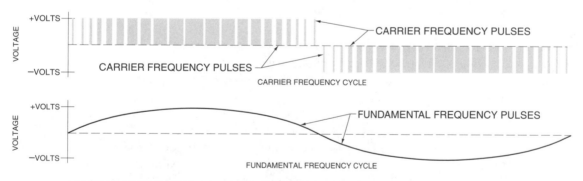

ELECTRIC MOTOR DRIVE CARRIER FREQUENCY PULSES				
FUNDAMENTAL FREQUENCY*	CARRIER FREQUENCY†		NUMBER CARRIER PULSES PER FUNDAMENTAL CYCLE	
60	1	HIGHER VOLTAGE SPIKES AT MOTOR	16.66	SMALLER HEAT SINKS
60	2		33.33	
60	6		100	
60	8		133.33	
60	10		166.66	
60	12		200	
60	14	LESS NOISE	233.33	
60	16		266.66	

* in Hz
† in kHz

ELECTRIC MOTOR DRIVE FREQUENCIES

Fundamental frequency is the frequency of the voltage a motor uses, but the carrier frequency actually delivers the fundamental frequency voltage to the motor. The carrier frequency of most electric motor drives can range from 2 Hz to about 16 kHz. The higher the carrier frequency, the closer the output sine wave is to a pure fundamental frequency sine wave.

Increasing the fundamental frequency to a motor above the standard 60 Hz also increases the noise produced by the motor. Noise is noticeable in the 1 kHz to 2 kHz range because it is within the range of human hearing and is amplified by the motor. A motor connected to an electric motor drive delivering a 60 Hz fundamental frequency with a carrier frequency of 2 kHz is about three times louder than the same motor connected directly to a pure 60 Hz sine wave with a magnetic motor starter. Motor noise is a problem in electric motor drive applications such as HVAC systems in which the noise can carry throughout a building.

Higher carrier frequencies cause greater power losses (thermal losses) in an electric motor drive because of the solid-state switches in the inverter section. Electric motor drives must be slightly derated or the size of heat sinks increased because of the increased thermal losses. Derating an electric motor drive decreases the power rating of a drive and increasing heat sink size adds additional cost to a drive. **See Carrier Frequency Power Derating Curve.**

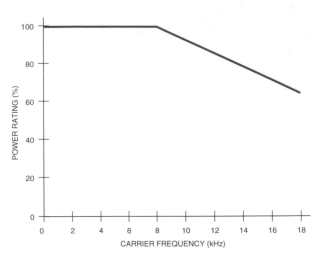

**CARRIER FREQUENCY
POWER DERATING CURVE**

Higher carrier frequencies are preferred, but only up to a point. A 6 kHz to 8 kHz carrier frequency simulates a pure sine wave better than a 1 kHz to 3 kHz carrier frequency and reduces heating in a motor. The more closely the voltage delivered to a motor simulates a pure sine wave, the cooler a motor operates. Even slightly reducing the temperature in a motor increases insulation life.

Carrier frequency can be changed at an electric motor drive to meet particular requirements. The factory default value is usually the highest frequency, and changing to a lower frequency is done through a parameter change, such as changing 12 kHz to 2.2 kHz. One effect of high carrier frequencies is that the fast switching of the inverter produces larger voltage spikes that damage motor insulation. The voltage spikes become more of a problem as cable length between an electric motor drive and motor increases.

Motor Lead Length

In any electrical system, the distance between components affects operation. The primary limit to the distance between a magnetic motor starter and the motor is the voltage drop of the conductors. The voltage drop of conductors should not exceed 3% for any type of motor circuit.

Conductors between an electric motor drive and motor have line-to-line (phase-to-phase) capacitance and line-to-ground (phase-to-ground) capacitance. The capacitance produced by conductors causes high voltage spikes in the voltage to a motor. Since voltage spikes are reflected into the system, the voltage spikes are often called reflective wave spikes, or reflective waves (reflected waves). As the length of conductors increases and/or an electric motor drive carrier frequency increases, the voltage spikes become larger. **See Conductor Length Voltage Spikes.**

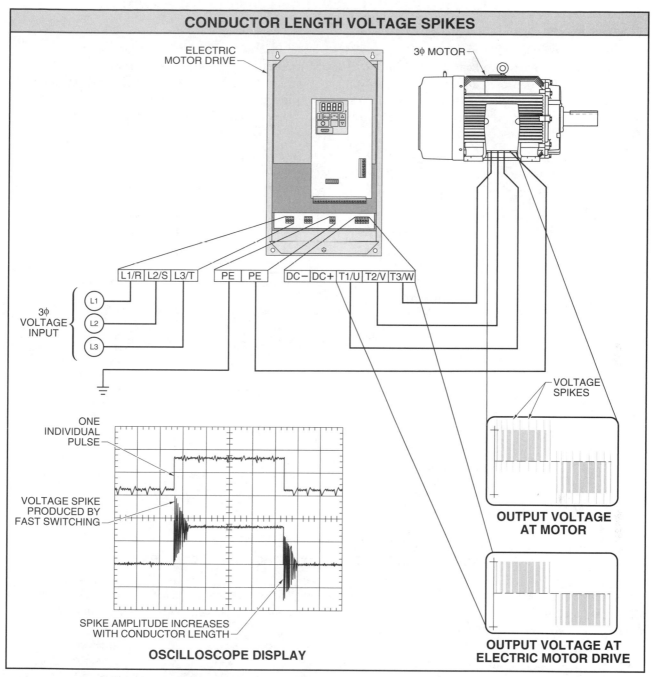

CONDUCTOR LENGTH VOLTAGE SPIKES

ELECTRIC MOTOR DRIVE

3φ MOTOR

L1/R | L2/S | L3/T | PE | PE | DC− | DC+ | T1/U | T2/V | T3/W

3φ VOLTAGE INPUT

L1
L2
L3

VOLTAGE SPIKES

OUTPUT VOLTAGE AT MOTOR

ONE INDIVIDUAL PULSE

VOLTAGE SPIKE PRODUCED BY FAST SWITCHING

SPIKE AMPLITUDE INCREASES WITH CONDUCTOR LENGTH

OSCILLOSCOPE DISPLAY

OUTPUT VOLTAGE AT ELECTRIC MOTOR DRIVE

 Refer to **Activity 17-2—Pulse Width Modulation** ....

APPLICATION 17-3—Electric Motor Drive Types

AC drives control motor speed and torque by converting incoming AC voltage to DC voltage and then converting the DC voltage to a variable-frequency AC voltage. The inverter section of an electric motor drive converts the DC voltage of the DC bus to AC voltage. AC drives control the voltage and frequency at the motor by switching the DC voltage ON and OFF at the proper time. Electronic switches such as SCRs or IGBTs are used to switch the DC voltage ON and OFF. A microprocessor circuit located within an electric motor drive controls the electronic switching.

Inverter Drives

Inverter drives are the oldest type of AC drive. Older six-step inverter drives used SCRs in the inverter section of the AC drive to produce the AC sine wave. PWM inverters use transistors in the inverter section to produce the AC sine wave. Transistors operate at much faster speeds than SCRs, allowing higher switching frequencies to produce electronically reproduced sine waves that closely resemble a pure sine wave. The improved sine wave produces less heating in a motor than six-step inverters. **See SCR and Transistor Output Voltages and Currents.**

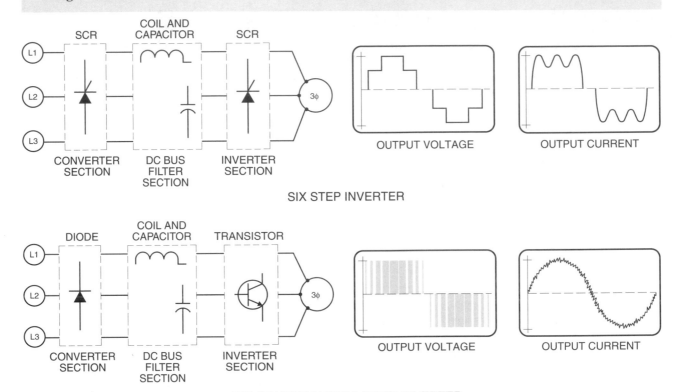

SCR AND TRANSISTOR OUTPUT VOLTAGES AND CURRENTS

Inverter drives are referred to according to the method used to change the frequency of the voltage. Inverter drives include variable voltage inverters (VVI), current source inverters (CSI), and pulse width modulated inverters (PWM). **See Output Voltages and Output Currents.**

Variable voltage inverter drives control an AC motor but produce a square wave instead of a sine wave. Square waves produce high-torque pulsations at the motor that cause a motor to have jerky movements at low speeds and operate at high temperatures. To overcome the square wave problem, electric motor drives use IGBTs to produce a more accurate sine wave.

In addition to naming electric motor drives by the type of inverter used (VVI, CSI, or PWM), drives are also referred to by the technology used by the drive to control motor torque.

Vector Drives

Applications that required better motor torque control at different speeds used DC motors. To overcome the problem of having to use DC motors with their higher maintenance costs, vector drives were developed. Vector electric motor drive (closed loop vector drive or open loop vector drive) is a name given to certain drives to better describe the operation of the drive.

OUTPUT VOLTAGES AND OUTPUT CURRENTS

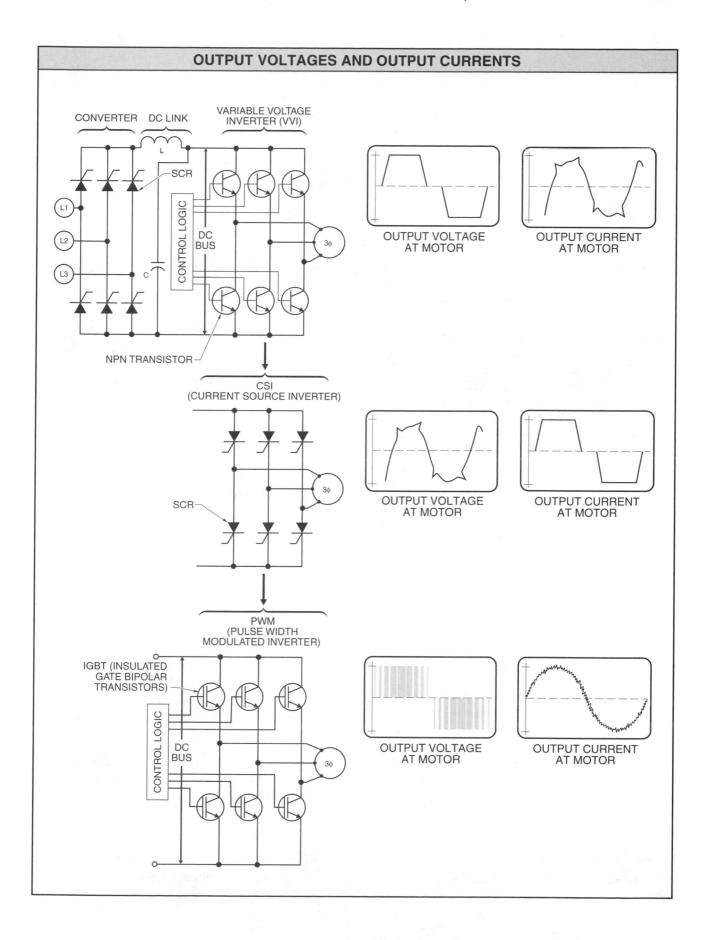

CONVERTER   DC LINK   VARIABLE VOLTAGE INVERTER (VVI)

L

SCR

L1
L2
L3

C

CONTROL LOGIC

DC BUS

3φ

NPN TRANSISTOR

OUTPUT VOLTAGE AT MOTOR

OUTPUT CURRENT AT MOTOR

CSI
(CURRENT SOURCE INVERTER)

SCR

3φ

OUTPUT VOLTAGE AT MOTOR

OUTPUT CURRENT AT MOTOR

PWM
(PULSE WIDTH MODULATED INVERTER)

IGBT (INSULATED GATE BIPOLAR TRANSISTORS)

CONTROL LOGIC

DC BUS

3φ

OUTPUT VOLTAGE AT MOTOR

OUTPUT CURRENT AT MOTOR

A *closed loop drive* is an electric motor drive that operates using a feedback sensor such as an encoder or tachometer connected to the shaft of the motor to send information about motor speed back to the drive. A *closed loop system* is a system with feedback from the motor sensors to the electric motor drive. Monitoring information from sensors allows an electric motor drive to automatically make adjustments to better meet the needs of the motor.

Vector control (also called flux vector or field orientation) drives are designed to operate AC motors at the same performance as DC motors. Vector control drives achieve the performance by measuring the current drawn by a motor at a known speed (encoder feedback) and comparing the current draw to the applied voltage. Induction motors produce torque at the rotor and shaft when the rotor turns at a slower speed than the rotating magnetic field of the stator. **See Motor Torque Characteristics.** AC motor performance using a vector drive is comparable with that of a DC motor. Vector drives are a good selection for many applications, but inverter drives are more economical and better for applications such as centrifugal pumps, fans, and blowers or any applications that do not require full torque control.

The main disadvantage of a vector drive is that an encoder or tachometer is required. Mounting and maintaining the encoder adds cost to the system. To achieve the improved performance of vector drives when compared to inverter drives without the cost penalty, a vector drive that does not require an encoder was developed. Electric motor drives without encoders are called open loop vector drives, or sensorless vector drives.

An *open loop drive* is an electric motor drive that operates without any feedback to the drive about motor speed. An *open loop system* is a system that has no feedback method. Programming the conditions believed necessary to achieve the desired results and accepting the results of the motor control is required in the open loop system. Open loop vector drives can be programmed using motor and application data to anticipate how to control the motor and load.

Open loop vector drives are used where speed regulation is important, but not important enough to require encoder feedback. Open loop vector drives typically can control speed within .1% of the drive setting and a closed loop vector drive typically can control speed within .01% of the drive setting.

Direct Torque Control (DTC) Drives

As new technologies emerge, manufacturers describe and advertise electric motor drives using different names to better express the technology used, such as a type of improved open loop vector drive called direct torque control (DTC) that is designed to provide better torque control. DTC drives perform the same function of controlling motor speed and torque as inverter drives and open loop vector drives, but apply technologies that improve motor torque control without requiring any feedback from motor sensors.

MOTOR TORQUE CHARACTERISTICS

INVERTER MOTOR TORQUE CHARACTERISTICS

VECTOR MOTOR TORQUE CHARACTERISTICS

 Refer to **Activity 17-3—Electric Motor Drive Types**..

 Refer to Quick Quiz® on CD-ROM

 Refer to Chapter 17 in the **Troubleshooting Electrical/Electronic Systems Workbook** *for additional questions.*

Electric Motor Drives

Name_____ Date _____

ACTIVITY 17-1—Electric Motor Drive Components

Electric motor drive components enable a drive to be programmed using parameters to customize the drive to the application as long as the parameter setting does not damage the motor or cause a safety problem. An AC motor can be operated at higher frequencies when powered by an electric motor drive. The higher the frequency, the faster the motor speed. The upper limits of a motor are based on the motor voltage and mechanical balancing limits. In general, a motor can be operated at a speed that does not exceed 10% more than the motor rated (nameplate) speed. The maximum frequency drive parameter setting limits the operating speed. Determine the maximum frequency drive parameter setting (MFS) that limits motor operating speed to no more than 10% above nameplate rated speed.

TITLE	PARAMETERS	ADJUSTABLE RANGE	DEFAULT VALUE
AU1	Automatic acceleration and decleration	0: NO 1: YES	0
AU2	Automatic torque boost*	0: NO 1: AUTOMATIC 50 Hz MOTOR 2: SENSORLESS VECTOR CONTROL 3: SENSORLESS VECTOR CONTROL AUTOMATIC TUNING	0
AU3	Automatic enviornment setting*	0: NO 1: AUTOMATIC 50 Hz MOTOR 2: AUTOMATIC 60 Hz MOTOR	0
FMC	FM terminal function selection	0: FREQUENCY METER 1: OUTPUT CURRENT METER	0
SSM	Standard mode selection*	0: NO ACTION 1: 50 Hz STANDARD 2: 60 Hz STANDARD 3: DEFAULT SETTING 4: CLEARING LOG ERRORS 5: CLEARING ACCUMULATED OPERATION TIME 6: INITIALIZE INVERTER TYPEFORM	3
FRS	Forward/reverse selection (panel)	0: FORWARD 1: REVERSE	0
ACC	Acceleration time #1 (sec)	0.1 ~ 3600	10.0
DAC	Deceleration time #1 (sec)	0.1 ~ 3600	10.0
MFS	Maximum frequency (Hz)*	30.0 ~ 320.0	80.0
BFS	Base frequency (Hz)	25.0 ~ 320.0	60.0

* Theses parameters cannot be changed while running

_____ 1. What is the factory default MFS setting?

_____ 2. What is the setting of the MFS for a 60 Hz rated motor so that the motor does not operate 10% above rated motor speed?

_____ 3. What is the maximum motor speed based on the factory default setting of the MFS if the 60 Hz motor is rated at 1800 RPM?

_____ **4.** What is the setting of the MFS for a 50 Hz rated motor so that the motor does not operate 10% above rated motor speed?

Understanding motor drive components and wiring is required when installing, troubleshooting, and replacing a drive. For example, electric motor drives can be set for controlling a motor from the drive panel located on the drive, or by external switches and external signals connected to the drive. Manufacturers provide wiring diagrams that show electric motor drive control wiring. These wiring guides are used to determine the connections between individual components and electric motor drives.

5. Wire the three-position selector switch to the electric motor drive terminal strip to control motor direction. Wire the reset pushbutton to the electric motor drive terminal strip to reset the drive and drive error messages. Wire the potentiometer to the electric motor drive terminal strip to control motor speed.

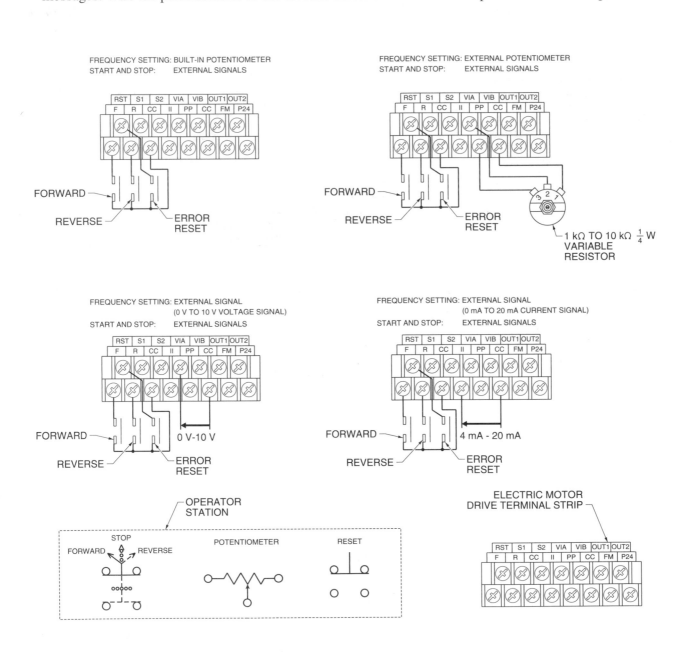

ACTIVITY 17-2—Pulse Width Modulation

Electric motor drive carrier frequency (pulse width modulation frequency) affects the distance a drive can be located away from a motor. Electric motor drive manufacturers list the recommended maximum cable length between the drive and motor. Depending on the electric motor drive manufacturer, cable lengths may be listed in meters (m), or feet (ft). Convert each of the following measurements.

1 m = 3.3′ (meters × 3.2808 = feet).

1′ = 0.3 m (feet × 0.3048 = meters).

MAXIMUM MOTOR CABLE LENGTHS			
Drive Part Number	**MAXIMUM MOTOR CABLE LENGTHS***		
	4 kHz	8 kHz	16 kHz
01	30	20	10
02	40	25	15
03	30	20	15
04	40	30	20

* in M

_____ **1.** Circuit 1 maximum cable length is ___′.

_____ **2.** Circuit 2 maximum cable length is ___′.

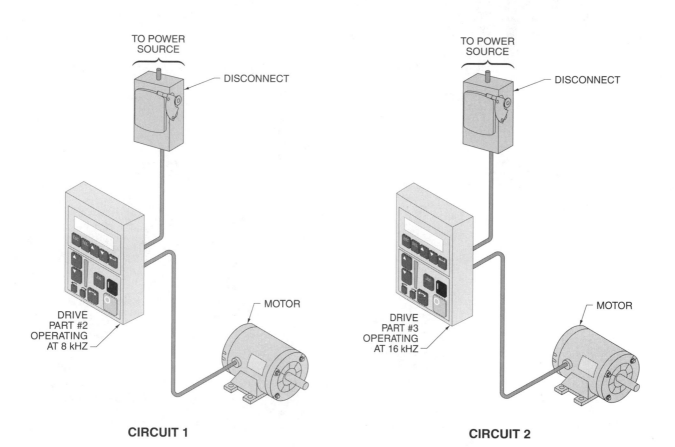

CIRCUIT 1   CIRCUIT 2

ACTIVITY 17-3—Electric Motor Drive Types

Motor torque is the force that causes a motor shaft to rotate. The standard units of torque are pound-inches (lb-in.) or pound-feet (lb-ft). The exact amount of torque a motor can produce depends on the motor design, applied voltage, and motor current. General motor torque rules can be applied for most motor applications. Using the general motor torque rule table and motor nameplate, determine the motor torque.

GENERAL MOTOR TORQUE		
Number of Motor Poles	Motor rated speed*	Motor torque†
2	3600 (3450)	1.5
4	1800 (1725)	3
6	1200 (1140)	4.5
8	900 (860)	6

* Synchronous (typical running)
† in lb-ft per HP

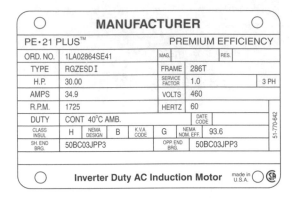

1. _____ **1.** What is the developed motor torque for the motor?

2. _____ **2.** How much torque does the motor develop if it is nameplate rated at 1140 rpm?

3. _____ **3.** How much torque does the motor develop if it is nameplate rated at 3450 rpm?

Troubleshooting Electric Motor Drives

APPLICATION 18-1—Troubleshooting Electric Motor Drive Safety

Troubleshooting electric motor drives and electrical equipment is inherently dangerous. Troubleshooting normally involves removing covers from electric motor drives, exposing internal parts with dangerous voltages present. The motor and driven load can be running, exposing personnel to machine hazards. Unexpected events occur during troubleshooting procedures such as an electric motor drive stopping unexpectedly due to a drive undervoltage fault.

The following safety guidelines must be observed at all times when troubleshooting electric motor drives:
- Only qualified personnel should troubleshoot electric motor drives.

- Always refer to recommendations and instructions of the manufacturer and applicable federal, state, and local regulations. Failure to follow manufacturer recommendations can result in serious physical injury and/or property damage.

- Technicians must understand the machinery and the process that the electric motor drive controls, plus the consequences of starting, stopping, and running a drive. For example, running an electric motor drive can cause product to fall off a conveyor.

- Avoid using two-way radios around electric motor drives, especially when the drive covers are removed. Electric motor drives are susceptible to radio frequency interference (RFI). Using two-way radios in close proximity to an electric motor drive can result in the drive running unexpectedly.

- The proper personal protective equipment must be used at all times. For example, electrical gloves with cover gloves and insulating matting are used to provide maximum insulation from electrical shock hazards.

- After a problem is identified, technicians must make every effort to find the cause of the problem.

- After a problem is corrected, a technician must verify that the electric motor drive and motor operate as designed for the application.

Gathering Information

The initial task of a technician is to gather information about the electric motor drive application problem. Technicians are sent to unfamiliar locations to troubleshoot electric motor drive application problems without the aid of engineering or maintenance shop records. Machine operators and other technicians are valuable sources of information about the electric motor drive application. Technicians should ask a series of questions in order to gather useful troubleshooting information:
- What function was the electric motor drive performing when it failed, such as accelerating, decelerating, running at speed?

- Did the electric motor drive display a fault code or error message? If so, what was the fault code?

- How long has the problem been occurring?

- Does the problem occur all the time, at a particular time, or randomly?

- Is the problem linked to a time of day, a specific event, or a specific process?

- Has anyone worked on the electric motor drive or motor recently? If so, what was done and who did the work? Did the problem start after the work was finished?

- Have there been any recent changes to the load, system, or electric motor drive programming?

Technicians must obtain all appropriate electric motor drive manuals and programming parameters. Electric motor drive manuals include installation, operation, and troubleshooting procedures. The manuals also contain drive schematics, fault code explanations, and parameter descriptions. Troubleshooting an electric motor drive without the manuals is extremely difficult.

Electric motor drive parameters are saved as hard copies or as electronic files that are downloaded to clear text display units or personal computers (PCs). Electric motor drive parameter record systems guarantee that electric motor drive parameters are not lost or destroyed when a drive is reset to factory default settings. Some electric motor drive manufacturers include DIP switches that are used to set some of the drive parameters. Understanding how to set the DIP switches is important for proper electric motor drive/motor performance.

Inspecting Electric Motor Drive Applications

After gathering information, technicians must inspect the electric motor drive application. An inspection allows technicians to become familiar with the physical layout and operation of an application. Inspections normally yield clues as to the cause of an electric motor drive application problem. **See Application Inspection Points.**

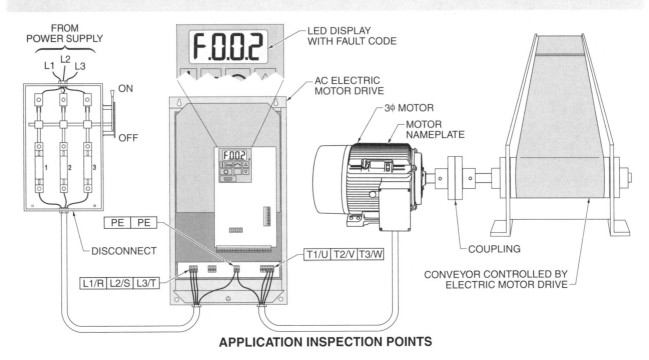

APPLICATION INSPECTION POINTS

To inspect an electric motor drive application, apply the procedure:
1. Check that all power disconnects are ON.
2. Access the fault history of the electric motor drive for information on possible causes and record the software version number.
3. Inspect the electric motor drive for physical damage and signs of overheating or fire.
4. Record the electric motor drive nameplate model number, serial number, input voltage, input current, output current, and horsepower rating.
5. Check the exterior of the motor and the area adjacent to the motor for debris to ensure proper ventilation to cool the motor.
6. Check that the motor power rating corresponds to the electric motor drive power rating.
7. Check that the motor is correctly aligned with the driven load.

8. Check that the coupling or other connection method between the motor and driven load is not loose or broken.

9. Check that the motor and the driven load are securely fastened in place.

10. Check that an object is not preventing the motor or load from turning.

11. Determine if any special equipment is required to work on the electric motor drive application.

 Refer to **Activity 18-1—Troubleshooting Electric Motor Drive Safety**..

APPLICATION 18-2—Troubleshooting Incoming Power

In order for an electric motor drive to function properly, the incoming power must be ON, the power must be within the voltage operating range of the drive, and the power must have sufficient kVA capacity and be free of quality problems. Common incoming power problems are high input voltage, low input voltage, no input voltage, voltage unbalance, improper grounding, and harmonics. Momentary power problems that may also occur are voltage sag or voltage swell.

Technicians must frequently test the incoming power supplies to electric motor drives. The tests are performed at the power terminal strip of electric motor drives using digital multimeters (DMMs) set to measure AC voltage. **See Measuring Electric Motor Drive Line Voltage.**

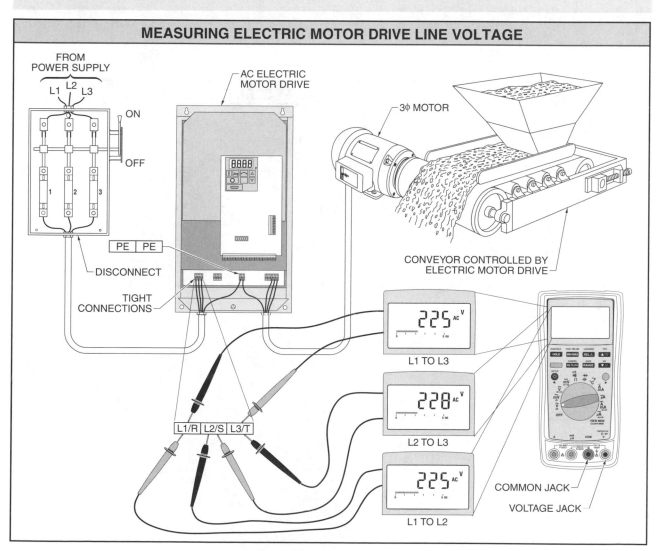

MEASURING ELECTRIC MOTOR DRIVE LINE VOLTAGE

To test the incoming power supply, apply the procedure:

1. Verify that all disconnects are ON and that fuses or circuit breakers are operational.

2. Verify that the line conductors from the local disconnect terminate at the correct spot on the electric motor drive.

3. Check that the line conductors are shielded cable or are in separate metal conduits with no other conductors.

4. Check that the line conductors are the proper AWG size.

5. Check that the grounding conductor is the proper AWG size and terminates at the correct spots.

6. Check that all connections at the power supply terminal strip are tight.

7. Measure and record the line voltage with no load (electric motor drive not running), L1 to L2, L1 to L3, and L2 to L3. Verify that the voltage is within the operating range of the electric motor drive. When a measurement of no voltage (0 VAC) is present, technicians must perform additional electrical distribution system tests.

8. Measure and record the line voltage under full-load operating conditions, L1 to L2, L1 to L3, and L2 to L3. Compare full-load readings with the no-load readings from Step 7. A voltage difference greater than 3% between no load and full load indicates that the electric motor drive does not have sufficient capacity (kVA) or the drive is overloaded.

9. Use the readings from Step 7 to calculate voltage unbalance. A value greater than 2% is not acceptable.

When no voltage is found at Step 7, technicians must determine the cause. The local disconnect can be OFF, a fuse may be blown, or a circuit breaker may be tripped. **See Determining Cause of No Voltage at Electric Motor Drive.**

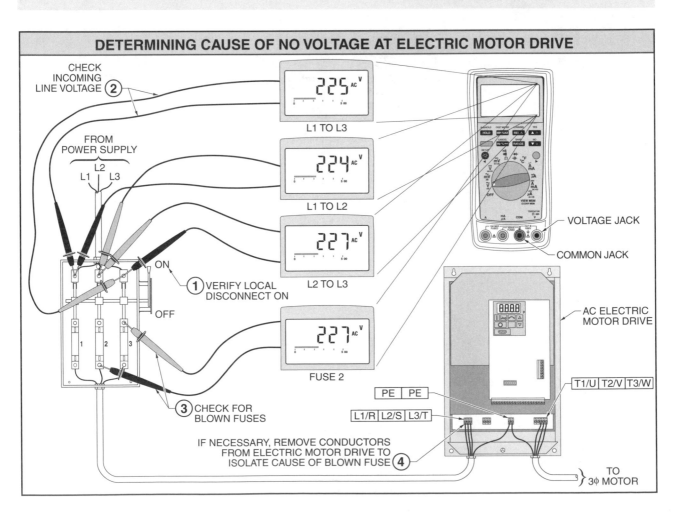

DETERMINING CAUSE OF NO VOLTAGE AT ELECTRIC MOTOR DRIVE

To identify the cause of a 0 VAC reading, continue with procedure:

10. Turn ON the local disconnect if it is in OFF position. Stand to the side of the disconnect and electric motor drive when energizing, in case of a major failure. Return to Step 7.

11. Verify that voltage is present at the disconnect, if the local disconnect is ON.

   A. Open the disconnect cover and measure the line voltage, L1 to L2, L1 to L3, and L2 to L3. If any of the measurements are 0 VAC or significantly less than the known line voltage, a problem exists in the electrical distribution system.

   B. Use a DMM to check fuses and circuit breakers. Verify that fuses or circuit breakers have the correct voltage rating, current rating, and trip characteristic for the electric motor drive. Replace any blown fuses or reset any tripped breakers. Do not remove or install fuses with the disconnect ON.

   C. Do not turn circuit breakers ON while the internal cover of the disconnect is removed.

   D. Close the disconnect cover. Stand to the side of the disconnect and electric motor drive when energizing, in case of a major failure. Turn the local disconnect ON. Return to Step 7.

12. Turn the local disconnect OFF if a fuse blows or circuit breaker trips again when power is applied to an electric motor drive.

   A. Use a DMM to verify that the AC line voltage is not present at the electric motor drive power terminal strip. Disconnect the line conductors at the power terminal strip of the electric motor drive and insulate the conductors.

   B. Replace any blown fuses or reset any tripped breakers. Turn the local disconnect ON.

   C. If fuses do not blow or circuit breakers do not trip, the electric motor drive has a problem. If fuses blow or circuit breakers trip, there is a problem with the wiring to the electric motor drive.

Incoming Power Solutions

When a problem with the incoming power is identified, the appropriate solution is applied. The solution may require modifications to the power source that supplies the electric motor drive, or troubleshooting the drive. Electric motor drives can be the cause or victim of harmonics and related problems. **See Incoming Power Troubleshooting Matrix. See Appendix.**

INCOMING POWER TROUBLESHOOTING MATRIX

SYMPTOM/FAULT CODE	PROBLEM	CAUSE	SOLUTION
ELECTRIC MOTOR DRIVE OVERVOLTAGE FAULTS. BLOWN CONVERTER (RECTIFIER) SEMICONDUCTOR	HIGH INPUT VOLTAGE/VOLTAGE SWELL	SWITCHING OF POWER FACTOR CORRECTION CAPACITORS	STOP SWITCHING POWER FACTOR CORRECTION CAPACITORS. INSTALL ELECTRIC MOTOR DRIVE ON ANOTHER FEEDER
		UTILITY SWITCHING TRANSFORMER TAPS FOR LOAD ADJUSTMENT	INSTALL A LINE REACTOR, OR INSTALL ELECTRIC MOTOR DRIVE ON ANOTHER FEEDER
		PROXIMITY TO LOW IMPEDANCE VOLTAGE SOURCE	INSTALL LINE REACTOR
		TRANSFORMER SECONDARY VOLTAGE IS HIGH	ADJUST TAPS ON THE TRANSFORMER
ELECTRIC MOTOR DRIVE OVERVOLTAGE FAULTS	VOLTAGE UNBALANCE GREATER THAN 2%	UNBALANCE FROM UTILITY	CONTACT UTILITY
		SINGLE PHASE LOADS DROPPING ON AND OFF THE SAME FEEDER AS THE ELECTRIC MOTOR DRIVE	INSTALL ELECTRIC MOTOR DRIVE ON SEPARATE FEEDER
HARMONICS	HARMONICS PRESENT ON ELECTRICAL DISTRIBUTION SYSTEM	ELECTRIC MOTOR DRIVE OR EXISTING NONLINEAR LOADS ARE POSSIBLE SOURCE	INSTALL LINE REACTOR

 Refer to **Activity 18-2—Troubleshooting Incoming Power** ....

APPLICATION 18-3—Troubleshooting an Electric Motor Drive

When the incoming power is eliminated as the source of a problem, the electric motor drive is the next element to test. A series of tests is used to eliminate the source of possible problems within the electric motor drive application. The possible sources of problems are the electric motor drive, drive parameters, input and output devices, motor, and load. Common electric motor drive problems are component failure, incorrect parameter settings, and input and output devices that are incorrectly connected or that have failed. Motor problems and load problems are mistaken for electric motor drive problems. Electric motor drive fault codes aid in identifying problems. Tests must be performed in the proper sequence to correctly identify a problem in the least amount of time.

An *electric motor drive test* is an initial test that verifies if an electric motor drive is operational. A partial failure of an electric motor drive is uncommon. Electric motor drives normally work or do not work. An electric motor drive set to factory default settings and controlled by an integral keypad is tested with the motor disconnected. At this point parameter settings, inputs and outputs, the motor, and the load are not tested as the source of the problem. When the control mode is sensorless vector control or closed-loop vector control, it may not be possible to run the vector control drive with the motor disconnected. When possible, technicians should change the control mode to constant torque or variable torque in order to perform electric motor drive tests. **See Electric Motor Drive Initial Test.**

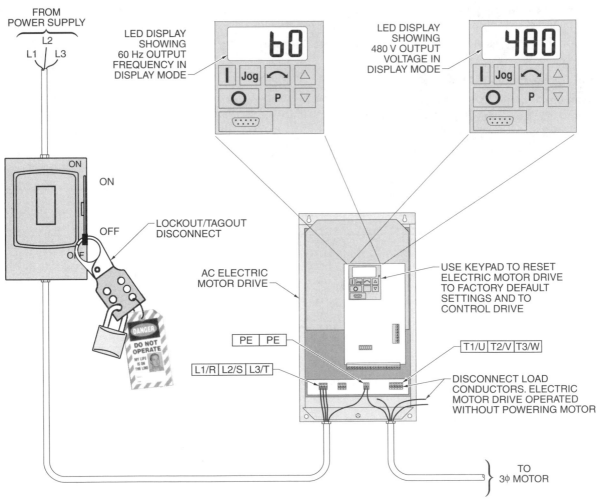

ELECTRIC MOTOR DRIVE INITIAL TEST

To test an electric motor drive, apply the procedure:

1. Push the stop (O) button if the electric motor drive is ON.

2. Turn disconnect OFF. Lockout/tagout disconnect.

3. Wait for the DC bus capacitors to discharge. Do not manually discharge the capacitors by shorting + to –. Remove the electric motor drive cover. Use a DMM to verify that the AC line voltage is not present. Use a DMM to verify that the DC bus capacitors have discharged. Do not rely on the DC bus charge LEDs because LEDs can burn out, giving a false indication.

4. Disconnect the load conductors from the electric motor drive power terminal strip. Note where motor wires are connected in order to maintain the correct rotation upon reconnection. Reinstall the electric motor drive cover.

5. Remove the lockout/tagout from the local disconnect.

6. Stand to the side of the disconnect and electric motor drive when energizing, in case of a major failure. Turn the local disconnect ON. Do not push the start (I) button. The electric motor drive LED display or clear text display activates. The electric motor drive cooling fan(s) may or may not start when power is applied, depending on the drive model. If the fans do not start at this point, a technician must check that they start when the start (I) button is pushed. If there are any loud noises, smoke, or explosions, immediately turn the local disconnect OFF and proceed to electric motor drive component tests.

7. Record or download electric motor drive parameter values. Reset parameters to factory default settings.

8. Program the input mode to keypad to control the electric motor drive by the integral keypad.

9. Program speed reference to internal.

10. Program the display mode to show electric motor drive output frequency in hertz (Hz).

11. Stand to the side of the electric motor drive when pushing the start (I) button, in case of a major drive failure. Push the start (I) button. The cooling fan(s) should start, if cooling fan(s) did not start when power was applied. The LED display should ramp up to a low speed. If the LED display shows 0 Hz, push the ramp up (↑) button until 5 Hz is shown.

12. Push the ramp up (↑) button until 60 Hz is shown on the LED display.

13. Program display mode to show the electric motor drive output voltage. The electric motor drive output voltage should be approximately the same as the 60 Hz drive input voltage, such as 480 VAC displayed when the input voltage is 485 VAC.

14. Push the stop (O) button. The voltage should decrease to 0 VAC.

15. Proceed to the next test. If the electric motor drive performed without any problem, it is not the source of the problem.

16. Proceed to Electric Motor Drive Component Tests if the electric motor drive did not perform correctly.

An *electric motor drive, motor, and load test* is a test used to verify that a drive and motor properly rotate the driven load. An electric motor drive set to factory defaults, and controlled by the integral keypad, is tested with the motor connected. At this point inputs and outputs are not tested as the source of the problem. If the control mode was changed to perform the electric motor drive test, return the control mode to its original setting. **See Electric Motor Drive, Motor, and Load Test.**

ELECTRIC MOTOR DRIVE, MOTOR, AND LOAD TEST

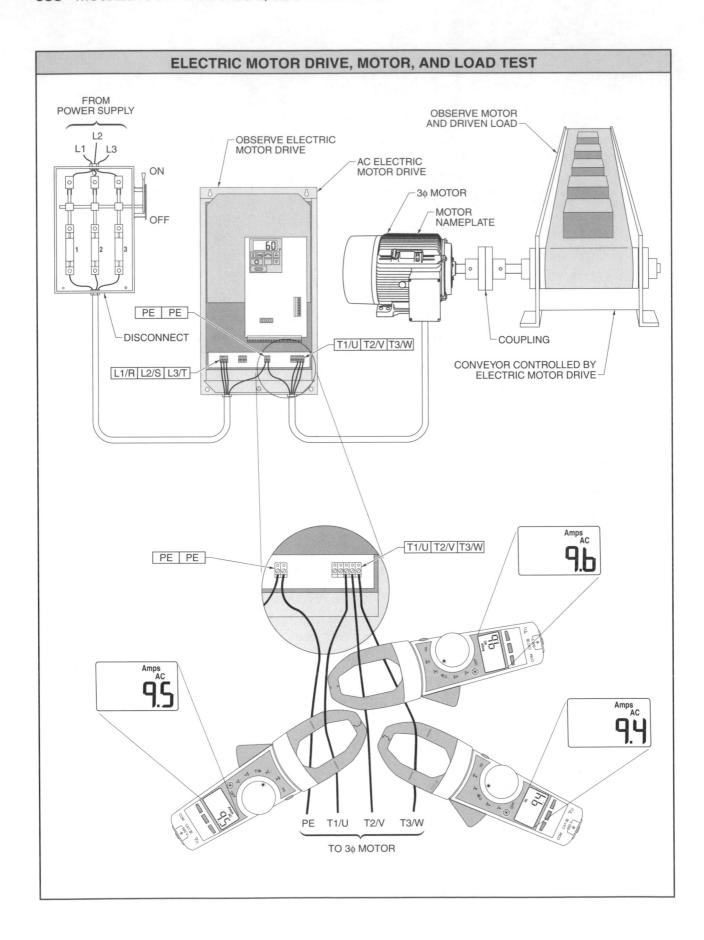

FROM POWER SUPPLY

L2
L1 L3

ON

OFF

OBSERVE ELECTRIC MOTOR DRIVE

AC ELECTRIC MOTOR DRIVE

3φ MOTOR

MOTOR NAMEPLATE

OBSERVE MOTOR AND DRIVEN LOAD

PE | PE

DISCONNECT

T1/U | T2/V | T3/W

COUPLING

L1/R | L2/S | L3/T

CONVEYOR CONTROLLED BY ELECTRIC MOTOR DRIVE

PE | PE

T1/U | T2/V | T3/W

Amps AC
9.6

Amps AC
9.5

Amps AC
9.4

PE   T1/U   T2/V   T3/W

TO 3φ MOTOR

To test an electric motor drive, motor, and load apply the procedure:

1. Push the stop (O) button if the electric motor drive is ON.

2. Turn disconnect OFF. Lockout/tagout disconnect.

3. Wait for the DC bus capacitors to discharge. Do not manually discharge the capacitors by shorting + to −. Remove the electric motor drive cover. Use a DMM to verify that the AC line voltage is not present. Use a DMM to verify that the DC bus capacitors have discharged. Do not rely on the DC bus charge LEDs.

4. Reconnect the load conductors to their previous locations on the power terminal strip in order to maintain correct motor rotation because incorrect motor rotation causes damage in certain applications. Reinstall the electric motor drive cover.

5. Remove lockout/tagout from disconnect.

6. Stand to the side of the disconnect and electric motor drive when energizing, in case of a major failure. Turn disconnect ON. Do not push the start (I) button. The electric motor drive LED display or clear text display activates.

7. Program the appropriate parameters into the electric motor drive with motor nameplate data.

8. Program the display mode to show electric motor drive output frequency.

9. Do not start the electric motor drive until a check has been made that personnel are not at risk from the driven load.

10. Stand to the side of the disconnect and electric motor drive when pushing the start button, in case of a major drive failure. Push the start (I) button. The LED display should ramp up to a low speed. If the LED display shows 0 Hz, push the ramp up (↑) button until 5 Hz is shown.

11. Increase the speed of the motor to 60 Hz using the ramp up (↑) button. The motor and driven load must accelerate smoothly to 60 Hz. Any unusual noises or vibrations must be recorded and the frequency of the occurrence recorded. Unusual noises or vibrations indicate alignment problems or require the use of the skip frequency parameter to avoid unwanted mechanical resonance.

12. Remove the electric motor drive cover. Dangerous voltage levels exist when the electric motor drive cover is removed and the drive is energized. Exercise extreme caution and use the appropriate personal protective equipment.

13. Measure and record the current in each of the three load conductors using a true-rms clamp-on ammeter. True-rms clamp-on ammeters are required because the current waveform of an electric motor drive is non-sinusoidal. Current readings are taken at 60 Hz because the motor nameplate current is based on 60 Hz.

    A. Current readings of the three load conductors must be equal or very close to each other—for example, T1 = 9.5 A, T2 = 9.6 A, and T3 = 9.4 A. A problem with the load conductors or motor is present if the current readings of the load conductors are not equal or very close.

    B. An *overloaded motor* is a motor with a current reading greater than 105% of nameplate current rating. There is a problem with the motor or the load if the current readings are greater than 105% of the nameplate current rating.

    C. Reinstall the electric motor drive cover.

14. Decrease the speed of the motor to 0 Hz using the ramp down (↓) button. The motor and driven load should decelerate smoothly to 0 Hz. Any unusual noises or vibrations must be recorded and the frequency of the occurrence recorded. Unusual noises or vibrations indicate alignment problems, or can require the use of the skip frequency parameter to avoid unwanted mechanical resonance.

15. Push the stop (O) button.

If the electric motor drive, motor, and load performed without any problems, the drive, motor, and load are not the source of the problem. Proceed to Electric Motor Drive Input and Output Test. There is a problem if the electric motor drive, motor, and load did not perform correctly. The problem is with the electric motor drive parameters, motor, or load. The electric motor drive was eliminated as the problem in the Initial Test. Proceed to the Electric Motor Drive Solutions section.

An *electric motor drive input and output test* is a test used to verify that the inputs and outputs of a drive function properly when operated as designed. An electric motor drive programmed for a specific application, and controlled by inputs and outputs specific to the application, is tested with the motor connected. The complexity of an electric motor drive application can require more than one technician or a technician and qualified person to complete the test. A qualified person must be able to verify that the electric motor drive is working properly in relation to the entire process. The qualified person must also be able to make suggestions for optimizing an electric motor drive, such as slowing down a mixing motor on a tank or vat in a batch process. **See Electric Motor Drive Input and Output Test.**

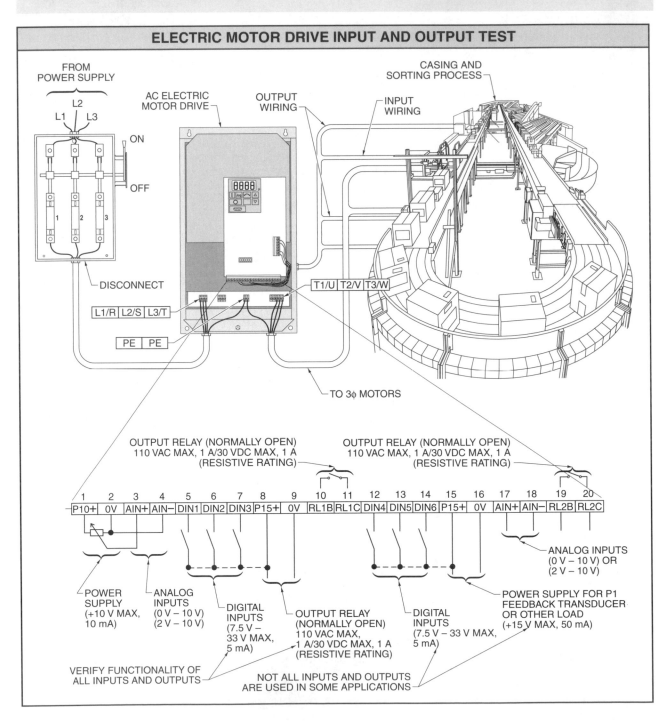

ELECTRIC MOTOR DRIVE INPUT AND OUTPUT TEST

To test electric motor drive inputs and outputs, apply the procedure:

1. Push the stop (O) button if the electric motor drive is ON.

2. Return the electric motor drive parameters to the application values copied or downloaded in Step 7 of the Initial Test.

3. Do not start an electric motor drive until a check has been made that personnel are not at risk from the driven load.

4. Stand to the side of the electric motor drive when pushing the start (I) button, in case of a major drive failure. Push the electric motor drive start (I) button.

5. Monitor the electric motor drive, motor, and driven load under full-load condition.

6. Verify that the electric motor drive application works properly. Adjust parameters as needed to optimize performance of the electric motor drive and the controlled process.

7. Verify the functionality of all inputs and outputs connected to the electric motor drive.

8. Remove the electric motor drive cover. Dangerous voltage levels exist when the electric motor drive cover is removed and the drive is energized. Exercise extreme caution and use the appropriate personal protective equipment.

9. Measure and record the current in each of the load conductors using a true-rms clamp-on ammeter. Verify that the motor is not overloaded.

10. Reinstall the electric motor drive cover.

11. Record or download the electric motor drive application parameter values. Store this information in a safe location.

12. Enable parameter protection to prevent unauthorized personnel from adjusting parameters. If the electric motor drive, motor, load, inputs, and outputs performed without any problems, the inputs and outputs are not the source of the problem. The electric motor drive application is ready for use.

If the electric motor drive, motor, load, inputs, and outputs did not perform correctly, there is a problem. The problem is an electric motor drive parameter problem or an input and output problem. The electric motor drive, motor, and load were eliminated as possible problems in the electric motor drive, motor, and load test.

Electric Motor Drive Component Tests

An electric motor drive component test identifies which components in a drive are defective. The DC bus capacitors, bus capacitor balancing resistors, electric motor drive cooling fan(s), converter semiconductors, and inverter semiconductors are tested. The tests consist of visual inspections, converter test, and an inverter test. The tests must be performed in sequence. An electric motor drive component test does not test every possible component that can fail. Technicians that are unable to identify a defective component have technical support available from electric motor drive manufacturers.

Electric Motor Drive Solutions

When the problem with an electric motor drive is identified, the appropriate solution is applied. The solution requires changing a parameter value, replacing a component, or troubleshooting the motor and the load. **See Electric Motor Drive Troubleshooting. See Appendix.** Items that must be followed when applying a solution to an electric motor drive are:

• Replace a soldered component where cost-effective. Normally, replacing multiple components is not cost-effective because on small horsepower and newer electric motor drives, all components are soldered in place using special equipment.

• Follow the instructions of the electric motor drive manufacturer when replacing a component.

• Exercise electrostatic discharge (ESD) precautions when working with circuit boards and components.

• Mark and record locations of wires and cables before disconnecting control cables and circuit boards to prevent mistakes during reconnection.

- Check pin and socket alignment before reconnecting cables to circuit boards.
- Replace all DC bus capacitors when any DC bus capacitor is defective because good DC bus capacitors are the same age as the failed capacitor and may fail at any time. Also, the defective DC bus capacitor can cause other capacitors to be overstressed.
- Electrify spare DC bus capacitors periodically per electric motor drive manufacturer recommendations. Follow the instructions of the manufacturer regarding DC bus capacitor storage and replacement.
- Remove old heat sink compound when replacing semiconductors. Apply new heat sink compound before installing new semiconductors.
- Replace both semiconductors when one semiconductor of a paralleled pair is defective.

ELECTRIC MOTOR DRIVE TROUBLESHOOTING

FAULTS

SYMPTOM/FAULT CODE	PROBLEM	CAUSE	SOLUTION
ELECTRIC MOTOR DRIVE OVERVOLTAGE FAULT	ELECTRIC MOTOR DRIVE OVERVOLTAGE	DECELERATION TIME IS TOO SHORT	INCREASE DECELERATION TIME
		HIGH INPUT VOLTAGE (VOLTAGE SWELL)	SEE INCOMING POWER TROUBLESHOOTING MATRIX
		LOAD IS OVERHAULING MOTOR	ADD DYNAMIC BRAKING RESISTOR AND/OR INCREASE DECELERATION TIME

COMPONENT FAILURES

SYMPTOM/FAULT CODE	PROBLEM	CAUSE	SOLUTION
ELECTRIC MOTOR DRIVE DOES NOT TURN ON. BLOWN FUSE OR TRIPPED BREAKER	DEFECTIVE CONVERTER SECTION (RECTIFIER SEMICONDUCTOR)	HIGH INPUT VOLTAGE (VOLTAGE SWELL)	REPLACE CONVERTER SECTION SEMICONDUCTOR OR REPLACE ELECTRIC MOTOR DRIVE
			SEE ALSO INCOMING POWER MATRIX
		ELECTRIC MOTOR DRIVE COOLING FAN IS DEFECTIVE	REPLACE CONVERTER SECTION SEMICONDUCTOR AND COOLING FAN OR REPLACE ELECTRIC MOTOR DRIVE

PARAMETER PROBLEMS

SYMPTOM/FAULT CODE	PROBLEM	CAUSE	SOLUTION
UNUSUAL NOISES OR VIBRATIONS WHEN ELECTRIC MOTOR DRIVE POWERING MOTOR	PARAMETERS INCORRECT	PARAMETER(S) INCORRECTLY PROGRAMMED	ADJUST SKIP FREQUENCY PARAMETER
	PROBLEM WITH MOTOR AND/OR LOAD	PROBLEM WITH MOTOR AND/OR LOAD	SEE MOTOR AND LOAD TROUBLESHOOTING MATRIX

INPUT AND OUTPUT PROBLEMS

SYMPTOM/FAULT CODE	PROBLEM	CAUSE	SOLUTION
ELECTRIC MOTOR DRIVE DOES NOT OPERATE CORRECTLY WHEN INPUT MODE IS OTHER THAN KEYPAD, MOTOR AND LOAD ARE CONNECTED, AND DRIVE IS OPERATED AS DESIGNED	EXTERNALLY CONNECTED INPUTS AND OUTPUTS INCORRECT	INPUT(S) AND/OR OUTPUT(S) INCORRECTLY WIRED. INPUT OR OUTPUT DEVICES NOT FUNCTIONAL	TIGHTEN LOOSE WIRES AND/OR REPLACE NON FUNCTIONAL OR INCORRECT DEVICES FOR APPLICATION
		PROBLEM WITH INPUTS THAT SUPPLY START, STOP, REFERENCE, OR FEEDBACK SIGNALS	CHECK INPUT SYSTEM FOR PROPER INPUT
	PARAMETERS INCORRECT	PARAMETERS INCORRECTLY PROGRAMMED	SEE ELECTRIC MOTOR DRIVE PARAMETER PROBLEMS

OPERATIONAL PROBLEMS

SYMPTOM/FAULT CODE	PROBLEM	CAUSE	SOLUTION
MOTOR ROTATION INCORRECT WHEN POWERED BY ELECTRIC MOTOR DRIVE	INCORRECT PHASING	WIRING	INTERCHANGE TWO OF THE LOAD CONDUCTORS AT THE ELECTRIC MOTOR DRIVE LOAD TERMINAL STRIP
MOTOR ROTATION INCORRECT WHEN IN BYPASS MODE	WIRING INCORRECT	WIRING	INTERCHANGE TWO LINE CONDUCTORS AT DISCONNECT NOTE: ASSUMES ELECTRIC MOTOR DRIVE AND BYPASS SHARE COMMON FEED

Motor and Load Tests

After the incoming power and the electric motor drive have been eliminated as sources of problems, the motor and the load are the next elements to test. A series of tests is used to eliminate the source of possible problems. Motor and load tests must be performed in sequence to identify a problem. Common motor and load problems are a grounded motor stator, defective motor bearings, and motor-to-load misalignment. Motor and load problems can require the assistance of the manufacturer to solve. Electric motor drive fault codes aid in identifying motor and load problems.

 Refer to **Activity 18-3—Troubleshooting an Electric Motor Drive** ....

APPLICATION 18-4—Troubleshooting the Motor and Load

An *insulation spot test* is a test that checks motor insulation over the life of a motor. Megohmmeters are used to perform insulation spot tests. Motor insulation is damaged by moisture, oil, dirt, excessive heat, excessive cold, corrosive vapors, and vibration. Good motor insulation has a high resistance reading. Poor motor insulation has a low resistance reading.

The ideal megohmmeter reading is infinite resistance (resistance) between the conductor or winding being tested, and ground. Megohmmeter readings of less than infinite resistance are common. The rule of thumb states that for every 1000 V of insulation rating, 1 MΩ of resistance should exist. Wires used in 240 VAC or 480 VAC distribution systems have a rating of 600 V. During insulation spot-tests, consider the wires to have 1000 V insulation. The stator winding insulation for inverter duty motors is rated at approximately 1500 V. The motor windings and the load conductors are simultaneously tested because the megohmmeter readings are normally measured from the electric motor drive end of the motor conductors. To perform an insulation spot test, apply the procedure:

1. Push the stop (O) button when an electric motor drive is ON.

2. Turn disconnect OFF. Lockout/tagout disconnect.

3. Wait for the DC bus capacitors to discharge. Do not manually discharge the capacitors by shorting + to –. Remove the electric motor drive cover. Use a DMM to verify that AC line voltage is not present. Use a DMM to verify that the DC bus capacitors have discharged.

4. Disconnect the load conductors from the electric motor drive at the power terminal strip.

5. Set the megohmmeter to the selected test voltage level. The test voltage is normally set higher than the voltage rating of the insulation under test in order to stress the insulation. The 1000 V setting is normally used for motors and conductors operating at 480 VAC or less. Some megohmmeters do not have a 1000 V setting and use a voltage setting close to 1000 V, but not exceeding 1000 V.

6. Connect the megohmmeter to measure the resistance of each winding lead to ground. Record the readings after 60 sec. Service the motor if a reading does meet the minimum acceptable resistance.

7. Discharge the motor windings.

8. Interpret the readings taken.

Megohmmeter readings must be a minimum of 2 MΩ and be relatively close to each other, such as T1 to ground measuring 10 MΩ, T2 to ground measuring 9.7 MΩ, and T3 to ground measuring 9.8 MΩ. A reading less than 2 MΩ, or a large difference between the readings (greater than 50%), is cause for additional testing.

When additional testing is necessary, the cause of the low readings or large difference between readings must be determined. Possible causes are the load conductors from the electric motor drive to the motor or the motor windings. **See Isolation Insulation Spot Test.**

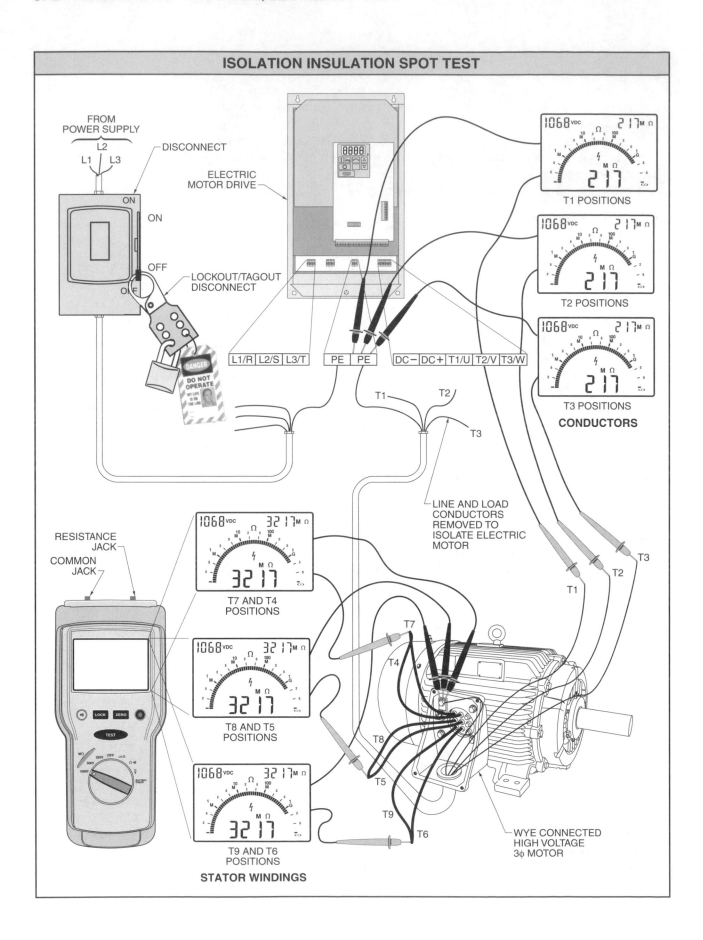

ISOLATION INSULATION SPOT TEST

FROM POWER SUPPLY
L2
L1 L3

DISCONNECT

ELECTRIC MOTOR DRIVE

ON
ON
OFF
OFF

LOCKOUT/TAGOUT DISCONNECT

DANGER
DO NOT OPERATE
MY LIFE IS ON THE LINE

L1/R | L2/S | L3/T   PE | PE   DC- | DC+ | T1/U | T2/V | T3/W

1068 VDC   217 M Ω
217
T1 POSITIONS

1068 VDC   217 M Ω
217
T2 POSITIONS

1068 VDC   217 M Ω
217
T3 POSITIONS

CONDUCTORS

T1   T2
T3

LINE AND LOAD CONDUCTORS REMOVED TO ISOLATE ELECTRIC MOTOR

T3
T2
T1

RESISTANCE JACK
COMMON JACK

1068 VDC   3217 M Ω
3217
T7 AND T4 POSITIONS

1068 VDC   3217 M Ω
3217
T8 AND T5 POSITIONS

1068 VDC   3217 M Ω
3217
T9 AND T6 POSITIONS

LOCK ZERO
TEST
MΩ OFF
500V 250V
1000V Ω LoΩ
BATTERY FAULT

T7
T4
T8
T5
T9
T6

WYE CONNECTED HIGH VOLTAGE 3φ MOTOR

STATOR WINDINGS

To isolate the cause of the low readings or large difference between readings, apply the procedure:

1. Open the motor termination box. Disconnect the load conductors from the motor leads. Do not disconnect the connections between windings.

2. Perform the insulation spot test to test the load conductors. Technicians must ensure that the load conductors are disconnected from the electric motor drive and are clear of ground and personnel. Megohmmeters produce voltages that can injure personnel and damage electric motor drives.

3. Perform the insulation spot test to test the motor windings. Technicians must ensure that the motor leads not under test are clear of ground and personnel. Megohmmeter voltage is present at the motor leads not under test because of the internal winding connections inside a motor.

4. When the cause of the low readings or large difference between readings is determined, the motor and/or load conductors must be repaired or replaced.

5. The load conductors and motor windings must be insulation spot-tested with a megohmmeter after repair or replacement work has been performed.

6. Reconnect the load conductors to the motor leads and replace the motor termination box cover. Reconnect the load conductors to the appropriate positions on the electric motor drive terminal strip.

Motor Mechanical Test

A *motor mechanical test* is a test that checks the mechanical operation of a motor. A mechanical problem with a motor results in electric motor drive faults. **See Motor Mechanical Test.**

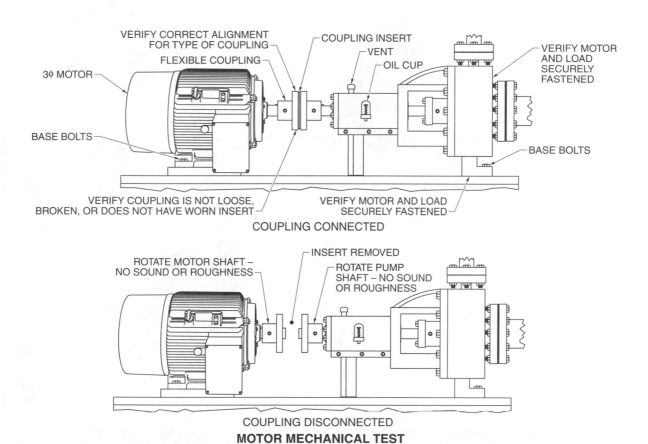

MOTOR MECHANICAL TEST

To perform a motor mechanical test, apply the procedure:

1. Check that the motor and driven load are aligned correctly for the type of coupling used.
2. Check that the coupling connecting the motor and driven load is not loose or broken.
3. Check that the motor is securely bolted in place.
4. Check that an object is not preventing the motor from rotating.
5. Disconnect the motor from the driven load. Turn the motor shaft by hand. The shaft must rotate freely and not be noisy when rotated. A bind in the rotation of the motor shaft or noise when the shaft is rotating indicates a problem with the motor.

Motor Current Test

A *motor current test* is a test used to find hidden motor problems not found with the motor mechanical test. A possibility exists that a motor mechanical test did not detect a problem that is just starting to develop within a motor. The motor current test is designed to catch hidden problems in motors. A motor that is disconnected from the driven load draws less than its motor nameplate current when running at 60 Hz. **See Load Current Test.**

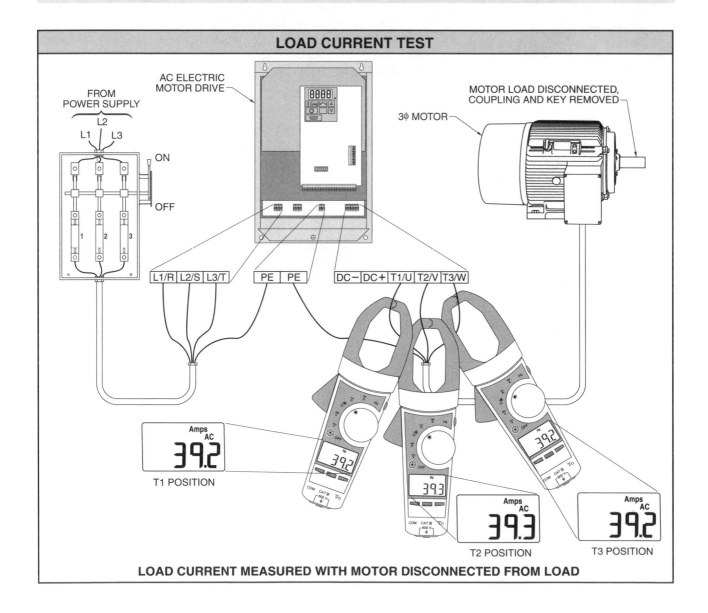

LOAD CURRENT TEST

FROM POWER SUPPLY
L2
L1   L3
ON
OFF

AC ELECTRIC MOTOR DRIVE

3φ MOTOR

MOTOR LOAD DISCONNECTED, COUPLING AND KEY REMOVED

L1/R L2/S L3/T   PE PE   DC− DC+ T1/U T2/V T3/W

Amps AC
39.2
T1 POSITION

39.3

Amps AC
39.3
T2 POSITION

Amps AC
39.2
T3 POSITION

LOAD CURRENT MEASURED WITH MOTOR DISCONNECTED FROM LOAD

To perform a motor current test, apply the procedure:

1. Remove the lockout/tagout from the disconnect.

2. Stand to the side of the disconnect and electric motor drive when energizing, in case of a major failure. Turn disconnect ON.

3. Do not start an electric motor drive until a check has been made that personnel are not at risk from the driven load.

4. Program the display mode to show the electric motor drive output frequency.

5. Stand to the side of the electric motor drive when pushing the start button, in case of a major drive failure. Push the start (I) button.

6. Increase the speed of the motor to 60 Hz using the ramp up (↑) button.

7. Remove the electric motor drive cover. Dangerous voltage levels exist when the electric motor drive cover is removed and the drive is energized. Exercise extreme caution and use the appropriate personal protective equipment.

8. Measure and record the current in each of the load conductors using a true-rms clamp-on ammeter. Current readings equal to or greater than the motor nameplate current indicate a problem with the motor. The current readings must be equal or very close to each other.

9. Reinstall the electric motor drive cover.

10. Push the stop (O) button.

11. Turn disconnect OFF. Lockout/tagout disconnect.

If problems were not found using the insulation spot test, the motor mechanical test, or the motor current test, the motor is not the source of the problem. The load is the problem. If problems were found using the motor insulation test, the motor mechanical test, or the motor current test, the motor is the source of the problem. Proceed to the Motor and Load Solutions section.

Load Current Test

A possibility exists that a load mechanical test did not detect a problem that is just starting to develop. The load current test is designed to catch this type of problem. To perform a load current test, apply the procedure:

1. Reconnect the motor with the driven load. Check that the motor is correctly aligned with the driven load.

2. Remove lockout/tagout from disconnect.

3. Stand to the side of the disconnect and electric motor drive when energizing, in case of a major failure. Turn the disconnect ON.

4. Do not start the electric motor drive until a check has been made that personnel are not at risk from the driven load.

5. Program the display mode to show electric motor drive output frequency.

6. Stand to the side of the electric motor drive when pushing the start button, in case of a major drive failure. Push the start (I) button.

7. Increase the speed of the motor to 60 Hz using the ramp up (↑) button.

8. Remove the electric motor drive cover. Dangerous voltage levels exist when the electric motor drive cover is removed and the drive is energized. Exercise extreme caution and use the appropriate personal protective equipment.

9. Measure and record the current in each of the load conductors using a true-rms clamp-on ammeter. Verify that the motor is not overloaded. The current readings must be equal or very close to each other.

10. Push the stop (O) button.

11. Reinstall the electric motor drive cover.

If problems were not found using the load mechanical test or the load current test, the load is not the source of the problem. If problems were found using the load mechanical test or the load current test, the load is the source of the problem. Proceed to Motor and Load Solutions section.

Motor and Load Solutions

When the problem with a motor and/or the load is identified, the appropriate solution is applied. The solution requires changing a parameter value, repairing the motor, or replacing the motor. A technician may require the assistance of other trades in order to correct a motor and/or load problem. **See Motor and Load Troubleshooting Matrix. See Appendix.**

MOTOR AND LOAD TROUBLESHOOTING MATRIX			
MOTOR AND LOAD PROBLEMS			
SYMPTOM/FAULT CODE	PROBLEM	CAUSE	SOLUTION
DRIVE OVERVOLTAGE FAULT	ELECTRIC MOTOR DRIVE OVERVOLTAGE	LOAD IS OVERHAULING MOTOR	CONTACT ELECTRIC MOTOR DRIVE MANUFACTURER
			SEE ELECTRIC MOTOR DRIVE TROUBLESHOOTING MATRIX
UNUSUAL NOISES OR VIBRATIONS WHEN ELECTRIC MOTOR DRIVE IS POWERING LOAD	PROBLEM WITH MOTOR AND/OR LOAD	MISALIGNMENT OF MOTOR AND LOAD	ALIGN MOTOR AND LOAD
		MOTOR AND/OR LOAD NOT SECURELY FASTENED IN PLACE	SECURELY FASTEN MOTOR AND LOAD TO BASE
		DEFECTIVE BEARING(S) IN MOTOR AND/OR LOAD	REPLACE DEFECTIVE BEARING(S) OR REPLACE MOTOR OR LOAD
		MOTOR AND ELECTRIC MOTOR DRIVE NOT PROPERLY SIZED FOR LOAD	CONTACT ELECTRIC MOTOR DRIVE MANUFACTURER
	PARAMETERS INCORRECT	PARAMETER(S) INCORRECTLY PROGRAMMED	SEE ELECTRIC MOTOR DRIVE TROUBLESHOOTING MATRIX
RELATED PROBLEMS			
ELECTRIC MOTOR DRIVE DOES NOT OPERATE CORRECTLY WHEN SET TO DEFAULT PARAMETERS, INPUT MODE IS KEYPAD WITH MOTOR AND LOAD CONNECTED	PROBLEM WITH MOTOR AND/OR LOAD	PROBLEM WITH MOTOR AND/OR LOAD	SEE MOTOR AND LOAD PROBLEMS
	PARAMETERS INCORRECT	PARAMETER(S) INCORRECTLY PROGRAMMED	SEE ELECTRIC MOTOR DRIVE TROUBLESHOOTING MATRIX

Items to be aware of when applying a solution to a motor and/or load problem are the following:
- Motors connected to an electric motor drive should be inverter rated or inverter duty, NEMA MG-1 Section IV Part 31 compliant.
- Improper installation of an electric motor drive and motor can result in motor problems and failures. For example, long distances between an electric motor drive and a motor result in destructive voltage spikes at the terminals of the motor.

- Dirty motors cause excessive motor heating.
- Bearings fail from lack of lubrication or overlubrication.
- Rotor and/or stator damage results from overlubrication.

 Refer to **Activity 18-4—Troubleshooting the Motor and Load**..

APPLICATION 18-5—Electric Motor Drive Maintenance

After successful installation and startup, electric motor drives and motors do not require a large amount of maintenance. In most electric motor drive and motor applications, an annual maintenance inspection is all that is required. Turn power OFF to the electric motor drive during an annual maintenance inspection. An annual maintenance inspection must be scheduled so the inspection does not conflict with the process that an electric motor drive controls. Problems found during an annual maintenance inspection must be corrected before the problems lead to electric motor drive failure and equipment downtime.

Electric Motor Drive Maintenance Data

Technicians must take measurements and record information from an electric motor drive. The maintenance data collected is used to establish a baseline. **See Electric Motor Drive Record Sheet.** A significant deviation from the prior annual maintenance inspection data is a warning of imminent failure. Technicians must determine the cause of the deviations and take appropriate corrective actions.

To collect maintenance data, apply the procedure:
1. Record the electric motor drive model number, serial number, input voltage, input current, output current, and horsepower rating from the nameplate of the drive.
2. Record motor nameplate data.
3. Measure and record line voltages, L1 to L2, L1 to L3, and L2 to L3.
4. Measure and record the DC bus voltage with the electric motor drive running at a constant speed. Unstable DC bus voltage when the electric motor drive is running at a constant speed is an indication of imminent failure or defective DC bus capacitors.
5. Measure and record the currents in each line conductor using a true-rms clamp-on ammeter with the electric motor drive powering a motor at 60 Hz.
6. Measure and record the currents in each load conductor using a true-rms clamp-on ammeter with the drive powering a motor at 60 Hz. The reading must be 105% or less than the current listed on the nameplate of the motor.
7. Perform an insulation spot test of the load conductors and motor windings. Record the lowest reading on an insulation spot test graph if all readings are above the minimum acceptable reading.
8. Record the heat sink temperature. An increase in heat sink temperature between annual inspections indicates a problem.

Electric Motor Drive Replacement Parts

Cooling fans and DC bus capacitors of electric motor drives are two components that have fixed life expectancies. Cooling fans and capacitors must be replaced periodically to avoid major electric motor drive failures. Always follow the recommended replacement intervals and procedures given by the electric motor drive manufacturer.

ELECTRIC MOTOR DRIVE RECORD SHEET

ELECTRIC MOTOR DRIVE INFORMATION

Drive Name:		DRIVE LOCATION	
Drive Manufacturer:		Building:	Floor:
Drive Model:		System:	
Drive Serial Number:		Machine:	
Drive Horsepower:			

INPUT	OUTPUT	PARAMETERS CHANGED FROM DEFAULT	
Voltage:	Voltage:	P013 - 60 Hz TO (90 Hz)	P -
Current:	Current:	P -	P -
Frequency:	Frequency:	P -	P -
Phase:	Phase:	P -	P -
kVA:	kVA:	P -	P -

MOTOR INFORMATION

Horsepower or Kilowatts:	Service Factor:	
Voltage:	Current:	
Speed:	Frequency:	
Magnetizing Current:	Stator Resistance:	
NEMA Design:	NEMA Efficiency:	
Duty:	Frame:	
Motor is NEMA MG-1 Section IV part 31 compliant	☐ YES	☐ NO

DRIVE MAINTENANCE INFORMATION

LINE VOLTAGE			LINE CURRENT	LOAD CURRENT
L1 to L2:	L1 to L3:	L2 to L3:	L1:	T1:
DC Bus Voltage:	Heat Sink temp:		L2:	T2:
			L3:	T3:
MEGOHMETER – MOTOR AND LOAD CONDUCTOR TEST				
T1 to Ground:		T2 to Ground:		T3 to Ground:

MAINTENANCE LOG

DATE	PROBLEM FOUND	SOLUTION

To maintain the cooling fan(s) of an electric motor drive, apply the procedure:
1. The cooling fan must operate correctly. The fan must not be noisy and should rotate freely by hand when not powered. A fan that is noisy or does not rotate freely requires replacement.
2. The fan housing, fan blades, and air intake must be free of dirt, dust, and obstructions. Replace the air intake filter per manufacturer schedule or when dirty.
3. An increase in heat sink temperature indicates a fan is beginning to fail or that the intake filter is dirty.
4. Normally, fans are replaced every 3 years to 5 years.

To maintain the DC bus capacitors of an electric motor drive, apply the procedure:
1. Inspect the DC bus capacitors. DC bus capacitors that are swollen and/or have a protruding pressure relief valve are failing and must be replaced. If one DC bus capacitor is defective, replace all DC bus capacitors.
2. When DC bus capacitors are replaced, also check the capacitor balancing resistors. Defective capacitor balancing resistors damage DC bus capacitors.
3. DC bus capacitors must be replaced every 5 years to 10 years. Ambient temperatures and electric motor drive loading affect the life expectancy of capacitors.

Visual Inspection

A thorough visual inspection of an electric motor drive and motor must be made periodically. Components are checked for wear and loose connections. To perform a visual inspection, apply the procedure:

1. Inspect an electric motor drive for physical damage and signs of overheating.
2. Inspect an electric motor drive heat sink to ensure that the heat sink is clean and air can flow freely across the heat sink.
3. Inspect the input contactor, output contactor, and bypass contactor for loose connections, worn or damaged parts, and burned or pitted contacts. Power must be OFF and proper lockout/tagout procedures observed.
4. Tighten all connections on the control terminal strip and power terminal strip per the recommendations of the manufacturer.
5. Inspect the motor for physical damage and signs of overheating. Check that the exterior of the motor and the area next to the motor are free of debris to ensure proper ventilation. Check that the motor and load are fastened securely in place correctly aligned, and that the coupling method is not loose or broken.

 Refer to **Activity 18-5—Electric Motor Drive Maintenance**..

 *Refer to Quick Quiz® on CD-ROM*

 Refer to Chapter 18 in the **Troubleshooting Electrical/Electronic Systems Workbook** *for additional questions.*

Name _____   **Date** _____

ACTIVITY 18-1—Troubleshooting Electric Motor Drive Safety

One of the most important safety steps to prevent an overloaded motor, motor damage, and a fire hazard is understanding when a drive will remove power from the motor when a fault occurs. Electric motor drives must be programmed (using a parameter setting) to trip if the drive detects the motor is overheated. The time it takes before the electric motor drive trips the overloads and turns the motor OFF depends on the extent of the overload and the motor operating frequency. Electric motor drive manufacturers use graphs to explain the approximate overload trip time based on the amount of overload and the electric motor drive's output frequency (motor operating frequency). The percent of motor overload is found by applying the formula:

$$\% \, OL = \frac{I_{RUN}}{I_{FL}} \times 100$$

where
% *OL* = percent motor overload
I_{RUN} = motor current draw (in A)
I_{FL} = motor nameplate listed current (in A)

_____ **1.** Using the manufacturer supplied graph, determine the approximate amount of trip time, in seconds, before the electric motor drive would automatically stop the overloaded motor.

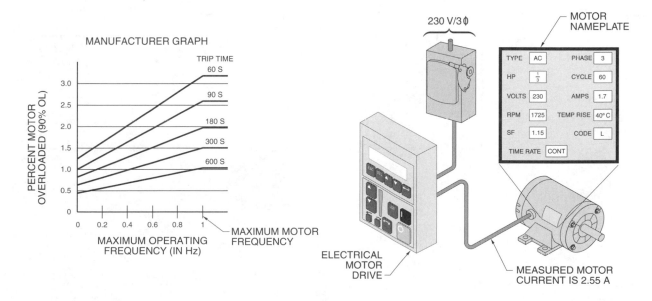

ACTIVITY 18-2—Troubleshooting Incoming Power

Electric motor drives are supplied with power from the power distribution system and deliver a modified output power to the motor. Understanding a manufacturers rating aids in proper electric motor drive installation, operation, and maintenance. Electric motor drive manufacturers also provide recommendations for fuse and wire size that can be used as a guide during motor installation. Using the manufacturers drive specifications and recommendations, determine the required data.

DRIVE SPECIFICATIONS										
Drive	Rated input current*		Output current		Maximum rated motor power		Recommended input fuse		Recommended wire size (AWG)†	
Part #	1φ 230 V	3φ 230 V	Rated	Short Term overload current‡	HP	kW	1φ	3φ	1φ	3φ
01	6.6	——	3	4.5	1.5	.55	10	——	14	14
02	8.9	——	4.3	6.5	1	.75	10	——	14	14
03	12.2	——	5.5	8.3	1.5	1.1	16	——	12	14
04	12.2	8.4	5.5	8.3	1.5	1.1	16	10	12	14
05	15.7	——	7.1	10.7	2	1.5	16	——	12	14
06	15.7	9.8	7.1	10.7	2	1.5	16	10	12	14
07	22.4	——	10.7	13	3	2.2	32	——	8	14
08	22.4	12.9	10.7	13	3	2.2	32	16	8	12

* In Amps
† Always apply local, state, and national codes
‡ Allowed overcurrent for 1 min even/10 min at 50°C ambient (or less)

_____ **1.** Circuit 1 motor wire size is ___ AWG.

_____ **2.** Circuit 1 drive part number is ___.

_____ **3.** Circuit disconnect fuse size A is ___ A.

_____ **4.** Circuit 2 motor wire size is ___ AWG.

_____ **5.** Circuit 2 drive part number is ___.

_____ **6.** Circuit 2 disconnect fuse size A is ___ A.

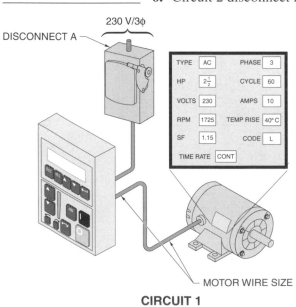

CIRCUIT 1

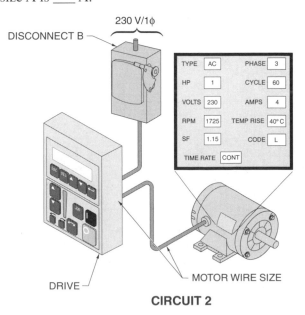

CIRCUIT 2

ACTIVITY 18-3—Troubleshooting an Electric Motor Drive

An electric motor drive can operate without a ground, but an ungrounded drive can cause electrical shock and drive damage. An electric motor drive may operate a motor if the DC bus voltage is too high or too low, but will damage the drive and motor over time. Voltage measurements can be taken to ensure an electric motor drive is grounded and the DC bus voltage is correct.

1. Set the correct position of Meter 1 selector switch to test the electric motor drive for ground. Connect Meter 1 test leads to test if the drive is grounded when the drive is powered but the motor is not operating. Set the correct position of Meter 2 selector switch to test the drive bus voltage. Connect Meter 2 test leads to test the drive bus voltage.

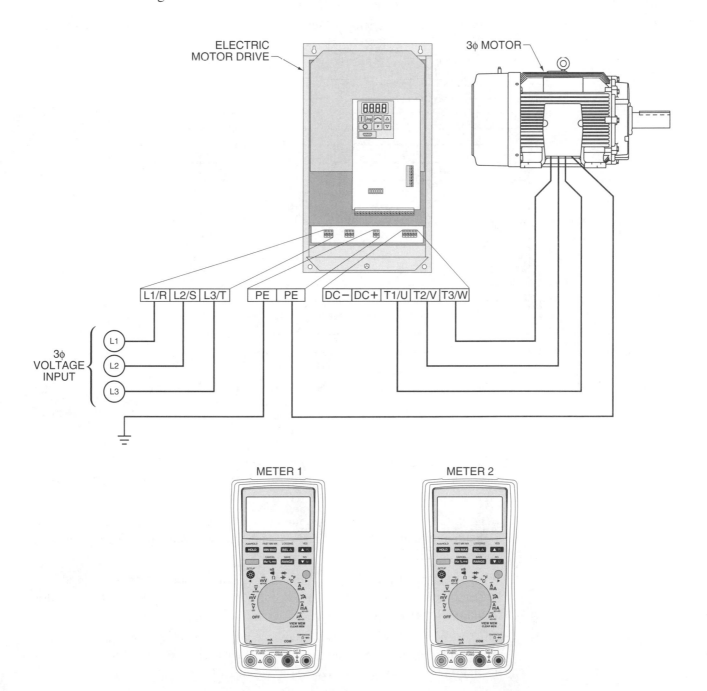

ACTIVITY 18-4—Troubleshooting the Motor and Load

Motor insulation life is based on the type of insulation used (A, B, F, or H) and the maximum temperature the insulation can withstand. Motor insulation is normally rated for an average life of 20,000 hr to 60,000 hr. Insulation life can last longer at lower temperatures and is reduced by higher temperatures. Temperature increases are caused within motors by motor overloading, higher (or lower) motor applied voltage, unbalanced voltage lines, high ambient temperatures, blocked ventilation, and frequent starts.

The temperature rise of the insulation is based on the ambient temperature (40°C standard), maximum insulation rated temperature rise, and a hot spot allowance that takes into consideration the center of the motor's winding where the temperature is higher. For example, the maximum temperature rating of class A insulation is 105°C (40°C assumed ambient +60°C rise for class A insulation +5°C hot spot allowance).

Manufacturers use charts to show the expected insulation life based on the type of insulation and the maximum temperature the insulation can withstand. Using the manufacturer insulation life chart, determine the approximate insulation life for each motor/application.

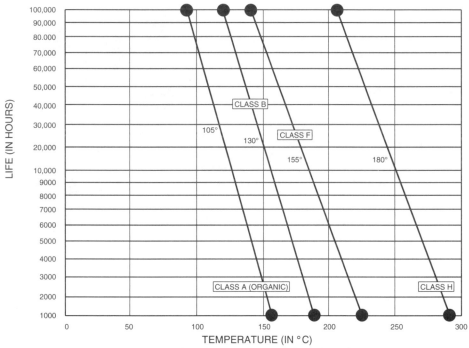

MANUFACTURER INSULATION LIFE CHART

_____ 1. What is the rated life of Class F motor insulation when the motor is used in a high ambient temperature area that raises the motor insulation temperature to an average of about 200°C?

_____ 2. What is the rated life of Class H motor insulation when the motor is used in a high ambient temperature area that raises the motor insulation temperature to an average of about 200°C?

_____ 3. What is the rated life of Class F motor insulation when the motor is used in a low ambient temperature area (inside freezer warehouse), that holds the motor insulation temperature to an average of about 150°C?

_____ 4. What is the rated life of Class B motor insulation when the motor is used in a low ambient temperature area (inside freezer warehouse), that holds the motor insulation temperature to an average of about 150°C?

ACTIVITY 18-5—Electric Motor Drive Maintenance

As part of any drive maintenance service call, the drive operating parameter setting should be checked. Some electric motor drive manufacturers include DIP switches that are used for setting some of the drive settings. DIP switches must be set correctly for proper electric motor drive/motor performance.

1. Using the manufacturer information, determine the position of each DIP switch on setting A-2 and setting B-2 to match the given motor for operating the motor at a speed no greater than the nameplate rating with an automatic boost at low speed and a standard coast to a stop mode.

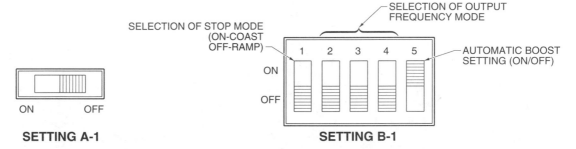

MANUFACTURER			
PE·21 PLUS™		PREMIUM EFFICIENCY	
ORD. NO.	1LA02864SE41	MAG. 21.8	
TYPE	RGZESDI	FRAME	286T
HP	30.00	SERVICE FACTOR 1.0	3 PH
AMPS	34.9	VOLTS	460
RPM	1765	HERTZ	60
DUTY	CONT 40°C AMB.	DATE CODE	
CLASS INSUL H	NEMA DESIGN B	KVA CODE G	NEMA NOM. EFF. 93.6
SH. END BRG.	50BC03JPP3	OPP. END BRG.	50BC03JPP3

Inverter Duty AC Induction Motor — made in U.S.A. (SA)

51-770-642

SETTING A-1

ON   OFF

SETTING B-1

SELECTION OF STOP MODE
(ON-COAST
OFF-RAMP)

SELECTION OF OUTPUT
FREQUENCY MODE

AUTOMATIC BOOST
SETTING (ON/OFF)

Setting A Information		
Input Voltage	Switch	Drive Output
230 V	OFF	3.83 V/Hz (230 V at 60 Hz)
	ON	4.6 V/Hz (230 V at 50 Hz)
208 V	OFF	3.46 V/Hz (208 V at 60 Hz)
	ON	4.16 V/Hz (208 V at 50 Hz)
460 V	OFF	7.66 V/Hz (460 V at 60 Hz)
	ON	9.2 V/Hz (460 V at 50 Hz)
415 V	OFF	6.92 V/Hz (415 V at 60 Hz)
	ON	8.3 V/Hz (415 V at 50 Hz)
380 V	OFF	6.3 V/Hz (380 V at 60 Hz)
	ON	7.6 V/Hz (380 V at 50 Hz)

Setting B Information			
SW2	SW3	SW4	Description
OFF	OFF	OFF	0 Hz TO 60 Hz
ON	OFF	OFF	0 Hz TO 50 Hz
OFF	ON	OFF	0 Hz TO 120 Hz
OFF	OFF	ON	0 Hz TO 240 Hz

SETTING A-2

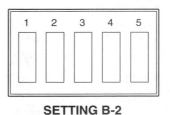

SETTING B-2

APPLICATION 19-1—Lighting Terminology

Lamp output is expressed in lumens. A *lumen (lm)* is the unit used to measure the total amount of light produced by a light source. Lamps are rated in lumens, initial lumens, and mean lumens. *Initial lumen* is the amount of light produced when a lamp is new. *Mean lumen* is the average light produced after a lamp has operated about 40% of its rated life. The lumen rating is normally the most common general rating. **See Lamp Ratings.**

LAMP RATINGS			
Ratings	**Bulb Type**		
	STANDARD INCANDESCENT	COMPACT FLUORESCENT (CFL)	LIGHT EMITTING DIODE (LED)
Light Output (in lumens)	800	800	800
Power Rated (in W)	60	60	60
Power Used (in W)	60	13	8
Lifetime (in hr)	750–1000	10,000–12,000	40,000–80,000

Illumination

Illumination is the effect that occurs when light falls on a surface. The unit of measure of illumination is the footcandle. A *footcandle (fc)* is the amount of light produced by a lamp (lumens) divided by the area that is illuminated.

The light falling on a surface may be direct or indirect light. *Direct light* is light that travels directly from a lamp to the surface being illuminated. *Indirect light* is light that is reflected from one or more objects to the surface being illuminated. In indoor applications, both direct and indirect light is considered. In outdoor applications, only direct light is considered. **See Direct Light** and **Indirect Light.**

The amount of illumination produced by a lamp may be measured or calculated. Footcandle level is measured using a light meter. A *light meter* is a portable instrument that measures light. A light meter measures from a fraction of a footcandle to several thousand footcandles.

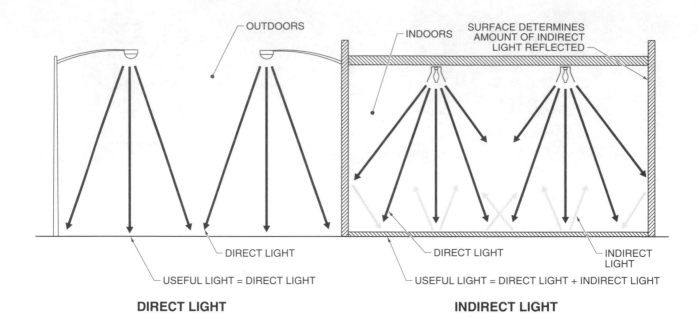

DIRECT LIGHT

INDIRECT LIGHT

A number of readings are taken at different locations and an average of the readings is used because a light meter indicates the level of illumination only at the location of the light-sensitive measuring element. When measuring the amount of illumination, both horizontal and vertical readings are taken to measure both direct and indirect light. **See Light Meter Use.**

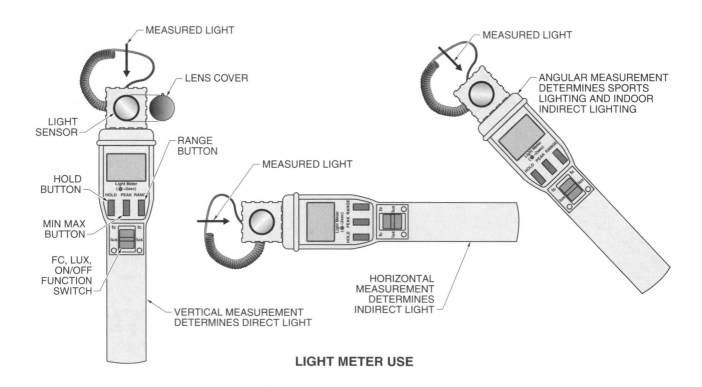

LIGHT METER USE

The intensity of a light source in a given direction depends on the angle the light travels from the lamp. Some manufacturers rate a lamp's output in candela. *Candela (cd)* is the unit of luminous intensity produced by a light source in a given direction. For example, a 175 W mercury-vapor lamp's bulb may be rated at 181 cd at a 45° vertical angle, 135 cd at 60°, and 11 cd at 90°.

Illumination is calculated before lamps are installed or measured after lamps are in operation to determine the number of footcandles that are produced on a given surface. This is checked against recommended light levels. For example, the recommended light level for an accounting office is 150 fc. **See Recommended Light Levels.**

RECOMMENDED LIGHT LEVELS

INTERIOR LIGHTING

Area	Light Level*
Assembly	
Rough, easy seeing	30
Medium	100
Fine	500
Auditorium	
Exhibitions	30
Banks	
Lobby, general	50
Writing areas	70
Teller station	150
Canning	
Cutting, sorting	100
Clothing Manufacturing	
Pattern making	50
Shops	100
Garages (Auto)	
Parking	10
Repair	50
Hospital/Medical	
Lobby	30
Dental chair	1000
Operating table	2500
Machine Shop	
Rough bench	50
Medium bench	100
Materials handling	
Picking stock	30
Packing, labeling	50
Offices	
Regular office work	100
Accounting	150
Detailed work	200
Printing	
Proofreading	150
Color inspecting	200
Schools	
Auditoriums	20
Classrooms	60-100
Indoor gyms	30-40
Stores	
Stockroom	30
Service area	100
Warehousing, storage	
Inactive	5
Active	30

* in footcandles

EXTERIOR LIGHTING

Area	Light Level*
Building	
Light surface	15
Dark surface	50
Loading/unloading area	20
Parking areas	
Industrial	2
Shopping	5
Storage yards (Active)	20
Street	
Local	0.9
Expressway	1.4
Car lots	
Front line	100-500
Remaining area	20-75

* in footcandles

SPORTS LIGHTING (PROFESSIONAL)

Area	Light Level*
Baseball	
Outfield	100
Infield	150
Boxing	
Professional	200
Championship	500
Football	100
Hockey	50
Racing	
Auto, horse	20
Dog	30
Skating	
Rink	5
Soccer	100
Tennis Courts	
Recreational	15
Tournament	30
Volleyball	
Recreational	10
Tournament	20

* in footcandles

FOOTCANDLE - A UNIT OF MEASURE OF THE INTENSITY OF LIGHT FALLING ON A SURFACE, EQUAL TO ONE LUMEN PER SQUARE FOOT

Manufacturers use footcandle levels to aid in selecting and installing lamps. Manufacturers' charts list the expected footcandle rating for an area given the mounting height and spacing of the lamps. For example, about 30 fc is produced when lamps are mounted 25′ high and are spaced 30′ apart. Thirty footcandles satisfy the aisle floor lighting requirements of an active warehouse storage area. **See Aisle Floor Footcandles.**

The location of a lamp is considered when determining the amount of light produced on a surface. Manufacturers provide light distribution data that is used to determine the amount of light produced for different lamp types. **See Determining Footcandles.**

AISLE FLOOR FOOTCANDLES

Spacing*	Mounting Height*							
	15	20	25	30	35	40	45	50
20	—	—	46	39	33	63	26	24
25	—	45	36	30	26	24	21	19
30	—	37	30	25	22	19	18	15
35	—	31	26	22	19	17	15	14
40	33	28	23	19	17	14	13	12
45	30	25	20	17	14	13	12	10

* in feet

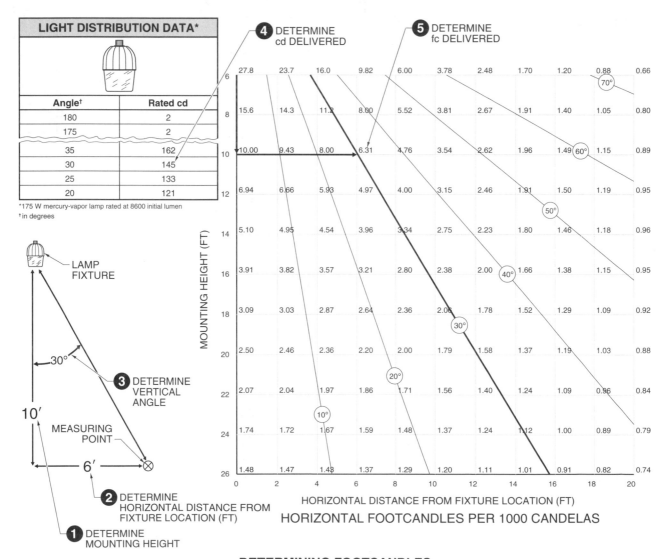

DETERMINING FOOTCANDLES

To determine the light delivered to an indoor surface, apply the procedure:

1. Determine the mounting height of the lamp fixture.
2. Determine the horizontal distance from the fixture location to the measuring point.
3. Determine the vertical angle. See Step 4 if the vertical angle is given. The vertical angle is found on Horizontal Footcandles Per 1000 Candelas. The vertical angle is determined by the sloping line that intersects the mounting height and horizontal distance. The sloping line is followed to the vertical angle.
4. Determine the candelas delivered to the measuring point. The candelas delivered are found on Light Distribution Data* for the lamp used. To determine the candelas delivered to the measuring point, apply the formula:

$$cd = \frac{rcd \times rlm}{1000}$$

where

cd = candelas

rcd = rated candela at given vertical angle (from Light Distribution Data* for lamp used)

rlm = rated lamp lumens

1000 = constant

5. Determine the footcandles delivered per 1000 candelas at the given vertical angle. The footcandles delivered per 1000 candelas at the given angle are located at the intersection of the mounting height and horizontal distance on Horizontal Footcandles Per 1000 Candelas.

To determine the fc delivered to the measuring point, convert cd to fc. To convert cd to fc, apply the formula:

$$fc = \frac{rfc \times cd}{1000}$$

where

fc = footcandles at measuring point

rfc = rated footcandles per 1000 candelas (from Horizontal Footcandles Per 1000 Candelas)

cd = candelas at measuring point

1000 = constant

Example: Determining Footcandles

A 400 W high-pressure sodium lamp rated at 50,000 initial lumens is mounted 12′ high. Determine the footcandles delivered to a surface 10′ away from the lamp. **See Surface Lighting.**

1. Determine the mounting height of the lamp fixture. Mounting height equals **12′**.
2. Determine the horizontal distance from the fixture location to the measuring point. Horizontal distance equals **10′**.
3. Determine the vertical angle. Vertical angle equals **40°**.
4. Determine the candelas delivered to the measuring point (from Light Distribution Data*).

$$cd = \frac{rcd \times rlm}{1000}$$

$$cd = \frac{181 \times 50,000}{1000}$$

cd = **9050 cd**

5. Determine the footcandles delivered per 1000 candelas at the given vertical angle. The footcandles delivered per 1000 candelas equals **3.15** (from Horizontal Footcandles Per 1000 Candelas).

Convert cd to fc.

$$fc = \frac{rfc \times cd}{1000}$$

$$fc = \frac{3.15 \times 9050}{1000}$$

fc = **28.5 fc**

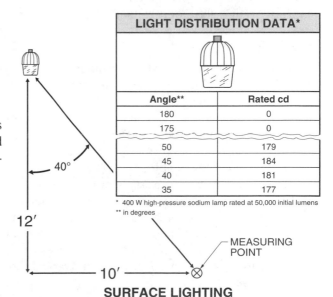

LIGHT DISTRIBUTION DATA*

Angle**	Rated cd
180	0
175	0
50	179
45	184
40	181
35	177

* 400 W high-pressure sodium lamp rated at 50,000 initial lumens
** in degrees

MEASURING POINT

12′

10′

SURFACE LIGHTING

Lamps should be spaced so that the light produced on a surface is a combination of more than one light source. This helps to reduce glare and shadows. When determining the amount of light on a surface, all lamp sources that produce light are considered. To determine the total light on a surface, the light from several properly spaced lamps is calculated. **See Lamp Spacing.**

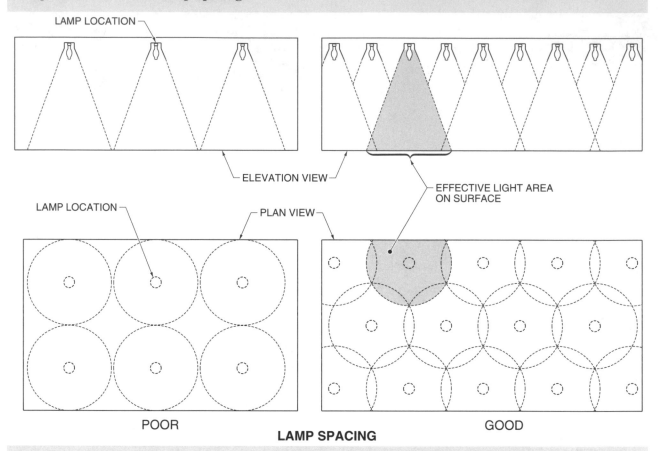

LAMP SPACING

Lumen per watt (lm/W) is the number of lumens produced per watt of electrical power. Lumen per watt is the measure of the effectiveness (efficacy) of the light source in producing light from electrical energy. The higher the lm/W rating, the more economical the light source is to operate.

Lamp Efficiency

Most of the energy produced by a lamp (65% to 90%) is in the form of ultraviolet energy, infrared energy, or heat. A small percentage of the total energy produced by a lamp is visible. Lamps are rated by the electrical efficiency (lm/W) of light produced. The electrical efficiency of lamps determines the cost to light a given area. **See Lamp Electrical Efficiency.**

Total lamp efficiency consists of electrical efficiency and economical efficiency. Economical efficiency depends on the initial cost of the lamp, lamp life, and lamp usage. The frequency of lamp service is important from a maintenance standpoint. Lamp service includes repair and changing burnt-out bulbs. **See Lamp Average Life Comparison.**

| LAMP ELECTRICAL EFFICIENCY ||
Lamp	Lumen Per Watt*
Incandescent	15–25
Mercury-vapor	40–50
Metal-halide	60–100
High-pressure sodium	60–110
Low-pressure sodium	80–180
Standard fluorescent	33–77
T-Fluorescent	85–105
CFL	50–80
LED	90–110

* exact lumen output depends on size and type of lamp used

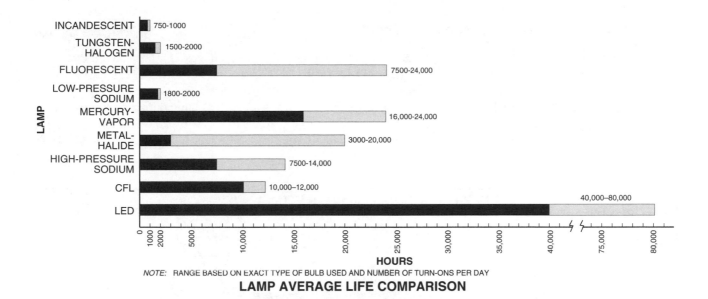

LAMP AVERAGE LIFE COMPARISON

NOTE: RANGE BASED ON EXACT TYPE OF BULB USED AND NUMBER OF TURN-ONS PER DAY

Many different lamps are available. Some lamps, such as incandescent lamps, have a low initial cost, but are more expensive to operate than other lamps, such as light emitting diode (LED) lamps and compact fluorescent lamps (CFLs). **See Lamp Advantages and Disadvantages.**

LAMP ADVANTAGES AND DISADVANTAGES		
Lamp	**Advantages**	**Disadvantages**
Compact Fluorescent (CFL)	Much more light output than incandescent; Long life; Lower bulb temperature than incandescent; Available in warm or cool light	No bright point lighting; Some have poor color rendering; Glare in some applications; Heat sensitive; Larger size than incandescent
Light Emitting Diode (LED)	More light output than incandescent or CFL; Longest life of any bulb type; Available in many colors; Can be dimmed; Cool bulb temperature; Shock and vibration resistant	Light output in only one direction; Expensive
Incandescent, tungsten-halogen	Low initial cost; Simple construction; No ballast required; Available in many shapes and sizes; Requires no warm-up or restart time; Inexpensively dimmed; Simple maintenance	Low electrical efficiency; High operating temperature; Short life; Bright light source in small space; Does not allow large distribution of light
Fluorescent	Available in many shapes and sizes; Moderate cost; Good electrical efficiency; Long life; Low shadowing; Low operating temperature; Short turn-ON delay	Not suited for high-level light in small, highly-concentrated applications; Requires ballast; Higher initial cost than incandescent lamps; Light output and color affected by ambient temperature; Expensive to dim
Low-pressure sodium, mercury-vapor, metal-halide, high-pressure sodium	Good electrical efficiency; Long life; High light output; Slightly affected by ambient temperature	May cause color distortion; Long start and restart time; High initial cost; High replacement cost; Requires ballast; Expensive or not possible to dim; Problem starting in cold weather; High-socket voltage required

 Refer to **Activity 19-1—Lighting Terminology**...

APPLICATION 19-2—Light Emitting Diode Lamps

A *light emitting diode (LED) lamp* is a solid-state semiconductor device that produces light when a DC current passes through it. Because LEDs require direct current, an AC to DC converter (normally diodes) must be included as part of the lamp base on lamps designed to be connected to AC power. LEDs are extremely long lasting (some rated over 100,000 hr), but the AC to DC converter may not last as long. LED lamp output decreases over time. **See Light Emitting Diode (LED) Lamps.**

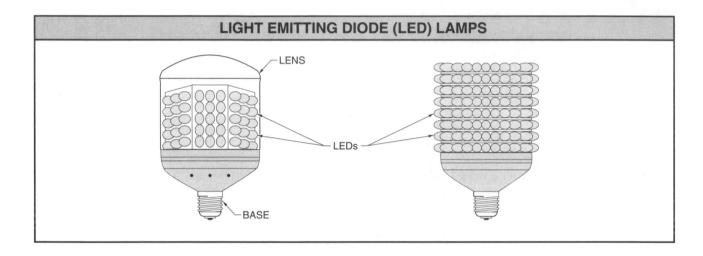

LIGHT EMITTING DIODE (LED) LAMPS

LED lamps consist of many individual LEDs, each with a diffuser lens to help direct the light produced. LED lamps are available based on the number of individual LEDs that they contain and are categorized based on the direction in which they direct the light. Because they have such a long life span and are available in many colors, they are used in special applications such as airport runway lighting and tower warning lights in addition to general use. LED lamps are not designed to be repaired and should be replaced when they fail. However, premature failure normally results from poor power quality, such as an excessively high voltage and/or transients (high-voltage spikes).

 Refer to **Activity 19-2—Light Emitting Diode Lamps**...

APPLICATION 19-3—Incandescent Lamps

An *incandescent lamp* is an electric lamp that produces light by the flow of current through a tungsten filament inside a gas-filled, sealed glass bulb. An incandescent lamp has a low initial cost and is simple to install and service. An incandescent lamp has a lower electrical efficiency and shorter life than other lamps. **See Incandescent Lamp.**

An incandescent lamp operates as current flows through a filament. A *filament* is a conductor that has a resistance high enough to cause the conductor to heat. The filament glows white-hot and produces light. The filament is made of tungsten and limits the current to a safe operating level.

Inrush current is higher than operating current because the filament has a low resistance when cold. The high inrush current causes most incandescent lamps to fail when first switched ON.

The air inside the bulb is removed before it is sealed to prevent oxidation of the filament. The filament burns out quickly if oxygen is present. A gas mixture (nitrogen and argon) is placed inside most incandescent bulbs to increase the life of the lamp.

Incandescent Lamp Bases

The base of an incandescent lamp holds the lamp firmly in the socket and connects electricity from the outside circuit to the filament. The ends of the filament are brought out to the base. The base simplifies the replacement of an incandescent lamp. Most incandescent lamp bases contain threads. Some smaller lamps are fitted with bayonet bases. The standard screw base has right-hand threads that allow removal of the lamp by turning counterclockwise (CCW) and installation by turning clockwise (CW). **See Incandescent Lamp Bases.**

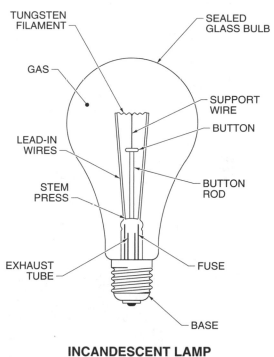

INCANDESCENT LAMP

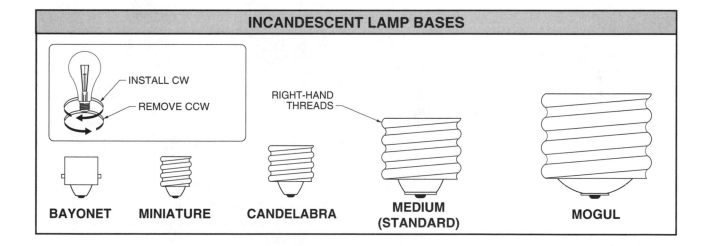

Incandescent Bulb Shapes

Incandescent lamp bulbs are available in a variety of shapes and sizes. Bulb shapes are designated by letters. The most common bulb shape is the A bulb. The A bulb is used for most residential and commercial indoor lighting applications. The A bulb is normally available in sizes from 15 W to 200 W. A common variation of the A bulb is the PS bulb. The PS bulb has a long neck. The PS bulb is available in sizes from 150 W to 2000 W. The PS bulb is used for commercial, school, industrial, and some old street lighting applications. **See Incandescent Bulb Shapes.**

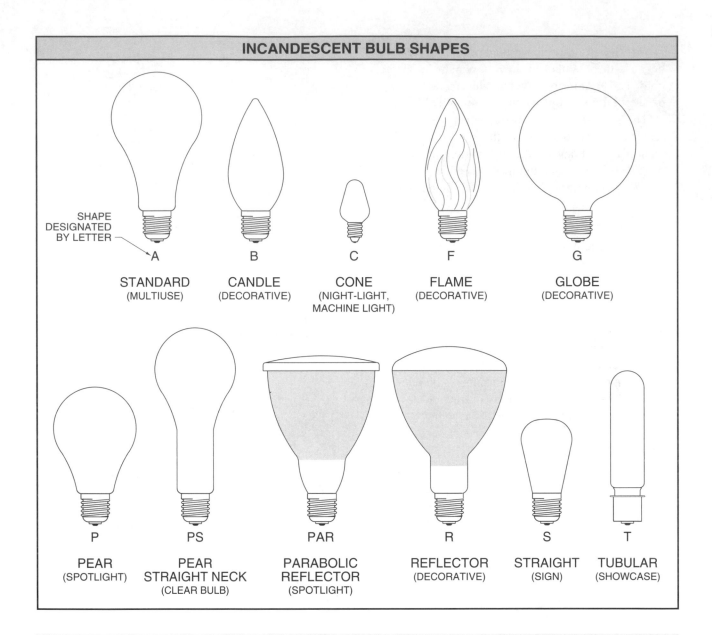

INCANDESCENT BULB SHAPES

SHAPE DESIGNATED BY LETTER

A
STANDARD
(MULTIUSE)

B
CANDLE
(DECORATIVE)

C
CONE
(NIGHT-LIGHT,
MACHINE LIGHT)

F
FLAME
(DECORATIVE)

G
GLOBE
(DECORATIVE)

P
PEAR
(SPOTLIGHT)

PS
PEAR
STRAIGHT NECK
(CLEAR BULB)

PAR
PARABOLIC
REFLECTOR
(SPOTLIGHT)

R
REFLECTOR
(DECORATIVE)

S
STRAIGHT
(SIGN)

T
TUBULAR
(SHOWCASE)

Other bulbs are used for decorative or special-purpose applications. The C bulb is designed to withstand moderate vibration and is used in sewing, washing, and other machines. The B, F, and G bulbs are used for decorative applications and are normally of low wattage. The R bulb is also used as a decorative light, but is one of high wattage. The PAR bulb is normally used as an outdoor spotlight.

Incandescent Lamp Size

The size of an incandescent lamp is determined by the diameter of the lamp's bulb. **See Common Incandescent Lamps.** The diameter of a bulb is expressed in eighths of an inch ($\frac{1}{8}''$). For example, an A-21 bulb is 21 eighths ($\frac{21}{8}''$) or $2\frac{5}{8}''$ in diameter at the bulb's maximum dimension.

COMMON INCANDESCENT LAMPS			
Type	Watts	Size*	Volts
A-19	40-100	$2\frac{3}{8}$	120-130
F-15	25-60	$1\frac{7}{8}$	120
PS-35	300-500	$4\frac{3}{8}$	120
PAR-38	115-150	$4\frac{3}{4}$	120-130
T-19	40-100	$2\frac{3}{8}$	120

* in in.

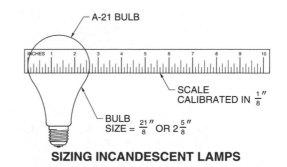

SIZING INCANDESCENT LAMPS

A scale calibrated in eighths is used to measure the size of a lamp. The scale is placed at the largest diameter of the bulb. **See Sizing Incandescent Lamps.**

Tungsten-Halogen Lamps

A *tungsten-halogen lamp* is an incandescent lamp that is filled with a halogen gas (iodine or bromine). The gas combines with tungsten that is evaporated from the filament as the lamp burns. Tungsten-halogen lamp wattages range from 15 W to 1500 W. Tungsten-halogen lamps are available in a limited selection of bulb sizes and wattages. **See Tungsten-Halogen Lamp.**

Tungsten-halogen lamps are used for display lighting, outdoor lighting, and in photocopy machines because they produce a large amount of light instantly. The lamp lasts about twice as long as a standard incandescent lamp.

Care must be taken when replacing a burned-out tungsten-halogen lamp. Ensure that all power is OFF because a high temperature is produced when the lamp is turned ON. The bulb is made of quartz because of the high temperature of the lamp. High-wattage tungsten-halogen lamps are available in T or PAR bulbs. The replacement cost is about three times the cost of an incandescent lamp.

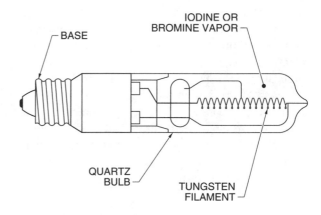

TUNGSTEN-HALOGEN LAMP

 Refer to **Activity 19-3—Incandescent Lamps** ....

APPLICATION 19-4—*Fluorescent Lamps*

A *fluorescent lamp* is a low-pressure discharge lamp in which ionization of mercury-vapor transforms ultraviolet energy generated by the discharge into light. The bulb contains a mixture of inert gas (normally argon and mercury-vapor), which is bombarded by electrons from the cathode. This provides ultraviolet light. The ultraviolet light causes the fluorescent material on the inner surface of the bulb to emit visible light.

A fluorescent lamp consists of a cylindrical glass tube (bulb) sealed at both ends. Each end (base) includes a cathode that supplies the current to start and maintain the arc. Both cathodes are constructed to allow conduction in either direction. Conduction in either direction allows the lamp to be operated on alternating current. The cathode is normally made of coiled tungsten wire and coated with a material that emits electrons. **See Fluorescent Lamp Construction.**

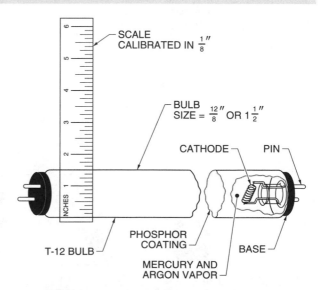

FLUORESCENT LAMP CONSTRUCTION

Standard fluorescent lamps vary in diameter from ⅝" to 2⅛" and in length from 6" to 96". The overall diameter of the lamp is designated by a number that indicates eighths of an inch. The most common size is the 40 W lamp, which is designated T-12 (1½" in diameter) and has a length of 48". *Compact fluorescent lamps (CFLs)* are small fluorescent lamps designed to use less energy than incandescent bulbs and screw into standard lamp sockets. CFLs are about four times more energy efficient than standard incandescent bulbs and last about ten times longer. **See Common Fluorescent Lamps.**

Fluorescent lamp holders are designed for the various types of lamp bases. The base is used to support a fluorescent lamp and provide the required electrical connections. Fluorescent lamps that use preheat or rapid-start cathodes require four electrical contacts (pins). The four connections are made by using a bi-pin base at either end. The three standard types of bi-pin bases are miniature bi-pin, medium bi-pin, and mogul bi-pin. Instant-start fluorescent lamps require a single pin at each end of the lamp. **See Fluorescent Lamp Bases.**

COMMON FLUORESCENT LAMPS

Type	Watts	Length*	Size*	Base
CFL	4–250	1–8	Depends on shape	Standard Incandescent Screw-In
T-5	4–13	6–21	⅝	Miniature Bi-pin
T-5	13	21	⅝	Miniature Bi-pin
T-8	15–13	18–36	1	Medium Bi-pin
T-9	20–40	6–16 OD	1⅛	4-pin
T-10	40	48	1¼	Medium Bi-pin
T-12	14–75	15–96	1½	Medium Bi-pin
T-12	60–75	96	1½	Single Pin

* in in.

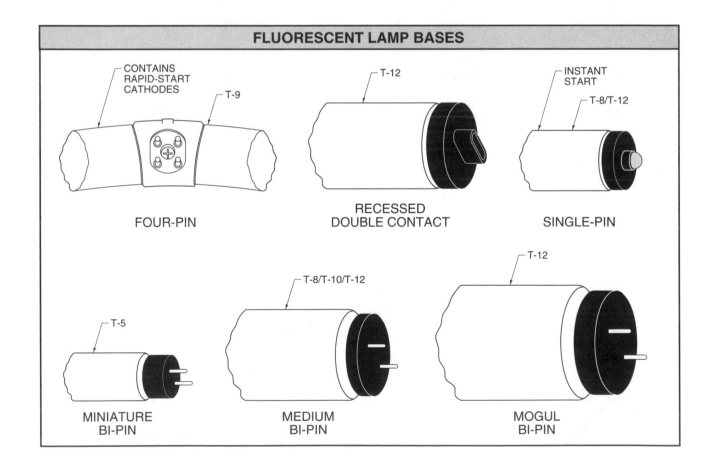

FLUORESCENT LAMP BASES

CONTAINS RAPID-START CATHODES
T-9
FOUR-PIN

T-12
RECESSED DOUBLE CONTACT

INSTANT START
T-8/T-12
SINGLE-PIN

T-5
MINIATURE BI-PIN

T-8/T-10/T-12
MEDIUM BI-PIN

T-12
MOGUL BI-PIN

Voltage Fluctuations

A fluorescent lamp is not connected directly to power lines. The high inrush current would destroy the lamp. A ballast is used to limit and control the current flow in a fluorescent lamp. A *ballast* is a transformer or solid-state circuit that limits current flow and supplies the starting voltage for a fluorescent lamp. **See Fluorescent Lamp Fixture.**

A supply voltage higher or lower than the rating of the ballast affects the life of the lamp and ballast. The voltage supplied to a fluorescent lamp should be within ±7% of the lamp's rating.

A supply voltage lower than the rated voltage but within 7% of the rated voltage normally has no serious effect on the lamp or ballast. A supply voltage deviation of more than 7% below the rated voltage shortens lamp life and reduces the light output. The reduced voltage may cause frequent flashing during starting.

A supply voltage higher than the rated voltage of the ballast increases light output and shortens lamp and ballast life. A higher-than-rated voltage also increases the lamp temperature. High temperatures break down insulation and reduce ballast life. **See Fluorescent Lamp Voltage Change Characteristics.**

Ballast Sound

A slight hum is present in fluorescent lamps. The hum originates from the inherent magnetic action in the core and coil assembly of the ballast. The sound is normally not a problem. The sound may be annoying if amplified. Reasons ballast sound may be amplified include:

- Improper ballast mounting. Use all mounting holes to securely mount a ballast.
- Loose parts in the fixture. Ensure no loose parts are present including loose screws and extra screws.
- Sound resonating through ceilings, walls, and furniture. Avoid mounting fluorescent lamps near air ducts, in loosely assembled metal display cabinets, or other locations that may amplify sound.

A fluorescent lamp ballast is selected based on the quietest rating for a specific location. Most ballasts are sound-rated by a letter code. An A ballast has the least hum and an F ballast has the most hum. The greater the number of fixtures used in an area, the higher the total sound output. **See Ballast Sound Rating.**

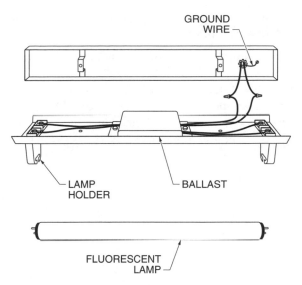

FLUORESCENT LAMP FIXTURE

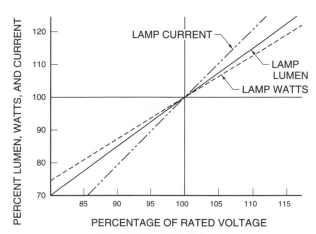

**FLUORESCENT LAMP
VOLTAGE CHANGE CHARACTERISTICS**

BALLAST SOUND RATING		
Sound Rating	Noise Level*	Recommended Application
A	20-24	Broadcasting booths, churches, study areas
B	25-30	Libraries, classrooms, quiet office areas
C	31-36	General office areas, commercial buildings
D	37-42	Retail stores, stock rooms
E	43-48	Light production areas, general outdoor lighting
F	Over 48	Street lighting, heavy production areas

* in decibels

Ballast Ventilation

A fluorescent lamp ballast generates heat during normal operation. A ballast using Class A insulation should have a maximum ballast coil temperature of 105°C (221°F) and a maximum ballast case temperature of 90°C (194°F) at its hottest point. Ballast life is reduced and an unsafe condition is produced when a ballast is operated at temperatures above these limits. A ballast should be positioned at a distance great enough to allow for the dissipation of the combined heating effect in multiple ballast enclosures. Per the NEC®, thermal insulation for recessed lighting must be at least 3″ from the fixture. **See Ballast Ventilation.**

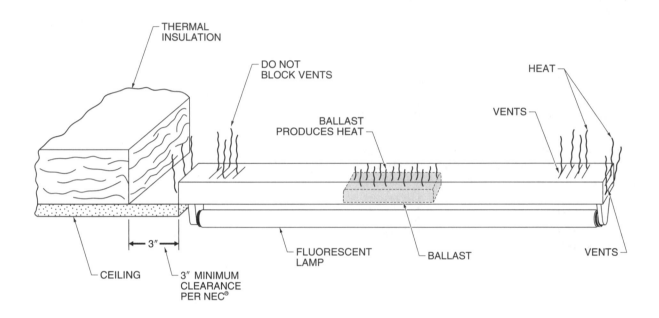

BALLAST VENTILATION

The temperature rise of ballasts may be reduced by:
- Mounting ballasts with the maximum number of sides in direct contact with the metal channel of the fixture to help dissipate the heat.
- Providing fixture ventilation.
- Painting unpainted fixture channels with a non-metallic finish to increase thermal radiation.
- Placing the ballast in a cool location outside the fixture when fixture ventilation is restricted.
- Installing the fixture for maximum heat dissipation by conduction, convection, or radiation.

Cold Weather Operation

Lumen rating of a fluorescent lamp applies to operation in still air at an ambient temperature of 77°F. Fluorescent lamps and fluorescent lamp ballasts provide good light output down to 50°F. Further decreases in ambient temperature result in decreased light output.

Variations in humidity, line voltage, fixture design, and within the design of a lamp and ballast affect the low-temperature starting limit of a fluorescent lamp. A fluorescent lamp with a high wall-bulb temperature is recommended for efficient cold weather operation. *Wall-bulb temperature* is the operating temperature of the lamp's wall, which affects the mercury-vapor pressure inside the bulb. Variations in mercury-vapor pressure change the light output of the lamp. The colder the bulb, the less light output.

Radio and Sound Interference

Radio waves are created at the arc of fluorescent lamp electrodes. The radio waves may interfere with radio reception by direct radiation from the fluorescent lamp to the radio, feedback from the lamp through the power line to the radio, and direct radiation from the power source line to the radio. **See Fluorescent Lamp Interference.**

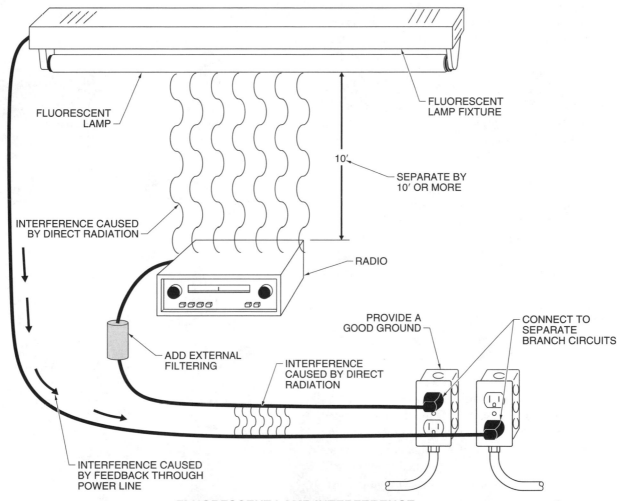

FLUORESCENT LAMP INTERFERENCE

Interference caused by fluorescent lamps is eliminated by:
* Separating radio and fluorescent lamps by at least 10' or more
* Providing a good ground for the radio or sound system
* Adding external circuit filtering
* Connecting radio and fluorescent lamp fixtures to separate branch circuits

Fluorescent Lamp Starting

A fluorescent lamp cannot be operated directly from a standard electrical circuit. A fluorescent lamp connected directly to a standard electrical circuit does not light. All fluorescent lamps require a means of starting. The method used to start the lamp determines the circuit name. The methods of fluorescent lamp-starting include preheat, instant-start, and rapid-start.

Preheat Circuits

A *preheat circuit* is a fluorescent lamp-starting circuit that heats the cathode before an arc is created. Preheat circuits are used in some low-wattage lamps (4 W to 20 W). Preheat circuits are common in desk lamps.

A preheat circuit uses a separate starting switch that is connected in series with the ballast. The electrodes in a preheat circuit require a few seconds to reach the proper temperature to start the lamp after the starting switch is pressed. **See Preheat Circuit.**

The opening of the starting switch breaks the path of current flow through the starting switch. This leaves the gas in the lamp as the only path for current to travel. A high-voltage surge is produced by the collapsing magnetic field as the starting switch is no longer connected to power. This high-voltage surge starts the flow of current through the gas in the lamp. Once started, the ballast limits the flow of current through the lamp.

Instant-Start Circuits

An *instant-start circuit* is a fluorescent lamp-starting circuit that provides sufficient voltage to strike an arc instantly. The instant-start circuit was developed to eliminate the starting switch and overcome the starting delay of preheat circuits. An arc strikes without preheating the cathodes when a high enough voltage is applied across a fluorescent lamp.

The high initial voltage requires a large autotransformer as an integral part of the ballast. The autotransformer delivers an instant voltage of 270 V to 600 V, depending on bulb size and voltage rating. The larger the bulb, the higher the voltage. Instant-start lamps require only one pin at each base because no preheating of the electrode is required. **See Instant-Start Circuit.**

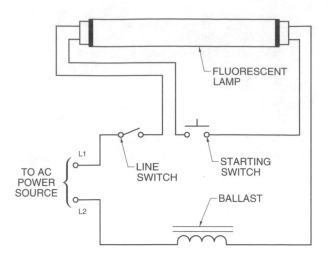

PREHEAT CIRCUIT

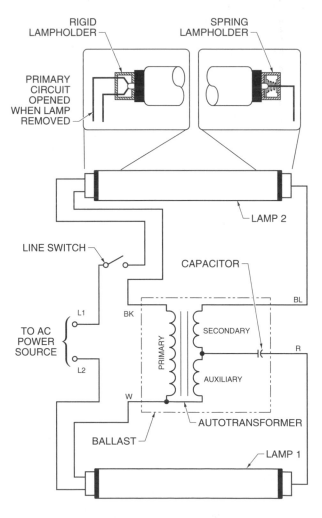

INSTANT-START CIRCUIT

A safety circuit is used with an instant-start lamp to prevent electrical shock due to the high starting voltage. When the lamp is removed, the base pin acts as a switch, interrupting the circuit to the ballast. The lamp is replaced by pushing the lamp into the spring lampholder at the high-voltage end of the fixture and then inserting it into the rigid lampholder at the low-voltage end of the fixture.

Both lamp ends must be in place before current can flow through the ballast winding. The circuit does not operate when the lamp is not in place.

Rapid-Start Circuits

A *rapid-start circuit* is a fluorescent lamp-starting circuit that brings the lamp to full brightness in about two seconds. A rapid-start circuit is the most common circuit used in fluorescent lighting.

A rapid-start circuit uses lamps that have short, low-voltage electrodes that are automatically preheated by the lamp ballast. The rapid-start ballast preheats the cathodes by means of a heater winding. The heater winding continues to provide current to the lamp after ignition. **See Rapid-Start Circuit.**

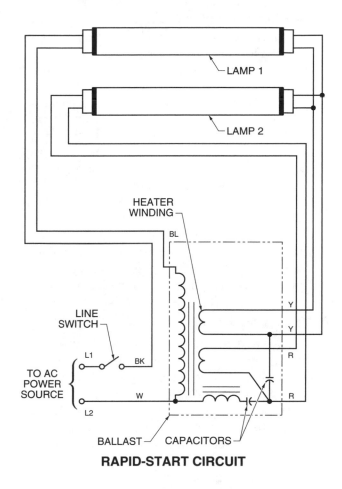

RAPID-START CIRCUIT

 Refer to **Activity 19-4—Fluorescent Lamps**...

APPLICATION 19-5—High-Intensity Discharge Lamps

High-Intensity Discharge Lamps

A *high-intensity discharge (HID) lamp* is a lamp that produces light from an arc tube. An *arc tube* is the light-producing element of an HID lamp. An arc tube contains metallic and gaseous vapors and electrodes inside the tube. An arc is produced in the tube between the electrodes. The arc tube is enclosed in a bulb which may contain phosphor or diffusing coating that improves color rendering, increases light output, and reduces surface brightness.

HID lamps include low-pressure sodium, mercury-vapor, metal-halide, and high-pressure sodium lamps. All HID lamps are electric discharge lamps. An *electric discharge lamp* is a lamp that produces light by an arc discharged between two electrodes. High vapor pressure is used to convert a large percentage of the energy produced into visible light. Arc tube pressure for most HID lamps are normally from one to eight atmospheres.

HID lamps provide an efficient, long-lasting source of light and are used for street, parking lot, and general lighting applications.

Low-Pressure Sodium Lamps

A *low-pressure sodium lamp* is an HID lamp that operates at a low vapor pressure and uses sodium as the vapor. A low-pressure sodium lamp has a U-shaped arc tube. The arc tube has both electrodes located at the same end. The arc tube is placed inside a glass bulb and contains a mixture of neon, argon, and sodium-metal. On start-up, an arc is discharged through the neon, argon, and sodium-metal. As the sodium-metal heats and vaporizes, the amber color of sodium is produced. **See Low-Pressure Sodium Lamp.**

A low-pressure sodium lamp gets its name because it uses sodium inside the arc tube. A low-pressure sodium lamp has the highest efficiency rating of any lamp. Some low-pressure sodium lamps deliver up to 200 lm/W of power. This is 10 times the output of an incandescent lamp.

Mercury-Vapor Lamps

A *mercury-vapor lamp* is an HID lamp that produces light by an electric discharge through mercury-vapor. Mercury-vapor lamps are used for general lighting applications. Phosphor coating is added to the inside of the bulb to improve color-rendering characteristics. **See Mercury-Vapor Lamp.**

A mercury-vapor lamp contains a starting electrode and two main electrodes. An electrical field is set up between the starting electrode and one main electrode when power is first applied to the lamp. The electrical field causes current to flow and an arc to strike. Current flows between the two main electrodes as the heat vaporizes the mercury.

Metal-Halide Lamps

A *metal-halide lamp* is an HID lamp that produces light by an electric discharge through mercury-vapor and metal-halides in the arc tube. A *metal-halide* is an element (normally sodium and scandium iodide) which is added to the mercury in small amounts. The metal-halide improves the light output of the lamp. A metal-halide lamp produces more lumen per watt than a mercury-vapor lamp. **See Metal-Halide Lamp.**

The light produced by a metal-halide lamp does not produce as much color distortion as a mercury-vapor lamp. A metal-halide lamp is an efficient source of white light. A metal-halide lamp has a shorter bulb life than the other HID lamps.

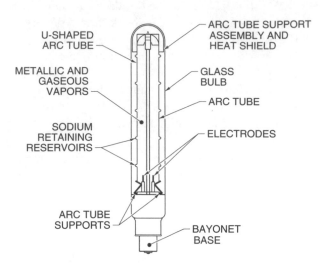

LOW-PRESSURE SODIUM LAMP

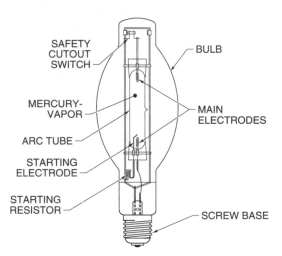

MERCURY-VAPOR LAMP

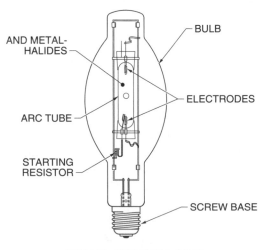

METAL-HALIDE LAMP

High-Pressure Sodium Lamps

A *high-pressure sodium lamp* is an HID lamp that produces light when current flows through sodium-vapor under high pressure and high temperature. A high-pressure sodium lamp is a more efficient lamp than a mercury-vapor or metal-halide lamp. The light produced from a high-pressure sodium lamp appears as a golden-white color. **See High-Pressure Sodium Lamp.**

A high-pressure sodium lamp is constructed with a bulb and an arc tube. The arc tube is made of ceramic to withstand high temperature. The bulb is made of weather-resistant glass to prevent heat loss and protect the arc tube.

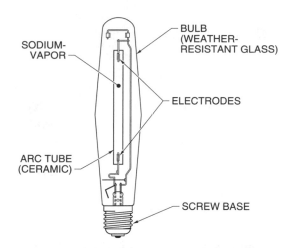

HIGH-PRESSURE SODIUM LAMP

HID Lamp Bases

HID lamps are designed to be easily connected and removed from the lamp fixture. Mercury-vapor, metal-halide, and high-pressure sodium lamps use the same medium and mogul brass screw bases as incandescent lamps. Low-pressure sodium lamps normally use a bayonet type base. Many lamp bases include a date code on the base to record the month and year of installation.

HID Lamp Operating Characteristics

HID lamps take several minutes to warm up before full light output is reached. Any short interruption in the power supply may extinguish the arc. An HID lamp cannot restart immediately after being turned OFF. The lamp must cool enough to reduce the vapor pressure in the tube to a point where the arc can re-strike. **See HID Lamp Operating Characteristics.**

HID LAMP OPERATING CHARACTERISTICS		
Lamp	Start Time*	Restart Time
Low-pressure sodium	6-12	4-12 sec
Mercury-vapor	5-6	3-5 min
Metal-halide	2-5	10-15 min
High-pressure sodium	3-4	30-60 sec

* in minutes

HID Lamp Color Rendering

Color rendering is the appearance of a color when illuminated by a light source. For example, a red color may be rendered light, dark, pinkish, or yellowish depending on the light source under which it is viewed. Color rendering of HID lamps varies depending on the lamp used.

Low-pressure sodium lamps produce yellow to yellow-orange light. The color of the lamp distorts the true color of objects viewed under the lamp. The yellow light is produced by the sodium in a low-pressure sodium lamp. Low-pressure sodium lamps are normally not used where the appearance of people and colors are important. The yellow light produces severe color distortion on most light-colored objects.

Mercury-vapor lamps with clear bulbs have poor rendering of reds. Blue colors appear purplish with most other colors appearing normal. Phosphor-coated bulbs improve color rendering. Blue colors viewed under phosphor-coated bulbs still have a slight purplish hue and yellow colors take on a greenish overtone.

Metal-halide lamps produce good overall color rendering. Red colors appear slightly muted with some pinkish overtones. High-pressure sodium lamps have good color rendering but reds, blues, greens, and violets are slightly muted.

HID Lamp Ballasts

HID lamps require ballasts to limit the current in the lamps to the correct operating level and provide the proper starting voltage to strike and maintain the arc. Each HID ballast is designed for a specific lamp, bulb size, voltage range, and line frequency.

Different lamps are not interchangeable because lamp wattage is controlled by the ballast and not the lamp. For example, when a 250 W mercury-vapor lamp is connected to a 1000 W ballast, the lamp tries to operate at 1000 W. The lamp does not operate correctly or is destroyed because the 250 W lamp is not designed to operate at 1000 W.

HID lamp sizes cannot be interchanged without changing the ballast. For example, a 100 W lamp cannot be replaced with a 175 W lamp to produce more light. Also, a 175 W lamp cannot be replaced with a 100 W lamp to produce less light.

Low-Pressure Sodium Ballasts

Low-pressure sodium lamps must be operated on a ballast designed to meet the lamp's starting and running requirements. Low-pressure sodium lamps do not have a starting electrode or ignitor. The ballast must provide an open-circuit voltage of approximately three to seven times the lamp's rated voltage to start and sustain the arc. **See Low-Pressure Sodium Ballast.**

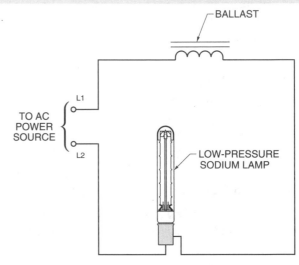

LOW-PRESSURE SODIUM BALLAST

Mercury-Vapor Ballasts

Mercury-vapor ballasts include reactor, high-reactance autotransformer, constant-wattage autotransformer, and two-winding constant-wattage ballasts. The ballast used normally depends on economics. **See Mercury-Vapor Ballasts.**

A *reactor ballast* is a ballast that connects a coil (reactor) in series with the power line leading to the lamp. A reactor ballast is used when the incoming supply voltage meets the starting voltage requirements of the lamp. This is common when the incoming supply voltage is 240 V or 277 V. Both 240 V and 277 V mercury-vapor lamps are standard. A capacitor is added to some reactor ballasts to improve the power factor. A reactor ballast costs the least, but has poor lamp-wattage regulation. A reactor ballast should be used only when line voltage regulation is good because a 5% change in line voltage produces a 10% change in lamp wattage in a reactor ballast.

A *high-reactance autotransformer ballast* is a ballast that uses two coils (primary and secondary) to regulate both voltage and current. A high-reactance autotransformer ballast is used when the incoming supply voltage does not meet the starting requirements of the lamp. Incoming voltages of 115 V, 208 V, and 460 V require a voltage change to the lamp.

A *constant-wattage autotransformer ballast* is a high-reactance autotransformer ballast with a capacitor added to the circuit. The capacitor improves the power factor. A constant-wattage autotransformer ballast is the most commonly used ballast.

A *two-winding constant-wattage ballast* is a ballast that uses a transformer that provides isolation between the primary and secondary circuits. A two-winding constant-wattage ballast has excellent lamp-wattage regulation. A 13% change in line voltage produces only a 2% to 3% change in lamp wattage.

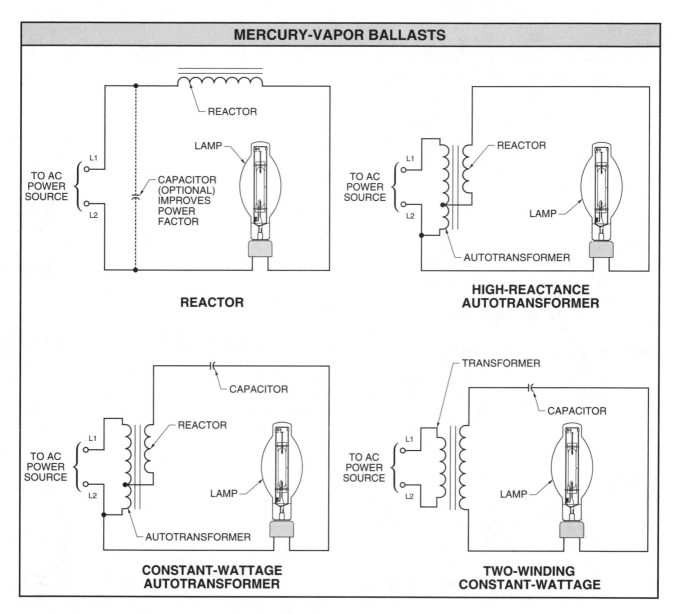

MERCURY-VAPOR BALLASTS

REACTOR

HIGH-REACTANCE
AUTOTRANSFORMER

CONSTANT-WATTAGE
AUTOTRANSFORMER

TWO-WINDING
CONSTANT-WATTAGE

Metal-Halide Ballasts

A metal-halide ballast uses the same basic circuit as the constant-wattage autotransformer mercury-vapor ballast. The ballast is modified to provide high starting voltage required by metal-halide lamps. **See Metal-Halide Ballast.**

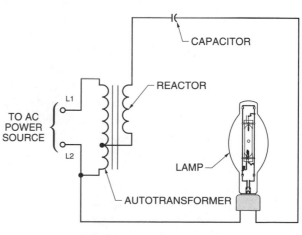

METAL-HALIDE BALLAST

High-Pressure Sodium Ballasts

High-pressure sodium lamps do not have a starting electrode. The ballast must deliver a voltage pulse high enough to start and maintain the arc. This voltage pulse must be delivered every cycle and must be 4000 V to 6000 V for 1000 W lamps and 2500 V to 4000 V for smaller lamps. The starter (ignitor) is the device inside the ballast that produces the high starting voltage. **See High-Pressure Sodium Ballast.**

A high-pressure sodium ballast is similar to a mercury-vapor reactor ballast. The main difference is the added starter. The reactor ballast is used where the input voltage meets the lamp requirements. A transformer or autotransformer is added to the ballast circuit when the incoming voltage does not meet the lamp requirements.

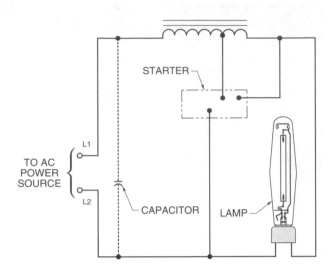

HIGH-PRESSURE SODIUM BALLAST

HID Ballast Loss

All ballasts consume a certain amount of power. Depending on the size of the lamp used, the amount of power consumed by the ballast may be a large percentage of the total lamp power.

With HID lamps, the larger the lamp, the less power the ballast loss is as a percent of total power used by the lamp. For example, a 40 W mercury-vapor lamp has a 32% ballast loss and a 1000 W mercury-vapor lamp has a 7% ballast loss. **See Ballast Loss in Appendix.**

HID Lamp Identification

ANSI lists a standard code that is used to identify HID lamps. This code is used among manufacturers and allows for interchangeability of lamps. **See Lamp Identification in Appendix.**

HID Lamp Selection

When selecting an HID lamp for an application, consider using low-pressure sodium lamps first because they are the most efficient. Low-pressure sodium lamps are good for outdoor lighting installations. The yellow-orange color is acceptable for street, highway, parking lot, and floodlight applications.

Low-pressure sodium lamps are also used for some indoor applications such as warehouse lighting and other areas where color distortion is not critical to an operation. The long start-up time is not a problem for outdoor lighting because the lights are normally turned ON by a photoelectric cell at dusk. Photoelectric-cell switches may be adjusted to turn ON at different light levels.

Consider using metal-halide lamps when the color rendering of low-pressure sodium lamps is not acceptable for the application. Metal-halide lamps are used for sport, street, highway, parking lot, and floodlighting applications. Metal-halide lamps are normally specified for most sport, indoor, and outdoor lighting applications.

Consider using mercury-vapor lamps when the initial installation cost is of major importance. Mercury-vapor lamps are used in all types of applications. Use metal-halide lamps when true color appearance is important such as in a car dealer lot.

Consider using high-pressure sodium lamps when lower operating cost is important and some color distortion is acceptable. High-pressure sodium lamps may be used in parking lots, street lighting, shopping centers, exterior buildings, and most storage areas. **See Lamp Characteristics Summary.**

LAMP CHARACTERISTICS SUMMARY				
Lamp	Lm/W	Rated Bulb Life*	Color Rendition	Operating Cost
Incandescent	15-25	750-1000	Excellent	Very high
Tungsten-halogen	20-25	1500-2000	Excellent	High
Fluorescent	55-100	7500-24,000	Very good	Average
Low-pressure sodium	190-200	1800	Poor	Low
Mercury-vapor	50-60	16,000-24,000	Depends on type used	Average
Metal-halide	80-125	3000-20,000	Very good	Average
High-pressure sodium	65-115	7500-14,000	Good (golden white)	Low

* in hours

Refer to **Activity 19-5—High-Intensity Discharge Lamps** ....

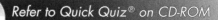

Refer to Quick Quiz® on CD-ROM

Refer to Chapter 19 in the **Troubleshooting Electrical/Electronic Systems Workbook** *for additional questions.*

Lighting Circuits

Name_____   Date_____

ACTIVITY 19-1—Lighting Terminology

Using Horizontal Footcandles Per 1000 Candelas and Light Distribution Data*, determine the footcandles delivered at each measuring point.

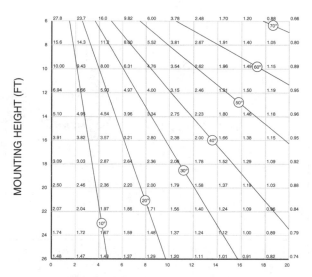

MOUNTING HEIGHT (FT)

HORIZONTAL DISTANCE FROM FIXTURE LOCATION (FT)

HORIZONTAL FOOTCANDLES PER 1000 CANDELAS

A 175 W mercury-vapor lamp rated at 8600 initial lumens is used for an application.

LIGHT DISTRIBUTION DATA*

Angle†	Rated cd	Angle†	Rated cd
180	2	85	20
175	2	80	33
170	2	75	53
165	2	70	79
160	2	65	108
155	2	60	135
150	2	55	157
145	2	50	173
140	2	45	181
135	2	40	177
130	2	35	162
125	2	30	145
120	2	25	133
115	2	20	121
110	2	15	107
105	3	10	97
100	4	5	93
95	7	0	93
90	11		

* 175 W mercury-vapor lamp rated at 8600 initial lumen
† in degrees

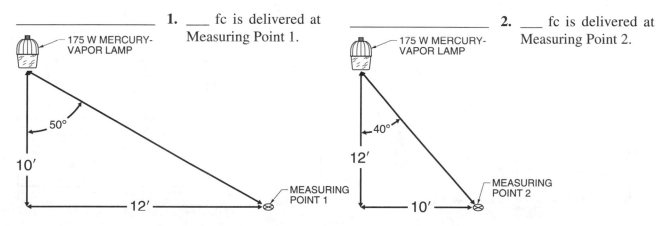

_____   **1.** ___ fc is delivered at Measuring Point 1.

175 W MERCURY-VAPOR LAMP

50°

10′

12′

MEASURING POINT 1

_____   **2.** ___ fc is delivered at Measuring Point 2.

175 W MERCURY-VAPOR LAMP

40°

12′

10′

MEASURING POINT 2

_____ **3.** The vertical angle between the lamp and Measuring Point 3 equals ___°.

_____ **4.** ___ fc is delivered at Measuring Point 3.

_____ **5.** The vertical angle between the lamp and Measuring Point 4 equals ___°.

_____ **6.** ___ fc is delivered at Measuring Point 4.

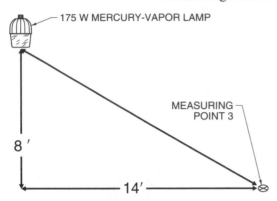

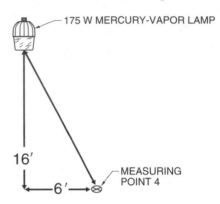

A 400 W high-pressure sodium lamp rated at 50,000 initial lumens is used for an application.

_____ **7.** The vertical angle between the lamp and Measuring Point 5 equals ___°.

_____ **8.** ___ fc from Lamp 1 is delivered at Measuring Point 5.

_____ **9.** ___ fc from Lamp 2 is delivered at Measuring Point 5.

_____ **10.** The total fc delivered at Measuring Point 5 equals ___ fc.

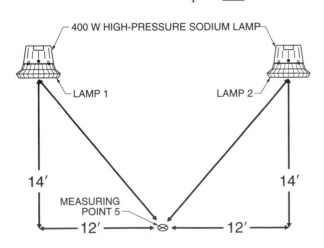

LIGHT DISTRIBUTION DATA*			
Angle†	**Rated cd**	**Angle†**	**Rated cd**
180	0	80	18
175	0	85	28
170	0	75	42
165	0	70	63
160	0	65	928
155	0	60	128
150	0	55	160
145	0	50	179
140	0	45	184
135	0	40	181
130	2	35	177
125	5	30	176
120	7	25	174
115	8	20	166
110	8	15	153
105	8	10	142
100	7	5	138
95	9	0	137
90	11		

* 400 W high-pressure sodium lamp rated at 50,000 initial lumen
† in degrees

_____ **11.** The vertical angle between Lamp 1 and Measuring Point 6 equals ___°.

_____ **12.** ___ fc from Lamp 1 is delivered at Measuring Point 6.

_____ **13.** The total fc delivered at Measuring Point 6 equals ___ fc.

_____ **14.** The vertical angle between Lamp 1 and Measuring Point 7 equals ___°.

_____ **15.** ___ fc from Lamp 1 is delivered at Measuring Point 7.

_____ **16.** ___ fc from Lamp 3 is delivered at Measuring Point 7.

_____ **17.** The total fc delivered at Measuring Point 7 equals ___ fc.

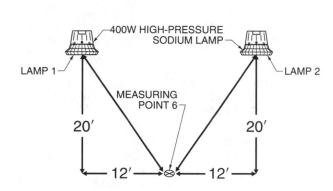

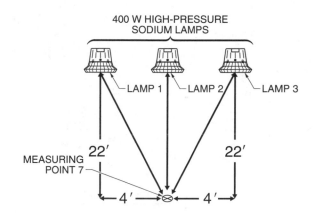

ACTIVITY 19-2—Light Emitting Diode Lamps

Using Lamp Specifications, answer the following questions.

1. All models are rated to produce a light output of 25 W. How much energy (in W) is saved by using Model 1 over using Model 4?

2. How much energy (in %) is saved using Model 1 over using Model 4?

3. Can both Model 1 and Model 2 be used on either a 115 V or 230 V circuit?

4. Why can Model 3 not be used in the same light fixture as Model 1?

5. Which lamp type continues to provide good light output at temperatures as low as 50°F?

6. Which model is best for flood lighting of a given area?

LAMP SPECIFICATIONS

MODEL 1
- LED, White
- Power Consumed: 3.5 W
- Light Output: 25 W
- Voltage: 85–260
- Lumens: 200–250
- Beam Width: 360 degrees
- Operating Temperature: –30°C to 55°C
- Base: Standard screw

MODEL 2
- LED, (A) Red, (B) Green, (C) Yellow
- Power Consumed: 3.5 W
- Light Output: 25 W
- Voltage: 85–260
- Lumens: 150–175
- Beam Width: 210 degrees
- Operating Temperature: –30°C to 55°C
- Base: Standard screw

MODEL 3
- LED, White
- Power Consumed: 3.5 W
- Light Output: 25 W
- Voltage: 85–260
- Lumens: 200–250
- Beam Width: 360 degrees
- Operating Temperature: –30°C to 55°C
- Base: Miniature screw

MODEL 4
- CFL
- Power Consumed: 6 W
- Light Output: 25 W
- Voltage: 100–130
- Lumens: 125–155
- Beam Width: 360 degrees
- Operating Temperature: –20°C to 55°C
- Base: Standard screw

MODEL 5
- CFL
- Power Consumed: 6 W
- Light Output: 25 W
- Voltage: 100–130
- Lumens: 125–155
- Beam Width: 120 degrees
- Operating Temperature: –20°C to 55°C
- Base: Standard screw

ACTIVITY 19-3—Incandescent Lamps

An incandescent lamp is the most costly lamp to operate because it is the least efficient. Energy is saved and operating costs reduced when standard incandescent lamps are replaced with energy-efficient lamps. An energy-efficient lamp delivers the same number of lumen at a slightly reduced wattage rating than a standard lamp. Energy-efficient lamps initially cost more than standard lamps. The amount of energy used is always reduced when using energy-efficient lamps. To determine the operating cost of an energy-efficient lamp, apply the formula:

$$oc = (W - Wr) \times rll \times \frac{er}{1000}$$

where

oc = operating cost (in $)

W = lamp rating (in W)

Wr = watt reduction

rll = rated lamp life (in hr)

er = electric rate (in $/kWh)

1000 = constant

To determine the operating cost of a standard lamp, apply the formula:

$$oc = W \times rll \times \frac{er}{1000}$$

STANDARD INCANDESCENT LAMPS					
Lamp Rating (in W)	Initial Cost (in $)	Rated Lamp Life (in hr)	Electric Rate (in $/kWh)		
			#1 (low)	#2 (med)	#3 (high)
40	1.44	1500	.09	.15	.20
60	1.15	1000	.09	.15	.20
75	1.15	750	.09	.15	.20
100	1.15	750	.09	.15	.20
150	1.64	750	.09	.15	.20

ENERGY-EFFICIENT INCANDESCENT LAMPS			
Lamp Rating (in W)	Initial Cost (in $)	Rated Lamp Life (in hr)	Watt Reduction
40	1.44	1500	6
60	1.29	1000	8
75	1.29	750	8
100	1.29	750	10
150	2.28	750	15

Using Standard Incandescent Lamps and Energy-Efficient Incandescent Lamps, answer the questions.

A 60 W energy-efficient incandescent lamp is used to replace a standard incandescent lamp.

_____ **1.** The watt reduction of using an energy-efficient incandescent lamp is ___ W.

_____ **2.** The rated lamp life of a 60 W standard incandescent lamp is ___ hr.

_____ **3.** The operating cost of a standard incandescent lamp at electric rate #1 is $___.

_____ **4.** The operating cost of an energy-efficient incandescent lamp at electric rate #1 is $___.

_____ **5.** The cost savings using rate #1 (including initial cost) is $___.

_____ **6.** The cost savings using rate #2 (including initial cost) is $___.

_____ **7.** The cost savings using rate #3 (including initial cost) is $___.

A 100 W energy-efficient incandescent lamp is used to replace a standard incandescent lamp.

_____ **8.** The watt reduction of using an energy-efficient incandescent lamp is ___ W.

_____ **9.** The rated lamp life of a 100 W standard incandescent lamp is ___ hr.

_____ **10.** The operating cost of a standard incandescent lamp at electric rate #1 is $___.

_____ **11.** The operating cost of an energy-efficient incandescent lamp at electric rate #1 is $___.

_____ **12.** The cost savings using rate #1 (including initial cost) is $___.

_____ **13.** The cost savings using rate #2 (including initial cost) is $___.

_____ **14.** The cost savings using rate #3 (including initial cost) is $___.

ACTIVITY 19-4—Fluorescent Lamps

Fluorescent lamps are available in energy-efficient lamps. The high lamp life (10 to 20 times longer than incandescent lamps) makes great savings possible. Using Standard Fluorescent Lamps and Energy-Efficient Fluorescent Lamps, answer the questions.

STANDARD FLUORESCENT LAMPS					
Lamp Rating (in W)	Initial Cost (in $)	Rated Lamp Life (in hr)	Electric Rate (in $/kWh)		
			#1 (low)	#2 (med)	#3 (high)
30	5.98	18,000	.09	.15	.20
40	2.53	20,000	.09	.15	.20
75	9.96	12,000	.09	.15	.20
110	7.83	12,000	.09	.15	.20

ENERGY-EFFICIENT FLUORESCENT LAMPS			
Lamp Rating (in W)	Initial Cost (in $)	Rated Lamp Life (in hr)*	Watt Reduction
30	7.14	18,000	5
40	6.29	15,000	6
75	10.65	12,000	15
110	9.21	12,000	15

* at 3 hours average run time per start

A 30 W energy-efficient fluorescent lamp is used to replace a standard fluorescent lamp.

_____ 1. The watt reduction of using an energy-efficient fluorescent lamp is ___ W.

_____ 2. The rated lamp life of a 30 W standard fluorescent lamp is ___ hr.

_____ 3. The operating cost of a standard fluorescent lamp at electric rate #1 is $___.

_____ 4. The operating cost of an energy-efficient fluorescent lamp at electric rate #1 is $___.

_____ 5. The cost savings using rate #1 (including initial cost) is $___.

_____ 6. The cost savings using rate #2 (including initial cost) is $___.

_____ 7. The cost savings using rate #3 (including initial cost) is $___.

A 75 W energy-efficient fluorescent lamp is used to replace a standard fluorescent lamp.

_____ 8. The watt reduction of using an energy-efficient fluorescent lamp is ___ W.

_____ 9. The rated lamp life of a 75 W standard fluorescent lamp is ___ hr.

_____ 10. The operating cost of a standard fluorescent lamp at electric rate #1 is $___.

_____ 11. The operating cost of an energy-efficient fluorescent lamp at electric rate #1 is $___.

_____ 12. The cost savings using rate #1 (including initial cost) is $___.

_____ 13. The cost savings using rate #2 (including initial cost) is $___.

_____ 14. The cost savings using rate #3 (including initial cost) is $___.

ACTIVITY 19-5—High-Intensity Discharge Lamps

Using High Mast Lighting, Footcandles Delivered, and Minimum Recommended Light Levels, answer the questions.

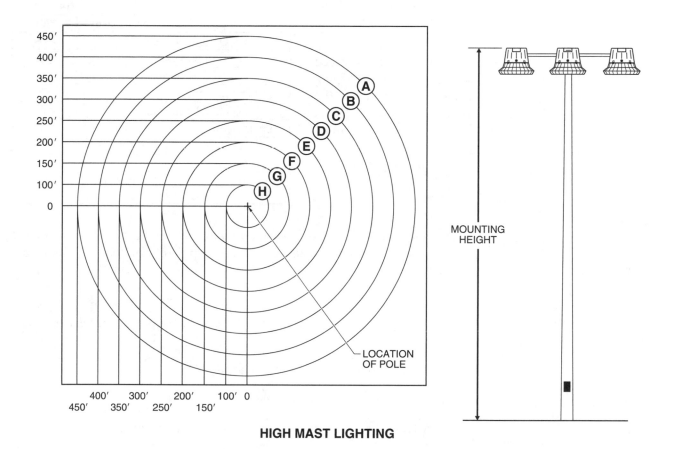

HIGH MAST LIGHTING

Area	FOOTCANDLES DELIVERED											
	Mounting Height*											
	80			100			120			150		
	Lamp Number			Lamp Number			Lamp Number			Lamp Number		
	4	8	12	4	8	12	4	8	12	4	8	12
A	.4	3.2	6	.3	.9	1	.2	.4	.8	.1	.3	.5
B	2.2	8	15	1.2	2.2	4.2	.9	1.5	2.7	.7	1	1.8
C	10	16	31	3	5.5	10	2.3	3	7	1.5	2.3	4.5
D	21	31	62	7	12	20	5	7	14	3.2	4.5	9
E	52	76	90	15	21	42	9	12	26	6	9	17
F	100	125	165	32	50	95	22	35	68	15	21	45
G	120	160	300	65	98	200	46	68	145	30	45	86
H	160	240	480	110	160	310	70	100	220	45	68	135

* in feet

_____ **1.** A 120′ pole with 8 lamps is placed in the middle of a railroad switching yard. The maximum distance from the pole that has an acceptable minimum light level is ___′.

_____ **2.** A high mast light is placed in the middle of a shopping center parking lot on a 100′ pole. The parking lot is 350′ by 350′. ___ lamps are required on the pole to provide an acceptable light level.

MINIMUM RECOMMENDED LIGHT LEVELS	
Area	Minimum fc (per sq ft)
Race track (auto, horse, motorcycle)	20
Outdoor ice rink	5
Prison yard	5
Quarry	5
Shopping center parking area	5
Industrial parking area	2
Lumberyard, storage (protective)	1
Railroad yard (switching point)	2
Railroad yard (protective)	.3
General park and garden lighting (display)	5

_____ **3.** A 100′ pole in a 200′ by 200′ industrial parking area could not be placed on the parking lot surface. Determine how many lamps are required. ___ lamps are required.

Determine the maximum distance that a mast pole 120′ high with eight lamps provides as an acceptable minimum light level for each location.

_____ **4.** Prison yard minimum light distance equals ___′.

_____ **5.** Industrial parking area minimum light distance equals ___′.

_____ **6.** Lumberyard storage (protective) area minimum light distance equals ___′.

_____ **7.** Railroad yard (protective) area minimum light distance equals ___′.

_____ **8.** General park and garden (protective) minimum light distance equals ___′.

_____ **9.** Auto racing minimum light distance equals ___′.

APPLICATION 20-1—Troubleshooting Incandescent Lamps

Incandescent Lamps

An incandescent lamp is the simplest light used. No auxiliary equipment, such as a ballast, is required for operation. The lamp is always replaced if it is burned out. Frequent lamp replacement is cause for concern. Problems that require frequent lamp replacement are normally caused by using the wrong lamp type or a voltage problem.

Incandescent Lamp Voltage Ratings

Incandescent lamps are rated based on their voltage and power. The voltage rating is the maximum voltage that may be applied to the lamp. All incandescent lamps should be operated as close to their voltage rating as possible. Operating a lamp at its voltage rating assures the lamp is operating at its highest efficiency and life. Increasing the applied voltage increases the light output of the lamp and decreases lamp life.

All incandescent lamps produce less light as voltage is decreased. This is observed when large current-consuming appliances, such as furnaces, are connected to the same circuit as a lamp. The heavy current draw when the appliance is turned ON may cause a temporary voltage drop in the circuit. This voltage drop dims the lamp until the voltage returns to full level. Temporary voltage fluctuations do not significantly reduce the life of an incandescent lamp. Consistent operation at a voltage other than the lamp rating reduces lamp efficiency, lamp life, and increases the overall cost of lighting. Light output is reduced with a decrease in applied voltage and increased with an increase in applied voltage. **See Incandescent Lamp Characteristics.**

An increase in voltage causes more current to flow through the lamp and increase the brightness of the lamp. This also increases bulb temperature and shortens lamp life.

For example, operating a 120 V lamp on 125 V (+4%) produces approximately 15% more lumen,

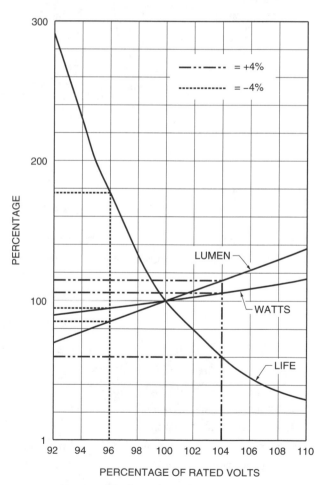

INCANDESCENT LAMP CHARACTERISTICS

uses 7% more watts, and shortens lamp life 40%. Operating the same 120 V lamp on 115 V (−4%) produces approximately 15% less lumen, uses 7% less watts, and increases lamp life 70%.

Measuring Applied Voltage

A voltmeter is used to measure the voltage applied to a lamp. A standard voltmeter may be used to measure the applied voltage when troubleshooting, replacing, or during lamp installation. A voltmeter with a min/max function may be used to record voltage levels over time. **See Measuring Applied Voltage.**

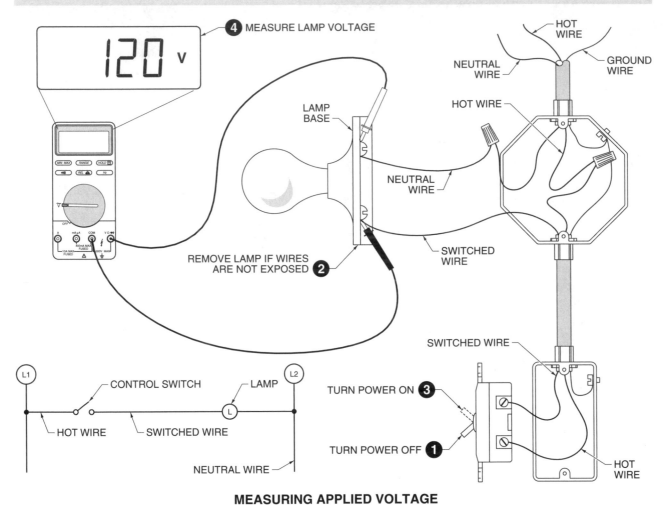

MEASURING APPLIED VOLTAGE

To measure applied voltage, apply the procedure:
1. Turn power to the lamp OFF.
2. Remove the lamp from the socket when the wires are not exposed.
3. Turn power to the lamp ON.
4. Measure the lamp voltage. Set the voltmeter to the correct voltage setting and connect the voltmeter to the lamp socket.

Warning: The socket has power. Care must be taken to ensure that no body part comes in contact with the circuit.

Standard Voltage Ratings

The standard voltage ratings of the majority of incandescent lamps are 115 V and 120 V. Incandescent lamps are also designed for use with 105 V, 110 V, and 130 V. A lamp with a high voltage rating may be used for an application when a lamp is not lasting as long as expected and the voltage is equal to or higher than the rating of the lamp.

Lamp Problems

Replacing burned-out lamps is generally routine for incandescent lamps. There may be a problem when lamp replacement becomes excessive. Consider lamp position, bulb type, service conditions, and operation time.

Bulb Blackening

Standard incandescent lamps may generally be placed in any position. The best position for most lamps is base up. Reduction of light may occur in lamps that are positioned horizontally because the bulb tends to blacken.

Blackening of the bulb is normal as a lamp ages. The blackening is the result of the normal evaporation of the tungsten filament. Mount a lamp base up when bulb blackening is a problem. In this position, the hot gas inside the bulb carries the burnt particles up to the base. **See Bulb Positioning.**

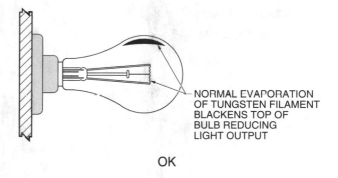

NORMAL EVAPORATION OF TUNGSTEN FILAMENT BLACKENS TOP OF BULB REDUCING LIGHT OUTPUT

OK

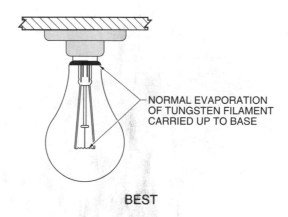

NORMAL EVAPORATION OF TUNGSTEN FILAMENT CARRIED UP TO BASE

BEST

BULB POSITIONING

Rough Service Lamps

Standard incandescent lamps cannot receive much shock without burning out. A rough service lamp should be used in applications that include shocks, bumps, or rough handling. A *rough service lamp* is a lamp made of a specially processed tungsten filament that has extra supports to resist breakage. Rough service lamp applications include extension cord lights, service areas, and machine shops.

Vibration-Resistant Lamps

High-speed machinery produce high-frequency vibrations that destroy standard incandescent lamp filaments. In such applications, vibration-resistant lamps may be used. Vibration-resistant lamps have a special tungsten structure that resists the effect of vibration by allowing the filament to move.

Special lamps have decreased lumen ratings. The lower light output is a result of designing the lamp for long life, vibration, or rough service. **See 100 W Lamp Ratings.**

100 W LAMP RATINGS			
Voltage	Service Rating	Rated Lumen Per Watt	Rated Life*
120	General lighting	19.5	750
120	Long life	15	2500
120	Vibration resistant	14	1000
120	Rough service	12.5	1000

* in hours

Incandescent Lamp Testing

An incandescent lamp may be tested for continuity when it is removed from a circuit. **See Incandescent Lamp Test.**

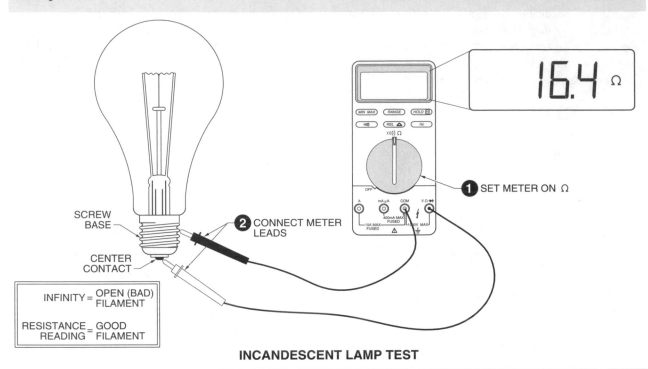

SCREW BASE

CENTER CONTACT

① SET METER ON Ω

② CONNECT METER LEADS

INFINITY = OPEN (BAD) FILAMENT

RESISTANCE READING = GOOD FILAMENT

INCANDESCENT LAMP TEST

To test an incandescent lamp, apply the procedure:
1. Set the meter on the lowest resistance scale.
2. Connect the meter leads to the center contact and screw base of the lamp. Polarity of the meter leads is not important. A meter reading of infinity indicates an open (bad) filament. Any resistance reading indicates a good filament.

The resistance reading varies directly with the filament temperature and wattage rating. A cold filament has a low resistance reading. The higher the lamp wattage, the lower the resistance reading. The lower the lamp wattage, the higher the resistance reading.

Group Lamp Replacement

Incandescent lamps may be replaced individually as they burn out or in groups. As a lamp ages, its life expectancy and output decrease. Lamps begin to fail after reaching about 70% of their rated life. Light output also decreases as a bulb blackens with age. Lamps are replaced in groups in applications such as schools and commercial buildings. Group replacement is performed when a lamp reaches between 70% and 85% of its rated life. This is accomplished by establishing a maintenance schedule for lamp replacement. **See Incandescent Lamp Life Expectancy.**

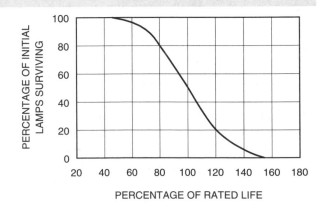

INCANDESCENT LAMP LIFE EXPECTANCY

 Refer to **Activity 20-1 — Troubleshooting Incandescent Lamps** ....

APPLICATION 20-2—*Troubleshooting Fluorescent Lamps*

Bulb Discoloration

A uniform blackening throughout the length of a fluorescent bulb is normal as the lamp ages. The blackening starts gradually and is not normally detectable until the ends of the bulb blacken heavily. Heavy end blackening soon after a new lamp is installed indicates a problem. The problem may be a defective lamp, but is normally caused by excessive or low voltage. Replace any lamp with more than 2″ of end blackening. Light spotting may occur on new lamps at any point on the bulb. Light spotting is due to mercury condensation in the bulb. **See Fluorescent Bulb Discoloration.**

Lamp Blinking

A fluorescent lamp that blinks ON and OFF is an indication of normal lamp failure that requires replacement. Other possible causes of lamp blinking include low voltage or low ambient temperature. Blinking lamps should be replaced as soon as they are detected because a blinking lamp may ruin the lamp and the starter.

Lamp Seating

Fluorescent lamps may turn ON before they are properly seated in the lamp holder. The lamp may fall out and lamp life is normally reduced when a lamp is not properly seated. Ensure that a fluorescent lamp is seated properly by aligning the base mark on the lamp with the center of the lamp holder. Both ends of the lamp must be checked for correct seating. **See Fluorescent Lamp Seating.**

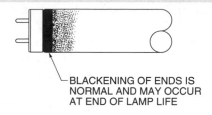
BLACKENING OF ENDS IS NORMAL AND MAY OCCUR AT END OF LAMP LIFE

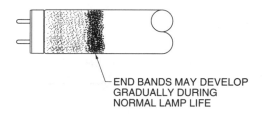
HEAVY SPOTTING MAY DEVELOP GRADUALLY DURING NORMAL LAMP LIFE

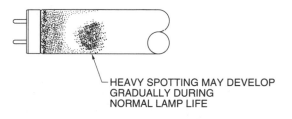
END BANDS MAY DEVELOP GRADUALLY DURING NORMAL LAMP LIFE

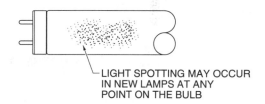
LIGHT SPOTTING MAY OCCUR IN NEW LAMPS AT ANY POINT ON THE BULB

FLUORESCENT BULB DISCOLORATION

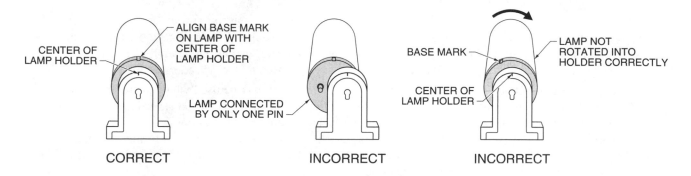

CENTER OF LAMP HOLDER — ALIGN BASE MARK ON LAMP WITH CENTER OF LAMP HOLDER — LAMP CONNECTED BY ONLY ONE PIN

CORRECT   INCORRECT

BASE MARK — LAMP NOT ROTATED INTO HOLDER CORRECTLY — CENTER OF LAMP HOLDER

INCORRECT

FLUORESCENT LAMP SEATING

Preheat Circuit Testing

A preheat circuit requires a few seconds to heat the filament before light is produced. A starting switch is used to preheat the lamp filament. As the lamp ages, the starting switch has to be pressed to start and restart the lamp several times before the lamp stays ON. Ballast life is reduced when a hard-to-start lamp is not replaced. Replace the lamp any time a hard-to-start lamp is allowed to remain in operation for more than one week. In two-lamp circuits, premature ballast failure occurs when only one lamp is operating. Replace any inoperative lamp immediately. The ballast may require replacement on any circuit in which only one lamp is allowed to operate for longer than a few days. **See Preheat Circuit Testing.**

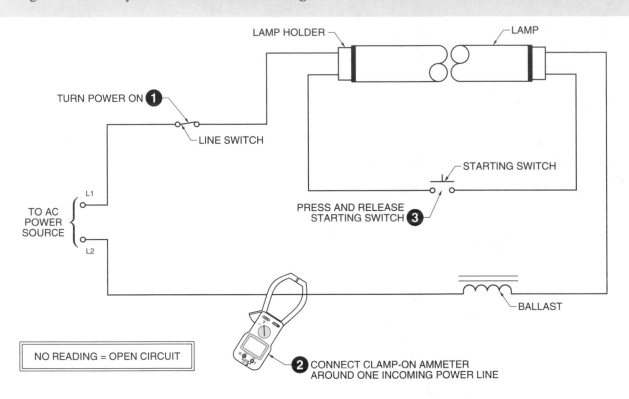

PREHEAT CIRCUIT TESTING

To test a preheat circuit, apply the procedure:

1. Turn power ON by placing the line switch in the ON position.

2. Connect a clamp-on ammeter around one of the incoming power lines.

3. Press and release the starting switch after two seconds. Record the starting current as soon as the switch is released. The starting current is the current value measured just after the starting switch is released. The starting current should be about 20% to 80% higher than the operating current when the circuit is working properly. The operating current is the current value measured after the lamp is energized.

No current reading indicates an open circuit. The same procedure may be used when an automatic starting switch is used. The starting current is read immediately after the line switch is closed. There is a problem with the starter or lamp when the lamp does not ignite properly and continues to try to start.

A lamp does not start when it is at the end of its normal life. The starter continues to try to start the lamp when the lamp is not replaced. This damages the starter. An old lamp that cycles during starting must be replaced. The starter requires replacement when a new lamp continues to cycle.

Instant-Start Circuit Testing

An instant-start circuit produces light instantly. Short lamp life is normally due to low supply voltage, improper lamp and lamp holder contact, or miswiring of the ballast. In two-lamp circuits, immediately replace any lamp that burns out. Replace both lamps when the good lamp shows signs of blackening. Look for ballast problems when there is a problem shortly after replacing a lamp in any circuit in which one lamp was allowed to operate alone for more than a few weeks.

Warning: Care must be taken to prevent electrical shock when measuring the high voltage. The voltmeter must be set to the high-voltage range. **See Instant-Start Circuit Testing.**

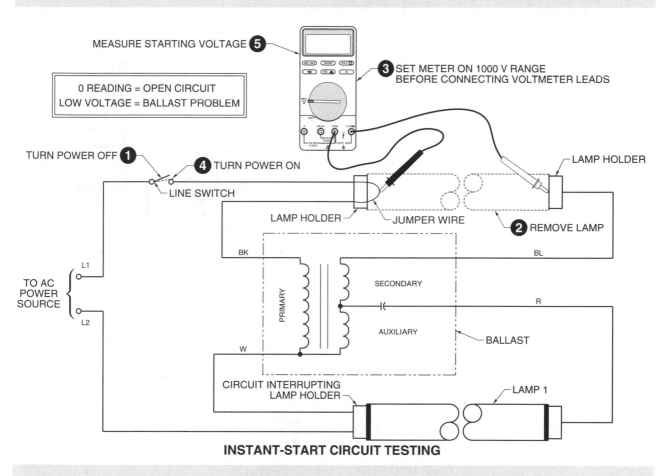

INSTANT-START CIRCUIT TESTING

To test an instant-start circuit, apply the procedure:

1. Turn the power to the circuit OFF by placing the line switch in the OFF position.

2. Remove the lamp.

3. Set the voltmeter to the 1000 V range. Connect the voltmeter leads to each end of the lamp holder. The voltmeter must complete the primary circuit at the lamp holder having two conductors. The two conductors may be shorted using a jumper wire.

Warning: Ensure that power is OFF before shorting the pins.

4. Turn the power to the circuit ON.

5. Measure the starting voltage. The ballast must provide an open circuit voltage of about three times the operating voltage to produce the arc required to light the lamp. There is a problem with the circuit when there is not a high starting voltage present. A zero voltage reading indicates an open circuit. A low voltage normally indicates a ballast problem.

Rapid-Start Circuit Testing

A rapid-start circuit produces light in less than one second. Long start times normally indicate a lamp problem. There is normally a ballast problem when start time is still long after replacing a lamp. In two-lamp circuits, immediately replace any lamp that burns out. Replace both lamps when the good lamp shows signs of blackening. **See Rapid-Start Circuit Testing.**

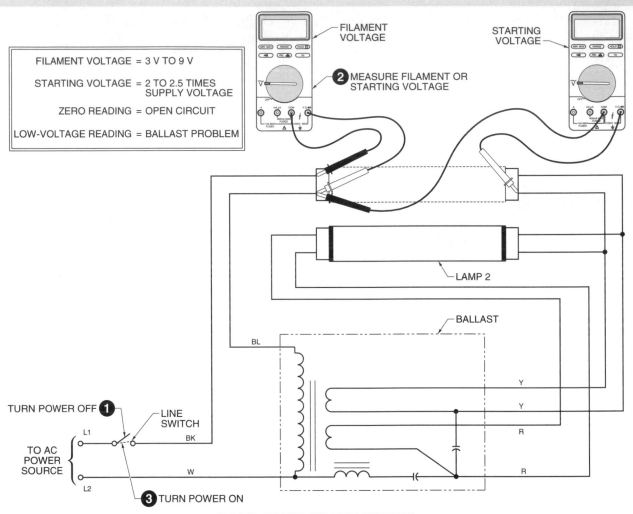

RAPID-START CIRCUIT TESTING

To test a rapid-start circuit, apply the procedure:

1. Turn the power to the circuit OFF by placing the line switch in the OFF position.

2. Measure filament or starting voltage. Remove the lamp and connect a voltmeter across the opposing lamp holders to measure starting voltage. Connect the voltmeter across the end of each holder (normally red-red, blue-blue, and yellow-yellow) to measure the filament voltage.

3. Turn power to the circuit ON.

Filament voltage should equal 3 V to 9 V, depending on the lamp. Starting voltage is given by the manufacturer and is normally about 2 to 2.5 times the supply voltage. A zero voltage reading indicates an open circuit. A low-voltage reading normally indicates a ballast problem.

 Refer to **Activity 20-2 — Troubleshooting Fluorescent Lamps** ....

APPLICATION 20-3—Troubleshooting High-Intensity Discharge Lamps

Measuring Line Voltage

The voltage at the input of an HID lamp fixture should be checked to ensure it is at the proper level. Non-starting, non-cycling, and lamp problems result when the voltage is not within a given limit. For constant wattage-rated ballasts, the line voltage should be within ±10% of the ballast (or fixture) nameplate rating. For reactor or high-reactance ballasts, the line voltage should be within ±5% of the ballast (or fixture) nameplate rating. **See Measuring Line Voltage.**

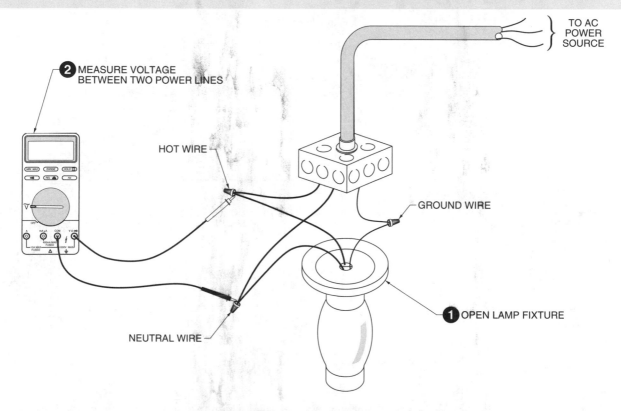

MEASURING LINE VOLTAGE

To measure the line voltage at an HID fixture, apply the procedure:
1. Open the lamp fixture so that the voltage may be measured directly at the lamp input.
2. Measure the voltage between the two incoming power lines. Ensure that the voltmeter is set to an AC voltage setting high enough to measure the voltage between the two incoming power lines. Check the measured voltage against the lamp's rated voltage. The voltage should be within ±10% for a constant-wattage ballast and ±5% for a reactor or high-reactance ballast.

Warning: Mercury-vapor and metal-halide lamps may cause serious skin burns and eye inflammation when the lamp's bulb is broken. The burns and eye inflammation result from the release of shortwave ultraviolet radiation. The lamp's bulb normally absorbs the shorter ultraviolet waves. The bulb should contain a fusible filament that extinguishes the arc when the bulb breaks. Always turn the power to the circuit OFF before troubleshooting the lamp. Replace any lamp that is cracked or damaged.

Ballast Output Voltage

The voltage output of a ballast should be checked when the incoming voltage is correct. An open circuit voltage measurement is required to see when the ballast is delivering the proper starting voltage to the lamp. **See Measuring Ballast Output Voltage.**

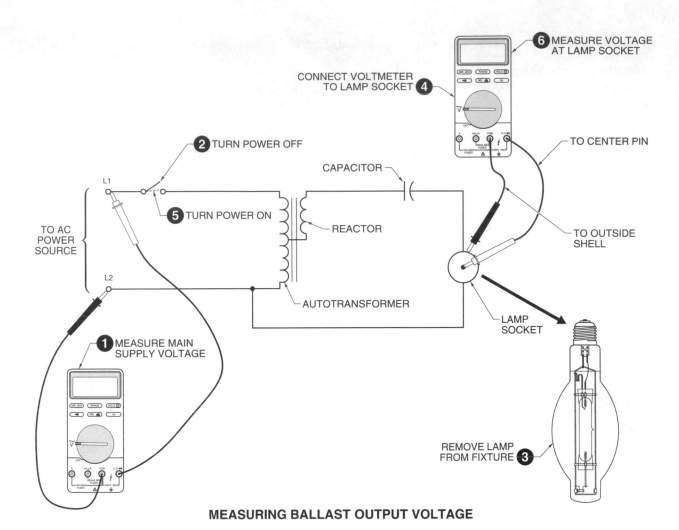

MEASURING BALLAST OUTPUT VOLTAGE

Warning: Never remove or insert an HID lamp when power is ON. Ensure that power is turned OFF. Measure the voltage to ensure power is OFF.

To measure the ballast output voltage, apply the procedure:
1. Measure the main supply voltage to ensure it is within the proper range.
2. Turn the power to the fixture OFF.
3. Remove the lamp from the fixture.
4. Connect the voltmeter to the lamp socket. Ensure that the voltmeter is set to a voltage setting high enough when connecting a voltmeter to the lamp socket (between the center pin and outside shell).

Caution: Disconnect the starter before measuring the ballast output voltage on a high-pressure sodium lamp. The high starting voltage pulse may damage some voltmeters.
5. Turn the power to the fixture ON.
6. Measure the voltage at the lamp socket. The voltage must be within the set limits. **See Recommended Ballast Output Voltage Limits in Appendix.**

There is a problem with the fixture when the voltage at the socket is not correct (or no voltage is present), and the incoming supply voltage is correct. The problem may be in the lamp socket, capacitor, ballast, or wiring. The problem is not with the lamp because the lamp is not receiving the correct voltage.

Warning: Metalarc lamps must be installed in fixtures that are enclosed with tempered glass or other manufacturer-approved materials. Metalarc lamps operating at high pressure (50 psi) may burst and shatter the lamp's bulb when they fail. This results in the discharge of hot quartz particles. These particles may be as hot as 1000°C.

Ballast Output Current

A ballast must deliver the proper voltage and current to a lamp. The output current of a ballast should be measured to ensure the ballast is delivering the correct starting current. A short-circuit lamp test may be used to measure ballast output current. In a short-circuit lamp test, the lamp is removed and an AC ammeter is connected across the socket. **See Measuring Ballast Output Current.**

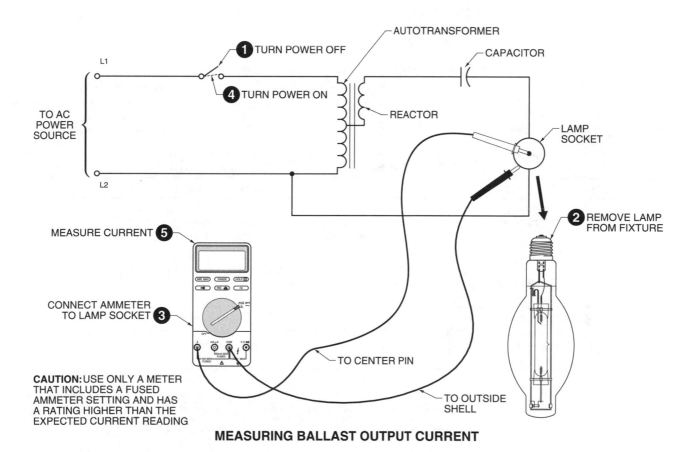

MEASURING BALLAST OUTPUT CURRENT

To measure ballast output current, apply the procedure:
1. Turn the power to the fixture OFF.
2. Remove the lamp from the fixture.
3. Connect the ammeter to the lamp socket. Ensure that the ammeter is set to a current setting high enough when connecting an in-line ammeter to the lamp socket (between the center pin and outside shell).
4. Turn the power to the fixture ON.
5. Measure the current at the lamp socket. The current must be within the set limits. **See Recommended Short-Circuit Current Test Limits in Appendix.**

There is a problem with the fixture when the current at the lamp socket is not correct (or no current is present) and the incoming supply voltage is correct. The problem may be in the lamp socket, capacitor, ballast, or wiring. The problem is not with the lamp because the lamp is not receiving the correct voltage.

Normal End of Lamp Life

Low-pressure sodium lamps maintain their light output even at the end of their lamp life. At the end of their lamp life, starting is intermittent and finally impossible. There may be some blackening of the arc tube at the end of the lamp life.

Mercury-vapor and metal-halide lamps have a low light output at the end of their lamp life. They also start having intermittent starting problems. Some blackening of the arc tube occurs.

High-pressure sodium lamps start to cycle ON and OFF at the end of their lamp life. Some blackening of the bulb occurs. Most lamps have a brownish color on the arc tube due to sodium deposits.

Troubleshooting Guides

Troubleshooting guides state a problem, possible cause(s) and corrective action(s) that may be taken. These guides may be used to quickly determine potential problems and possible courses of action. **See Troubleshooting Guides.**

INCANDESCENT LAMP TROUBLESHOOTING GUIDE		
Problem	**Possible Cause**	**Corrective Action**
Short lamp life	Voltage higher than lamp rating	Ensure that voltage is equal to or less than lamp's rated voltage. Lamp life is decreased when the voltage is higher than the rated voltage. A 5% higher voltage shortens lamp life by 40%.
	Lamp exposed to rough service conditions or vibration	Replace lamp with one rated for rough service or resistant to vibration
Lamp does not turn ON after new lamp is installed	Fuse blown, poor electrical connection, faulty control switch	Check circuit fuse or CB. Check electrical connections and voltage out of control switch. Replace switch when voltage is present into, but not out of the control switch.

FLUORESCENT LAMP TROUBLESHOOTING GUIDE			
Problem	**Circuit Type**	**Possible Cause**	**Corrective Action**
Lamp blinks and has shimmering effect during lighting period	All types	Depletion of emission material on electrodes	Replace with new lamp
New lamp blinks	All types	Loose lamp connection	Reseat lamp in socket securely. Ensure lamp holders are rigidly mounted and properly spaced
		Low voltage applied to circuit	Ensure that voltage is within ±7% of rated voltage
		Cold area or draft hitting lamp	Protect lamp with enclosure. Use a low temperature-rated ballast
	Preheat	Defective or old starter	Replace starter. Lamp life is reduced when starter is not replaced

Lamp does not light or is slow starting	All types	Lamp failure	Replace lamp. Test faulty lamp in another fixture. Check circuit fuse or CB. Check voltage at fixture
		Loss of power to the fixture or low voltage	Troubleshoot fluorescent fixture when voltage is present at the fixture. Replace broken or cracked holder. Check for poor wire connection
	Preheat	Normal end of starter life	Replace starter
	Rapid-start	Failed capacitor in ballast	Replace ballast
		Lamp not seated in holder	Seat lamp properly in holder. In rapid-start circuits, the holder includes a switch that removes power when the lamp is removed due to high voltage present
Bulb ends remain lighted after switch is turned OFF	Preheat	Starter contacts stuck together	Replace starter. Lamp life is reduced when starter is not replaced
Short lamp life	All types	Frequent turning ON and OFF of lamps	Normal operation is based on one start per three hour period of operation time. Short lamp life must be expected when frequent starting cannot be avoided
		Supply voltage excessive or low	Check the supply voltage against the ballast rating. Short lamp life must be expected when supply voltage is not within ±7% of lamp rating
		Low ambient temperature. Low temperature causes a slow start	Protect lamp with enclosure. Use a low temperature-rated ballast
	Instant-start	One lamp burned out and other burning dimly due to series-start ballast circuit	Replace burned-out lamp. Ballast is damaged when lamp is not replaced
		Wrong lamp type. May be using rapid-start or preheat lamp instead of instant-start	Replace lamp with correct type
Light output decreases	All types	Light output decreases over first 100 hours of operation	Rated light output is based on output after 100 hours of lamp operation. Before 100 hours of operation, light output may be as much as 10% higher than normal
		Low circuit voltage	Check the supply voltage against the ballast rating. Short lamp life and low light output must be expected when supply voltage is not within ±7% of lamp rating
		Dirt build-up on lamp and fixture	Clean bulb and fixture

HID LAMP TROUBLESHOOTING GUIDE		
Problem	**Possible Cause**	**Corrective Action**
Lamp does not start	Normal end of life operating characteristic	Replace with new lamp
	Loose lamp connection	Re-seat lamp in socket securely. Ensure lamp holder is rigidly mounted and properly spaced
	Defective photocell used for automatic turn ON	Replace defective photocell
	Low line voltage applied to circuit	Ensure that voltage is within ±7% of rated voltage
	Cold area or draft hitting lamp	Protect lamp with enclosure. Use a low temperature-rated ballast
	Defective ballast	Replace ballast
Lamp cycles ON and OFF or flickers when ON	Normal end of life operating characteristic	Replace with new lamp
	Poor electrical connection or loose bulb	Check electrical connections and socket contacts
	Line voltage variations	Ensure that voltage is within ±7% of rated voltage. Move lamps to separate circuit when lamps are on same circuit as high-power loads. High-power loads cause a voltage dip when turned ON. This voltage dip may cause the lamp to turn OFF
Short lamp life	Wrong wattage lamp or ballast	HID lamps and ballast must be matched in size. Lamp life is shortened when a large-wattage lamp is used. The same size and type ballast must be installed when replacing a ballast
	Power tap set too low for line voltage	Check ballast taps and ensure they are set for the correct supply voltage
	Defective sodium ballast starter	Replace starter
Low light output	Normal end of life operating characteristic	Replace with new lamp
	Dirty lamp or lamp fixture	Keep lamp fixture clean
	Early blackening of bulb caused by incorrect lamp size or ballast	Ensure that lamp size, lamp type, and ballast match

 *Refer to **Activity 20-3—Troubleshooting High-Intensity Discharge Lamps** ....*

 *Refer to Quick Quiz® on CD-ROM*

 *Refer to Chapter 20 in the **Troubleshooting Electrical/Electronic Systems Workbook** for additional questions.*

Troubleshooting Lighting Circuits

Name_____ Date _____

ACTIVITY 20-1—Troubleshooting Incandescent Lamps

1. Using Incandescent Lamp Circuit 1 and Incandescent Lamp Characteristics, determine each value.

_____ **A.** The wattage of the lamp based on Meter Reading 1 is ___ W.

_____ **B.** The lumen of the lamp based on Meter Reading 1 is ___ lm.

_____ **C.** The lamp life changes by ___% based on Meter Reading 1.

_____ **D.** The wattage of the lamp based on Meter Reading 2 is ___ W.

_____ **E.** The lumen of the lamp based on Meter Reading 2 is ___ lm.

_____ **F.** The lamp life changes by ___% based on Meter Reading 2.

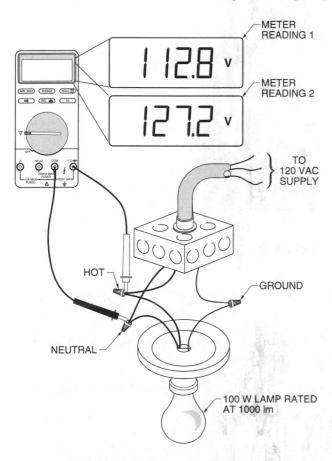

INCANDESCENT LAMP CIRCUIT 1

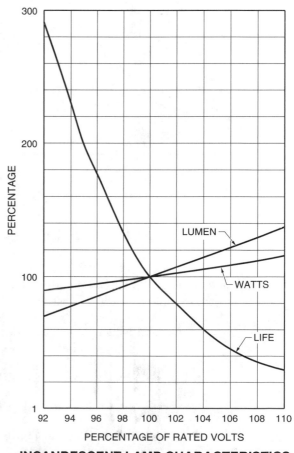

INCANDESCENT LAMP CHARACTERISTICS

2. Using Incandescent Lamp Circuit 2 and the meter readings, determine each value. *Note:* Assume all components in the circuit are good for questions A-D.

_____ **A.** Meter 1 reads ___ V when the switch is in the OFF position.

_____ **B.** Meter 2 reads ___ V when the switch is in the OFF position.

_____ **C.** Meter 1 reads ___ V when the switch is in the ON position.

_____ **D.** Meter 2 reads ___ V when the switch is in the ON position.

_____ **E.** The problem is a(n) ___ if Meter 1 reads 0 VAC when the switch is in the ON position and 0 VAC when the switch is in the OFF position and the lamp does not light.

_____ **F.** The problem is a(n) ___ if Meter 2 reads 115 VAC when the switch is in the ON position and 115 VAC when the switch is in the OFF position and the lamp is always ON.

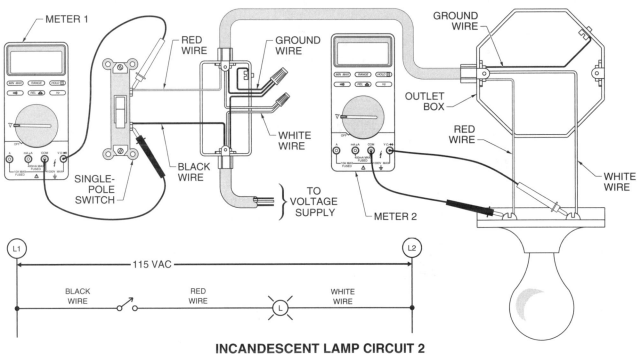

INCANDESCENT LAMP CIRCUIT 2

3. Using Incandescent Lamp Circuit 3 and the meter readings, determine each value. *Note:* Assume all components in the circuit are good for questions A-E.

_____ **A.** Meter 1 connected in Test Position 1 reads ___ V regardless of the position of the two switches.

_____ **B.** Meter 1 connected in Test Position 2 reads ___ V when both switches are up.

_____ **C.** Meter 1 connected in Test Position 2 reads ___ V when both switches are down.

_____ **D.** Meter 2 connected in Test Position 1 reads ___ V when Switch 1 is in the up position and Switch 2 is in the down position.

_____ **E.** Meter 2 connected in Test Position 1 reads ___ V when Switch 1 is in the up position and Switch 2 is in the up position.

_____ **F.** The problem is a(n) ___ if Meter 1 reads 115 VAC in Test Position 1, 0 VAC in Test Position 2 (at all times), and Meter 2 reads 0 VAC in either test position (regardless of either switch position).

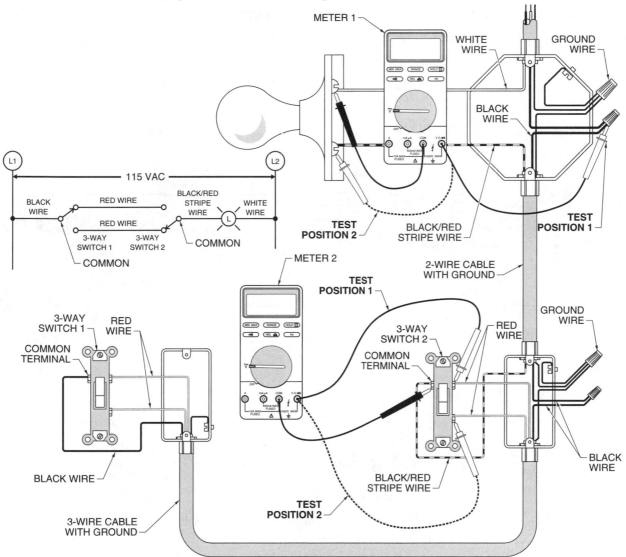

INCANDESCENT LAMP CIRCUIT 3

ACTIVITY 20-2—Troubleshooting Fluorescent Lamps

1. Using Fluorescent Lamp Circuit 1 and the meter readings, determine each value. *Note:* Assume all components in the circuit are good for questions A-I.

_____ **A.** Meter 1 reads ___ V before the Starting Switch is pressed.

_____ **B.** Meter 2 reads ___ A before the Starting Switch is pressed.

_____ **C.** Meter 3 reads ___ A before the Starting Switch is pressed.

_____ **D.** Meter 1 reads ___ V when the Starting Switch is held down.

_____ **E.** Meter 2 reads ___ A when the Starting Switch is held down.

_____ **F.** Meter 3 reads ___ A when the Starting Switch is held down.

_____ **G.** Meter 1 reads ___ V when the Starting Switch is open and the lamp is ON.

_____ **H.** Meter 2 reads ___ A when the Starting Switch is open and the lamp is ON.

_____ **I.** Meter 3 reads ___ A when the Starting Switch is open and the lamp is ON.

_____ **J.** Meter 1 reads ___ V when the Starting Switch is held down and the filament or ballast is open.

_____ **K.** Meter 2 reads ___ A when the Starting Switch is held down and the filament or ballast is open.

_____ **L.** Meter 3 reads ___ A when the Starting Switch is held down and the filament or ballast is open.

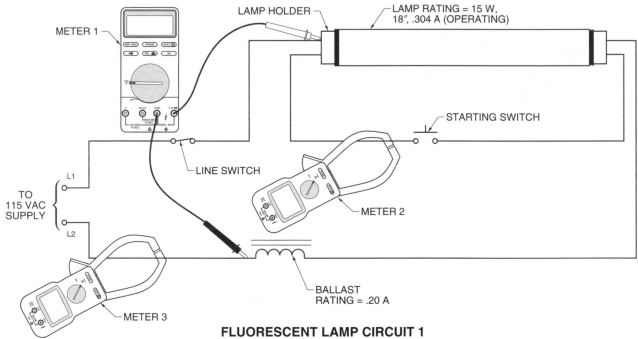

FLUORESCENT LAMP CIRCUIT 1

2. Using Fluorescent Lamp Circuit 2 and the meter readings, determine each value. *Note:* Assume all components in the circuit are good for questions A-C.

_____ **A.** Meter 1 reads ___ V when the lamps are ON.

_____ **B.** Meter 2 reads ___ V when the lamps are ON.

_____ **C.** Meter 3 reads ___ A when the lamps are ON.

_____ **D.** Meter 1 reads ___ V when the line switch is closed, one of the lamps is not properly seated (or removed), and neither lamp is ON.

_____ **E.** Meter 2 reads ___ V when the line switch is closed, one of the lamps is not properly seated (or removed), and neither lamp is ON.

_____ **F.** Meter 3 reads ___ A when the line switch is closed, one of the lamps is not properly seated (or removed), and neither lamp is ON.

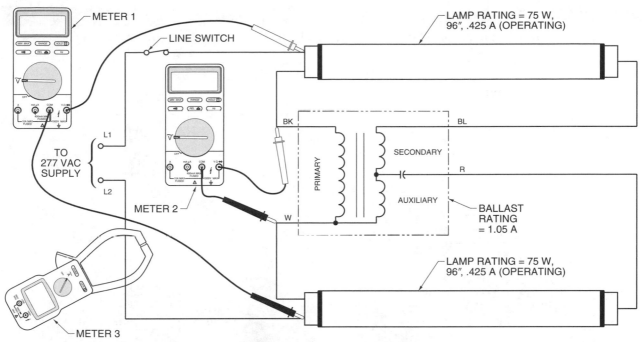

FLOURESCENT LAMP CIRCUIT 2

3. Using Fluorescent Lamp Circuit 3 and the meter readings, determine each value. *Note:* Assume all components in the circuit are good for questions A and B.

_____ **A.** Meter 1 reads ___ V when the lamps are ON.

_____ **B.** Meter 2 reads ___ A when the lamps are ON.

_____ **C.** Does Meter 2 reading increase or decrease if the line voltage is increased 5%?

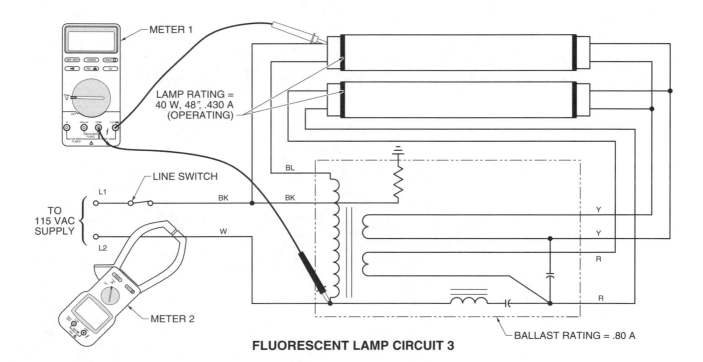

FLUORESCENT LAMP CIRCUIT 3

ACTIVITY 20-3—Troubleshooting High-Intensity Discharge Lamps

1. Using the wiring diagram, connect the devices in HID Lamp Circuit 1. Add a wire nut to insulate any wires not used.

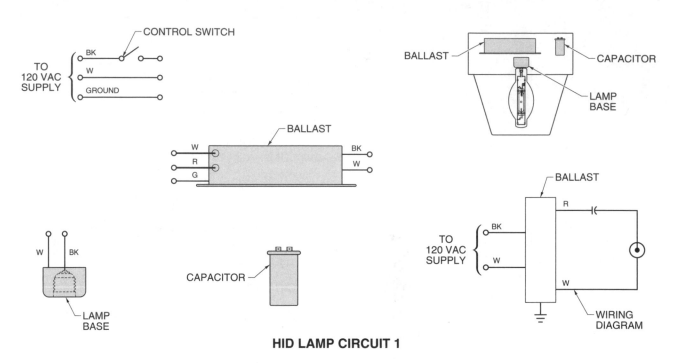

HID LAMP CIRCUIT 1

2. Using the wiring diagram, connect the devices in HID Lamp Circuit 2. Add a wire nut to insulate any wires not used.

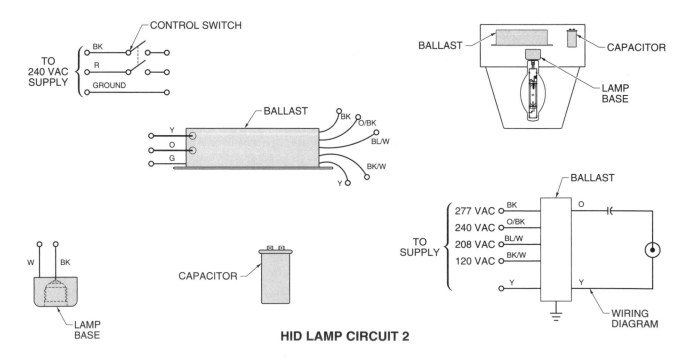

HID LAMP CIRCUIT 2

3. Using the wiring diagram, connect the devices in HID Lamp Circuit 3. Add a wire nut to insulate any wires not used.

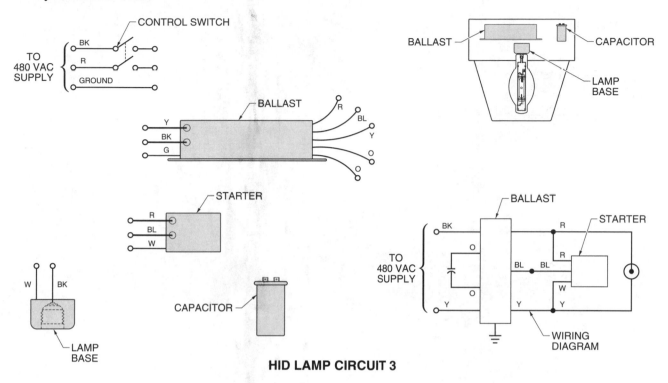

HID LAMP CIRCUIT 3

4. Using the wiring diagram, connect the devices in HID Lamp Circuit 4. Add a wire nut to insulate any wires not used.

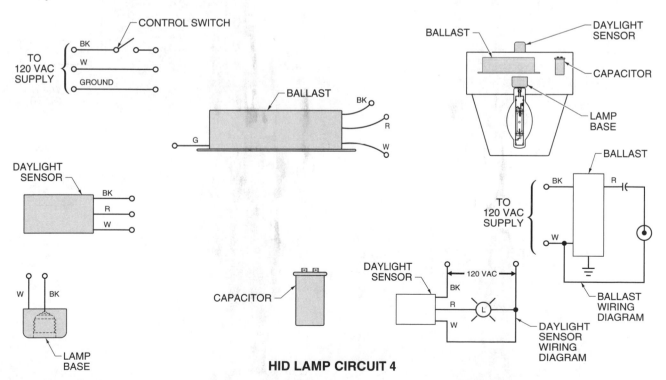

HID LAMP CIRCUIT 4

5. Using the wiring diagram, connect the devices in HID Lamp Circuit 5. Add a wire nut to insulate any wires not used.

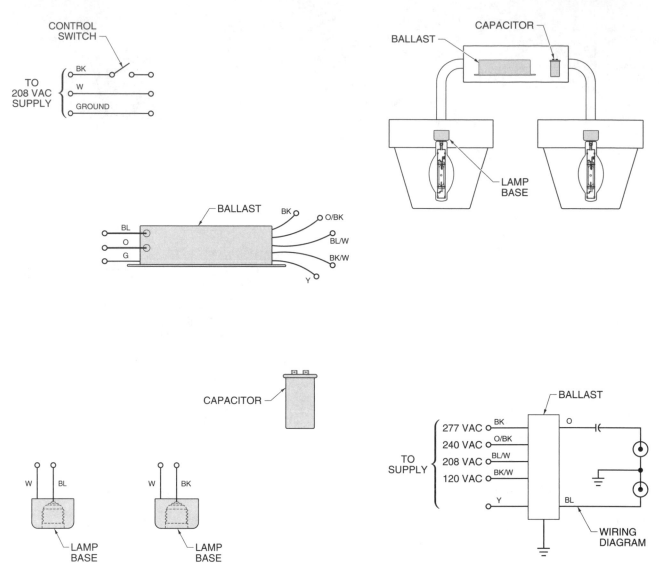

HID LAMP CIRCUIT 5

APPLICATION 21-1—Solenoid-Operated Valves

A *solenoid* is an electromechanical device that converts electrical energy to mechanical energy. Solenoids are used to operate motor starters, contactors, relays, valves, clutches, and many other applications. **See Solenoid.**

The amount of force a solenoid develops depends on the number of turns in the coil and the amount of applied current. The larger the coil, the greater the force. The current draw is the highest (inrush current) when a solenoid is first energized. The current draw is less (rated current) after the solenoid has moved through its rated travel distance.

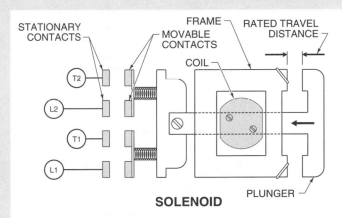

SOLENOID

Solenoid-Operated Valves

A *solenoid-operated valve* is a valve that uses electrical force to directly or indirectly change the position of a valve's spool. Solenoid-operated valves are used in most fluid power circuits to control the direction of fluid flow. **See Fluid Power Symbols in Appendix.** Directional control valves are identified by the ways, number of positions, and types of actuators. **See Solenoid-Actuated Directional Control Valves.**

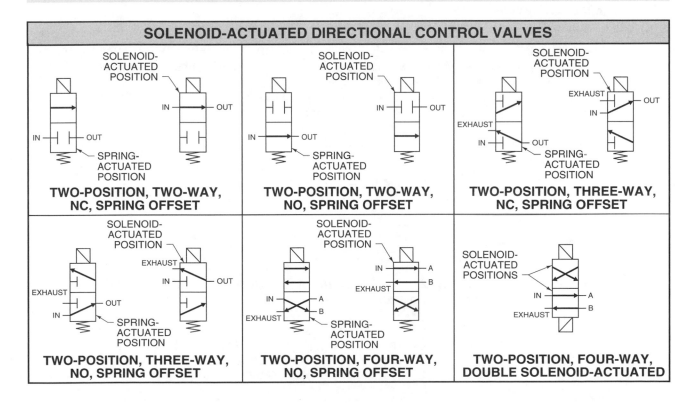

SOLENOID-ACTUATED DIRECTIONAL CONTROL VALVES

TWO-POSITION, TWO-WAY, NC, SPRING OFFSET

TWO-POSITION, TWO-WAY, NO, SPRING OFFSET

TWO-POSITION, THREE-WAY, NC, SPRING OFFSET

TWO-POSITION, THREE-WAY, NO, SPRING OFFSET

TWO-POSITION, FOUR-WAY, NO, SPRING OFFSET

TWO-POSITION, FOUR-WAY, DOUBLE SOLENOID-ACTUATED

Ways

A *way* is the flow path through the valve. Most directional control valves are either two-way, three-way, or four-way. The number of ways required depends on the application. Two-way directional control valves have two main ports that allow or stop the flow of fluid. Two-way valves are used as shutoff, check, and quick-exhaust valves. Three-way valves allow or stop fluid flow or exhaust. They are used to control single-acting cylinders, fill-and-drain tanks, and nonreversible fluid motors. Four-way directional control valves have four (or five) main ports that change fluid flow from one port to another. Four-way valves are used to control the direction of double-acting cylinders and reversible-fluid motors.

Positions

A directional control valve is placed in different positions to start, stop, or change the direction of fluid flow. A *position* is the number of positions within the valve that the spool is placed in to direct fluid flow through the valve. Two-way and three-way valves have two positions. Four-way valves may have two or three positions.

Actuators

Directional control valves must have a means to change the valve position. An *actuator* is a device that changes the position of a valve. Directional control valve actuators include pilots, manual levers, mechanical levers, springs, and solenoids. Solenoid-operated, directional control valves are either solenoid-operated/spring return, or double solenoid-operated.

Normally-Closed and Normally-Open Valves

A *normally-closed (NC) valve* is a valve that does not allow pressurized fluid to flow out of the valve in the spring-actuated position. A *normally-open (NO) valve* is a valve that allows pressurized fluid to flow out of the valve in the spring-actuated position. Two-way and three-way valves may be either normally open or normally closed. The valve must have a spring position to be normally open or normally closed.

Valve Numbering System

The ways, positions, port size, actuators, coil voltage, and current type must be specified when ordering a new or replacement valve. Valve manufacturers provide a numbering system to simplify valve selection and ordering. The numbering system normally consists of letters and numbers that represent the different valve models. **See Valve Selections.**

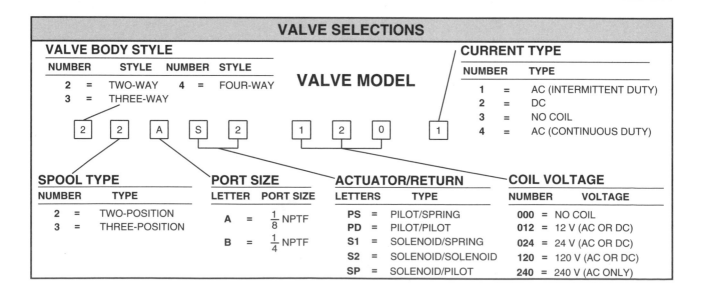

Incorrect Voltage

The voltage applied to a solenoid-operated valve should be ±10% of the rated solenoid value. The voltage is measured directly at the valve when the solenoid is energized. A voltmeter is set on an AC voltage range for AC solenoids. The voltmeter is set on a DC voltage range for DC solenoids. The range setting must be greater than the applied voltage. **See Solenoid Voltage Measurement.**

A solenoid overheats when the voltage is excessive. The heat destroys the insulation on the coil wire and burns out the solenoid. The solenoid has difficulty moving the spool inside the valve when the voltage is too low. The slow operation causes the solenoid to draw its high inrush current longer. Lasting high inrush current causes excessive heat.

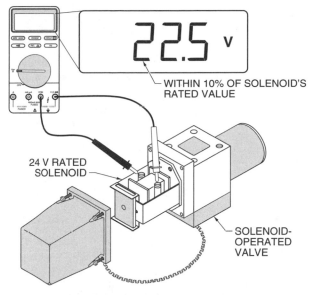

SOLENOID VOLTAGE MEASUREMENT

Incorrect Frequency

Solenoid valves are available with frequency ratings of 50 Hz or 60 Hz. Solenoids with a frequency rating of 50 Hz are used for export applications. Solenoids with a frequency rating of 60 Hz are used for domestic applications. Imported machines may have solenoid valves with a frequency rating of 50 Hz. A solenoid may operate if the frequency is not correct, but may have a higher failure rate and may produce noise.

Double-Acting Solenoid Valve

A *double-acting solenoid valve* is a valve that uses one solenoid to develop a force in one direction and a second solenoid to develop a force in the opposite direction. One solenoid moves the spool to one position and the other solenoid moves the spool to the other position. The proper operation of a double-acting solenoid is to energize one solenoid at a time. The solenoid overheats and burns out when both solenoids are energized simultaneously. **See Double-Acting Solenoid Valve.**

An *interlock* is a mechanical or electrical method used to prevent two different circuits from being energized at the same time. An interlock is included in applications that require double-acting solenoids. A voltmeter is connected to both solenoids when testing a circuit in operation. Both meters should never indicate a voltage simultaneously. A control circuit problem exists when both solenoids have power simultaneously. **See Solenoid Interlock.**

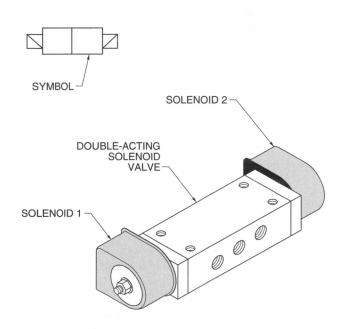

DOUBLE-ACTING SOLENOID VALVE

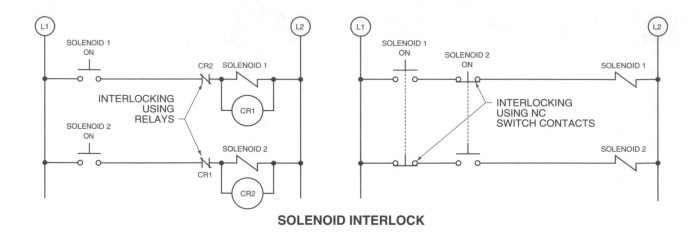

SOLENOID INTERLOCK

Rapid Cycling

A solenoid draws several times its rated current when first connected to power. This high inrush current produces heat. In normal applications, the heat is low and dissipates over time. Rapid cycling does not allow the heat to dissipate quickly. The heat buildup burns the coil insulation and causes solenoid failure. A high-temperature solenoid should be used in applications requiring a solenoid to be cycled more than 10 times per minute. A solenoid may be forced-cooled in some applications.

An AC solenoid heats faster than a DC solenoid. A DC solenoid should be used in fast cycling applications. An AC solenoid with a continuous duty rating should be used if circuit design does not provide for DC. Some manufacturers include a diode in series with the coil of AC solenoids to give the solenoid a continuous duty rating. This rating is normally marked on the valve as CD. The voltage after the diode in an AC-solenoid valve marked CD is pulsating DC, and the voltage before the diode is AC.

Excessive Valve Force

A solenoid moves the valve spool in a solenoid-operated valve. The spool has extra pressure applied to it when undersized valves are used. The solenoid has to apply a greater force to overcome the extra pressure. The greater force produces a higher-than-rated current in the solenoid. The higher-than-rated current is present longer than normal. To prevent this problem, ensure that the valve is correctly sized or slightly oversized.

Manufacturers provide specification curves that illustrate the amount of force a solenoid produces at its rated voltage. Most curves also illustrate the amount of force produced at less than rated voltage and the current draw of the solenoid. The lower the applied voltage, the less the delivered force and current draw. A solenoid must deliver enough force throughout its stroke to operate properly. **See Solenoid Specifications.**

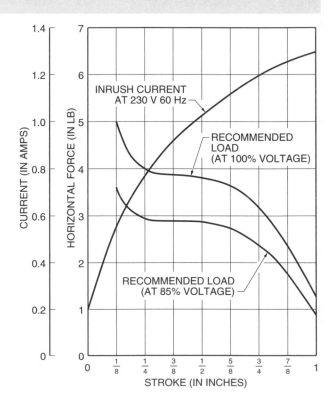

SOLENOID SPECIFICATIONS

Excessive valve force may be required at times even when the valve is correctly sized. This occurs in cold hydraulic systems. Cold temperatures cause oil to become thick, which may overload a solenoid-operated valve. Use a hydraulic oil that is suitable for low temperature when ambient temperature is low. A higher temperature coil or larger solenoid may also be used in low temperature applications.

Solenoid force should be as close to the load requirements as possible. Excessive force reduces solenoid life. Inadequate force keeps the solenoid from operating.

Transients

In most industrial applications, the power supplying a solenoid comes from the same power lines that supply electric motors and other solenoids. High transient voltages are placed on the power lines as these inductive loads are turned ON and OFF. Transient voltages may damage the insulation on the solenoid coil, nearby contacts, and other loads. The transient voltages may be suppressed by using snubber circuits.

Environmental Conditions

For proper operation, a solenoid must operate within its rating and not be mechanically damaged or damaged by the surrounding atmosphere. A solenoid coil is subject to heat during normal operation. This heat comes from the combination of fluid flowing through the valve, the temperature rise from the coil when energized, and the ambient temperature of the solenoid.

Solenoid-operated valves may be purchased with or without covers. Solenoid-operated valves that do not have covers are intended for mounting inside a dust-tight cabinet. Any valve that is not mounted inside a dust-tight cabinet must have a cover to protect the solenoid coils. The cover protects the solenoid coils from water, oil, dirt, and mechanical contact and should always be kept on the solenoid coils. The cover may be attached to the valve with a small chain to prevent it from being lost. **See Solenoid-Operated Valve Cover.**

A solenoid coil is damaged when heat increases the temperature of the coil insulation above its rating. The coil is damaged by the breakdown of the insulation. The insulation on the coil helps seal out moisture, dust, oil, corrosive material, and heat. The insulation also provides electrical insulation.

Solenoids that have high temperature-rated insulation are used in high-temperature applications. Solenoids use various classes of insulation. **See Solenoid Insulation Classes.**

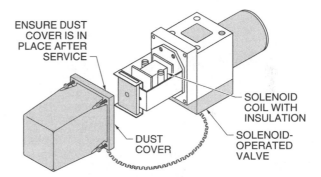

SOLENOID-OPERATED VALVE COVER

SOLENOID INSULATION CLASSES		
Class	Maximum Operating Temperature*	Insulation
A	105	Epoxy or paper
B	130	Epoxy
F	155	Epoxy
H	180	Epoxy

* in °C

High humidity and a frequently changing atmosphere cause condensation. Condensation may cause corrosion that hinders solenoid operation and aid in the deterioration of coil insulation. Condensation in low-temperature areas causes ice buildup. Ice buildup may prevent proper solenoid movement.

A solid molded coil or solenoid designed for oil immersion is used when condensation is a problem. This keeps the moisture away from the coil insulation. A solenoid-operated valve is placed where it cannot be mechanically dam-

 Refer to Activity 21-1—Solenoid-Operated Valves...

APPLICATION 21-2—Troubleshooting Solenoid–Operated Valves

aged. The valve is placed in an enclosure when required. Only solenoid valves that include tight covers are used. Four symptoms of a faulty solenoid include: failure to operate when energized, failure to operate when de-energized, noisy operation, and erratic operation. A solenoid that fails to operate when energized does not allow the valve to shift position or electrical contacts to operate. A solenoid that fails to operate when de-energized does not allow the valve to shift to the original position or electrical contacts to change to their normal condition.

A solenoid that operates noisily does not cause improper operation, although the noise may become a problem. A solenoid that operates erratically causes intermittent problems in the system and is a sign of a problem that normally gets worse. **See Faulty Solenoid Problems.**

FAULTY SOLENOID PROBLEMS		
Problem	**Possible Causes**	**Comments**
Failure to operate when energized	Complete loss of power to the solenoid	Normally caused by blown fuse or control circuit problem
	Low voltage applied to the solenoid	Voltage should be at least 85% of solenoid's rated value
	Burned out solenoid coil	Normally evident by pungent odor caused by burnt insulation
	Shorted coil	Normally a fuse is blown and continues to blow when changed
	Obstruction of plunger movement	Normally caused by a broken part, misalignment, or the presence of a foreign object
	Excessive pressure on solenoid plunger	Normally caused by excessive system pressure in solenoid-operated valves
Failure to operate spring-return solenoids when de-energized	Faulty control circuit	Normally a problem of the control circuit not disengaging the solenoid's hold or memory circuit.
	Obstruction of plunger movement	Normally caused by a broken part, misalignment, or the presence of a foreign object
	Excessive pressure on solenoid plunger	Normally caused by excessive system pressure in solenoid-operated valves
Failure to operate electrically-operated return solenoids when de-energized	Complete loss of power to solenoid	Normally caused by a blown fuse or control circuit problem
	Low voltage applied to solenoid	Voltage should be at least 85% of solenoid's rated value
	Burned out solenoid coil	Normally evident by pungent odor caused by burnt insulation
	Obstruction of plunger movement	Normally caused by broken part, misalignment, or presence of a foreign object
	Excessive pressure on solenoid plunger	Normally caused by excessive system pressure in solenoid-operated valves
Noisy operation	Solenoid housing vibrates	Normally caused by loose mounting screws
	Plunger pole pieces do not make flush contact	An air gap may be present causing the plunger to vibrate. These symptoms are normally caused by foreign matter
Erratic operation	Low voltage applied to the solenoid	Voltage should be at least 85% of the solenoid's rated voltage
	System pressure may be low or excessive	Solenoid size is inadequate for the application
	Control circuit is not operating properly	Conditions on the solenoid have increased to the point where the solenoid cannot deliver the the required force

Troubleshooting Solenoids

A voltmeter and ohmmeter are required when troubleshooting a solenoid. **See Troubleshooting Solenoids.**

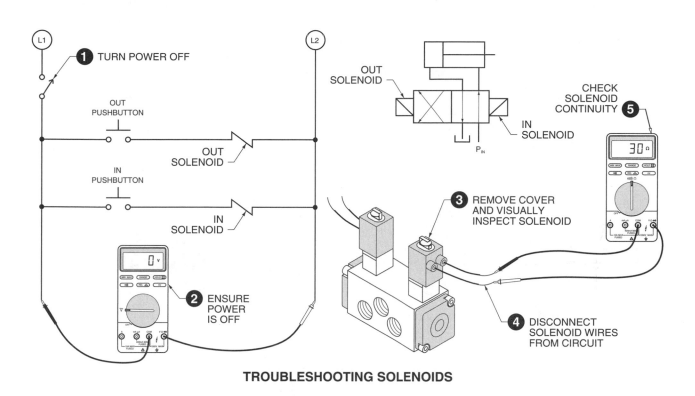

TROUBLESHOOTING SOLENOIDS

To troubleshoot a solenoid, apply the procedure:

1. Turn electrical power to the solenoid or solenoid circuit OFF.

2. Measure the voltage at the solenoid to ensure the power is OFF.

3. Remove the solenoid cover and visually inspect the solenoid. Look for a burnt coil, broken parts, or other problems. Replace the coil when burnt. Replace the broken parts when available. Replace the valve, contactor, starter, or solenoid-operated device when the parts are not available. *Note:* Determine the fault before installing a new coil when a solenoid has failed due to a burnt or shorted coil. The new coil burns out if the fault is not corrected. Always observe solenoid operation after a solenoid is replaced.

4. Disconnect the solenoid wires from the electrical circuit when no obvious problem is observed.

5. Check the solenoid continuity. Connect the meter leads to the solenoid wires with all power turned OFF. The meter should indicate a resistance reading of ±15% of the coil's normal reading.

Unknown readings are obtained by testing a good solenoid. A low or zero reading indicates a short or partial short. Replace the solenoid if there is a short. No movement of the needle on an analog meter or infinity resistance on a digital meter indicates the coil is open and defective. Replace the solenoid if the open is not obvious. *Note:* Set the ohmmeter to (R × 100) if the solenoid uses a small gauge wire. The small wire has high resistance.

 Refer to **Activity 21-2 — Troubleshooting Solenoid-Operated Valves** ....

APPLICATION 21-3—Heating Elements

A *heating element* is a heat-producing device that produces heat from the resistance to the flow of electric current. A heating element generally has a long life with little or no problems. Design, operation, or application problems are present if a heating element has early problems or does not last long. Most problems are caused by using the wrong type of heating element for the application or by using the wrong size heating element.

Heating Element Types

Many types of heating elements are available. The two main types are the open coil and enclosed coil. The open-coil heating element is the least expensive and presents a safety problem when not properly protected. The open-coil heating element is used in hot-air applications such as space heaters, hair dryers, and other small appliances that have a protective enclosure. **See Open and Enclosed-Coil Heating Elements.**

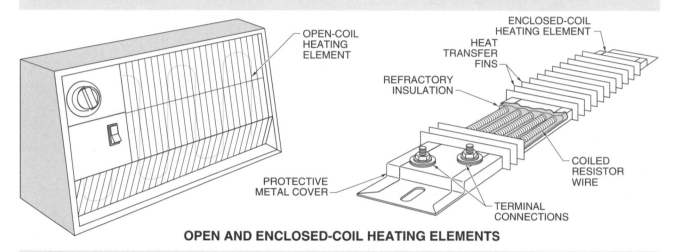

OPEN AND ENCLOSED-COIL HEATING ELEMENTS

The enclosed-coil heating element is the safest and most commonly used heating element in industrial applications. The *enclosed-coil heating element* has the resistance wire which produces the heat surrounded by a refractory insulating material. The refractory insulation is enclosed in a protective metal tube or cover. The cover protects the resistance wire from physical damage and improves thermal conductivity. The potential of electrical shock is also reduced because the live conductor is not exposed to the outside of the heating element. The most common enclosed-coil heating elements are the tubular, strip, ring, band, and combination types. **See Enclosed-Coil Heating Elements.**

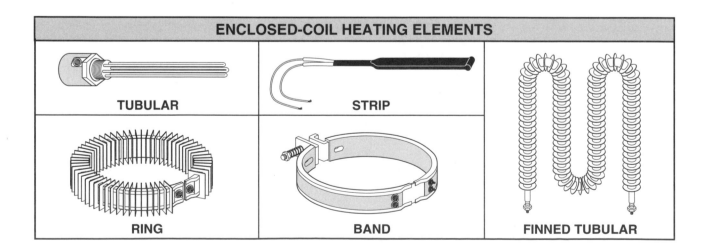

ENCLOSED-COIL HEATING ELEMENTS

Heating Element Ratings

Heating elements are rated by the manufacturer as to watt density. *Watt density* is the watts per square inch (W/sq in.) on the outer cover of the heating element. Excessive watt density wastes energy, shortens heating element life, and causes wide heat fluctuations. Low watt density causes long heating times or inadequate heat.

Heating a material to a high temperature quickly is accomplished by increasing the watt density of the heating element. This reduces the life of the heating element. The watt density of a heating element depends on the rate of heat absorption of the heated material. For example, water conducts (absorbs) heat away from a heating element faster than a heavy tar (asphalt). A heating element used to heat water may have a high watt density rating. Water conducts heat away from a heating element about five times faster than a heavy tar so the watt density of a water-heating element (approximately 50 W/in.) may be five times higher than a heavy tar-heating element (approximately 10 W/in.). The heavy tar may be heated faster by using a high-rated heating element, which reduces the life of the heating element.

Outer Cover Material

The type of material heated determines the type of outer cover material of the heating element. A copper jacket is normally used for water and cooking oil heating applications. A steel jacket is normally used for oil and kerosene heating applications. A stainless steel jacket or alloy is normally used for corrosive solutions and high-temperature air heating. **See Heating Element Specifications.**

HEATING ELEMENT SPECIFICATIONS				
Material State	Heated Material	Outer Cover Material	Maximum Watt Density*	Material Operating Temperature**
Liquid	Gasoline	Iron/steel	18-22	300
	Oil (low viscosity)	Steel/copper	20-25	Up to 180
	Oil (medium viscosity)	Steel/copper	12-18	Up to 180
	Oil (high viscosity)	Steel/copper	5-7	Up to 180
	Vegetable oil (cooking)	Copper	28-32	400
	Water (washroom)	Copper	70-90	140
	Water (process)	Copper	40-50	212
Air/Gas	Still air (ovens)	Steel/stainless steel	28-32 18-22 8-10 2-3	Up to 700 Up to 1000 Up to 1200 Up to 1500
	Air (moving at 10 fps)	Aluminum steel	30-34 23-26 14-16 2-3	Up to 700 Up to 1000 Up to 1200 Up to 1500
Solid	Asphalt	Iron/steel	9-12 8-10 6-8 5-6	Up to 200 Up to 300 Up to 400 Up to 500
	Molten tin (heated in pot)	Iron/steel	20-22	600
	Steel tubing (heated indirectly)	Iron/steel	45-48 50-52 54-56	500 750 1000

* in W/sq in.
** in °F

Terminal Contamination

Heating elements are connected to supply voltage at terminal points. The electrical resistance of the connection increases when the terminals are contaminated. Moisture, salt, oil, and acid are major causes of contamination.

Terminal connections are protected to prevent contamination. Moisture barriers or terminal seals are used to protect the connection. A variety of moisture barriers and terminal seals are available. These include rubber, epoxy, silicon, putty, and ceramic-to-metal hermetic seals. Rubber is used in applications up to 150°F. Epoxy is used in applications up to 300°F. Silicon is used in applications up to 500°F. Putty is used in applications up to 600°F. Ceramic-to-metal hermetic seals are used in applications up to 1000°F. **See Terminal Protection.**

TERMINAL SEALS	
Type	Max. Temp.*
Rubber	150
Epoxy	300
Silicon	500
Putty	600
Ceramic	1000

* in °F

TERMINAL PROTECTION

HEATING ELEMENT

TERMINALS COVERED TO PROTECT FROM CONTAMINATION

 Refer to **Activity 21-3—Heating Elements** ....

APPLICATION 21-4—Troubleshooting Heating Elements

The problems that occur in heating circuits include no heat, inadequate heat, excessive heat, or the system is giving electrical shocks. Inadequate or excessive heat normally indicates a control circuit problem. No heat normally indicates a blown fuse, a heating element problem, or control circuit problem. Electrical shocks are caused by a high leakage current that occurs when insulation breaks down. To determine the problem in a heating circuit, the circuit is tested for an open or short circuit, the power draw of the heating element is measured, and the insulation is tested.

Open or Short Circuit Test

Heating elements may open or short-circuit due to deterioration, physical damage, or misuse. No heat is produced when a heating element opens or shorts. The fuse (or CB) blows when a heating element shorts. The only sign that a partial short occurs in part of a heating element is a reduction of heat output. **See Open or Short Circuit Test.**

Warning: Avoid contact with any part of the heating circuit or element that may cause burns.

To test for an open or short circuit, apply the procedure:
1. Turn power to the system OFF when the heating element is suspected of having an open or short.
2. Measure the voltage at the heating element to ensure the power is OFF.
3. Disconnect the heating element from the circuit.
4. Measure the resistance of the heating element. The heating element should have some resistance. The higher the rated heater output, the lower the resistance reading. The lower the rated heater output, the higher the resistance reading. A heating element has a short when the meter measures 0 Ω. A heating element has an open when the meter measures infinity. The heating element should be removed and tested when there is a short or open indication.
5. Reconnect the heating element when there is a resistance reading. Ensure the ohmmeter is removed from the circuit.
6. Turn power to the system ON.

7. Set the temperature control unit to turn ON the heating element.

8. Measure the voltage into the heating element. The problem is located before the heating element when no power is present at the heating element. Turn power to the circuit OFF. The problem may be a blown fuse, faulty control circuit, or open connection. A power draw test is performed if there is correct voltage at the heating element but not the correct heat output.

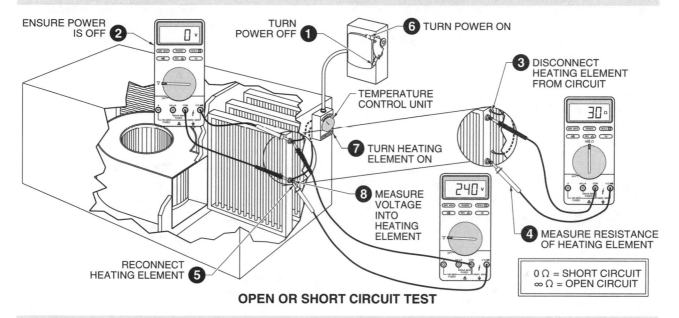

OPEN OR SHORT CIRCUIT TEST

Power Draw Test

A heating element converts electrical energy (watts) into heat. Most heaters are rated in watts. For example, a portable space heater is rated for 750 W to 1500 W for normal operation on a 15 A or 20 A circuit. The amount of electrical power a heating element draws is an indication of the condition of the heating element.

A wattmeter may be used to measure the electrical power draw of a heating element. A voltmeter and ammeter are normally used to measure the electrical power draw of a heating element because voltage and current are in phase in a heating element. **See Power Draw Test.**

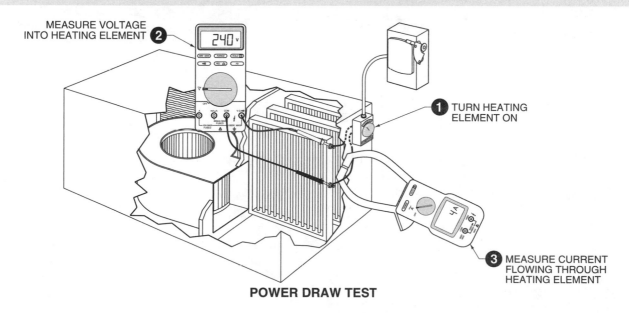

POWER DRAW TEST

To perform a power draw test on a heating element, apply the procedure:

1. Set the temperature control unit to turn the heating element ON.

2. Measure the voltage into the heating element.

3. Measure the current flowing through the heating element.

Multiply the measured voltage and current to obtain the amount of power draw. The amount of power draw should equal ±10% of the manufacturer's power rating. A reduction in heat output is produced when the power draw is less than 10% of the manufacturer's power rating. An increase in heat output is produced when the power draw is greater than 10% of the manufacturer's power rating. A reduced output normally indicates poor electrical connections or partially open elements. An increased output normally indicates partially shorted elements or high leakage current.

Insulation Breakdown Test

Heating elements absorb moisture over time. This causes current to leak from the current carrying conductors inside the heating element to the outer cover. Current leakage is a function of insulation resistance. Current leakage increases as insulation resistance decreases. Insulation resistance values on dry heating elements should be high. Dry insulation readings should equal several thousand megohms. Over time, the resistance decreases and the leakage current increases. A safety problem is present when the insulation value decreases greatly. **See Insulation Breakdown Test.**

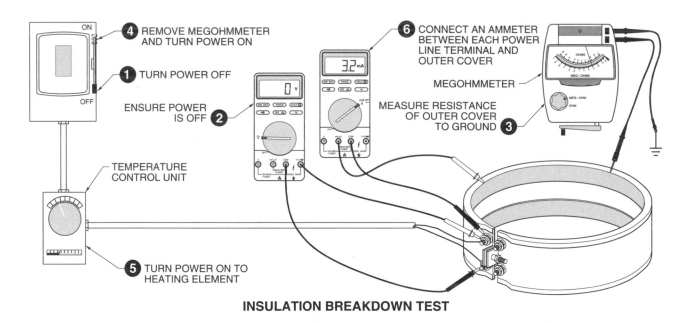

INSULATION BREAKDOWN TEST

To perform an insulation breakdown test, apply the procedure:

1. Turn power to the system OFF.

2. Measure the voltage at the heating element to ensure the power is OFF.

3. Measure the resistance of the outer cover of the heating element to ground. The resistance reading should be high. The resistance reading should equal 1000 times (or more) the rated voltage. For example, a 240 V heater should have a resistance reading of 240,000 Ω to ground. The heating element should be dried and retested if the reading is low.

4. Remove the megohmmeter and turn power ON when there is a high resistance reading.

5. Set the heating control circuit to turn ON the heating element.

6. Connect an ammeter between each power line terminal and the outer cover. Moisture is detected by an increase in normal electrical leakage from the element cover to ground. Leakage current should be less than 5 mA.

Constant Wattage Heating Element Drying

Any heating element that has an insulation resistance value of less than 250,000 Ω from the outer cover to ground should be dried before energizing. The preferred method of drying a heating element is to remove the element and bake it in an oven. The heating element should be baked at 250°F to 350°F for about 12 hours. Measure the insulation resistance after drying.

The low voltage drying method may be used when it is not practical to place a heating element in an oven. The low voltage method connects the heating element to 25% to 50% of the element's rated voltage. The low voltage is applied for 24 to 48 hours. The lower the resistance reading, the lower the applied voltage, and the longer the time period. Full voltage may be applied when the resistance reading equals 1000 times (or more) the rated voltage.

 Refer to **Activity 21-4—Troubleshooting Heating Elements**..

 Refer to Quick Quiz® on CD-ROM

Refer to Chapter 21 in the **Troubleshooting Electrical/Electronic Systems Workbook** *for additional questions.*

Name_____ Date _____

ACTIVITY 21-1—Solenoid–Operated Valves

1. Using the Glue Dispense Circuit, answer the questions.

_____ **A.** The valve has ___ positions.

_____ **B.** The valve is a ___-way valve.

_____ **C.** The valve is ___-actuated when glue is dispensed.

_____ **D.** The valve is ___-actuated when no glue is dispensed.

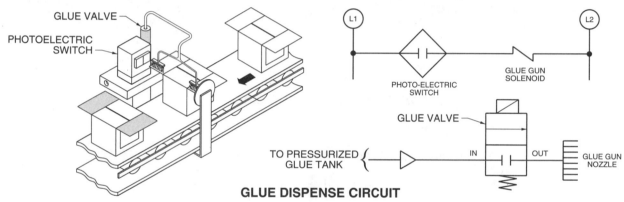

GLUE DISPENSE CIRCUIT

2. Using the Device Clamping Circuit, answer the questions.

_____ **A.** The valve has ___ positions.

_____ **B.** The valve is a ___-way valve.

_____ **C.** The valve is ___-actuated when the cylinder is advanced.

_____ **D.** The valve is ___-actuated when the cylinder is retracted.

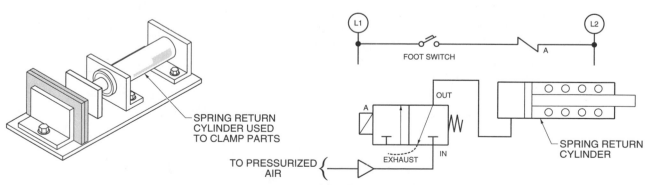

DEVICE CLAMPING CIRCUIT

3. Using the Conveyor Positioning Circuit, answer the questions.

_____ **A.** The valve has ___ positions.

_____ **B.** The valve is a ___-way valve.

_____ **C.** The valve is ___-actuated when the cylinder is advanced.

_____ **D.** The valve is ___-actuated when the cylinder is retracted.

CONVEYOR POSITIONING CIRCUIT

4. Using the Valve Control Circuit, answer the questions.

_____ **A.** The valve has ___ positions.

_____ **B.** The valve is a ___-way valve.

_____ **C.** The valve is ___-actuated when the cylinder is advanced.

_____ **D.** The valve is ___-actuated when the cylinder is retracted.

VALVE CONTROL CIRCUIT

5. In the Basic Glue Dispense Circuit, the glue gun dispenses glue anytime the photoelectric switch detects a box. The glue starts and stops on the edge of each box. This presents a maintenance problem because glue may be sprayed on the conveyor belt. An improved circuit is used to prevent the problem. Using the Improved Glue Dispense Circuit, answer the questions.

_____ **A.** The glue is dispensed ___ s after the photoelectric switch senses the edge of the box.

_____ **B.** The glue is dispensed for ___ s.

_____ **C.** When does the lamp turn ON?

_____ **D.** Does the Manual Test Button activate the glue gun even when the selector switch is in the OFF position?

BASIC GLUE DISPENSE CIRCUIT

IMPROVED GLUE DISPENSE CIRCUIT

6. In the Basic Device Clamping Circuit, the cylinder is extended anytime the Foot Switch is pressed. This may present a safety problem. Using the Improved Device Clamping Circuit, answer the questions.

_____ **A.** Does the system pressure have to be above or below 100 psi before the cylinder advances?

_____ **B.** How is the cylinder advanced after the pressure switch is closed?

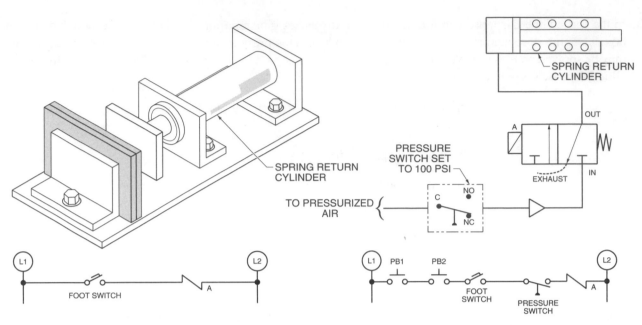

BASIC DEVICE CLAMPING CIRCUIT

IMPROVED DEVICE CLAMPING CIRCUIT

7. In the Basic Valve Control Circuit, the Product Flow Valve Cylinder is moved anytime a pushbutton is pressed. This may present a maintenance problem. Using the Basic and Improved Valve Control Circuit, answer the questions.

_____ **A.** In the basic circuit, can both solenoids be energized simultaneously?

_____ **B.** In the improved circuit, can both solenoids be energized simultaneously?

_____ **C.** In either circuit, can the cylinder be placed in a position other than fully-extended or fully-retracted?

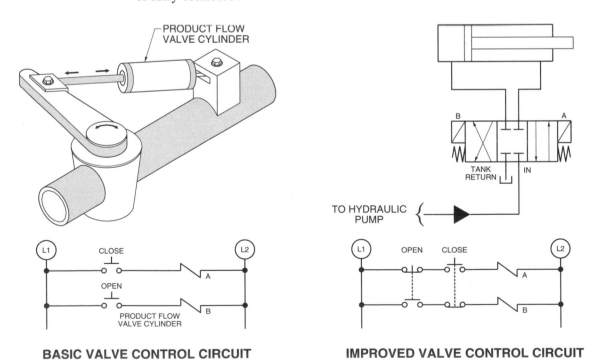

BASIC VALVE CONTROL CIRCUIT

IMPROVED VALVE CONTROL CIRCUIT

8. In the Basic Conveyor Positioning Circuit, the conveyor arm is moved anytime the selector switch is moved to a different position. This may present a safety problem. Using the Improved Conveyor Positioning Circuit, answer the questions.

_____ **A.** When does the alarm sound?

_____ **B.** The alarm sounds for ___ s.

_____ **C.** Does the gate change position as soon as the selector switch is moved to a different position?

_____ **D.** Is the gate moved before, during, or after the sounding of the alarm?

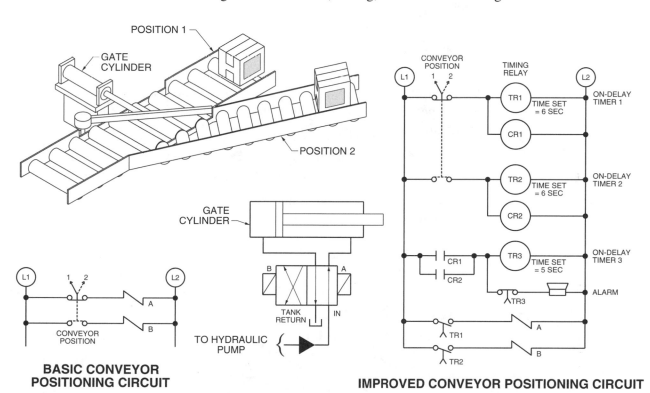

BASIC CONVEYOR POSITIONING CIRCUIT

IMPROVED CONVEYOR POSITIONING CIRCUIT

ACTIVITY 21-2—Troubleshooting Solenoid–Operated Valves

1. In the Glue Dispense Circuit, the timer is rated at .050 A, the solenoid at .60 A, and the lamp at .10 A. Using the Glue Dispense Circuit, answer the questions. *Note:* Assume all components in the circuit are good.

_____ **A.** Meter 1 reads ___ A when the Glue Gun Selector Switch is in the ON position and no box is in front of the photoelectric switch.

_____ **B.** Meter 2 in Test Position 1 reads ___ V when the Glue Gun Selector Switch is in the ON position and no box is in front of the photoelectric switch.

_____ **C.** Meter 2 in Test Position 2 reads ___ V when the Glue Gun Selector Switch is in the ON position and no box is in front of the photoelectric switch.

_____ **D.** Meter 2 in Test Position 3 reads ___ V when the Glue Gun Selector Switch is in the ON position and no box is in front of the photoelectric switch.

_____ **E.** Meter 3 reads ____ V when the Glue Gun Selector Switch is in the ON position and no box is in front of the photoelectric switch.

_____ **F.** Meter 4 reads ____ A when the Glue Gun Selector Switch is in the ON position and no box is in front of the photoelectric switch.

_____ **G.** Meter 1 reads ____ A when the Glue Gun Selector Switch is in the ON position, a box is in front of the photoelectric switch, and the glue gun is dispensing glue.

_____ **H.** Meter 2 in Test Position 1 reads ____ V when the Glue Gun Selector Switch is in the ON position, a box is in front of the photoelectric switch, and the glue gun is dispensing glue.

_____ **I.** Meter 2 in Test Position 2 reads ____ V when the Glue Gun Selector Switch is in the ON position, a box is in front of the photoelectric switch, and the glue gun is dispensing glue.

_____ **J.** Meter 2 in Test Position 3 reads ____ V when the Glue Gun Selector Switch is in the ON position, a box is in front of the photoelectric switch, and the glue gun is dispensing glue.

_____ **K.** Meter 3 reads ____ V when the Glue Gun Selector Switch is in the ON position, a box is in front of the photoelectric switch, and the glue gun is dispensing glue.

_____ **L.** Meter 4 reads ____ A when the Glue Gun Selector Switch is in the ON position, a box is in front of the photoelectric switch, and the glue gun is dispensing glue.

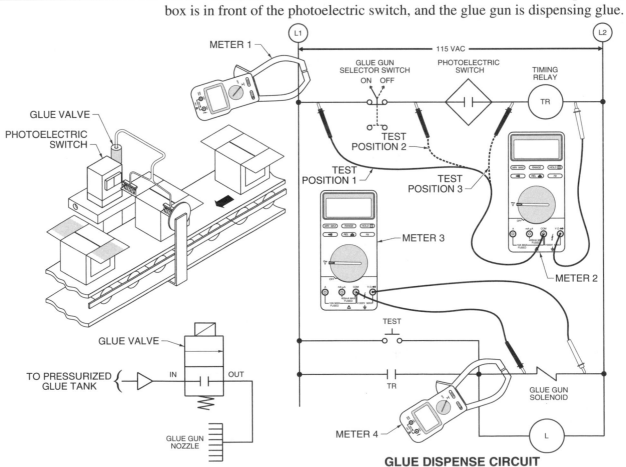

GLUE DISPENSE CIRCUIT

2. In the Device Clamping Circuit, the cylinder cannot be advanced unless the system pressure is at least 100 psi. This prevents the cylinder from advancing and not having enough pressure to hold the part in place. The problem may be in the pushbutton, foot switch, pressure switch, solenoid, electrical system, or fluid power system if the cylinder does not advance. To aid in troubleshooting a circuit, a push-to-test pushbutton is added to a circuit. Redraw the circuit adding a push-to-test pushbutton that overrides Pushbutton 1 and Pushbutton 2, the foot switch, and pressure switch. Add an alarm that sounds anytime the system pressure drops below 100 psi. *Note:* No additional components are required other than the push-to-test pushbutton and alarm. The components must be rearranged for the alarm to sound anytime pressure drops below 100 psi.

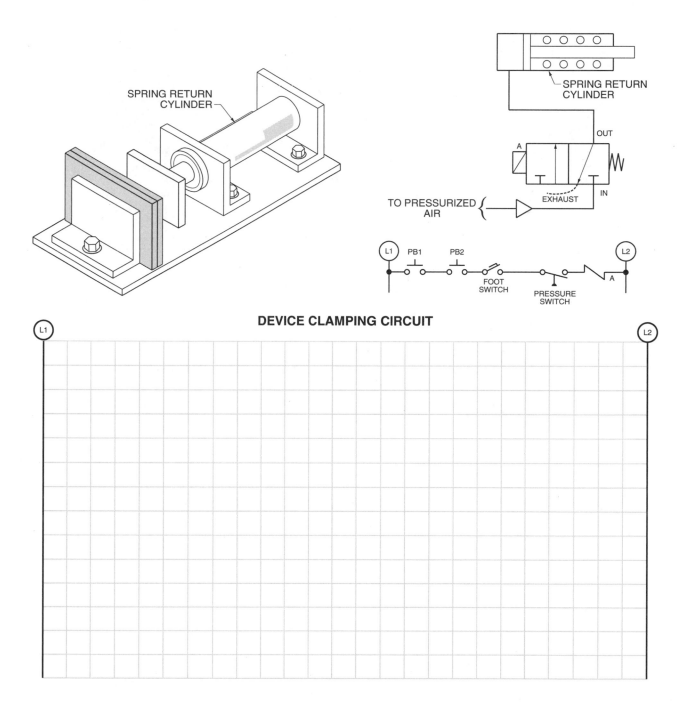

DEVICE CLAMPING CIRCUIT

3. In the Valve Control Circuit, the valve is opened or closed anytime a pushbutton is pressed. Redraw the circuit adding a selector switch that allows the valve to operate only when placed in the "valve operational" position. When the selector switch is placed in the OFF position, the valve cannot be moved. Add a red indicator lamp to the circuit to indicate the valve is operational and a green indicator lamp to indicate the valve is not operational.

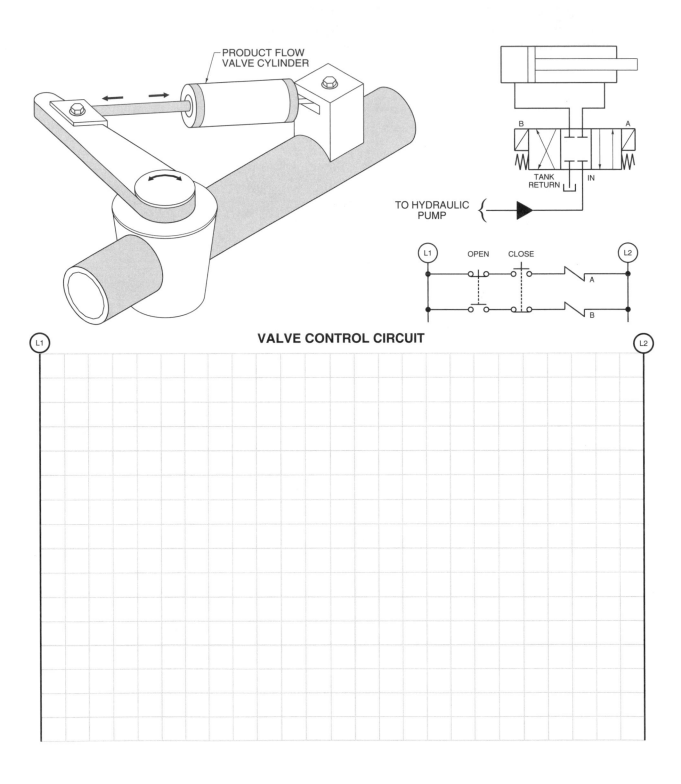

PRODUCT FLOW VALVE CYLINDER

TO HYDRAULIC PUMP

TANK RETURN   IN

B   A

OPEN   CLOSE

A

B

VALVE CONTROL CIRCUIT

ACTIVITY 21-3—Heating Elements

Power is consumed when an electric heating element produces heat. To calculate the amount of power required to heat air, apply the formula:

$$kW = \frac{cfm \times TR}{3000}$$

where

kW = power required to heat air (in kilowatts)

cfm = cubic feet per minute air flow
(measured at 70°F inlet air and 14 psi)

TR = temperature rise (in °F)

3000 = constant

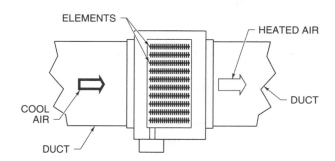

1. Complete Heating Air Power Requirements* by determining the power required to heat the air.

Air Flow Rate (in cfm)	Temperature Rise (°F)							
	50	100	150	200	250	300	350	400
100								
200								
300								
400								
500								
600								
700								
800								

HEATING AIR POWER REQUIREMENTS*

* in kilowatts

The power required to heat liquid in a tank depends on the amount of tank insulation. Uninsulated tanks require additional power due to heat loss. To calculate the amount of power required to heat water in a tank that is adequately insulated, apply the formula:

$$kW = \frac{gal. \times TR}{325 \times ht}$$

where

kW = power required to heat water (in kilowatts)

$gal.$ = amount of water (in gallons)

TR = temperature rise (in °F)

325 = constant

ht = heat-up time (in hours)

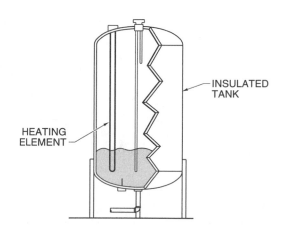

2. Complete Heating Water Power Requirements* by determining the amount of power required to heat the water. *Note:* Assume 1 hr heat-up time.

Liquid Amount (in gal.)	Temperature Rise (°F)				
	20	**40**	**80**	**100**	**140**
10					
20					
50					
100					
150					
200					
400					
500					

Table title: HEATING WATER POWER REQUIREMENTS*

* in kilowatts

To calculate the amount of current required to produce a given amount of power, apply the formulas:

For 1ϕ:

$$I = \frac{P}{E}$$

For 3ϕ:

$$I = \frac{P}{E \times \sqrt{3}}$$

where

I = current (in amperes)

P = power (in watts)

E = voltage (in volts)

THREE-PHASE VOLTAGE VALUES
For 208 V x 1.732, use 360
For 230 V x 1.732, use 398
For 240 V x 1.732, use 416
For 440 V x 1.732, use 762
For 460 V x 1.732, use 797
For 480 V x 1.732, use 831

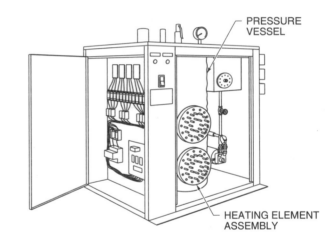

PRESSURE VESSEL

HEATING ELEMENT ASSEMBLY

3. Complete Power Heating Requirements* by determining the amount of current required to produce the given amount of power.

kW	1ϕ			3ϕ		
	120 V	**240 V**	**440 V**	**208 V**	**240 V**	**440 V**
1						
5						
10						
25						
50						
100						

Table title: POWER HEATING REQUIREMENTS*

* in amperes

ACTIVITY 21-4—Troubleshooting Heating Elements

Manufacturers specify a rated voltage and current for each heating element type. Heating elements are often used at voltages other than the rated voltage. When the voltage is not equal to the rated voltage, the amount of current draw from the power lines does not equal the rated current. A higher-than-rated voltage produces a higher-than-rated current and a higher-than-rated power output. A lower-than-rated voltage produces a lower-than-rated current and a lower-than-rated power output.

   To calculate power, apply the formulas:

For 1ϕ:	For 3ϕ:
$P = I \times E$	$P = I \times E \times \sqrt{3}$

1. Using the meter readings, determine the power.

_____ **A.** Power equals ___ W.

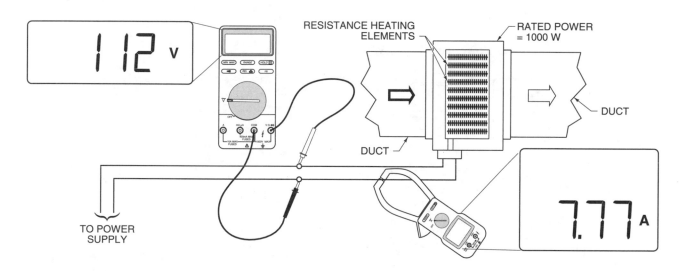

_____ **B.** Power equals ___ W.

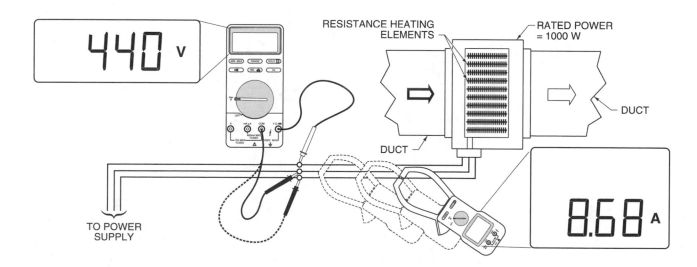

_____ **C.** Power equals ___ W.

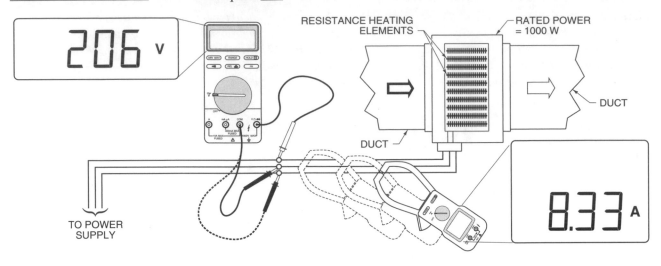

_____ **D.** Power equals ___ W.

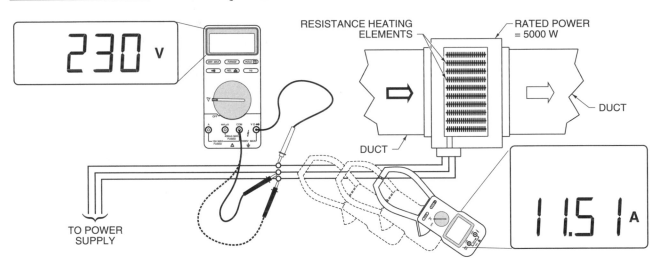

_____ **E.** Power equals ___ W.

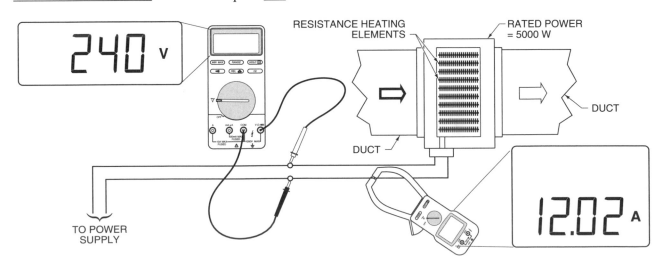

APPLICATION 22-1 — Troubleshooting Mechanical Switches

Mechanical and solid-state switches are used to switch ON and OFF the flow of electricity in a circuit. A *mechanical switch* is any switch that uses silver contacts to start and stop the flow of current in a circuit. Mechanical switches may include normally-open, normally-closed, or combination switching contacts. Mechanical switches may be manually, mechanically, or automatically operated.

Manually-operated switches are used when a circuit is controlled by a person. Manually-operated switches include pushbutton, foot switches, toggle switches, and keyboards.

Mechanically-operated switches are used when a circuit is controlled by the movement of an object. A limit switch is a mechanically-operated switch.

Automatically-operated switches are used when a circuit is controlled by the change in a given condition. Automatically-operated switches include flow, level, pressure, temperature, humidity, and gas-operated switches. **See Mechanical Switches.**

A suspected fault with a mechanical switch is tested using a voltmeter. The voltmeter is used to test the voltage into and out of the switch. **See Mechanical Switch Testing.**

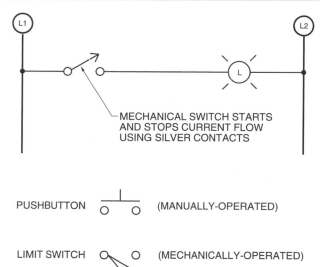

MECHANICAL SWITCH STARTS AND STOPS CURRENT FLOW USING SILVER CONTACTS

PUSHBUTTON (MANUALLY-OPERATED)

LIMIT SWITCH (MECHANICALLY-OPERATED)

LEVEL SWITCH (AUTOMATICALLY-OPERATED)

MECHANICAL SWITCHES

To test a mechanical switch, apply the procedure:

1. Measure the voltage into the switch. Connect the voltmeter between the neutral and hot conductor feeding the switch. The voltmeter lead may be connected to ground instead of neutral if the neutral conductor is not available in the same box in which the switch is located. The problem is located upstream from the switch when there is no voltage present or the voltage is not at the correct level. The problem may be a blown fuse or open circuit. Voltage must be reestablished to the switch before the switch may be tested.

2. Measure the voltage out of the switch. There should be a voltage reading when the switch contacts are closed. There should not be a voltage reading when the switch contacts are open. The switch has an open and must be replaced if there is no voltage reading in either switch position. The switch has a short and must be replaced if there is a voltage reading in both switch positions.

Warning: Always ensure power is OFF before changing a control switch. Use a voltmeter to ensure the power is OFF.

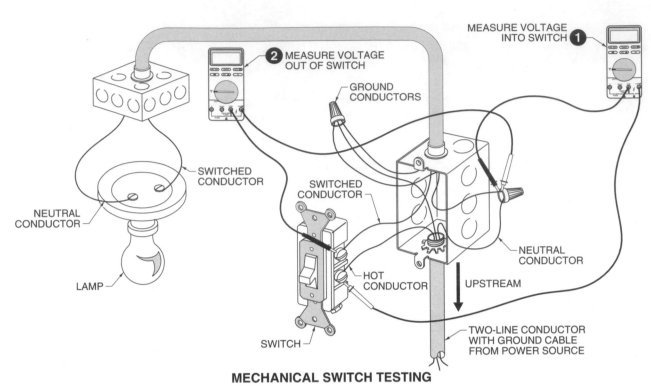

MECHANICAL SWITCH TESTING

 Refer to **Activity 22-1 — Troubleshooting Mechanical Switches** ....

APPLICATION 22-2 — Troubleshooting Solid-State Switches

A solid-state switch has no moving parts (contacts). A solid-state switch uses a triac, SCR, current sink (NPN) transistor, or current source (PNP) transistor output to perform the switching function. The triac output is used for switching AC loads. The SCR output is used for switching high-power DC loads. The current sink and current source outputs are used for switching low-power DC loads. Solid-state switches include normally-open, normally-closed, or combination switching outputs. **See Solid-State Switches.**

Two-Wire Solid-State Switches

A *two-wire solid-state switch* has two connecting terminals or wires (exclusive of ground). A two-wire switch is connected in series with the controlled load. A two-wire solid-state switch is also called a load-powered switch because it draws operating current through the load. The operating current flows through the load when the switch is not conducting (load OFF). This operating current is inadequate to energize most loads. Operating current is also called residual current or leakage current by some manufacturers.

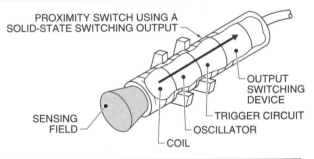

OUTPUT SWITCHING DEVICES	
Device	**Use**
TRIAC	SWITCH AC LOADS
SCR	SWITCH HIGH-POWER DC LOADS
NPN TRANSISTOR PNP TRANSISTOR	SWITCH LOW-POWER DC LOADS

SOLID-STATE SWITCHES

Operating current may be measured with an ammeter when the load is OFF. **See Operating Current** and **Load Current.**

The current in a circuit is a combination of the operating current and load current when a switch is conducting (load ON). A solid-state switching device must be rated high enough to carry the current of the load. Load current is measured with an ammeter when the load is ON.

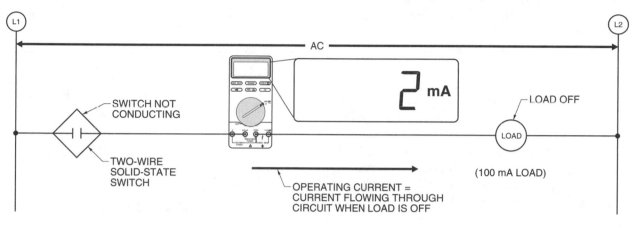

OPERATING CURRENT

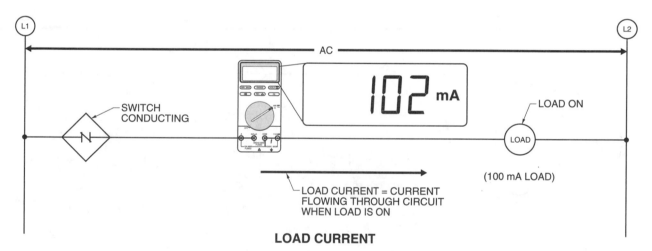

LOAD CURRENT

The current draw of a load must be sufficient to keep the solid-state switch operating when the switch is conducting (load ON). *Minimum holding current* is the minimum current that ensures proper operation of a solid-state switch. Minimum holding current values range from 2 mA to 20 mA.

Operating current and minimum holding current values are usually not a problem when a solid-state switch controls low-impedance loads, such as motor starters, relays, and solenoids. Operating current and minimum holding current values may be a problem when a switch controls high-impedance loads, such as programmable controllers and other solid-state devices. The operating current may be high enough to affect the load when the switch is not conducting. For example, a programmable controller may see the operating current as an input signal.

A load resistor must be added to the circuit to correct this problem. A load resistor is connected in parallel with the load. The load resistor acts as an additional load which increases the total current in the circuit.

Load resistors range in value from 4.5 kΩ to 7 kΩ. A 5 kΩ, 5 W resistor is used in most applications. **See Load Resistor.**

Two-wire solid-state switches connected in series affect the operation of the load because of the voltage drop across the switches. A two-wire switch drops about 3 V to 8 V. The total voltage drop across the switches equals the sum of the voltage drop across each switch. No more than three solid-state switches should be connected in series. **See Two-Wire Switches.**

Two-wire solid-state switches connected in parallel affect the operation of the load because each switch has its operating current flowing through the load. The load may turn ON if the current through the load becomes excessive. The total operating current equals the sum of the operating current of each switch. No more than three solid-state switches should be connected in parallel.

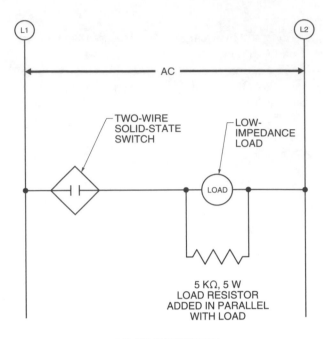

LOAD RESISTOR

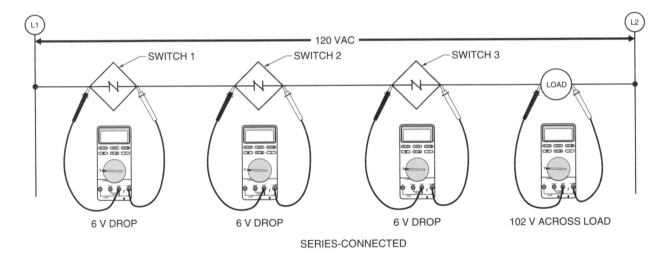

SERIES-CONNECTED

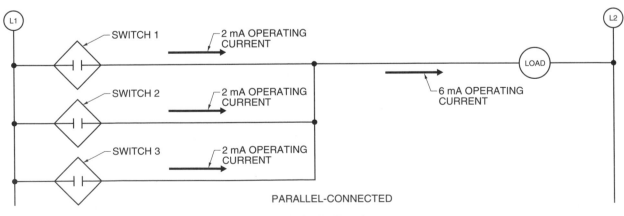

PARALLEL-CONNECTED

TWO-WIRE SWITCHES

A suspected fault in a two-wire solid-state switch may be tested using a voltmeter. The voltmeter is used to test the voltage into and out of the switch. **See Two-Wire Solid-State Switch Testing.**

To test a two-wire solid-state switch, apply the procedure:
1. Measure the supply voltage into the switch. The problem is located upstream from the switch when there is no voltage present or the voltage is not at the correct level. The problem may be a blown fuse or open circuit. Voltage must be reestablished to the switch before the switch may be tested.
2. Measure the voltage out of the switch. The voltage should equal the supply voltage minus the rated voltage drop (3 V to 8 V) of the switch when the switch is conducting (load ON). Replace the switch if the voltage output is not correct.

Warning: Always ensure power is OFF before changing a control switch. Use a voltmeter to ensure the power is OFF.

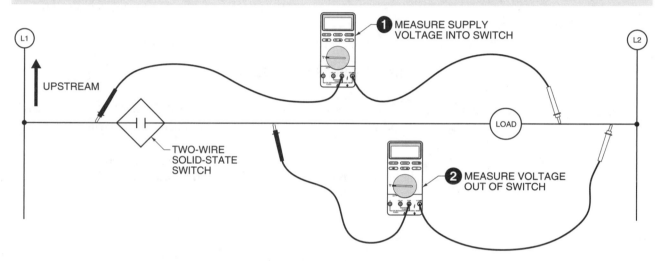

TWO-WIRE SOLID-STATE SWITCH TESTING

Three-Wire Solid-State Switches

A *three-wire solid-state switch* has three connecting terminals or wires (exclusive of ground). A three-wire solid-state switch is connected in series with the control load and also connected to the power lines. A three-wire solid-state switch draws operating current directly from the power lines. The operating current does not flow through the load.

A three-wire solid-state switch is also called a line-powered switch because it draws operating current from the power lines. The two types of three-wire solid-state switches are the current source (PNP) switch and the current sink (NPN) switch. **See PNP Switch** and **NPN Switch.**

Three-wire solid-state switches connected in series affect the operation of the load because each switch downstream from the last switch must carry the load current and the operating current of each switch. An ammeter may be used to measure operating and load current values. The measured values must not exceed the manufacturer's maximum rating. **See Three-Wire Switches.**

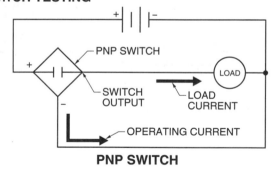

PNP SWITCH

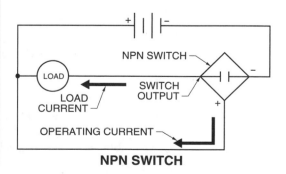

NPN SWITCH

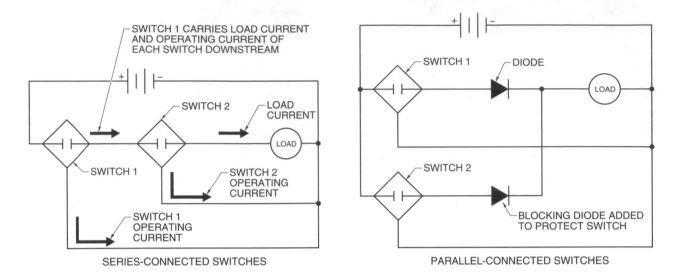

SWITCH 1 CARRIES LOAD CURRENT
AND OPERATING CURRENT OF
EACH SWITCH DOWNSTREAM

SWITCH 2

LOAD CURRENT

LOAD

SWITCH 1

SWITCH 2 OPERATING CURRENT

SWITCH 1 OPERATING CURRENT

SERIES-CONNECTED SWITCHES

SWITCH 1   DIODE

LOAD

SWITCH 2

BLOCKING DIODE ADDED TO PROTECT SWITCH

PARALLEL-CONNECTED SWITCHES

THREE-WIRE SWITCHES

Three-wire solid-state switches connected in parallel affect the operation of the load because the non-conducting switch may be damaged due to reverse polarity. A blocking diode should be added to each switch output to prevent reverse polarity on the switch.

A suspected fault with a three-wire solid-state switch may be tested using a voltmeter. The voltmeter is used to test the voltage into and out of the switch. **See Three-Wire Solid-State Switch Testing.**

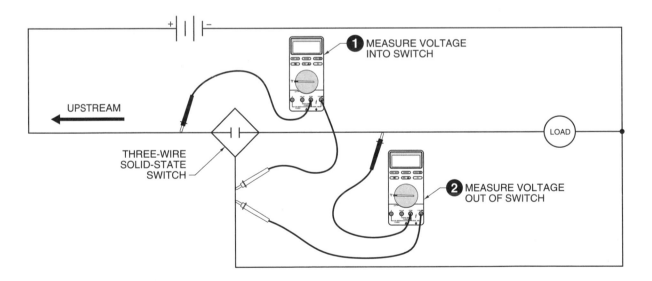

1 MEASURE VOLTAGE INTO SWITCH

UPSTREAM

THREE-WIRE SOLID-STATE SWITCH

2 MEASURE VOLTAGE OUT OF SWITCH

LOAD

THREE-WIRE SOLID-STATE SWITCH TESTING

To test a three-wire solid-state switch, apply the procedure:
1. Measure the voltage into the switch. The problem is located upstream from the switch when there is no voltage present or the voltage is not at the correct level. The problem may be a blown fuse or open circuit. Voltage must be reestablished to the switch before the switch may be tested.

2. Measure the voltage out of the switch. The voltage should be equal to the supply voltage when the switch is conducting (load ON). Replace the switch if the voltage out of the switch is not correct.

Warning: Always ensure power is OFF before changing a control switch. Use a voltmeter to ensure the power is OFF.

 Refer to **Activity 22-2—Troubleshooting Solid-State Switches** ....

APPLICATION 22-3—Protecting Switch Contacts

Switches are rated for the amount of current they may switch. The switch rating is usually specified for switching a resistive load. Resistive loads are the least destructive loads to switch. Most loads that are switched are inductive loads, such as solenoids, relays, and motors. Inductive loads are the most destructive loads to switch.

A large induced voltage appears across the switch contacts when inductive loads are turned OFF. The induced voltage causes arcing at the switch contacts. Arcing may cause the contacts to burn, stick, or weld together. Contact protection should be added when frequently switching inductive loads to prevent or reduce arcing. **See Contact Protection.**

A diode is added in parallel with the load to protect contacts that switch DC. The diode does not conduct when the load is ON. The diode conducts when the switch is open, providing a path for the induced voltage in the load to dissipate.

A resistor and capacitor are connected across the switch contacts to protect contacts that switch AC. The capacitor acts as a high impedance (resistor) load at 60 Hz, but becomes a short circuit at the high frequencies produced by the induced voltage of the load. This allows the induced voltage to dissipate across the resistor when the load is switched OFF.

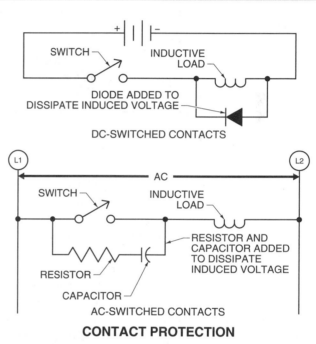

CONTACT PROTECTION

 Refer to **Activity 22-3—Protecting Switch Contacts**...

APPLICATION 22-4—Proximity Sensor Installation

Proximity sensors have a sensing head that produces a radiated sensing field. The sensing field detects the target of the sensor. The sensing field must be kept clear of interference for proper operation. *Interference* is any object other than the object to be detected that is sensed by the sensor. Interference may come from objects close to the sensor or from other sensors. General clearances are required for most proximity sensors.

Flush-Mounted Inductive and Capacitive Proximity Sensors

A distance equal to or greater than twice the diameter of the sensors is required between sensors when flush-mounting inductive and capacitive proximity sensors. The diameter of the largest sensor is used for installation when two sensors of different diameters are used. For example, at least 16 mm is required between sensors if two 8 mm inductive proximity sensors are flush-mounted. **See Flush-Mounted Proximity Sensor.**

Non-Flush-Mounted Inductive and Capacitive Proximity Sensors

A distance of three times the diameter of the sensor is required within or next to a material that may be detected when using non-flush-mounting inductive and capacitive proximity sensors. For example, at least 48 mm is required between sensors if two 16 mm capacitive proximity sensors are non-flush-mounted. **See Non-Flush-Mounted Proximity Sensor.**

Three times the diameter of the largest sensor is required when inductive and capacitive proximity sensors are installed next to each other. Spacing is measured from center to center of the sensors.

Six times the rated sensing distance is required for proper operation when inductive and capacitive proximity sensors are mounted opposite each other. Six times the rated sensing distance is required because the sensing field causes false readings on the other.

Mounting Photoelectric Sensors

A photoelectric sensor transmits a light beam. The light beam detects the presence (or absence) of an object. Only part of the light beam is effective when detecting the object. The *effective light beam* is the area of light that travels directly from the transmitter to the receiver. The object is not detected if the object does not completely block the effective light beam.

The receiver is positioned to receive as much light as possible from the transmitter when mounting photoelectric sensors. Greater operating distances are allowed and more power is available for the system to see through dirt in the air and on the transmitter and receiver lenses because more light is available at the receiver. The transmitter is mounted on the clean side of the detection zone because light scattered by dirt on the receiver lens affects the system less than light scattered by dirt on the transmitter lens. **See Correct Sensor Mounting** and **Incorrect Sensor Mounting.**

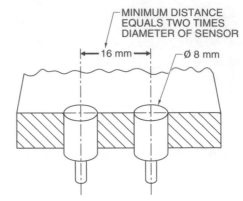

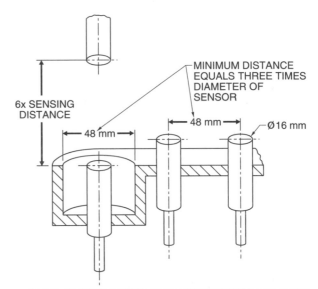

FLUSH-MOUNTED PROXIMITY SENSOR

NON-FLUSH-MOUNTED PROXIMITY SENSOR

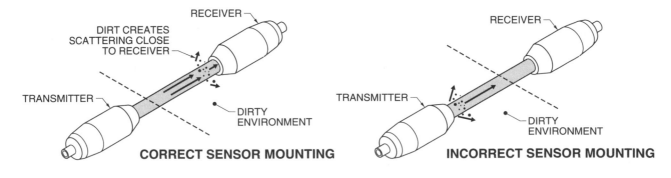

CORRECT SENSOR MOUNTING

INCORRECT SENSOR MOUNTING

 Refer to **Activity 22-4—Proximity Sensor Installation** ....

APPLICATION 22-5—Protecting Pressure Switches

A *pressure switch* is a switch that is designed to activate its contacts at a preset pressure. A pressure switch is rated according to its operating pressure range. A pressure switch may be damaged if its maximum pressure limit is exceeded. Protection for a pressure switch should be added in any system in which a higher pressure than the maximum limit is possible. A pressure relief valve is installed to protect the pressure switch. A pressure relief valve should be set just below the pressure switch's maximum limit. The valve opens when the system pressure increases to the setting of the relief valve.

Caution: The output of the relief valve must be connected to a proper drain (or return line) if the product under pressure is a gas or a fluid. **See Pressure Switch Protection.**

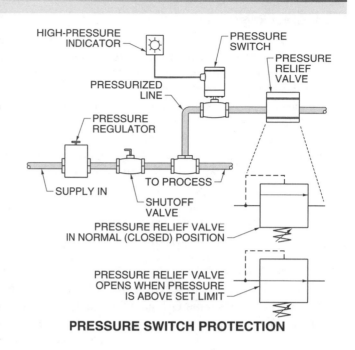

PRESSURE SWITCH PROTECTION

 Refer to **Activity 22-5—Protecting Pressure Switches**...

APPLICATION 22-6—Flow Switches

A *flow switch* is a switch designed to detect the movement (flow) of a product through a pipe or duct. Most flow switches use a paddle to detect the movement of the product. The paddle is designed to detect the product movement with the least possible pressure drop across the switch. A flow switch must be installed correctly to ensure it does not interfere with the movement of the product. Most flow switches are designed to operate in the vertical position. Allow at least three pipe diameters (ID) on each side of the flow switch when mounting a flow switch. **See Flow Switch Mounting.**

For example, an application moves a product through a 1½″ diameter pipe. The minimum horizontal distance of straight pipe required on each side of the flow switch is 4½″ (1½″ × 3 = 4½″).

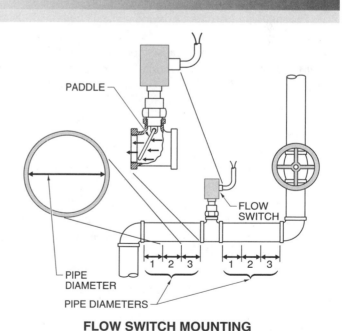

FLOW SWITCH MOUNTING

 Refer to **Activity 22-6—Flow Switches**..

APPLICATION 22-7—Level Control Problems

One-Level Registration

A *level switch* is a proximity sensor used to detect the presence of a liquid or solid in a tank. In level control applications, the distance between the maintained maximum and minimum levels must be determined. The distance may be small or large. Maintenance problems develop if the distance is not correct for the application.

One-level registration uses one sensor to determine the product level between the ON and OFF points of an operation. A small distance is maintained when using one-level registration. In one-level registration, the discharge pump motor starts when the sensor comes in contact with the product. The discharge pump motor stops when the sensor is no longer in contact with the product. **See One-Level Registration.**

Two-Level Registration

Two-level registration uses two sensors to determine the product level between the ON and OFF points of an operation. The distance may vary to almost any length when using two-level registration. In two-level registration, the discharge pump motor starts when the product comes in contact with the top sensor. The pump motor stays ON even when the top sensor is no longer in contact with the product. The pump motor stops when the bottom sensor is no longer in contact with the product. The discharge pump motor does not restart until the product comes in contact with the top sensor. **See Two-Level Registration.**

The smaller the distance to be maintained, the more times the pump motor must cycle ON and OFF. The motor heats faster when started frequently because motors draw high inrush current when starting. Consideration must be given to the speed of the product flowing from the tank. The faster the level in the tank drops, the faster the motor must cycle. A motor cycles faster if the product in the tank is turbulent. A pump does not have to cycle as often if two-level registration is used.

The type of product must be taken into consideration in determining the distance and time between the maximum and minimum level. For example, problems may develop with two-level registration if a common house paint is maintained in a fill tank. Problems develop because the paint accumulates as it is allowed to dry on the side of the tank. This may clog or impede the discharge or fill action.

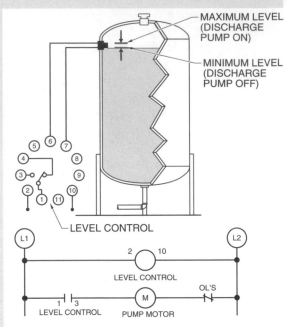

ONE-LEVEL REGISTRATION

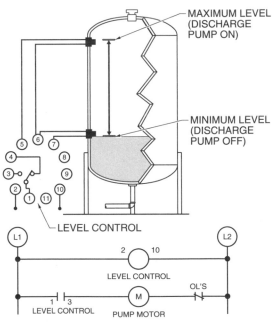

TWO-LEVEL REGISTRATION

 Refer to **Activity 22-7—Level Control Problems** ....

 Refer to Quick Quiz® *on CD-ROM*

 *Refer to Chapter 22 in the* **Troubleshooting Electrical/Electronic Systems Workbook** *for additional questions.*

Name_____ Date _____

ACTIVITY 22-1—Troubleshooting Mechanical Switches

1. Voltmeters are added to Light Circuit 1. Determine the meter readings for the given conditions.

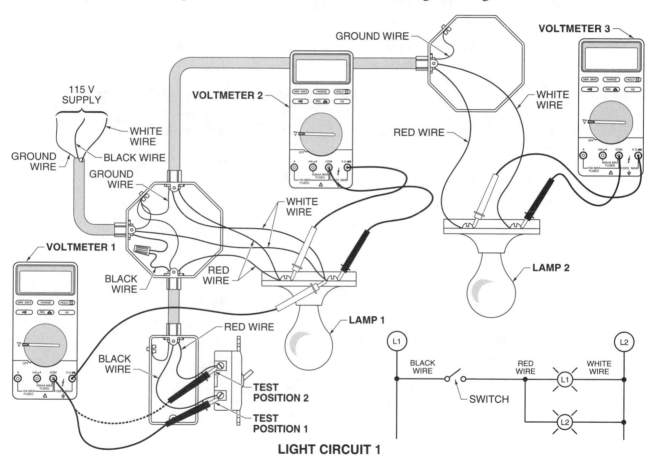

LIGHT CIRCUIT 1

_____ **A.** Voltmeter 1 in Test Position 1 reads ___ V when the switch is in the OFF position and the circuit is operating properly.

_____ **B.** Voltmeter 1 in Test Position 2 reads ___ V when the switch is in the OFF position and the circuit is operating properly.

_____ **C.** Voltmeter 1 in Test Position 1 reads ___ V when the switch is in the ON position and the circuit is operating properly.

_____ **D.** Voltmeter 1 in Test Position 2 reads ___ V when the switch is in the ON position and the circuit is operating properly.

_____ **E.** Voltmeter 1 in Test Position 1 reads ___ V when the switch is in the ON position and the switch has an open.

_____ **F.** Voltmeter 1 in Test Position 2 reads ___ V when the switch is in the ON position and the switch has an open.

_____ **G.** Voltmeter 1 in Test Position 1 reads ___ V when the switch is in the OFF position and the switch has a short.

_____ **H.** Voltmeter 1 in Test Position 2 reads ___ V when the switch is in the OFF position and the switch has a short.

_____ **I.** Voltmeter 2 reads ___ V when the switch is in the ON position and the circuit is operating properly.

_____ **J.** Voltmeter 3 reads ___ V when the switch is in the ON position and the circuit is operating properly.

2. Voltmeters are added to Light Circuit 2. Determine the meter readings for the given conditions.

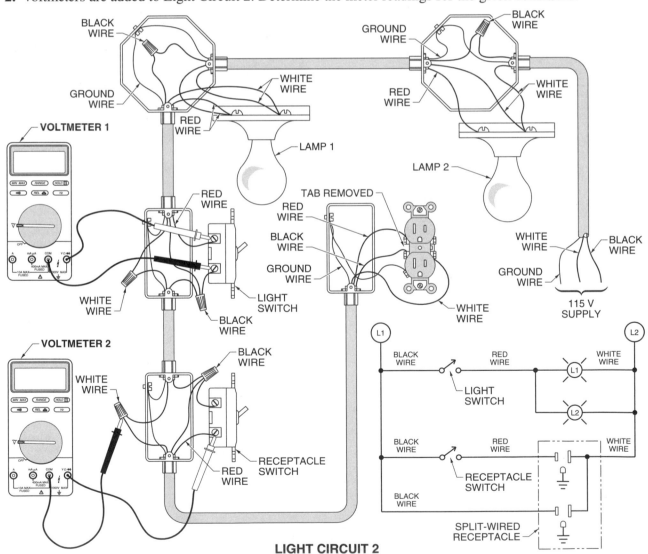

LIGHT CIRCUIT 2

_____ **A.** Voltmeter 1 reads ___ V when the Light Switch is in the OFF position and the circuit is operating properly.

_____ **B.** Voltmeter 1 reads ___ V when the Light Switch is in the ON position and the circuit is operating properly.

_____ **C.** Voltmeter 2 reads ___ V when the Receptacle Switch is in the OFF position and the circuit is operating properly.

_____ **D.** Voltmeter 2 reads ___ V when the Receptacle Switch is in the ON position and the circuit is operating properly.

_____ **E.** Voltmeter 1 reads ___ V when the Light Switch is in the OFF position and one lamp is burned out (open).

_____ **F.** Voltmeter 1 reads ___ V when the Light Switch is in the OFF position and both lamps are burned out (open).

_____ **G.** Voltmeter 1 reads ___ V when the Light Switch is in the ON position and one lamp is burned out (open).

_____ **H.** Voltmeter 1 reads ___ V when the Light Switch is in the ON position and both lamps are burned out (open).

ACTIVITY 22-2—Troubleshooting Solid-State Switches

1. Connect Voltmeter 1 to measure the voltage into the switch, Voltmeter 2 to measure the voltage out of the switch, and add a load resistor to the circuit.

_____ **2.** A(n) ___-powered switch is controlling the load in Test Circuit 1.

_____ **3.** A(n) ___ is used as the switching device in Test Circuit 1.

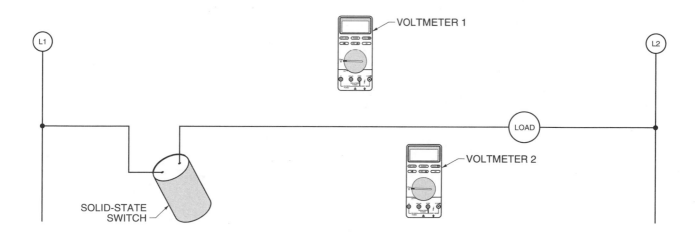

TEST CIRCUIT 1

4. Connect Voltmeter 1 to measure the voltage into the switch and Voltmeter 2 to measure the voltage out of the switch.

_____ **5.** A(n) ____-powered switch is controlling the load in Test Circuit 2.

_____ **6.** A(n) ____ is used as the switching device in Test Circuit 2.

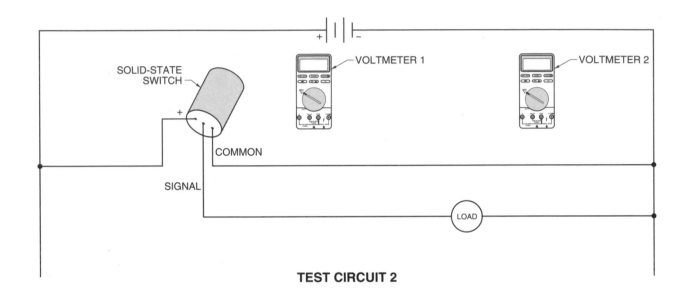

TEST CIRCUIT 2

7. Connect Voltmeter 1 to measure the voltage into the switch and Voltmeter 2 to measure the voltage out of the switch.

_____ **8.** A(n) ____-powered switch is controlling the load in Test Circuit 3.

_____ **9.** A(n) ____ is used as the switching device in Test Circuit 3.

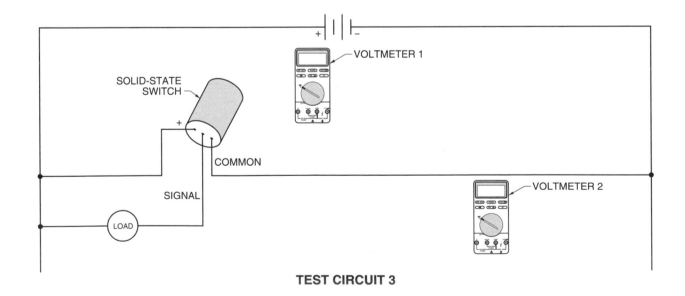

TEST CIRCUIT 3

ACTIVITY 22-3—Protecting Switch Contacts

1. Add a diode to the Automobile Starter Motor Circuit so that the induced voltage of the starter relay is dissipated to protect the electronic computer module circuits.

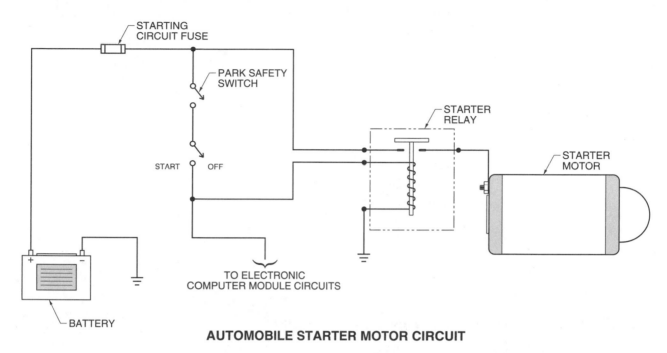

AUTOMOBILE STARTER MOTOR CIRCUIT

2. Add contact protection to the Air Conditioning Circuit to protect the thermostat contacts from the induced voltage of the evaporator blower relay and compressor motor relay. Add contact protection to protect the evaporator blower relay contacts and the compressor relay contacts.

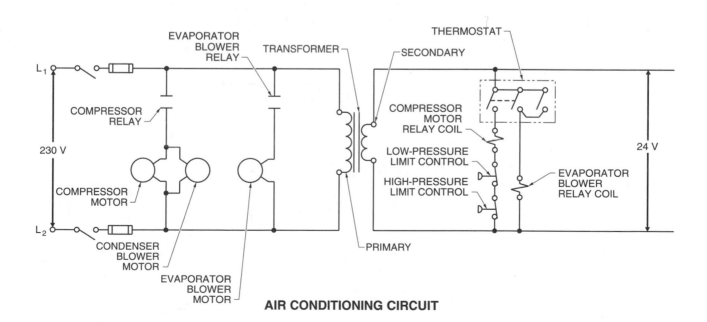

AIR CONDITIONING CIRCUIT

ACTIVITY 22-4—Proximity Sensor Installation

_____ **1.** The minimum distance required between the sensors for proper operation is ___ mm.

_____ **2.** The minimum distance required between the sensor and the surrounding material for proper operation is ___ mm.

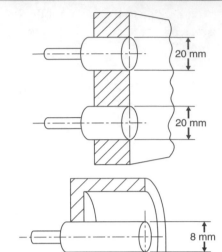

ACTIVITY 22-5—Protecting Pressure Switches

1. Add protection to the Fluid Power Circuit to protect the pressure switch from excessive fluid pressure. Add protection to the pressure switch contacts to protect them from induced voltage from the solenoid.

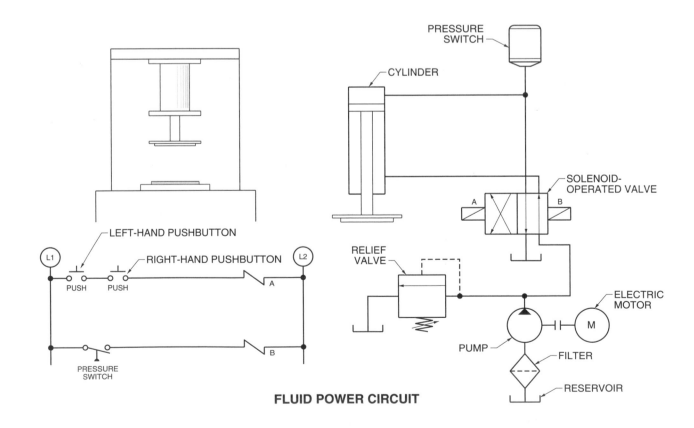

FLUID POWER CIRCUIT

ACTIVITY 22-6—Flow Switches

Determine the minimum length of straight pipe required on each side of the flow switch.

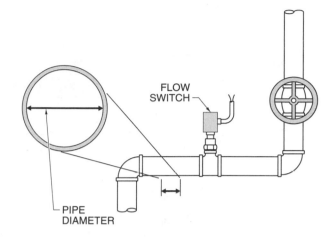

_____ 1. A ¼″ diameter pipe requires ___″ of straight pipe on either side of the flow switch.

_____ 2. A ½″ diameter pipe requires ___″ of straight pipe on either side of the flow switch.

_____ 3. A ¾″ diameter pipe requires ___″ of straight pipe on either side of the flow switch.

_____ 4. A 1″ diameter pipe requires ___″ of straight pipe on either side of the flow switch.

_____ 5. A 1¼″ diameter pipe requires ___″ of straight pipe on either side of the flow switch.

_____ 6. A 2″ diameter pipe requires ___″ of straight pipe on either side of the flow switch.

_____ 7. A 3″ diameter pipe requires ___″ of straight pipe on either side of the flow switch.

ACTIVITY 22-7—Level Control Problems

A level switch monitors the level of hot water in an automatic coffee brewer. The level switch sends a signal to the solenoid valve which opens to allow water to refill the reservoir.

_____ 1. The automatic coffee brewer uses ___-level registration.

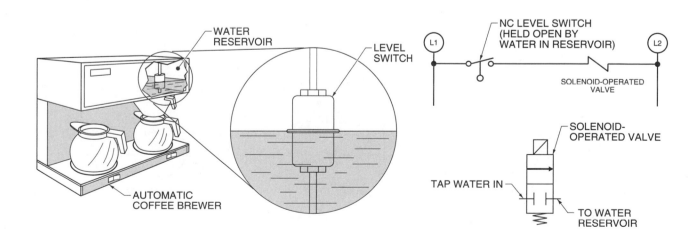

Two level switches maintain the product level in a tank. An additional level switch is used to sound an overfill alarm if the product rises too high.

_____ **2.** The tank uses ___-level registration.

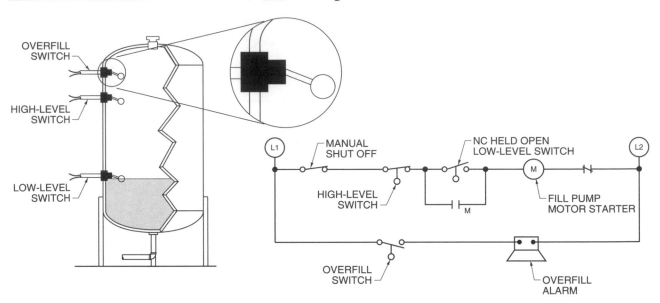

A high-level switch and low-level switch control the product level in a tank. The high-level switch sounds an alarm if the product rises too high. The low-level switch sounds an alarm if the product falls too low.

_____ **3.** The tank uses ___-level registration circuits.

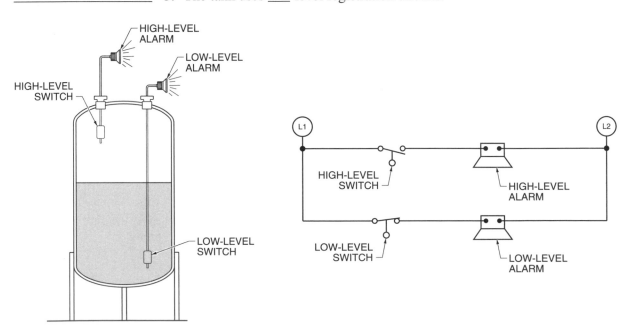

APPLICATION 23-1 — Diodes

A *diode* is an electronic device that allows current to flow in only one direction. A diode has an anode and a cathode. The *anode* is the positive lead. The *cathode* is the negative lead. **See Diode.**

A diode allows current to flow only when a positive voltage is applied to the anode (forward bias). The diode does not allow current to flow when the anode and cathode have the same polarity or if a positive voltage is applied to the cathode (reverse bias).

A diode is rated for the maximum forward current it can safely conduct and the maximum reverse voltage that can be applied. The maximum forward current rating applies to the amount of current the diode can withstand while conducting electrons in the forward direction. Exceeding the maximum forward current rating causes the diode to overheat and may cause the diode to fail. The maximum reverse voltage rating applies to the amount of voltage that may be applied in the reverse direction without allowing the diode to conduct (voltage breakover) in the opposite direction. **See Diode Current and Voltage.**

SCHEMATIC

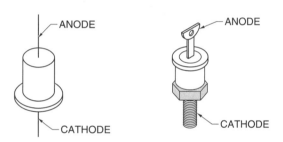

DIODE

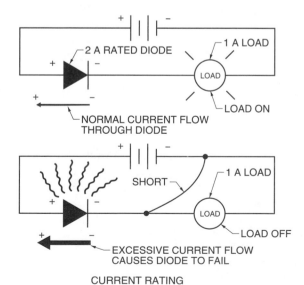

CURRENT RATING

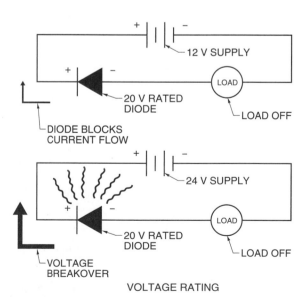

VOLTAGE RATING

DIODE CURRENT AND VOLTAGE

 Refer to **Activity 23-1 — Diodes**..

APPLICATION 23-2—Transistors

A *transistor* is a three-element device made of semiconductor material. A transistor is used in circuits as either a switch or an amplifier. Most transistors are used as current control devices. The three elements of a transistor are the emitter (E), base (B), and collector (C). The lead with the arrow is the emitter. A transistor is an NPN transistor if the arrow points out on the emitter. A transistor is a PNP transistor if the arrow points in on the emitter. **See Transistor Symbols.**

An *NPN transistor* has a thin layer of P-type material placed between two pieces of N-type material. N-type material carries current in the form of electrons because the material is manufactured to have extra electrons. P-type material carries current in the form of positive charges (holes) because the material is manufactured to have a deficiency of electrons. A *PNP transistor* has a thin layer of N-type material placed between two pieces of P-type material.

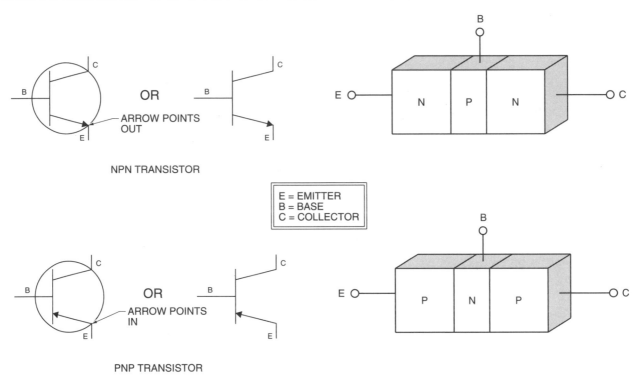

TRANSISTOR SYMBOLS

In a transistor, a very small current applied to one lead controls a large current flowing through the other two leads. A transistor may be used to start or stop current flow (switch) or to increase current flow (amplify). Any transistor may be used as a switch or an amplifier. Some transistors are designed to operate better as a switch while others are designed to operate better as an amplifier.

Transistor as a Switch

A transistor may be used as a solid-state switch in DC circuits. A transistor does not allow current to flow between the emitter and collector until a current is applied to the base of the transistor. Current flows through the emitter and collector when current is applied to the base of the transistor. A very low voltage may be used in the base circuit. A high voltage may be used in the emitter/collector circuit. For example, a 1.5 VDC base voltage can switch a 60 VDC emitter/collector circuit. **See Transistor Circuit Current Flow.**

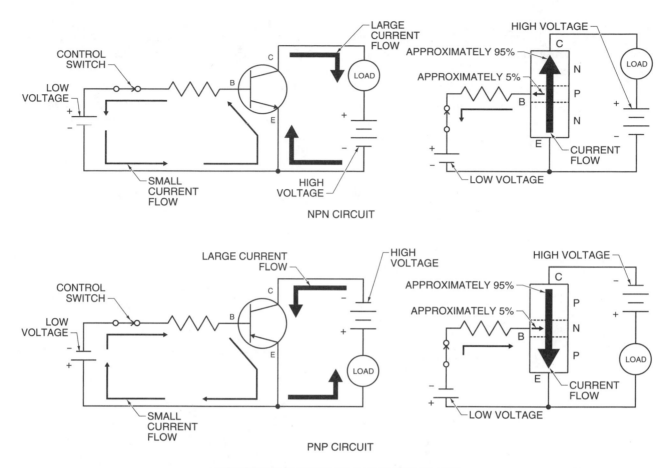

TRANSISTOR CIRCUIT CURRENT FLOW

Current flows in both the emitter/base circuit and the emitter/collector circuit when the transistor is switched ON. The emitter/base current is very low (about 5%) compared to the emitter/collector current. A transistor allows a small control current to control a high load current.

A transistor acts as a relay when used as a switch. A relay allows a small coil voltage and current to control a high load voltage and current. For example, a pushbutton on a beverage gun may be used to apply a low current to the base of a transistor. **See Transistor Switch Application.**

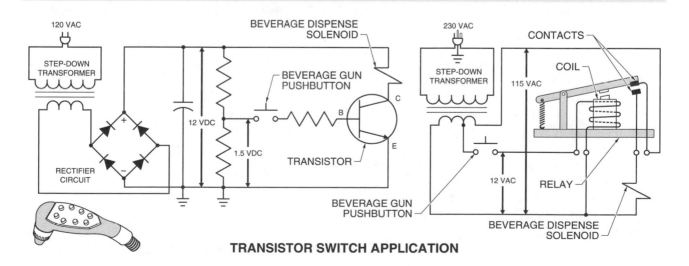

TRANSISTOR SWITCH APPLICATION

Current flows through the emitter/collector circuit when current is applied to the base of the transistor. The beverage-dispense solenoid is energized when current flows through the emitter/collector circuit. The solenoid is energized as long as current flows through the base circuit (pushbutton pressed).

A transistor used as a switch allows a very small input signal to control a high output. In the beverage gun application, the output current may be 100 times higher (or more) than the input current. In this application, the transistor has a current gain of 100. *Current gain* of a transistor (beta) is a measure of the ratio of the collector current to the base current. **See Transistor Gain.**

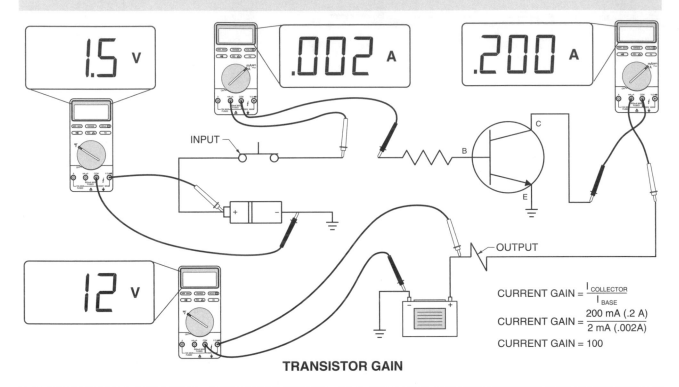

$$\text{CURRENT GAIN} = \frac{I_{COLLECTOR}}{I_{BASE}}$$

$$\text{CURRENT GAIN} = \frac{200 \text{ mA (.2 A)}}{2 \text{ mA (.002A)}}$$

$$\text{CURRENT GAIN} = 100$$

TRANSISTOR GAIN

Transistor as an Amplifier

The current in the emitter/collector circuit is switched completely ON or OFF when the transistor is used as a switch. This is accomplished by raising the base current high enough to allow full current flow through the emitter/collector circuit. The current flow through the emitter/collector circuit may be limited to any value less than full current flow. The amount of current flow is limited by limiting the base current to a value less than full current.

The transistor acts as a control valve allowing some of the current to flow through the emitter/collector circuit when the base current is limited. The amount of current flowing through the emitter/collector circuit is proportionately higher than the amount of current flowing through the emitter/base circuit.

Transistors amplify weak signals in radios, TVs, and industrial control equipment. For example, an NPN transistor may be used to amplify a signal from a microphone to a speaker. **See Transistor Amplification Application.**

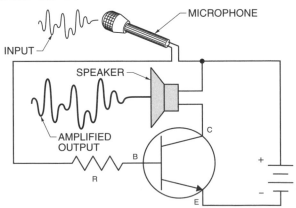

TRANSISTOR AMPLIFICATION APPLICATION

The microphone converts sound waves to electric current pulses. The electric current pulses from the microphone are fed to the base of the transistor. The transistor blocks the current flow through the speaker when there are no pulses (no sound at the microphone). A much larger but proportional quantity of current flows through the speaker when there is current at the base. A transistor is used because the current produced by the microphone is too small to drive the speaker.

 *Refer to **Activity 23-2—Transistors** ....*

APPLICATION 23-3—Thyristors

A *thyristor* is a solid-state switching device that switches current ON by a quick pulse of control current. A thyristor does not require control current to remain ON once the current is switched ON. The most common thyristors are the silicon-controlled rectifier (SCR) and the triac.

An *SCR* is a thyristor that is triggered into conduction in only one direction. A *triac* is a thyristor that is triggered into conduction in either direction. Thyristors have two terminals for load current and one terminal for control current. The three terminals of an SCR are the anode (A), cathode (K), and gate (G). The three terminals of a triac are the main terminal 1 (MT_1), main terminal 2 (MT_2), and gate (G). **See Thyristors.**

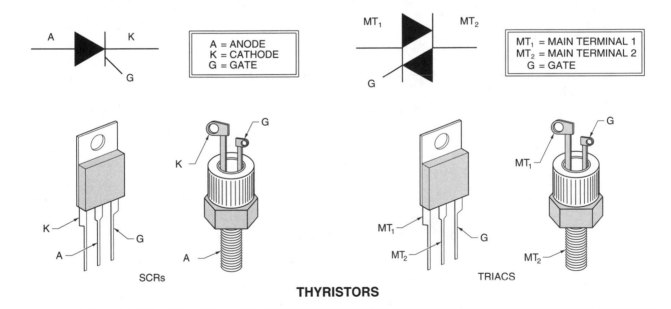

THYRISTORS

A thyristor is generally used to control large amounts of current. SCRs and triacs are used for most solid-state industrial switching circuits. SCRs and triacs control motor speed, heat output, brightness of lights, and other applications requiring a controlled solid-state switching.

SCR DC Voltage Control

An SCR is commonly used to control the voltage in a DC motor control circuit. Current flows between the anode and cathode when control current is applied to the gate. The gate control circuit turns the SCR ON once during each full cycle of the AC sine wave. The point at which the gate control circuit triggers the SCR determines the point on the AC voltage sine wave at which current starts to flow. This point also determines the amount of output voltage. **See SCR DC Voltage Control.**

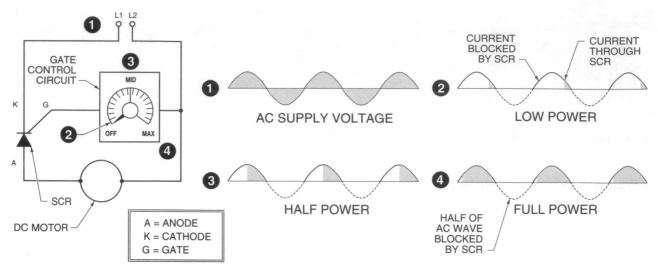

SCR DC VOLTAGE CONTROL

The gate control circuit may start current flowing at any point between points 2 and 4, varying the voltage to the motor from zero to full half-wave. Only full half-wave is possible because the SCR allows current to flow in only one direction.

Triac AC Voltage Control

A triac is commonly used to control the voltage in an AC heating control circuit. Current flows between MT_1 and MT_2 when control current is applied to the gate. The triac is triggered for each half-wave of the AC sine wave. The point at which the control circuit triggers the triac determines the point on the wave at which current begins to flow. This point also determines the amount of output voltage applied to the heating element. The higher the output voltage, the hotter the heating element. **See Triac AC Voltage Control.**

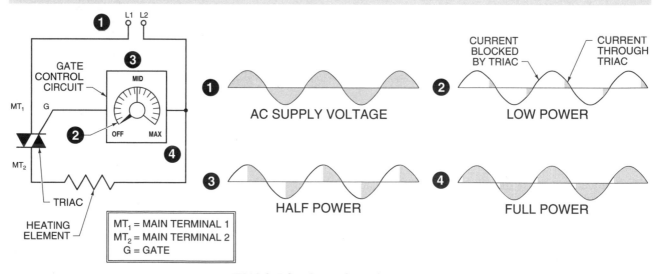

TRIAC AC VOLTAGE CONTROL

The gate control circuit may start current flowing at any point between points 2 and 4, varying the voltage to the heating element. When triggered at the beginning of each half-wave (point 4), there is full voltage output and the heating element is the hottest.

Thyristor Triggering Methods

A thyristor is triggered into conduction by applying a pulse of control current to the gate. Once turned ON, the thyristor remains ON as long as there is a minimum level of holding current flowing through the circuit. A thyristor is turned OFF by reducing the current flowing through the circuit below the holding current. **See Thyristor Triggering.**

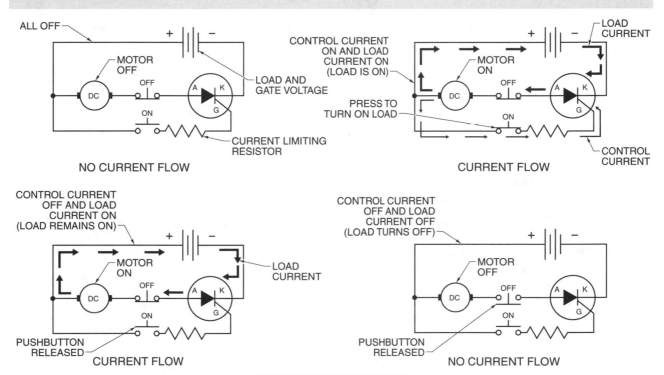

THYRISTOR TRIGGERING

The correct method of thyristor turn ON is applying a proper signal to the gate of the thyristor. The incorrect methods of thyristor turn ON are voltage breakover turn ON, static turn ON, and thermal turn ON.

Gate Turn ON

Gate turn ON is a method of thyristor turn ON that occurs when the proper signal is applied to the gate at the correct time. Gate turn ON is the only correct way to turn ON a thyristor. For an SCR, the gate signal must be positive with respect to the cathode polarity for the thyristor to turn ON. **See Gate Turn ON.**

A triac is triggered into conduction in either direction by a gate current. The simplest triac-triggering circuit uses two fixed resistors, a variable resistor and capacitor. The variable resistor determines the triggering point and the fixed resistor limits current flow. The capacitor is used to trigger the gate once it has charged to a value high enough to deliver sufficient gate current.

A diac is usually added to improve the gate triggering circuit. A *diac* is a two terminal bidirectional device used primarily to control the triggering of triacs. The advantage of using a diac is that it delivers a sharp pulse of gate current rather than a sinusoidal gate current. **See Triac Gate Triggering Methods.**

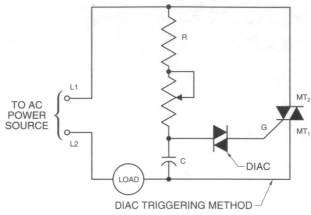

TRIAC GATE TRIGGERING METHODS

Voltage Breakover Turn ON

Voltage breakover turn ON is a method of thyristor turn ON that occurs when the voltage across the thyristor terminals exceeds the maximum voltage rating of the device. This causes localized heating in the thyristor and may damage the component. A varistor connected across the thyristor terminals prevents voltage breakover turn ON. **See Voltage Breakover Turn ON.**

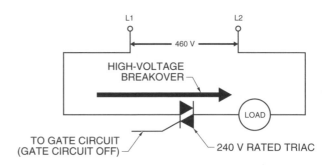

VOLTAGE BREAKOVER TURN ON

Static Turn ON

Static turn ON is a method of thyristor turn ON that occurs when a fast-rising voltage is applied across the anode and cathode terminals of an SCR or the main terminals of a triac. Manufacturers refer to this turn ON point as the dv/dt rating. This rating defines the amount of voltage over a given time period that turns ON the device. A typical rating is 250 V/1 s. Static turn ON does not damage the thyristor provided the surge current is limited. A snubber circuit is added across the thyristor terminals if static turn ON is a problem. **See Static Turn ON.**

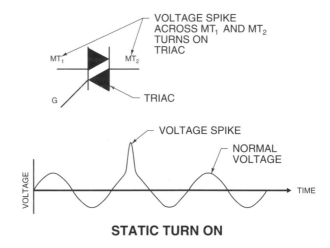

STATIC TURN ON

Thermal Turn ON

All solid-state components are heat sensitive. *Thermal turn ON* is a method of thyristor turn ON that occurs when the heat level exceeds the limit of a thyristor (normally 110°C). Once started, the component is usually damaged or destroyed. Thermal turn ON is eliminated by using the correct heat sink.

 Refer to **Activity 23-3—Thyristors** ....

 *Refer to Quick Quiz® on CD-ROM*

 Refer to Chapter 23 in the **Troubleshooting Electrical/Electronic Systems Workbook** *for additional questions.*

Name_____ **Date** _____

ACTIVITY 23-1—Diodes

An alternator *is a device that converts mechanical energy (spinning motion) to electrical energy. An alternator generates AC in its stator. Diodes convert the AC to DC. Draw the current path through the diodes for the given polarity of the transformers or windings. Determine the negative and positive polarity terminals of the alternator output.*

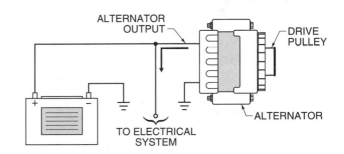

1.

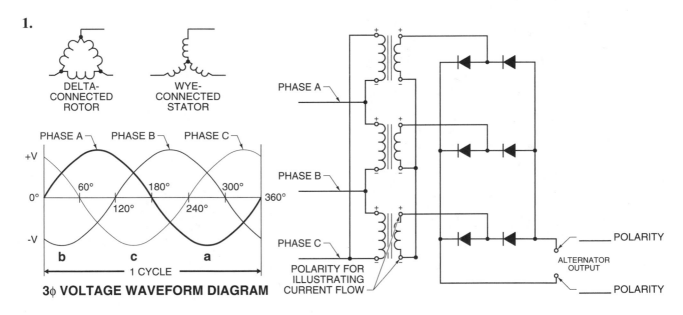

2.

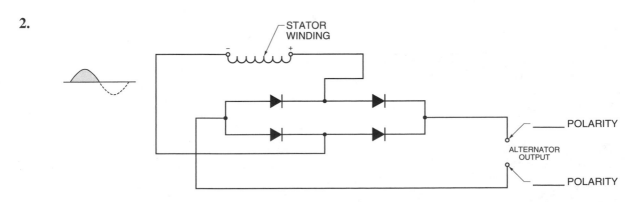

3.

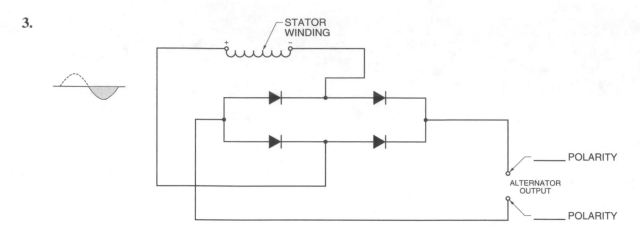

ACTIVITY 23-2—Transistors

Transistors may be connected in series or parallel to develop circuit logic. Complete Truth Table 1 by determining whether the load is ON or OFF for each input switch combination. Identify the equivalent logic function (AND, OR, etc.) of the transistor arrangement.

_____ **1.** The equivalent logic function of Circuit 1 is ___.

TRUTH TABLE 1			
Input			Output
Pushbutton	Temperature Switch	Level Switch	Load
Open	Open	Open	
Open	Open	Closed	
Open	Closed	Open	
Closed	Open	Open	
Closed	Closed	Open	
Open	Closed	Closed	
Closed	Closed	Closed	

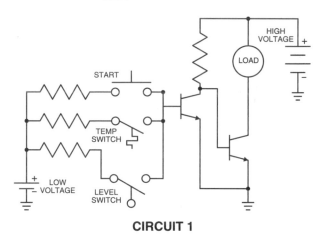

CIRCUIT 1

Complete Truth Table 2 by determining whether the load is ON or OFF for each input switch combination. Identify the equivalent logic function (AND, OR, etc.) of the transistor arrangement.

_____ **2.** The equivalent logic function of Circuit 2 is ___.

TRUTH TABLE 2		
Input		Output
Pushbutton	Limit Switch	Load
Closed	Open	
Open	Closed	
Open	Open	
Closed	Closed	

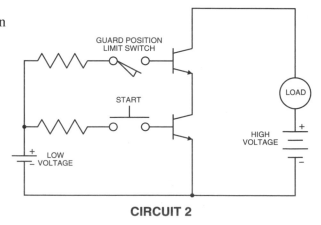

CIRCUIT 2

Diodes and transistors may be used to control the direction of rotation of a DC motor. Determine the positive and negative power bus. Draw the current path from the positive power bus to the negative power bus when Transistor 1 and Transistor 2 are switched ON. Mark the positive and negative side of the motor when Transistor 1 and Transistor 2 are switched ON.

_____ **3.** Power bus ___ is the negative power bus.

_____ **4.** Power bus ___ is the positive power bus.

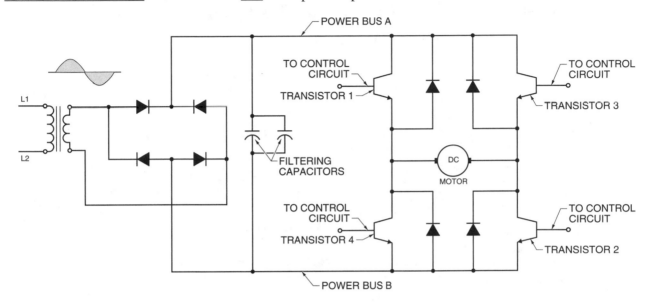

Determine the positive and negative power bus. Draw the current path from the positive power bus to the negative power bus when Transistor 3 and Transistor 4 are switched ON. Mark the positive and negative side of the motor when Transistor 3 and Transistor 4 are switched ON.

_____ **5.** Power bus ___ is the negative power bus.

_____ **6.** Power bus ___ is the positive power bus.

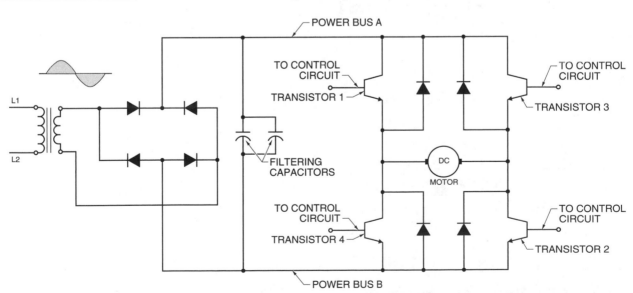

ACTIVITY 23-3—Thyristors

Phase control is the most common use of thyristor power control. A thyristor blocks current flow until it is turned ON by the control circuit. Voltage is applied to the load when the thyristor is turned ON. Conduction angle is the time during which the thyristor is ON. The conduction angle determines the peak voltage (V_{max}), rms voltage (V_{rms}), and average voltage (V_{avg}) applied to the load. Using Voltage/Conduction Angle, determine the peak voltage, rms voltage, and average voltage.

1. The conduction angle is 180°.

_____ **A.** V_{max} equals ___ V.

_____ **B.** V_{rms} equals ___ V.

_____ **C.** V_{avg} equals ___ V.

2. The conduction angle is 120°.

_____ **A.** V_{max} equals ___ V.

_____ **B.** V_{rms} equals ___ V.

_____ **C.** V_{avg} equals ___ V.

3. The conduction angle is 90°.

_____ **A.** V_{max} equals ___ V.

_____ **B.** V_{rms} equals ___ V.

_____ **C.** V_{avg} equals ___ V.

4. The conduction angle is 20°.

_____ **A.** V_{max} equals ___ V.

_____ **B.** V_{rms} equals ___ V.

_____ **C.** V_{avg} equals ___ V.

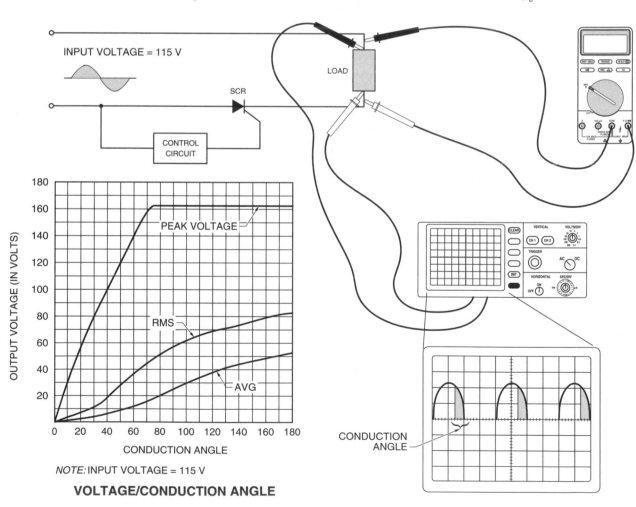

VOLTAGE/CONDUCTION ANGLE

5. The conduction angle is 180°.

_____ **A.** V_{max} equals ___ V.

_____ **B.** V_{rms} equals ___ V.

_____ **C.** V_{avg} equals ___ V.

7. The conduction angle is 90°.

_____ **A.** V_{max} equals ___ V.

_____ **B.** V_{rms} equals ___ V.

_____ **C.** V_{avg} equals ___ V.

9. The conduction angle is 20°.

_____ **A.** V_{max} equals ___ V.

_____ **B.** V_{rms} equals ___ V.

_____ **C.** V_{avg} equals ___ V.

6. The conduction angle is 160°.

_____ **A.** V_{max} equals ___ V.

_____ **B.** V_{rms} equals ___ V.

_____ **C.** V_{avg} equals ___ V.

8. The conduction angle is 60°.

_____ **A.** V_{max} equals ___ V.

_____ **B.** V_{rms} equals ___ V.

_____ **C.** V_{avg} equals ___ V.

10. The conduction angle is 0°.

_____ **A.** V_{max} equals ___ V.

_____ **B.** V_{rms} equals ___ V.

_____ **C.** V_{avg} equals ___ V.

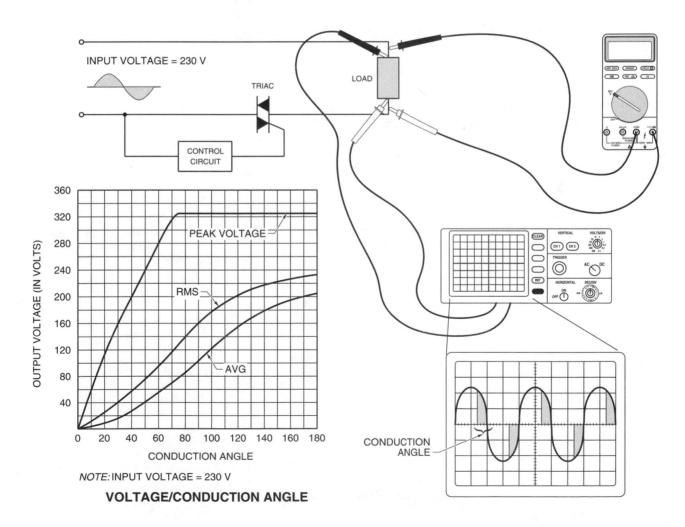

VOLTAGE/CONDUCTION ANGLE

11. Complete the R/C Triac Gate Triggering Circuit by adding a capacitor and conductors to the circuit. Connect the ammeter to measure the load current. Connect the voltmeter to measure the load voltage.

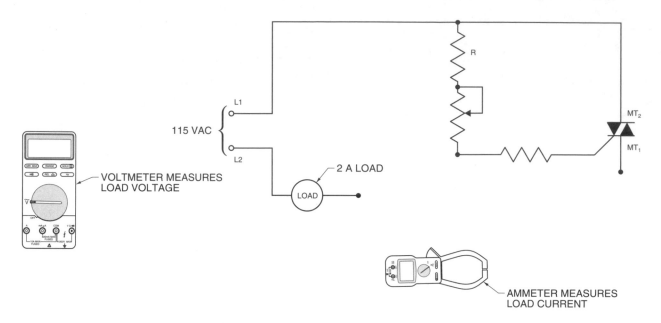

R/C TRIAC GATE TRIGGERING CIRCUIT

12. Complete the SCR Gate Triggering Circuit by adding an ON Pushbutton, OFF Pushbutton, and SCR to the circuit. Connect the ammeter to measure the load current. Connect the voltmeter to measure the load voltage.

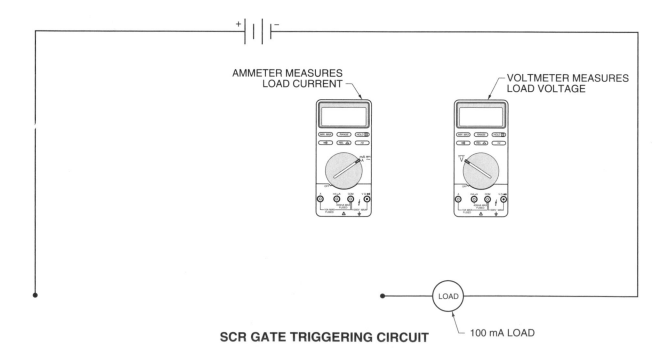

SCR GATE TRIGGERING CIRCUIT

APPLICATION 24-1—Diode Testing

Ohmmeter Diode Test

Forward bias is the condition of a diode while it conducts current. The anode in a forward-biased diode has a positive polarity compared to the cathode. *Reverse bias* is the condition of a diode when it acts as an insulator. The cathode of a reverse-biased diode has a positive polarity compared to the anode. A diode acts as a closed switch when it is forward-biased and as an open switch when it is reverse-biased. An ohmmeter is used to conduct a basic diode check. **See Ohmmeter Diode Test.**

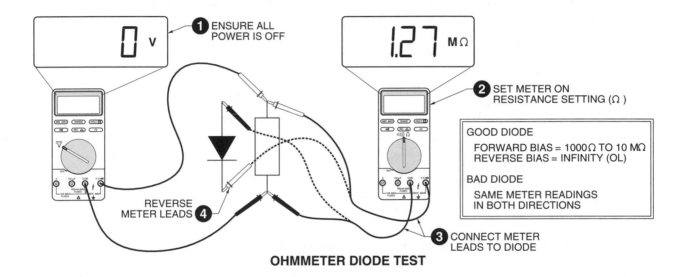

OHMMETER DIODE TEST

To use an ohmmeter to test a diode, apply the procedure:

1. Ensure that all power in the circuit is OFF. Test for voltage using a voltmeter to ensure power is OFF.

2. Set the meter on the resistance setting.

3. Connect the meter leads to the diode. Record the meter reading.

4. Reverse the meter leads. Record the meter reading.

The diode is forward-biased when the positive (red) probe is on the anode and the negative (black) probe is on the cathode. The forward-biased resistance of a good diode should be between 1000 Ω and 10 MΩ. The resistance reading is high when the diode is forward-biased because the current from the meter's voltage source flows through the diode and results in a resistance measurement.

The diode is reverse-biased when the positive (red) lead is on the cathode and the negative (black) lead is on the anode. The reverse-biased resistance of a good diode should equal infinity. An infinite resistance is displayed as an overload (OL) on a digital meter. The diode is bad if the meter readings are the same in both directions.

Multimeter Diode Test

Testing a diode using an ohmmeter does not always indicate whether a diode is good or bad. Testing a diode that is connected in a circuit with an ohmmeter may give false readings because other components may be connected in parallel with the diode under test. The best way to test a diode is to measure the voltage drop across the diode when it is forward-biased.

A good diode has a voltage drop across it when it is forward-biased and conducting current. The voltage drop is between .5 V and .8 V for the most commonly used silicon diodes. Some diodes are made of germanium and have a voltage drop between .2 V and .3 V.

A multimeter set on the diode test position is used to test the voltage drop across a diode. In this position, the meter produces a small voltage between the test leads. The meter displays the voltage drop when the leads are connected across a diode. **See Multimeter Diode Test.**

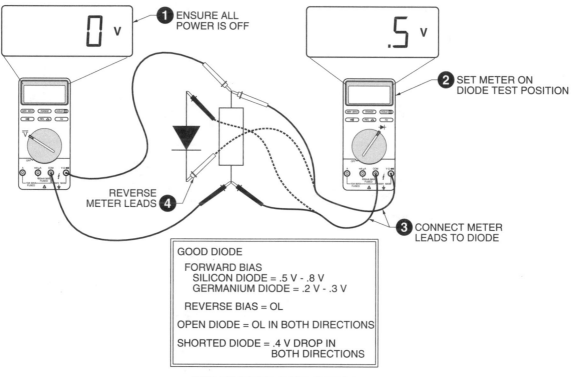

MULTIMETER DIODE TEST

To test a diode using the diode test position on a multimeter, apply the procedure:

1. Ensure that all power in the circuit is OFF. Test for voltage using a voltmeter to ensure power is OFF.

2. Set the meter on the diode test position.

3. Connect the meter leads to the diode. Record the meter reading.

4. Reverse the meter leads. Record the meter reading.

The meter displays a voltage drop between .5 V and .8 V (silicon diode) or .2 V and .3 V (germanium diode) when a good diode is forward-biased. The meter displays an OL when a good diode is reverse-biased. The OL reading indicates the diode is acting like an open switch. An open (bad) diode does not allow current to flow through it in either direction. The meter displays an OL reading in both directions when the diode is open. A shorted diode gives the same voltage drop reading in both directions. This reading is normally about .4 V.

 Refer to ***Activity 24-1—Diode Testing*** ....

APPLICATION 24-2—Transistor Testing

A transistor becomes defective from excessive current or temperature. A transistor normally fails due to an open or shorted junction. The two junctions of a transistor may be tested with an ohmmeter. **See Transistor Testing.**

To test an NPN transistor for an open or shorted junction, apply the procedure:
1. Connect a multimeter to the emitter and base of the transistor. Measure the resistance.
2. Reverse the meter leads and measure the resistance. The emitter/base junction is good when the resistance is high in one direction and low in the opposite direction.

   Note: The ratio of high to low resistance should be greater than 100 : 1. Typical resistance values are 1 kΩ (with the positive lead of the meter on the base) and 100 kΩ (with the positive lead of the meter on the emitter). The junction is shorted when both readings are low. The junction is open when both readings are high.
3. Connect the meter to the collector and base of the transistor. Measure the resistance.
4. Reverse the meter leads and measure the resistance. The collector/base junction is good when the resistance is high in one direction and low in the opposite direction.

   Note: The ratio of high to low resistance should be greater than 100 : 1. Typical resistance values are 1 kΩ (with the positive lead of the meter on the base) and 100 kΩ (with the positive lead of the meter on the collector).
5. Connect the meter to the collector and emitter of the transistor. Measure the resistance.
6. Reverse the meter leads and measure the resistance. The collector/emitter junction is good when the resistance reading is high in both directions.

   The same test used for an NPN transistor is used for testing a PNP transistor. The difference is that the meter test leads must be reversed to obtain the same results.

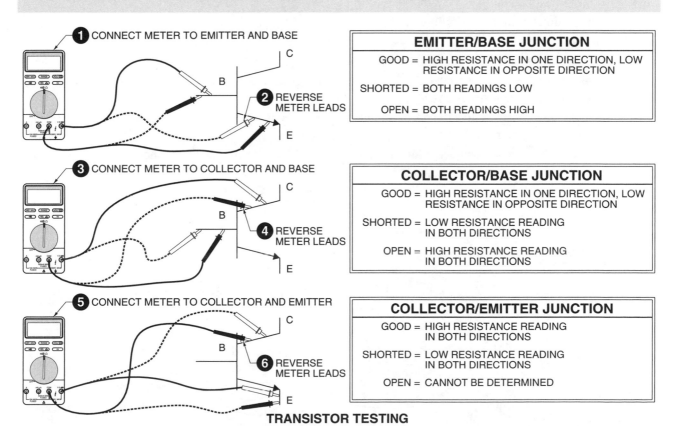

1 CONNECT METER TO EMITTER AND BASE

2 REVERSE METER LEADS

EMITTER/BASE JUNCTION
GOOD = HIGH RESISTANCE IN ONE DIRECTION, LOW RESISTANCE IN OPPOSITE DIRECTION
SHORTED = BOTH READINGS LOW
OPEN = BOTH READINGS HIGH

3 CONNECT METER TO COLLECTOR AND BASE

4 REVERSE METER LEADS

COLLECTOR/BASE JUNCTION
GOOD = HIGH RESISTANCE IN ONE DIRECTION, LOW RESISTANCE IN OPPOSITE DIRECTION
SHORTED = LOW RESISTANCE READING IN BOTH DIRECTIONS
OPEN = HIGH RESISTANCE READING IN BOTH DIRECTIONS

5 CONNECT METER TO COLLECTOR AND EMITTER

6 REVERSE METER LEADS

COLLECTOR/EMITTER JUNCTION
GOOD = HIGH RESISTANCE READING IN BOTH DIRECTIONS
SHORTED = LOW RESISTANCE READING IN BOTH DIRECTIONS
OPEN = CANNOT BE DETERMINED

TRANSISTOR TESTING

Clamping Diode

A clamping diode is used when transistors control inductive loads. A *clamping diode* is a diode that prevents voltage in one part of the circuit from exceeding the voltage in another part. A clamping diode protects a transistor from high-voltage surges. **See Clamping Diode.**

All coils have inductance. *Inductance* is the property of a circuit that causes it to oppose a change in current due to energy stored in a magnetic field. The collapsing magnetic field of a coil produces a high-voltage spike when switched OFF. This spike may damage or destroy a transistor. A diode prevents the spike from passing through the transistor because the diode conducts the excessive voltage away from the transistor.

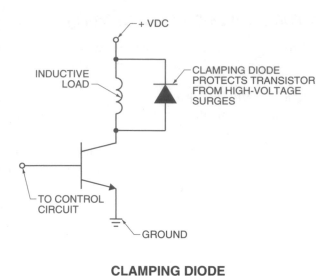

CLAMPING DIODE

 Refer to **Activity 24-2—Transistor Testing** ....

APPLICATION 24-3—Thyristor Testing

SCR Testing

High-power SCRs should be tested using a test circuit. Low-power and some high-power SCRs may be tested using a multimeter set on the resistance scale. **See SCR Testing.**

To test an SCR using a multimeter, apply the procedure:
1. Set the multimeter on the R × 100 scale.
2. Connect the negative lead of the meter to the cathode.
3. Connect the positive lead of the meter to the anode. The meter should read infinity.
4. Short circuit the gate to the anode using a jumper wire. The meter should read almost 0 Ω. Remove the jumper wire. The low-resistance reading should remain.
5. Reverse the meter leads so that the positive lead is on the cathode and the negative lead is on the anode. The meter should read almost infinity.
6. Short circuit the gate to the anode with a jumper wire. The resistance on the meter should remain high.

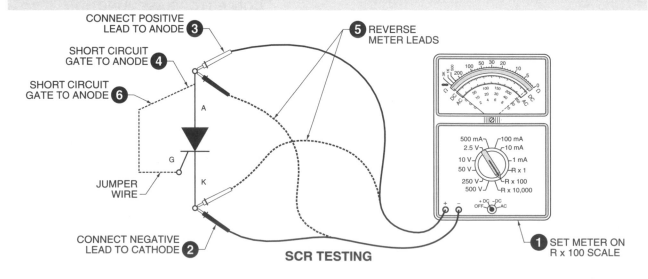

SCR TESTING

Triac Testing

Triacs should be tested under operating conditions using an oscilloscope. A multimeter may be used to make a rough test with the triac out of the circuit. **See Triac Testing.**

To test a triac using a multimeter, apply the procedure:
1. Set the multimeter on the R × 100 scale.
2. Connect the negative meter lead to main terminal 1.
3. Connect the positive meter lead to main terminal 2. The meter should read infinity.
4. Short circuit the gate to main terminal 2 using a jumper wire. The meter should read almost 0 Ω. The zero reading should remain when the lead is removed.
5. Reverse the meter leads so that the positive lead is on main terminal 1 and the negative lead is on main terminal 2. The meter should read infinity.
6. Short circuit the gate of the triac to main terminal 2 using a jumper wire. The meter should read almost 0 Ω. The zero reading should remain after the lead is removed.

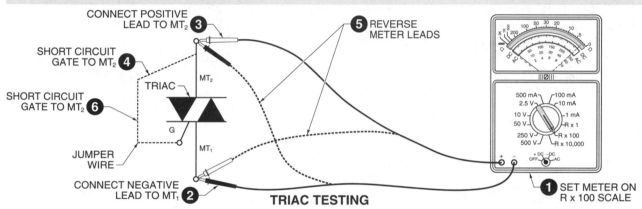

TRIAC TESTING

 Refer to **Activity 24-3—Thyristor Testing** ....

APPLICATION 24-4—Electronic Heat Problems

All solid-state switching components produce heat. Heat affects the operation, life, and reliability of solid-state switching components. Excessive heat destroys all solid-state components. Heat is produced by current passing through a solid-state component because of the friction of the electrons. Heat must be kept to a safe level. This is done by limiting the amount of current passing through the component or removing the heat. Many solid-state components control large amounts of power. This produces large amounts of excess heat that must be dissipated by heat sinks for proper operation of the solid-state components.

Heat Sinks

A *heat sink* is a device that conducts and dissipates heat away from a component. A heat sink is mounted on solid-state components to assure good heat conduction from the solid-state component to the heat sink. The heat sink must fit the solid-state component with no air space between for efficient heat conduction. Thermal grease or pads are used to ensure that no gap exists between the solid-state component and the heat sink. Solid-state components must be evenly spaced to assure uniform heat transfer when more than one is used on the same heat sink. Even spacing prevents hot spots that may damage or warp the mounting board.

Mounting Surface

The surface of the heat sink and solid-state component must be flat and compatible. Mounting holes should be only as large as required by the fasteners. The mounting surface must be free from all foreign material, film, and oxidation. The mounting surface may be polished with #000 steel wool. Clean with an alcohol based cleaner after polishing.

Thermal Compounds

Thermal grease or a thermal pad should be applied after cleaning. Thermal compounds fill air gaps between the metal surfaces. The thermal grease should cover the entire surface area between the solid-state component and the heat sink. Remove all excessive thermal grease after mounting. **See Mounting Heat Sinks.**

Natural convection occurs as warm air rises. The natural rising of warm air causes air to flow over a heated component. To take advantage of this natural airflow, always provide for both top and bottom vents in cabinets.

Allow the moving air to flow over each component with as little obstruction as possible. Use vertical fins and place high heat-producing components near the top. **See Airflow Over Heat Sinks.**

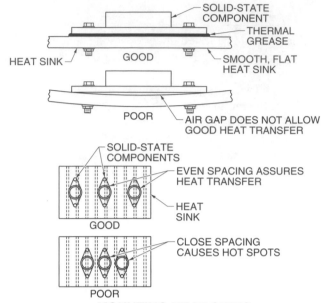

MOUNTING HEAT SINKS

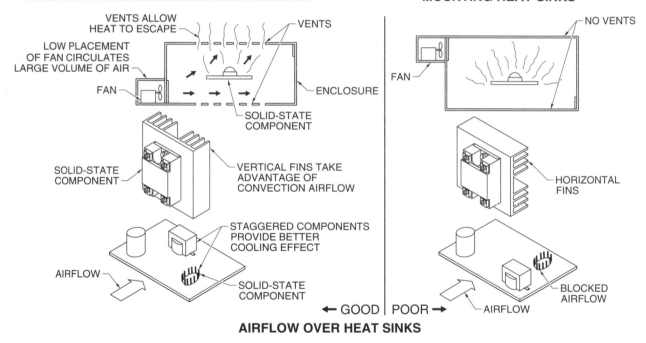

AIRFLOW OVER HEAT SINKS

To assure good airflow with forced convection, stagger the solid-state components so they have about the same amount of air flowing over them. The placement of a fan should be near the bottom, with the top vented. Add a filter to the intake of the fan to keep the fan moving as much air as possible. Change or clean the filter and fan as needed.

 Refer to **Activity 24-4—Electronic Heat Problems** ....

 *Refer to Quick Quiz® on CD-ROM*

 Refer to Chapter 24 in the **Troubleshooting Electrical/Electronic Systems Workbook** *for additional questions.*

Testing Diodes, Transistors, and Thyristors

Name_____   Date _____

ACTIVITY 24-1—Diode Testing

1. Using the meter readings, determine the anode and cathode lead of the diode. *Note:* Assume each diode is good.

_____ **A.** Lead 1 is the ___.

_____ **B.** Lead 2 is the ___.

_____ **C.** Lead 3 is the ___.

_____ **D.** Lead 4 is the ___.

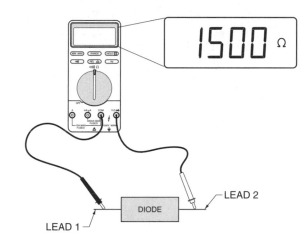

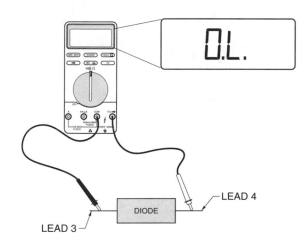

2. Using the meter readings, determine if the diode is forward-biased or reverse-biased.

_____ **A.** Diode 1 is ___-biased.

_____ **B.** Diode 2 is ___-biased.

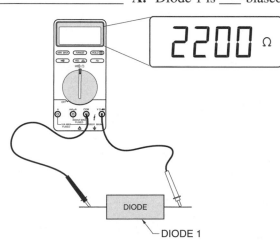

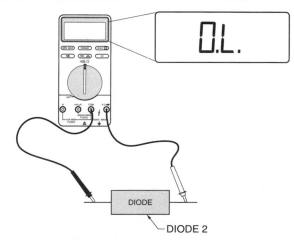

3. Using the meter readings, determine if the diode is good, open, or shorted.

_____ **A.** Diode 1 is ___.   _____ **B.** Diode 2 is ___.

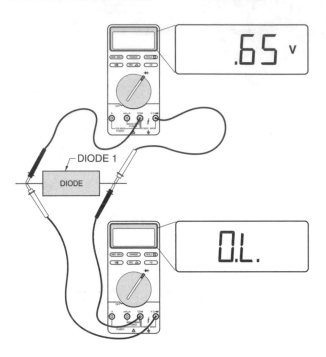   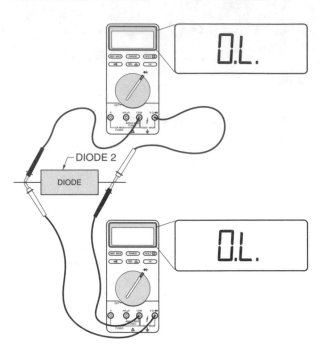

_____ **C.** Diode 3 is ___.   _____ **D.** Diode 4 is ___.

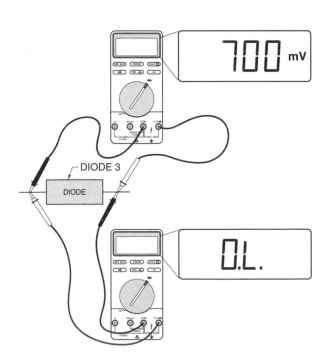   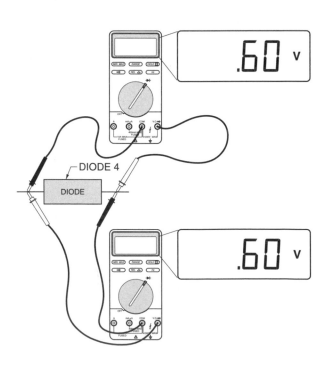

ACTIVITY 24-2—Transistor Testing

1. Using the meter readings, determine if the transistor junction is good, open, or shorted.

_____ **A.** Junction 1 is ___.

_____ **B.** Junction 2 is ___.

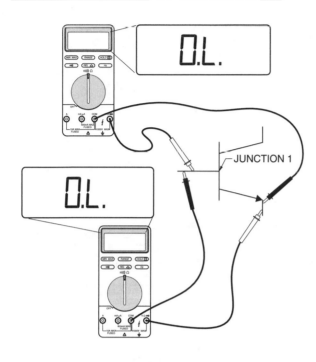

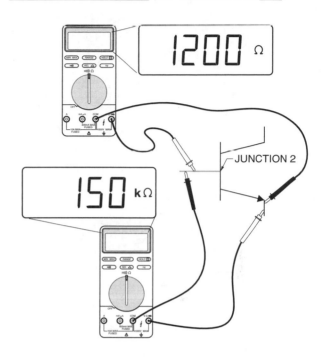

_____ **C.** Junction 3 is ___.

_____ **D.** Junction 4 is ___.

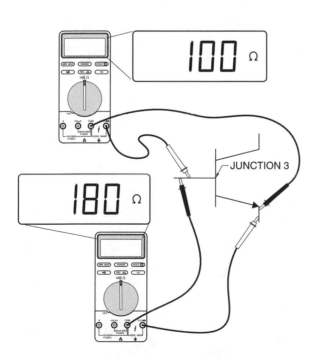

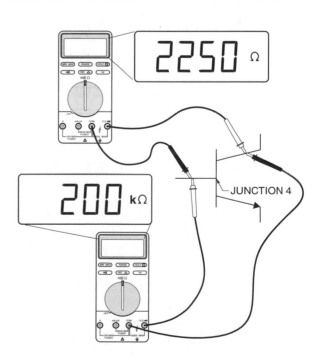

E. Junction 5 is ___.

F. Junction 6 is ___.

G. Junction 7 is ___.

H. Junction 8 is ___.

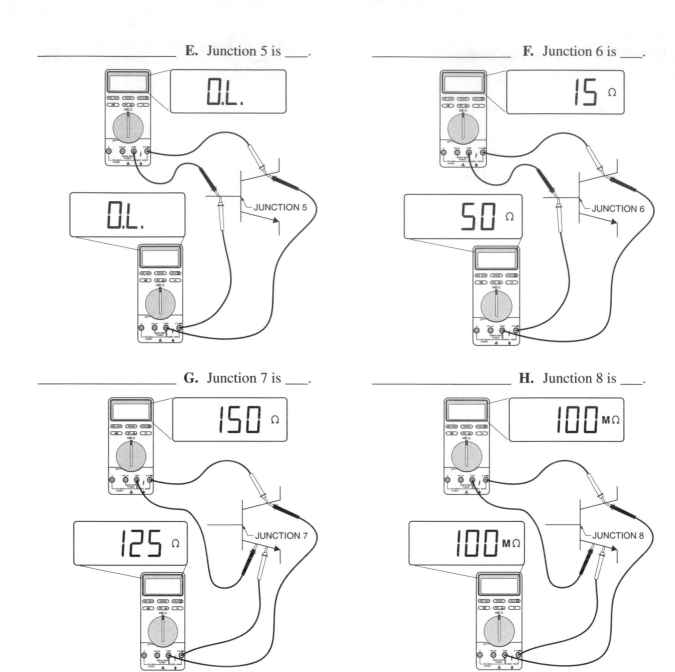

2. Add a clamping diode to each circuit. *Note:* Connect to correct polarity.

A.

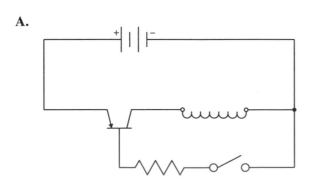

B.

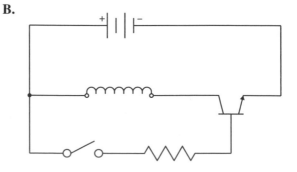

ACTIVITY 24-3—Thyristor Testing

Using the meter readings, determine if the SCR is good or bad.

_____ **1.** The SCR is ___.

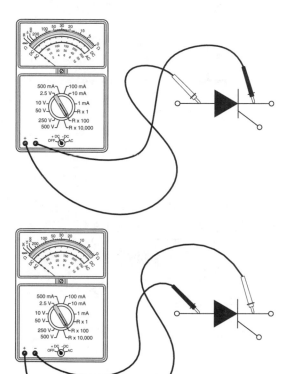

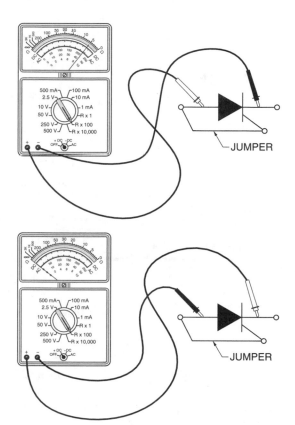

Using the meter readings, determine if the triac is good or bad.

_____ **2.** The triac is ___.

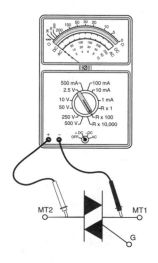

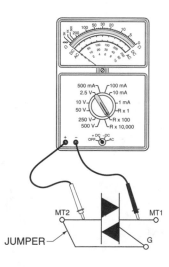

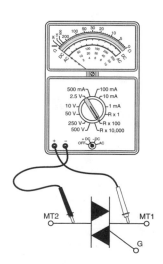

ACTIVITY 24-4—Electronic Heat Problems

Determine if the mounting of the heat sink and components is correct or incorrect.

_____ **1.** The mounting is ___.

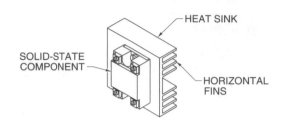

_____ **2.** The mounting is ___.

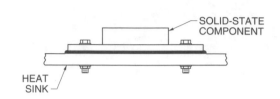

_____ **3.** The mounting is ___.

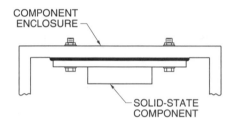

_____ **4.** The mounting is ___.

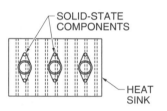

Determine if the airflow over the heat sink and components is good or poor.

_____ **5.** The airflow is ___.

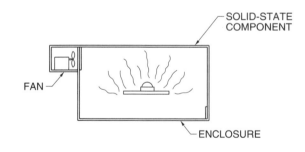

_____ **6.** The airflow is ___.

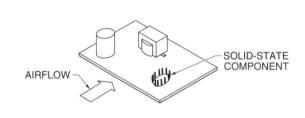

_____ **7.** The airflow is ___.

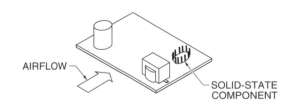

_____ **8.** The airflow is ___.

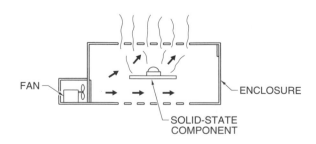

APPLICATION 25-1—Programmable Controllers

A *programmable controller* is a solid-state control device that is programmed and reprogrammed to automatically control machines and industrial processes. A programmable controller contains a power supply, programming terminal, processor, and input and output modules. The *power supply* is the device that provides all the voltages required for the internal operation of the programmable controller. **See Programmable Controller.**

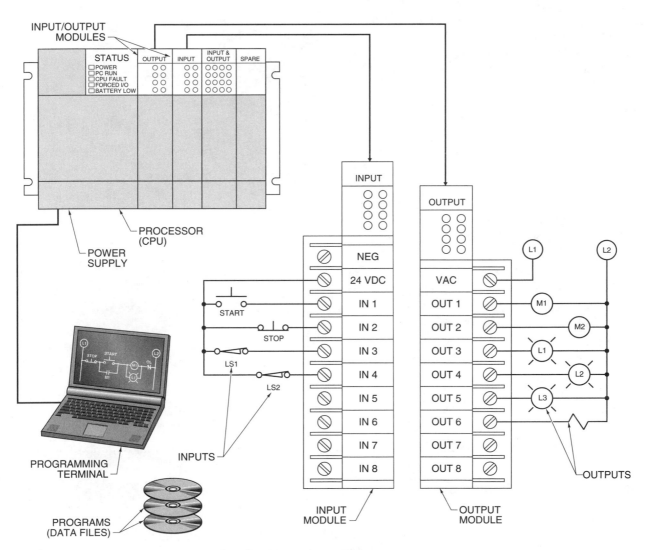

PROGRAMMABLE CONTROLLER

The *programming terminal* is a keyboard-operated device that is used to insert the control program into memory, make changes to the stored program, and display the program circuit and data. The programming terminal allows the circuit to be monitored before or after the program is placed into operation.

The *processor (CPU)* is part of the programmable controller that performs logic functions, program storage, and supervisory functions. The processor scans and monitors the status of all the input devices and output devices in response to the programmed instructions in memory. The processor energizes or de-energizes output devices as a result of the pre-programmed instructions.

An *input module* is the part of a programmable controller that receives information from pushbuttons, temperature switches, pressure switches, photo-electric and proximity switches, and other input devices. An *output module* is the part of a programmable controller that delivers the output voltage required to control alarms, lights, solenoids, motor starters, and other output devices.

Processor Modes

The processor of a programmable controller has different modes. The different modes allow the programmable controller to be taken on-line (system running), or off-line (system on stand-by). Processor modes include the program, run, and test modes. The program mode is used for developing the logic of the control circuit. In the program mode, the circuit is monitored and the program is edited, changed, saved, and transferred.

The run mode is used to execute the program. In the run mode, the circuit may be monitored and the inputs and outputs forced. Program changes cannot be made in the run mode. The test mode is used to check the program without energizing output circuits or devices. In the test mode, the circuit is monitored and inputs and outputs are forced (without actually energizing the load connected to the output). **See Processor Modes.**

Warning: A programmable controller is switched from the program mode to the run mode by placing the controller in the run mode. The machine or process is started when the controller is placed in the run mode. Extreme care must be taken to assure that no damage to personnel or equipment occurs when switching the controller to the run mode. Only qualified personnel should change processor modes and key-operated switches should always be used in any dangerous application.

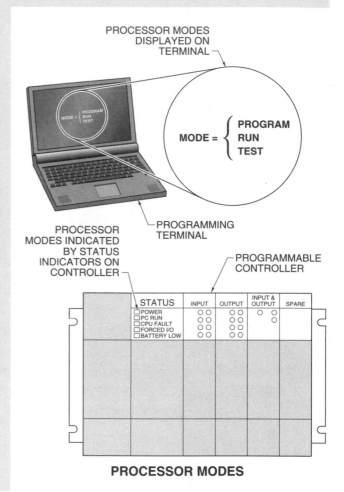

PROCESSOR MODES

 Refer to **Activity 25-1—Programmable Controllers** ....

APPLICATION 25-2—Status Indicators and Error Codes

A programmable controller is designed to withstand a more severe physical environment than a standard personal computer. The programmable controller's program is designed to keep production operating with as little downtime as possible. Manufacturers include status indicators and error messages (or codes) to help a troubleshooter find a problem that occurs.

Status Indicators

A *status indicator* is a light that shows the condition of the components in the programmable controller system. Each major part of the system normally has status indicators. The general location of a problem may be determined by observing the status indicators. Status indicators are a valuable troubleshooting aid when trying to determine a problem in the system. The two types of status indicators are the input/output status indicators, and the operating and fault status indicators. **See Status Indicators.**

Input/Output Status Indicators

An input/output (I/O) status indicator shows the status of the input and output devices. The status indicators on the input module are energized when an electrical signal is received at an input terminal. This occurs when an input contact is closed or signal is present. The status indicators on the output module are energized when an output device is energized. Each input device and output device has its own status indicator.

Operating and Fault Status Indicators

The processor of a programmable controller is programmed to look for potential problems (self diagnostics). The processor performs error checks as part of its normal operation and sends status information to the appropriate status indicators. Typical diagnostic checks include monitoring the input power, the processor's operating mode, CPU faults, forced inputs or outputs, and a low battery condition.

Operating and fault status indicators include the power status, PC run, CPU fault, forced I/O, and battery low indicators. The power status indicator turns ON to indicate that the processor is energized and power is being applied. This status indicator should normally be ON. The PC run indicator turns ON when the processor is in the run mode. Care must be taken when the run indicator is ON because the controller activates the loads as programmed. This indicator is OFF when the processor is placed in the program mode. The CPU fault indicator turns ON when the processor has detected an error in the controller. The processor automatically turns OFF all loads and stops operation when this indicator is ON. An error message is normally displayed on the monitor screen indicating the problem.

The forced I/O indicator turns ON when one or more input or output device has been forced ON or OFF. All force commands must be removed from the program before normal operation is resumed. A battery is used to provide back-up power for the processor memory in case of an external power failure. The battery-low indicator turns ON when the battery should be replaced or if the battery is not charging.

The condition of the status indicators must be checked when troubleshooting a problem in a system using a programmable controller. Potential problems may be determined based on the condition of the status indicators. **See Status Indicator Condition.**

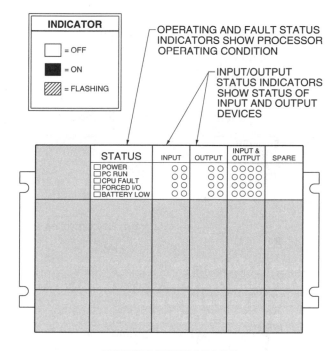

STATUS INDICATORS

STATUS INDICATOR CONDITION			
Status Indicator	**Problem**	**Possible Cause**	**Corrective Action**
☐ POWER ☐ PC RUN ☐ CPU FAULT ☐ FORCED I/O ☐ BATTERY LOW	No or low system power	Blown fuse, tripped CB, or open circuit	Test line voltage at power supply. Line voltage must be within 10% of the controller's rated voltage. Check for proper power supply jumper connections when voltage is correct. Replace the power supply module when the module has power coming into it but is not delivering the correct power
■ POWER ☐ PC RUN ☐ CPU FAULT ☐ FORCED I/O ☐ BATTERY LOW	Programmable controller not in run mode	Improper mode selection for system operation	Place in run mode. Ensure that all personnel are clear before placing the system in run mode
■ POWER ☐ PC RUN ■ CPU FAULT ☐ FORCED I/O ☐ BATTERY LOW	Fault in controller	Faulty memory module, memory loss, or memory error, normally caused by a high-voltage surge, short circuit, or improper grounding	Turn power OFF and re-start system. Remove power and replace the memory module when fault indicator is still ON. Load backup program on new memory module and reboot system
■ POWER ☐ PC RUN ☐ CPU FAULT ☐ FORCED I/O ▨ BATTERY LOW	Fault in controller due to inadequate or no power	Loss of memory when power was OFF and battery charge was inadequate to maintain memory	Replace battery and reload program
■ POWER ■ PC RUN ☐ CPU FAULT ■ FORCED I/O ☐ BATTERY LOW	System does not operate as programmed	Input or output device(s) in forced condition	Monitor program and determine forced input and output device(s). Disable forced input or output device(s) and test system
■ POWER ■ PC RUN ☐ CPU FAULT ☐ FORCED I/O ☐ BATTERY LOW	System does not operate	Defective input device, input module, output device, output module, or program	Monitor program and check condition of status indicators on the input and output modules. Reload program when there is a program error

Error Codes

The manufacturer normally provides a programmable controller which includes a self-troubleshooting program as part of the normal program. The monitor screen displays the problem when a pre-programmed error is detected in the system. The problem may be directly displayed as a message on the monitor screen or may be displayed as an error code number. Simple problems such as exceeding the counter's limit are normally displayed directly on the monitor screen.

Problems that require service are typically displayed by a coded number. The coded number must be cross-referenced to a listing in the service manual. Error code numbers and possible causes vary for each manufacturer. The corrective action taken for the problem is normally the same. **See Error Codes.**

ERROR CODES			
Error Code	**Problem**	**Possible Cause**	**Corrective Action**
01	Memory error occuring in run mode	Voltage surge, improper grounding, high noise interference on lines, inadequate power supply	Add surge suppression to incoming power lines and all inductive output lines. Check power supply voltage for correct level. Reload program with backup program and reboot system
02	Processor does not meet required software level	Software is not compatible with the hardware system. Normally occurs when updated software is loaded into an older system	Consult the software specifications or manufacturer to determine the required hardware level. Upgrade system hardware to meet the software requirements
03	Power failure of an expansion I/O module	Power was removed from module, inadequate power supply, or power has dipped below the minimum specification of the module	Measure voltage at the module power supply and correct any problems. Reboot the system to return to normal operation when message still appears after power is restored
04	Program is trying to address an I/O module in an empty slot on rack	Program is set to the wrong rack and/or slot number	Set the program to a new address number or insert the correct I/O module in the slot number being addressed
05	Module has been detected as being inserted under power	Power was not disabled before a module was inserted into a slot	Reboot the system to return to normal operation when message still appears after power is restored. Never insert a module while power is applied to the unit. Always turn power OFF before inserting or removing a module

 Refer to **Activity 25-3—Status Indicators and Error Codes** ....

APPLICATION 25-3—Force and Disable

The force command opens or closes an input device or turns ON or OFF an output device. The force command is designed for use when system troubleshooting. Forcing an input or output device allows checking the circuit using software. **See Force and Disable.**

An input device may be forced to test the circuit operation. Forcing an input device may also be used when service is required on a defective input device. The defective input device may be forced ON until the device may be serviced if the input device is not critical to production. The force command is removed after the device is fixed.

An output device turns ON regardless of the programmed circuit's logic when the force ON command is used. The output device remains ON until the force OFF command is used. Extra care must be taken when using the force command because the force command overrides all safety features designed for the program.

The disable command prevents an output device from operating. The disable command is the opposite of the force command. The disable command is used to prevent one or all of the output devices from operating. Ensure that all force and disable commands are removed before returning a system to normal operation.

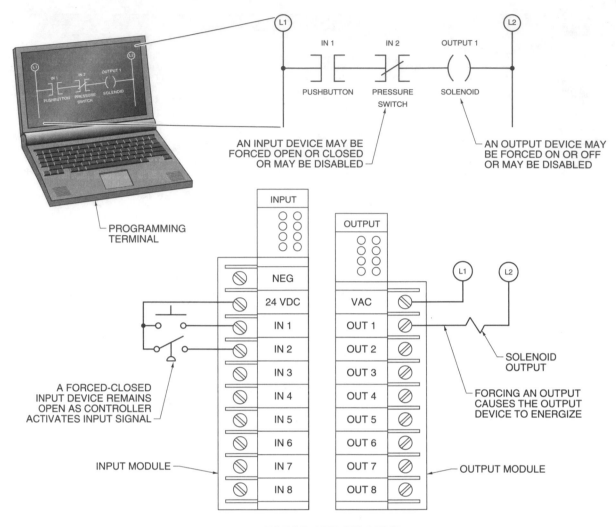

FORCE AND DISABLE

 Refer to **Activity 25-3—Force and Disable** ....

APPLICATION 25-4—Startup Procedure

A *startup procedure* is a logical, systematic procedure that isolates and eliminates potential problems that may cause personal injury and equipment damage. Common startup problems include equipment malfunctions, wiring mistakes, and programming errors.

A startup procedure should always be used when a system is first installed and after software or hardware changes are made. A startup procedure may be used any time troubleshooting is required. A programmable controller startup procedure includes:

• Inspecting the system

• Disabling motion producing output devices

• Verifying the program

• Troubleshooting input modules and devices

• Troubleshooting output modules and devices

• Testing the system

Inspecting the System

An inspection of all parts of a system is required before any power is applied to the system. Inspection is performed to ensure that each module is in the correct location, securely mounted, correctly wired, and properly programmed. **See System Inspection.**

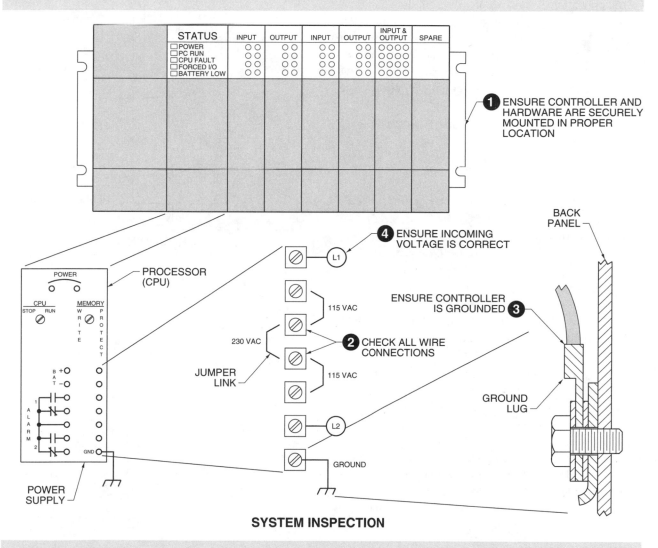

SYSTEM INSPECTION

To inspect a system, apply the procedure:

1. Ensure that the controller and all associated hardware is securely mounted in the proper location. Loose connections cause vibration which may cause plug-in circuit boards to loosen, open circuits, and arcing. Verify each module slot position.

2. Check all wire connections from the main disconnect to the controller input devices, output devices, expansion and communication cables, and jumper links. A *jumper link* is a dual-voltage unit used to set the power rating of the controller. A jumper link must be in the correct location.

Check to ensure that all wire connections are tight and that proper wiring procedures are followed. Proper wiring procedures include physically separating the main power lines from the input lines, input lines from the output lines, power lines from communication lines, and AC lines from DC lines. Separating different line types reduces the possibility of introducing electromagnetic interference on the lines. Electromagnetic interference may produce false signals and cause program errors.

3. Ensure that the controller is grounded. Proper grounding is an important safety precaution in any electrical installation. Proper grounding is especially important in a programmable controller application because improper grounding may lead to interference being induced into the controller. The induced interference may cause false turn ON of output devices, which may cause personnel or equipment damage. Refer to equipment manufacturer grounding recommendations. **See Grounding.**

To prevent problems, the grounding path must be:

• Permanent

• Continuous and uninterrupted

• Of minimum resistance

• Of sufficient size to carry any potential fault current

4. Ensure that incoming voltage is correct according to the programmable controller's specifications and wiring. Excessive voltage damages the controller. Inadequate voltage may cause malfunctions. One very damaging problem occurs when 230 V is applied to a controller set for 115 V. This normally occurs when the jumper links are not placed in the correct position.

Spare parts should be available because many programmable controller problems are solved by the substitution of PC boards and modules. Most manufacturers provide a list of recommended spare parts to stock in-house. This list normally includes a single replacement module for each CPU and power supply, and spare I/O modules equal to 10% of the total number used in the system. Output modules fail more often than input modules because they are connected to high current loads.

Disabling Motion Producing Output Devices

All motion producing output devices should be disabled to prevent damage to personnel and equipment. These include motors, solenoids, and valves. Output devices such as lights generally do not have to be disabled. The output devices may be enabled after the circuit input devices and programs are checked. Disabling the output devices does not prevent the system and outputs from being checked. Disabling output devices reduces the possibility of injury or damage. **See Disabling Output Devices.**

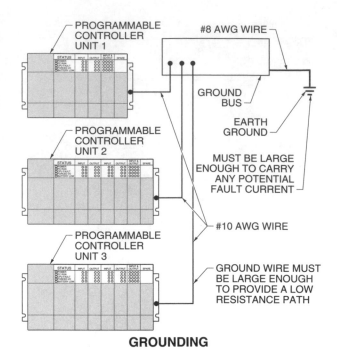

GROUNDING

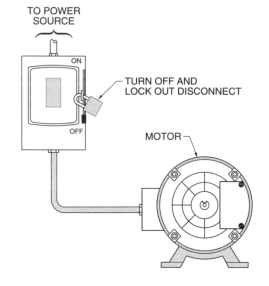

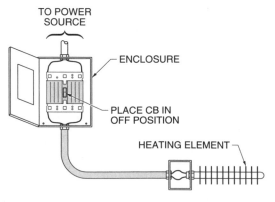

DISABLING OUTPUT DEVICES

Motors are disabled by turning OFF and locking out the disconnect switch feeding the main power to the motor. By locking out power at the disconnect, the motor starter is energized by the programmable controller but the motor is not capable of turning ON. Heating elements and solenoids that are on their own circuit are disabled by removing the main fuse or by placing the circuit breaker in the OFF position.

An output device may be disabled by disconnecting the wires directly at the output device. Ensure that the voltage is OFF by checking it with a voltmeter before disconnecting any wires. An output device may also be disabled by disconnecting the wires at the output module of the programmable controller. This is only recommended when there is no other practical way of disabling the output device. Ensure that the wire and terminal screws are clearly marked to prevent the wire from being re-connected to the wrong terminal.

Verifying the Program

Most programmable controllers include a test mode in addition to a program and run mode. In the test mode, the processor executes the program without actually turning ON the output devices. System operation is simulated without energizing the loads.

The program is tested because the program may have changed unexpectedly due to higher-than-rated temperatures, electromagnetic interference, high transient voltages, or unauthorized tampering. Check the program for changes any time a system does not seem to be operating properly and the input and output devices appear to be working.

Ensure that there is a backup program available. The program in the controller can always be reloaded using the backup program if there is a problem. **See Saving Programs.**

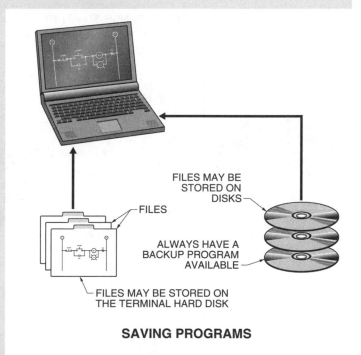

FILES MAY BE STORED ON DISKS

FILES

ALWAYS HAVE A BACKUP PROGRAM AVAILABLE

FILES MAY BE STORED ON THE TERMINAL HARD DISK

SAVING PROGRAMS

Troubleshooting Input Modules

Signals and information are sent to a programmable controller using input devices such as pushbuttons, limit switches, level switches, and pressure switches. The input devices are connected to the input module of the programmable controller. The controller does not receive the proper information if the input device or input module is not operating correctly. **See Troubleshooting Input Modules.**

To troubleshoot an input module, apply the procedure:
1. Measure the supply voltage at the input module to ensure that there is power supplied to the input device(s). Test the main power supply of the controller when there is no power.
2. Measure the voltage from the control switch. Connect the meter directly to the same terminal screw to which the input device is connected.

The voltmeter should read the supply voltage when the control switch is closed. The voltmeter should read the full supply voltage when the control device uses mechanical contacts. The voltmeter should read nearly the full supply voltage when the control device is solid-state. Full supply voltage is not read because .5 V to 6 V is dropped across the solid-state control device. The voltmeter should read zero or little voltage when the control switch is open.

3. Monitor the status indicators on the input module. The status indicators should illuminate when the voltmeter indicates the presence of supply voltage.

4. Monitor the input device symbol on the programming terminal monitor. The symbol should be highlighted when the voltmeter indicates the presence of supply voltage.

Replace the control device if the control device does not deliver the proper voltage. Replace the input module if the control device delivers the correct voltage, but the status indicator does not illuminate.

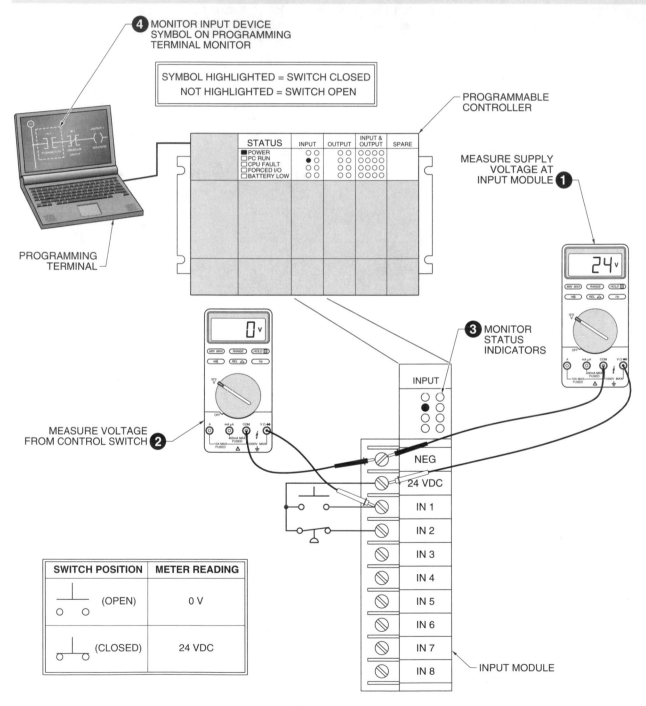

TROUBLESHOOTING INPUT MODULES

Troubleshooting Input Devices

Input devices such as pushbuttons, limit switches, pressure switches, and temperature switches are connected to the input module(s) of a programmable controller. Input devices send information and data concerning circuit and process conditions to the controller. The processor receives the information from the input devices and executes the program. All input devices must operate correctly for the circuit to operate properly. **See Troubleshooting Input Devices.**

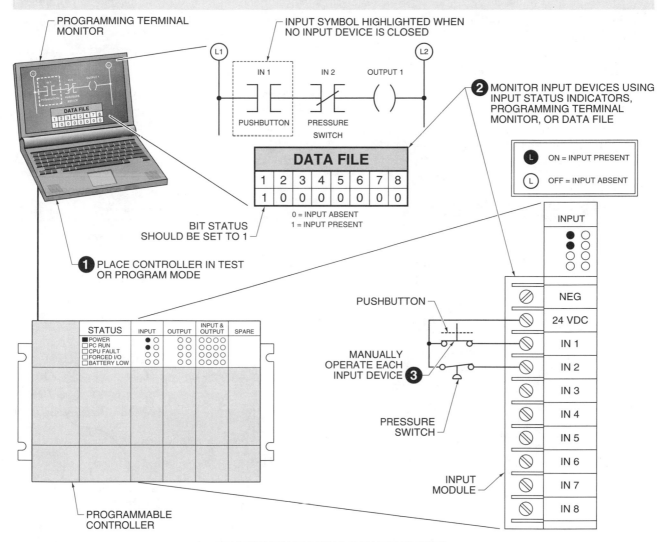

TROUBLESHOOTING INPUT DEVICES

To troubleshoot an input device, apply the procedure:
1. Place the controller in the test or program mode. This step prevents the output devices from turning ON. Output devices are turned ON when the controller is placed in the run mode.
2. Monitor the input devices using the input status indicators (located on each input module), the programming terminal monitor, or the data file. A *data file* is a group of data values (inputs, timers, counters, and outputs) that is displayed as a group and whose status may be monitored.
3. Manually operate each input device starting with the first input. Never reach into a machine when manually operating an input device. Always use a wood stick or other non-conductive device.

The input status indicator located on the input module should illuminate and the input symbol should be highlighted in the control circuit on the monitor screen when a normally-open input device is closed. The bit status on the programming terminal monitor screen should be set to 1 indicating a high or presence of voltage.

The input status indicator located on the input module should turn OFF and the input symbol should no longer be highlighted in the control circuit on the monitor screen when a normally-closed input device is open. The bit status on the programming terminal monitor screen should be set to 0 indicating a low or absence of voltage.

Select the next input device and test it when the status indicator and associated bit status match. Continue testing each input device until all inputs have been tested. Troubleshoot the input device and output device when the status indicator and associated bit status do not match.

Troubleshooting Output Modules

A programmable controller turns ON and OFF the output devices (loads) in the circuit according to the program. The output devices are connected to the output module of the programmable controller. No work is produced in the circuit when the output devices or output module are not operating correctly. The problem may lie in the output module, output device, or controller when an output device does not operate. **See Troubleshooting Output Modules.**

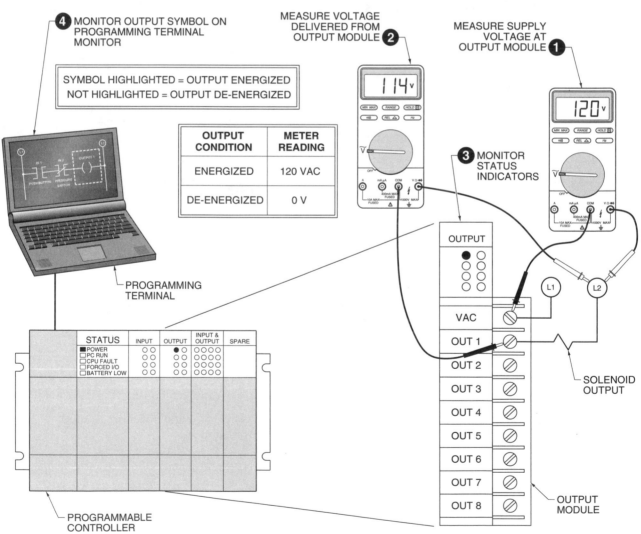

TROUBLESHOOTING OUTPUT MODULES

To troubleshoot an output module, apply the procedure:

1. Measure the supply voltage at the output module to ensure that there is power supplied to the output devices. Test the main power supply of the controller when there is no power.

2. Measure the voltage delivered from the output module. Connect the meter directly to the same terminal screw to which the output device is connected.

The voltmeter should read the supply voltage when the program energizes the output device. The voltmeter should read full supply voltage when the output module uses mechanical contacts. The voltmeter should read almost full supply voltage when the output module uses a solid-state switch. Full voltage is not read because .5 V to 6 V is dropped across the solid-state switch. The voltmeter should read zero or little voltage when the program de-energizes the output device.

3. Monitor the status indicators on the output module. The status indicators should be energized when the voltmeter indicates the presence of supply voltage.

4. Monitor the output device symbol on the programming terminal monitor. The output device symbol should be highlighted when the voltmeter indicates the presence of supply voltage. Replace the output module when the output module does not deliver the proper voltage. Troubleshoot the output device when the output module does deliver the correct voltage, but the output device does not operate.

Troubleshooting Output Devices

Output devices such as motor starters, solenoids, contactors, and lights are connected to the output modules of a programmable controller. An output device performs the work required for the application. The processor energizes and de-energizes the output devices according to the program. All output devices must operate correctly for the circuit to operate properly. **See Troubleshooting Output Devices.**

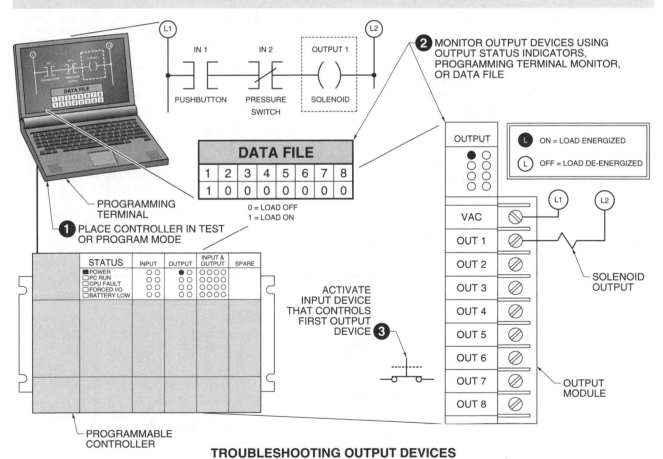

TROUBLESHOOTING OUTPUT DEVICES

To troubleshoot an output device using the test or program mode, apply the procedure:
1. Place the controller in the test or program mode. Placing the controller in the test or program mode prevents the output devices from turning ON. Output devices turn ON only when the controller is placed in the run mode.
2. Monitor the output devices using the output status indicators (located on each output module), the programming terminal monitor, or the data file.
3. Activate the input device that controls the first output device. Check the program displayed on the monitor screen to determine which input device activates which output device. Never reach into a machine to activate an input device.

Select the next output device and test it by checking if the outputs are highlighted on the monitor. Continue testing each output device until all output devices have been tested. Troubleshoot the input device and output device when the status indicator and associated bit status do not match.

Testing the System

The complete system may be tested after all input devices, output devices, and programs have been tested. Always have an individual ready to operate the emergency stop pushbutton when testing the system. The emergency stop pushbutton de-energizes the master control relay and removes power from the output devices. An emergency stop pushbutton that operates a master stop relay is required on all programmable controller installations. **See Emergency Stop Circuit.**

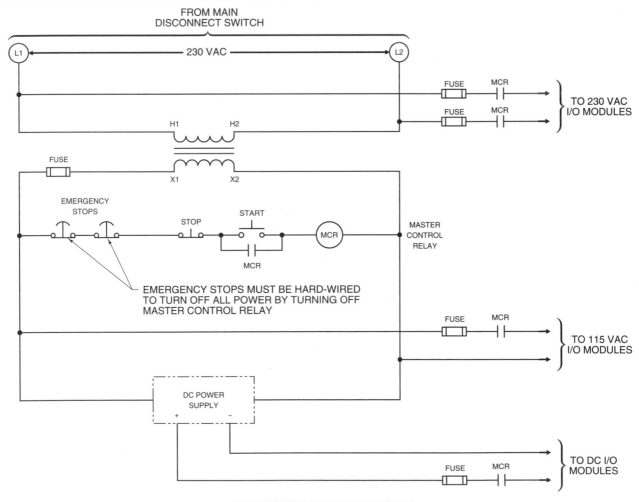

EMERGENCY STOP CIRCUIT

All personnel must remain clear of the controller and equipment when power is applied to test a system. There may be an intermittent problem and any sudden, unexpected machine motion could result in injury. **See Testing the System.**

To test the system, apply the procedure:
1. Apply power to the first output device. Leave the other output devices de-energized.
2. Place the controller in the run mode.
3. Cycle the control circuit until the first output device is activated.
4. Place the controller in the program or test mode when the first output device operates properly.
5. Apply power to the second output device with the controller in the program or test mode. Troubleshoot the output device when the device does not operate. Check the program for programming errors when the output device does not operate properly.
6. Place the controller in the run mode.
7. Cycle the control circuit until the second output device is activated.
8. Place the controller in the program or test mode when the second output device operates properly.
9. Apply power to the next output device with the controller in the program or test mode. Test the next output device and all other output devices using the same method as the first and second output devices.

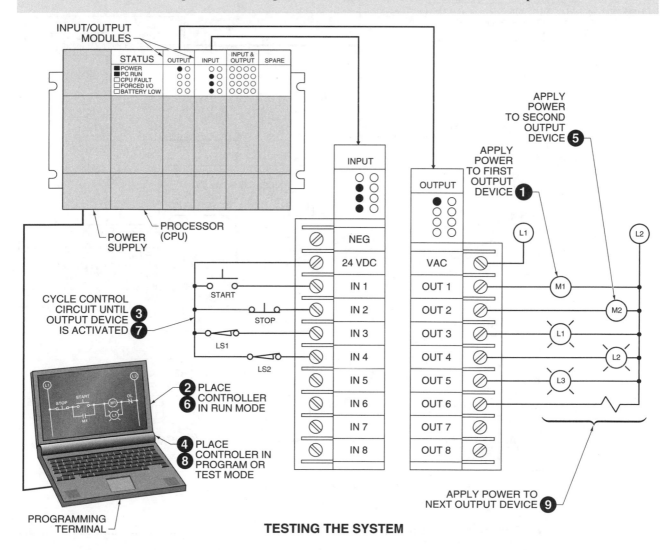

TESTING THE SYSTEM

Some manufacturers provide flow charts to aid in troubleshooting a programmable controller circuit. **See Output Troubleshooting Flow Chart.**

The manufacturer's service manual and flow chart may be used when a problem is not easily found. Using a flow chart aids in giving direction to the troubleshooting process.

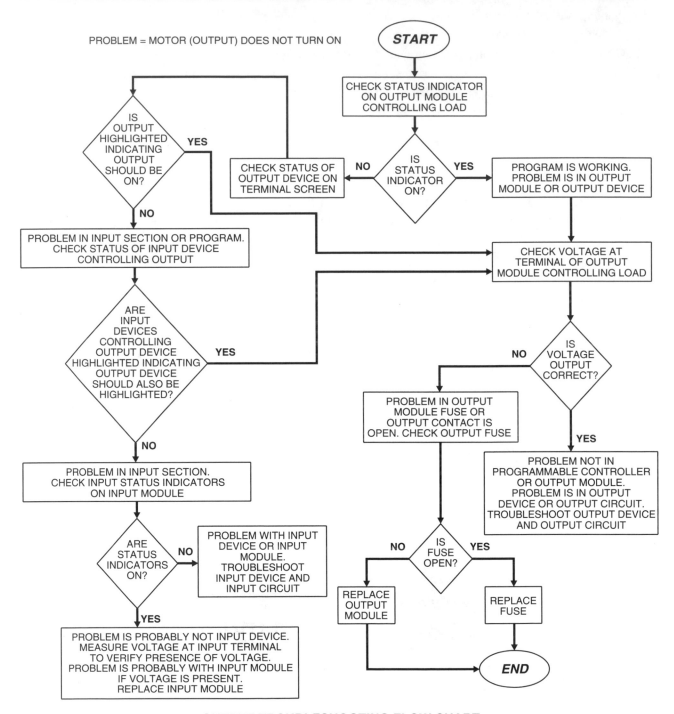

OUTPUT TROUBLESHOOTING FLOW CHART

 Refer to **Activity 25-4—Startup Procedure** ....

APPLICATION 25-5—Programmable Controller Preventive Maintenance

Programmable controllers are designed to operate for a long time with little maintenance. Most problems that occur are with the input devices and output devices connected to the controller. Preventive maintenance steps should be taken to keep the programmable controller working properly.

Enclosures

The PC boards used in the controller must be protected from dirt, oil, moisture, and other contaminants. To protect the PC boards, the controller must be installed in an enclosure suitable for the environment. The enclosure provides the main protection from atmospheric conditions. A NEMA 12 enclosure is used for most indoor general purpose areas. **See Enclosure Types in Appendix.**

A NEMA 12 enclosure is designed to be dust tight. A metal enclosure is used with programmable controllers because a metal enclosure helps minimize the effect of electromagnetic interference. Electromagnetic interference may cause false operations in the system. Ensure metal enclosures are properly grounded. Grounding provides a safe path for unwanted signals.

All programmable controllers produce heat that must be dissipated away from the controller. Proper spacing of components within the enclosure is required. Components are spaced so that the heat from one component does not cause problems in another. Use an enclosure that is large enough to allow room around each component to help dissipate the heat. The temperature inside the enclosure must not exceed the maximum operating temperature of the controller (normally 60°C). A fan is required in higher ambient temperatures.

A fan used to circulate air through the cabinet containing a programmable controller should pull the outside air into the cabinet. A filter must be used on the intake side of the fan to remove dirt from the air before it enters the fan and cabinet. This allows clean air to circulate over the components and out the vent holes. As air passes through the filter, the filter becomes dirty. The filter should be cleaned or replaced every three months (sooner in dirty environments).

Electrical Noise Suppression

Electrical noise is unwanted signals that are present on a power line. Electrical noise enters through input devices, output devices, and power supply lines. Unwanted noise pickup may be reduced by placing the controller away from noise-generating equipment such as motors, motor starters, welders, and drives.

Noise suppression should be included in every programmable controller installation because it is impossible to eliminate noise in an industrial environment. Certain sensitive input devices (analog, digital, and thermocouple) require a shielded cable to reduce electrical noise.

A shielded cable uses an outer conductive jacket (shield) to surround the two inner signal-carrying conductors. The shield blocks electromagnetic interference. The shield must be properly grounded to be effective. Proper grounding includes grounding the shield at only one point. A shield grounded at two points tends to conduct current between the two grounds. **See Input Shielding.**

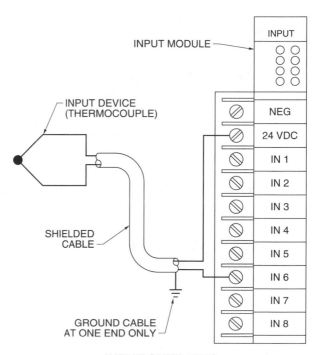

INPUT SHIELDING

A high-voltage spike is produced when inductive loads (motors, solenoids, coils) are turned OFF. These spikes may cause problems in a programmable controller system. High-voltage spikes should be suppressed to prevent problems. A snubber circuit is used to suppress a voltage spike. Typical snubber circuits use an RC (resistor/capacitor), MOV (metal oxide varistor), or a diode depending on the load. **See Snubber Circuits.**

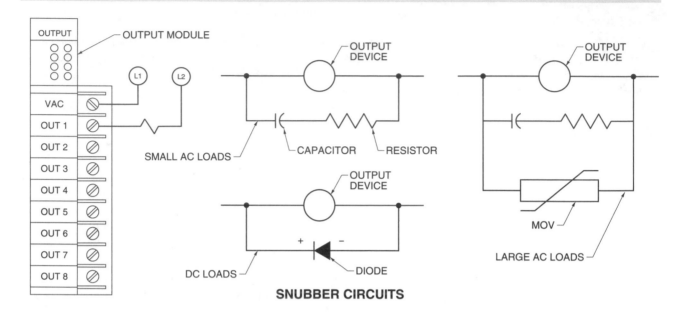

SNUBBER CIRCUITS

General Preventive Maintenance

General preventive maintenance steps are taken to prevent problems when servicing a programmable controller. Programmable controller general preventive maintenance steps include:

- Changing or cleaning filters. Programmable controllers that use a fan to help cool the circuits generally include a filter as part of the fan unit. The airflow is reduced or stopped as the filter gets dirty.
- Keeping the programmable controller modules clean, especially the fins on the power supply and I/O modules. The fins are designed to dissipate heat produced by the solid-state electronic devices. Wipe the fins with a clean cloth when dirty.
- Checking to ensure that the modules are placed tightly in their sockets. Modules are designed to be inserted and removed. The modules may loosen over time when vibration is present. A loose module often causes an intermittent problem, which is the hardest problem to troubleshoot.
- Keeping manufacturer service bulletins up-to-date and available.

 Refer to **Activity 25-5—Programmable Controller Preventive Maintenance** ....

Refer to Quick Quiz® on CD-ROM

 Refer to Chapter 25 in the **Troubleshooting Electrical/Electronic Systems Workbook** *for additional questions.*

Name_____   **Date**_____

ACTIVITY 25-1—Programmable Controllers

1. Using the Conveyor Circuit Line Diagram, complete the Conveyor Circuit table by identifying the type (input or output), name (master stop, pressure switch, etc.), and function (start Conveyor 1 motor, etc.) of the circuit's inputs and outputs.

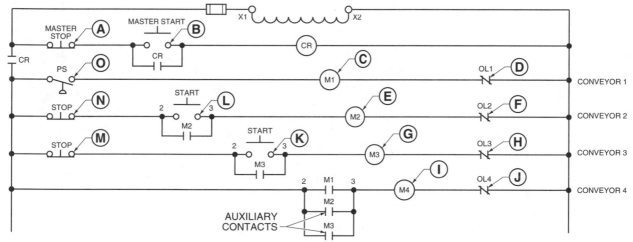

CONVEYOR CIRCUIT LINE DIAGRAM

CONVEYOR CIRCUIT			
Device	**Type**	**Name**	**Function**
A	Input	Master stop	Stop any or all motors that are running
B			
C			
D			
E			
F			
G			
H			
I			
J			
K			
L			
M			
N			
O			

2. Connect the inputs in the Conveyor Circuit to the input modules so they are listed A, B, C, etc. Mark any extras as spares.

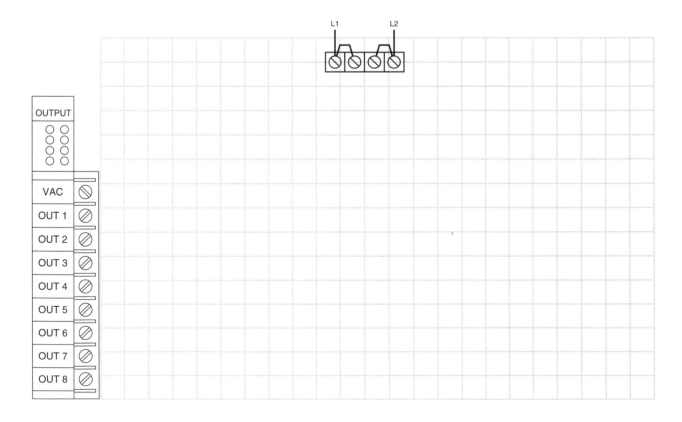

3. Connect the outputs in the Conveyor Circuit to the output module so they are listed A, B, C, etc. Mark any extras as spares.

4. Using the Tank Circuit Line Diagram, complete the Tank Circuit table by identifying the type (input or output), name (run pushbutton, temperature switch, etc.), and function (start pump motor, etc.) of the circuit's inputs and outputs.

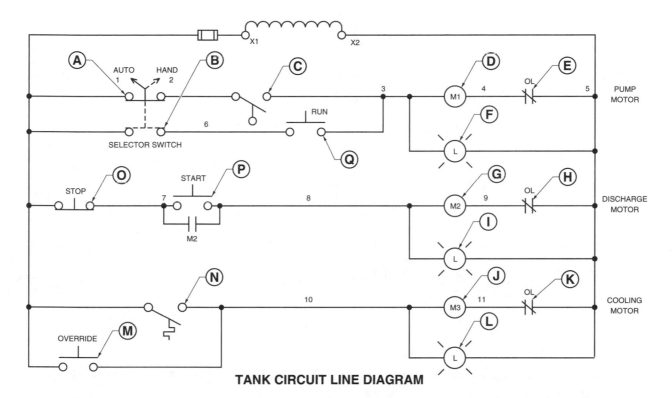

TANK CIRCUIT LINE DIAGRAM

		TANK CIRCUIT	
Device	Type	Name	Function
A	Input	Selector switch (automatic position)	Place pump circuit in automatic condition
B			
C			
D			
E			
F			
G			
H			
I			
J			
K			
L			
M			
N			
O			
P			
Q			

5. Connect the inputs in the Tank Circuit to the input modules so they are listed A, B, C, etc. Mark any extras as spares.

INPUT
NEG
24 VDC
IN 1
IN 2
IN 3
IN 4
IN 5
IN 6
IN 7
IN 8

INPUT
NEG
24 VDC
IN 1
IN 2
IN 3
IN 4
IN 5
IN 6
IN 7
IN 8

6. Connect the outputs in the Tank Circuit to the output module so they are listed A, B, C, etc. Mark any extras as spares.

L1   L2

OUTPUT
VAC
OUT 1
OUT 2
OUT 3
OUT 4
OUT 5
OUT 6
OUT 7
OUT 8

ACTIVITY 25-2—Status Indicators and Error Codes

The condition of the I/O status indicators is used when troubleshooting a control circuit. A troubleshooter must identify which status indicator(s) should be ON and which status indicator(s) should be OFF to interpret a problem. Using Pump Unloader Circuit Line Diagram, and Pump Unloader Circuit Input and Output Connections, determine the condition (ON or OFF) for each status indicator.

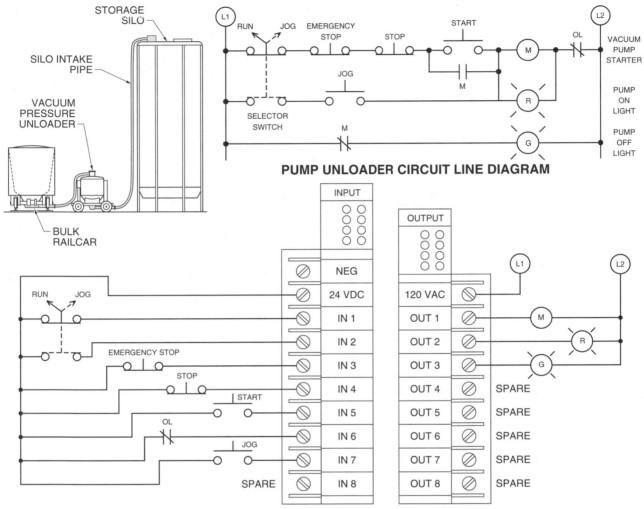

PUMP UNLOADER CIRCUIT LINE DIAGRAM

PUMP UNLOADER CIRCUIT INPUT AND OUTPUT CONNECTIONS

1. Circuit Condition 1 has normal (no problem) status indication when the programmable controller is in the run mode. The Selector Switch is in the run position and the Stop Pushbutton has been pressed and released.

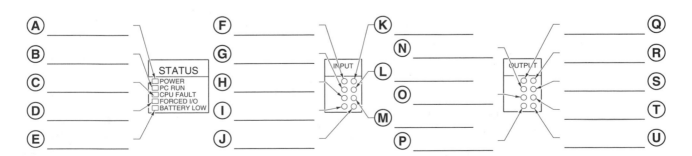

2. Circuit Condition 2 has normal (no problem) status indication when the programmable controller is in the run mode. The Selector Switch is in the run position and the Start Pushbutton has been pressed and released.

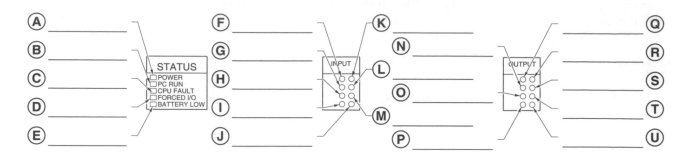

3. Circuit Condition 3 has normal (no problem) status indication when the programmable controller is in the run mode. The Selector Switch is in the jog position and the Jog Pushbutton is held down.

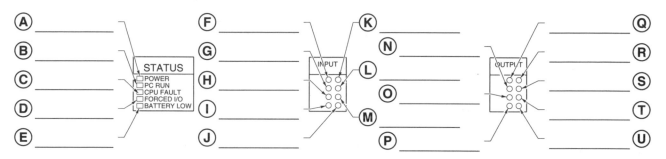

Using the status indicators, Pump Unloader Circuit Line Diagram, and Pump Unloader Circuit Input and Output Connections, determine the problem of each circuit condition.

4. The Start (or Jog) Pushbutton does not start the pump motor. The Selector Switch is in the run position.

_____ **A.** The problem is a(n) ___.

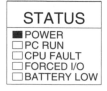
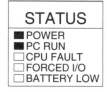

5. The Start (or Jog) Pushbutton does not start the pump motor. The Selector Switch is in the run position.

_____ **A.** The problem is ___.

ACTIVITY 25-3—Force and Disable

Forcing an input or output device allows troubleshooting a circuit using software. Using the Tank Circuit Line Diagram and Tank Circuit Input and Output Connections, determine the inputs and outputs that are forced when troubleshooting each problem.

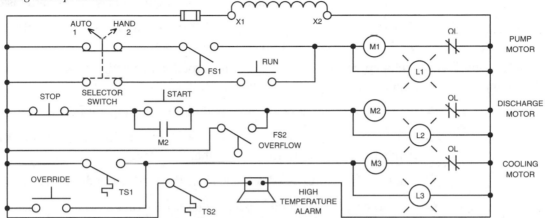

TANK CIRCUIT LINE DIAGRAM

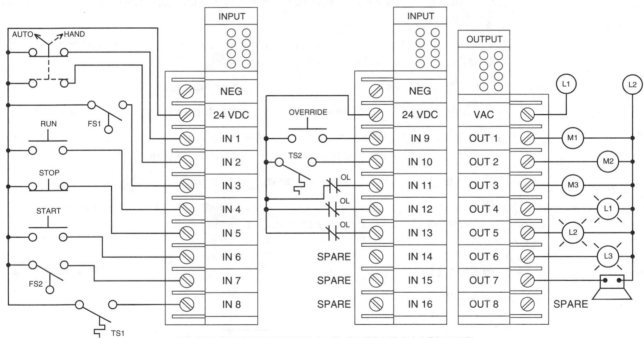

TANK CIRCUIT INPUT AND OUTPUT DIAGRAMS

1. The operator reports that the Discharge Motor does not start when the Start Pushbutton is pressed.

_____ **A.** ___ is forced to determine if the Start Pushbutton is operating.

_____ **B.** The force ___ (ON, OFF, open, or closed) instruction is used to determine if the Start Pushbutton is operating.

_____ **C.** ___ is forced to determine if the Overload Contact has tripped.

_____ **D.** The force ___ (ON, OFF, open, or closed) instruction is used to determine if the Overload Contact has tripped.

_____ **E.** Input ___ is forced to bypass the pushbutton station.

_____ **F.** The force ___ (ON, OFF, open, or closed) instruction is used to bypass the pushbutton station.

_____ **G.** ___ is forced to override all inputs in the circuit and test if the starter is operating.

_____ **H.** The force ___ (ON, OFF, open, or closed) instruction is used to override all inputs in the circuit and test if the starter is operating.

2. The Tank Circuit has been in operation for several years and a routine inspection is scheduled. The routine schedule requires a test of the circuit's visual and audible indicators.

_____ **A.** Output ___ is forced to determine if the High Temperature Alarm is operating.

_____ **B.** The force ___ (ON, OFF, open, or closed) instruction is used to determine if the High Temperature Alarm is operating.

_____ **C.** Output ___ is forced to determine if the Pump Motor Indicator Light is operating.

_____ **D.** The force ___ (ON, OFF, open, or closed) instruction is used to determine if the Pump Motor Indicator Light is operating.

_____ **E.** Output ___ is forced to determine if the Discharge Motor Indicator Light is operating.

_____ **F.** The force ___ (ON, OFF, open, or closed) instruction is used to determine if the Discharge Motor Indicator Light is operating.

3. The High Temperature Alarm has sounded. The Cooling Motor is not running.

_____ **A.** Output ___ is forced to shut the alarm OFF during the troubleshooting call.

_____ **B.** The force ___ (ON, OFF, open, or closed) instruction is used to force the output device to shut the alarm OFF during the troubleshooting call.

_____ **C.** Input ___ is forced to shut the alarm OFF during the troubleshooting call.

_____ **D.** The force ___ (ON, OFF, open, or closed) instruction is used to force the input to shut the alarm OFF during the troubleshooting call.

ACTIVITY 25-4—Startup Procedure

1. Connect the programmable controller to 115 V.

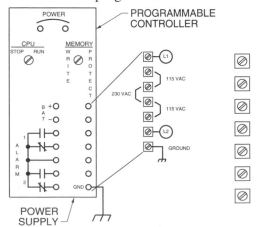

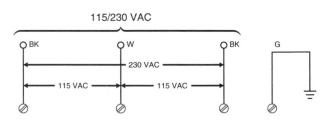

2. Connect the programmable controller to 230 V.

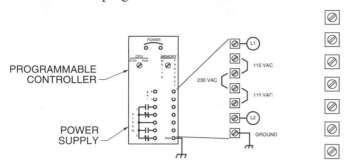

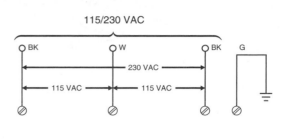

3. Determine the meter readings if the Tank Circuit is operating properly. The Selector Switch is in the auto position, FS1 is closed, the Start Pushbutton has been pressed and released, and TS1 is closed.

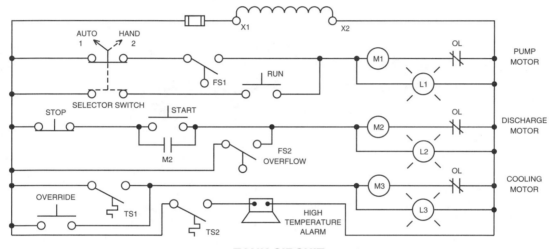

TANK CIRCUIT

_____ **A.** Meter 1 reading equals ___ V.

_____ **B.** Meter 2 reading equals ___ V.

_____ **C.** Meter 3 reading equals ___ V.

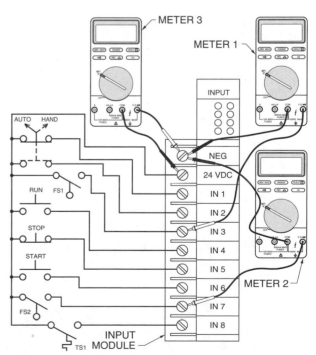

_____ **D.** Meter 4 reading equals
___ V.

_____ **E.** Meter 5 reading equals
___ V.

_____ **F.** Meter 6 reading equals
___ V.

_____ **G.** Meter 7 reading equals
___ V.

_____ **H.** Meter 8 reading equals
___ V.

_____ **I.** Meter 9 reading equals
___ V.

_____ **J.** Meter 10 reading equals
___ V.

ACTIVITY 25-5—Programmable Controller Preventive Maintenance

1. Add voltage spike suppression on each inductive load.

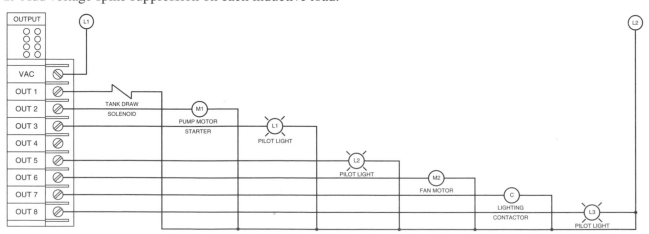

APPLICATION 26-1—Geothermal Systems

A *geothermal system* is a system that uses the temperature of the ground or water from a lake or ocean to produce temperature changes that can be used for heating or cooling applications. Depending on latitude, the temperature of the ground remains relatively constant (about 45°F to 70°F) at approximately 6′ to 8′ below grade level. Thus, in summer, the earth can be used to provide cooling for residential, commercial, and industrial buildings. In winter, the earth can be used to provide heat. The main components of a geothermal system are the piping system and ground source heat pump. **See Geothermal System.**

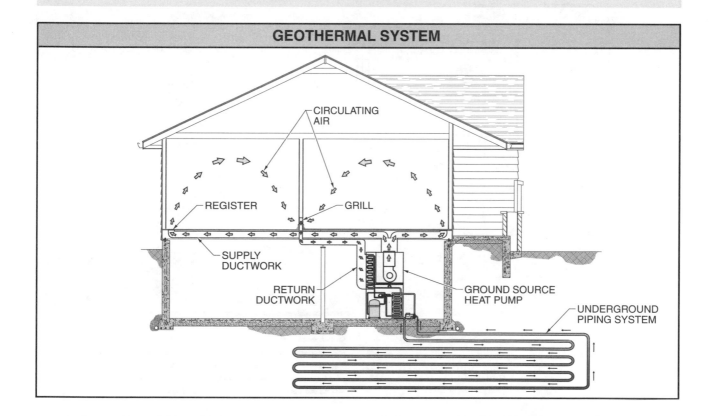

GEOTHERMAL SYSTEM

Piping System

The piping system consists of underground pipes that are buried far enough below the surface or underwater to be in contact with constant-temperature soil or water. The pipes may be configured horizontally, vertically, and/or in loops. The pipes absorb heat or cooling from the earth or water and transfer it to liquid inside the pipes. Most pipes are made of high-density polyethylene, which is strong and allows good heat transfer between the liquid in the pipe and the earth or water. The piping system is filled with water or, in cold climates, a water and antifreeze mixture.

Ground Source Heat Pumps

A *heat pump* is a mechanical compression refrigeration system that contains devices and controls that reverse the flow of refrigerant to move heat from one area to another area. A heat pump can absorb heat from inside a building (providing a cooling effect) and reject the heat outdoors, or the heat pump can absorb heat from the outdoors and reject the heat inside a building (providing a heating effect). A ground source heat pump system transfers heat or cooling from the ground through a piping system and to a heat exchanger. A ground source heat pump system consists of a compressor, reversing valve, heat exchanger(s), expansion device (valve), and refrigerant lines. **See Ground Source Heat Pump System.**

GROUND SOURCE HEAT PUMP SYSTEM

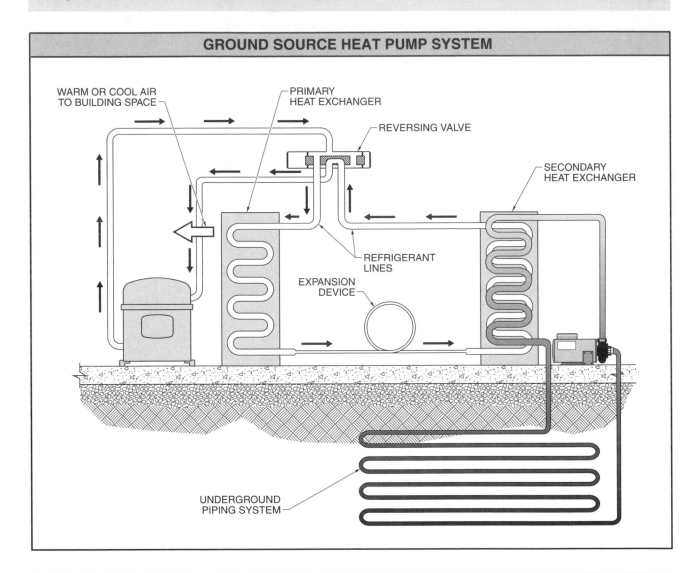

A *compressor* is a mechanical device that compresses refrigerant or other fluid. When refrigerant is compressed, its heat increases. A *reversing valve* is a four-way valve that reverses the flow of refrigerant in a heat pump. The primary heat exchanger transfers heat to low-pressure refrigerant liquid to cool the air in building spaces, or removes heat from high-pressure refrigerant vapor to heat the air in building spaces. The secondary heat exchanger is used to transfer heat or cooling from the underground piping system. An *expansion valve* is a device that reduces the pressure on liquid refrigerant by allowing the refrigerant to expand. The expansion of the refrigerant cools it, allowing a cooling effect to be transferred to the air in building spaces.

Ground Source Heat Pump System Heating Mode

In winter, a heat pump system transfers heat from the earth to the air in building spaces. When a ground source heat pump is in the heating mode, heat from the earth is transferred from the underground piping system to the secondary heat exchanger. In the secondary heat exchanger, heat from the earth is added to the system refrigerant. The system refrigerant then travels to the primary heat exchanger, where the heat is removed from the refrigerant as cool air from the building passes over the heat exchanger. This warms the air in the building spaces. **See Ground Source Heat Pump System Heating Mode.**

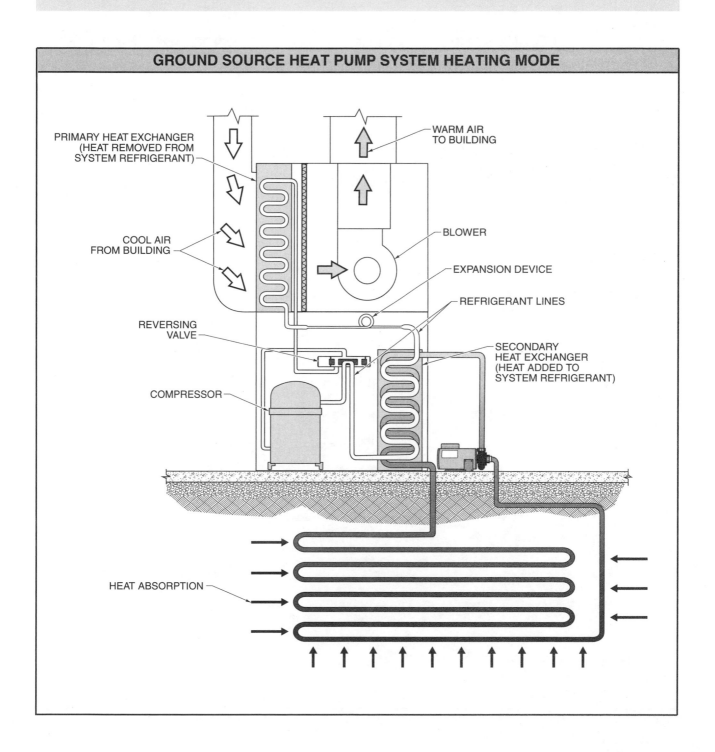

GROUND SOURCE HEAT PUMP SYSTEM HEATING MODE

Ground Source Heat Pump System Cooling Mode

In summer, a heat pump system transfers heat from the air in building spaces to the earth. When a ground source heat pump is in the cooling mode, cooling from the earth is transferred from the underground piping system to the secondary heat exchanger. In the secondary heat exchanger, cooling from the earth is added to the system refrigerant. The system refrigerant then travels to the primary heat exchanger, where the heat is added to the refrigerant as warm air from the building passes over the heat exchanger. This cools the air in the building spaces. **See Ground Source Heat Pump System Cooling Mode.**

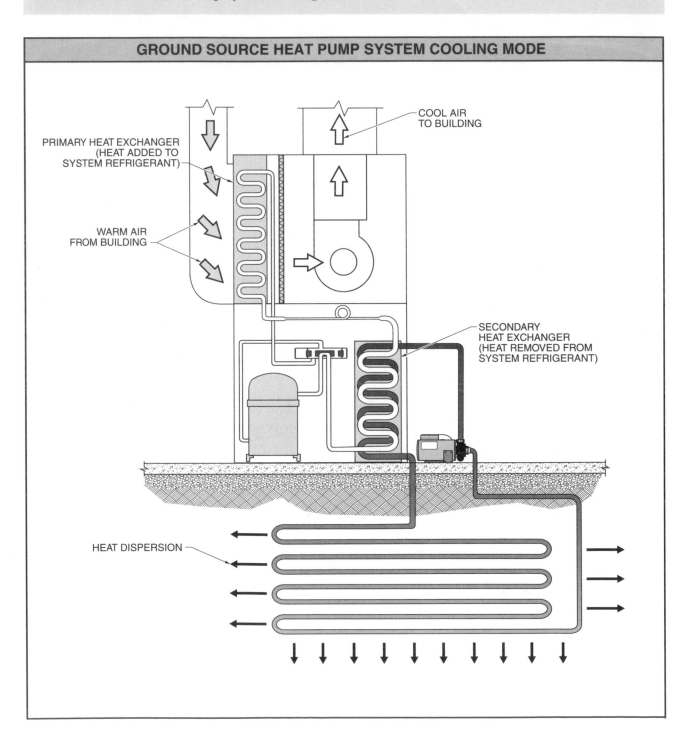

GROUND SOURCE HEAT PUMP SYSTEM COOLING MODE

COOL AIR
TO BUILDING

PRIMARY HEAT EXCHANGER
(HEAT ADDED TO
SYSTEM REFRIGERANT)

WARM AIR
FROM BUILDING

SECONDARY
HEAT EXCHANGER
(HEAT REMOVED FROM
SYSTEM REFRIGERANT)

HEAT DISPERSION

Troubleshooting Guides

Troubleshooting guides state a problem, possible cause(s), and corrective action(s) that may be taken. These guides may be used to quickly determine potential problems and possible courses of action. **See Geothermal Heat Pump System Troubleshooting Guide.**

GEOTHERMAL HEAT PUMP SYSTEM TROUBLESHOOTING GUIDE		
Problem	**Possible Cause**	**Corrective Action**
Reduced heating/cooling (system operating)	Dirty air filters	Check/clean/replace.
	Low refrigerant change (cooling problem)	Check charge. Check for system leaks. Recharge system.
	Poor water circulation	Check circulating pump, water flow (frozen pipe), or piping system for leaks.
System not operating	Ensure thermostat is set higher (heating) or lower (cooling) than room temperature	Correct thermostat setting.
	Tripped circuit breaker or blown fuse	Reset/replace circuit breaker/fuse and measure voltage and current level for correct operating level.
	Thermostat control transformer problem	Check/measure voltage into and out of transformer (usually 24 V output system voltage).
Part of system not operating	Electrical problem	Check voltage and current at each motor (pump, compressor, and circulating air blower).
System does not switch from cooling mode to heating mode or vice versa	Reversing valve stuck or faulty	Check reversing valve to verify that it is or is not operating. Replace if required.
Poor performance in extreme heat/cold	System undersized	Verify system operates properly under normal conditions. Check size of unit to size of building.

Geothermal System Basic System Checks

Before opening any panels or taking any measurements, the thermostat setting (COOL/OFF/HEAT) should be checked. If the system is to provide heat (heating mode), the thermostat should be set to HEAT at a setting higher than the room temperature. If the system is to provide cooling (cooling mode), the thermostat should be set to COOL at a setting lower than the room temperature.

The system power is checked by placing the fan switch in the ON position. The blower should turn ON even if there is no heating or cooling. If there seems to be no power, the system circuit breaker (or fuse) in the service panel supplying the system power should be checked. If the circuit breaker is tripped (or fuse blown), it should be replaced and the system operation observed. If the system seems to be operating, it should be monitored for proper operation. If the breaker trips or the fuse blows again, the system should be shut OFF until further troubleshooting can be done. Before doing any electrical troubleshooting, it is important to ensure the water lines are not blocked or frozen, or the valve shut off.

If the circuit breaker is not tripped (or fuse not blown) and the unit still does not turn ON, the disconnect switch located on the outside of the unit should be checked. The disconnect may have been switched OFF. If in the OFF position, it should be switched to the ON position. It may take a few minutes for the unit to cycle ON.

Before checking the electrical system, a troubleshooter must observe the system to see which components appear to be energized and/or operating. If no part of the system seems to have power, a thermostat or circuit breaker problem may have been caused by a short circuit or overloaded or jammed motor. If part of the system is working, such as the blower motor, compressor motor, or water circulating pump, the problem may be caused by one failing component in the system. If all of the systems seem to be operating, but there is little or no heating or cooling, it is normally not an electrical problem. Instead, it may be system coolant problem, mechanical problem, fluid circulating problem, or belt problem (squealing noise, especially during startup). Erratic system operation normally indicates a thermostat problem or thermostat programming problem.

Electrical System Troubleshooting

When checking an electrical system, a troubleshooter should always wear proper personal protective equipment (PPE) and stay clear of any rotating parts that may turn ON at any time. If power is not required to be ON for any checks or repairs, it is necessary to turn power OFF and lock out and tag out the main power feeding the system.

System troubleshooting should begin by looking for any problems that can be seen, such as motors that look worn or burnt, burnt wire insulation, broken motor/pump shaft/couplings, obstructions, or damage from flooding, rust, etc. If only one motor is not operating, the motor should be checked for a thermal overload switch. If there is a thermal overload switch, a troubleshooter should press the reset button on the motor. It may take several minutes for the system/motor to turn ON. If the problem is still not found and corrected, electrical measurements must be taken.

Voltage measurements are taken to determine if the correct voltage is present. Voltage coming into the system is measured to ensure that the problem is not somewhere outside the system. Voltage into and out of all fuses/circuit breakers is measured to ensure power is supplied to the fuses/circuit breakers and that the voltage is also coming out (fuse/circuit breaker not open). Voltage is also measured at all motors and solenoids. If voltage is present at the motors and solenoids but the device is not working, the motor or solenoid is probably faulty.

Current measurements are taken to determine if the device is operating and to determine its operating condition. A device that has voltage applied to it, but is not drawing current, is not producing power or operating (probably an open circuit within the device). A device that is drawing higher-than-rated current has a problem. Normally, the device is overloaded (belts too tight, frozen pipes, insulation breakdown, overheating, etc.). A device that is drawing normal current usually does not have a problem.

 Refer to **Activity 26-1—Geothermal Systems** ....

APPLICATION 26-2—Wind Turbine Systems

A *wind turbine* is a machine that converts the kinetic energy of wind into rotating mechanical energy. This rotating mechanical energy is used to rotate a generator that produces electrical energy. The electrical energy produced by a wind turbine can be stored in batteries until needed or can be used to directly operate electrical loads that are on or off the utility grid. The primary components of a wind turbine are the blades, hub, rotor, nacelle, gearbox, generator, power converter, controller, and tower. **See Wind Turbine Primary Components.**

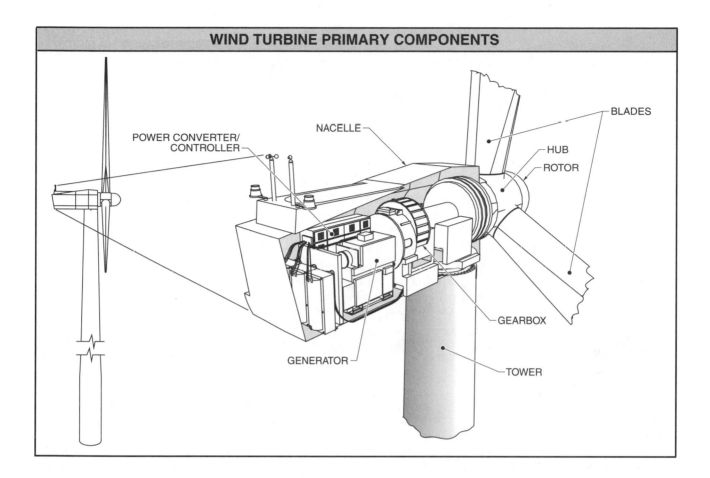

WIND TURBINE PRIMARY COMPONENTS

Blades

A *blade* is a long slender structure aerodynamically designed to capture the greatest amount of wind and produce the maximum rotating force. The blades are one of the most important parts of a wind turbine because they capture the kinetic energy of the wind and convert it to a rotating mechanical force that drives an electrical generator. Wind turbines may be designed with one, two, or multiple blades. Most wind turbines have three blades made of fiberglass-reinforced epoxy composites. Features such as lightning and de-icing protection can be included within the blade structure. On large wind turbines, lightning protection is required because of the height of the blades and the damage a lightning strike can cause to the blades and the electrical and electronic control systems.

Hub

A *hub* is the center part of a wind turbine rotor to which the blades are attached. The hub can have individual blades connected to it, or the hub and blades can be manufactured as one unit.

Rotor

A *rotor* is the central rotating shaft of a wind turbine that includes the blades and hub. The faster the rotor rotates, the higher the generated electrical power. However, slow rotor speeds are not efficient and excessive rotor speeds are dangerous. Large wind turbines are designed to produce an electrical power output when the rotor rotates within a set speed, such as 10 rpm to 60 rpm.

Nacelle

A *nacelle* is a housing that includes the gearbox, generator, power converter, electrical controls, brakes, and other components required to convert rotating mechanical force into electricity. The main function of the nacelle is to protect the system components from the elements and wildlife while also protecting individuals from the rotating and electrical components. The nacelle also provides protection from the elements for individuals working on the system.

Gearbox

A *gearbox* is a housing that contains several different sizes of gears that are used to change the rotational speed between its input shaft and output shaft. Gearboxes change the slow rotational speed of the rotor to a high rotational speed required by the generator. Normally, wind turbine rotors rotate between 10 rpm to 60 rpm during power generation. The generator rotates at between 1200 rpm to1800 rpm for 60 Hz systems, and between 1000 rpm to 1500 rpm for 50 Hz systems.

The gearbox has a low-speed shaft (blade side) and a high-speed shaft (generator side). *Gear ratio* is the number of revolutions per minute between the low-speed shaft and the high-speed shaft. For example, if the gear ratio is 1:50 and the low-speed shaft is rotating at 30 rpm, the high-speed shaft rotates at 1500 rpm. Although gear ratio describes the total gain (e.g., 1:50), the gearbox normally increases the shaft speed in several small stages using several internal gears instead of one gear. Proper lubrication and the limiting of the internal gearbox temperature are required for proper gearbox operation and for reducing noise and required maintenance.

Generator

A *generator (alternator)* is a device that converts mechanical energy into electrical energy by means of electromagnetic induction. Generators can be used to produce either DC or AC power. All generators operate on the principle that when a conductor (coil of wire) is rotated in a magnetic field, a voltage is produced. The magnetic field required to produce a voltage can be generated by permanent magnets or electromagnets. Permanent magnets are used in some wind turbine generators. Electromagnets can be made in any size and allow for large amounts of power to be generated. Electromagnets are used in most AC generators and in most wind turbine generators. **See Single-Phase AC Generator.**

AC generators can produce either single-phase AC power or three-phase AC power. A single-phase generator is less efficient than a three-phase generator and is normally limited to low-power applications, such as portable generators used on construction sites. Except for some very low power wind turbines, wind turbines use three-phase AC generators to produce electrical power. When a single-phase voltage output is required, a three-phase AC voltage can be still generated with the three-phase AC changed to single-phase AC. **See Three-Phase AC Generators.**

The two main types of AC generators used in wind turbines are induction generators (asynchronous generators) and synchronous generators. Both generate three-phase AC voltage that can be connected to the utility grid or used to directly operate electrical loads, and both have basically the same stator winding arrangements and generate the same type of three-phase voltage output. The main difference is in the way the electric field is produced in the rotor.

An *induction (asynchronous) generator* is an AC generator that rotates at a changing speed slightly below synchronous speed. *Synchronous speed* is the theoretical speed of a generator or motor based on the number of stationary poles and the line frequency. The rotor in an induction generator has a squirrel-cage winding design in which the current in the rotor bars is induced as the rotor bars cut through the stator magnetic field. Induction generators are the most common type of AC generator used in wind turbines.

A *synchronous generator* is an AC generator that rotates at a constant speed. The rotor in a synchronous generator has electromagnets mounted on the rotor, which cause the rotor magnetic field to lock in sync with the stator magnetic field, causing the generator to operate at synchronous speed. Synchronous speed equals 120 times the frequency (in Hz) divided by the number of poles. For example, in a two-pole generator operating at 60 Hz, the synchronous speed is 3600 rpm ($\frac{120 \times 60}{2} = 3600$ rpm). Synchronous generators are the most common type of AC generator used in utility power plants.

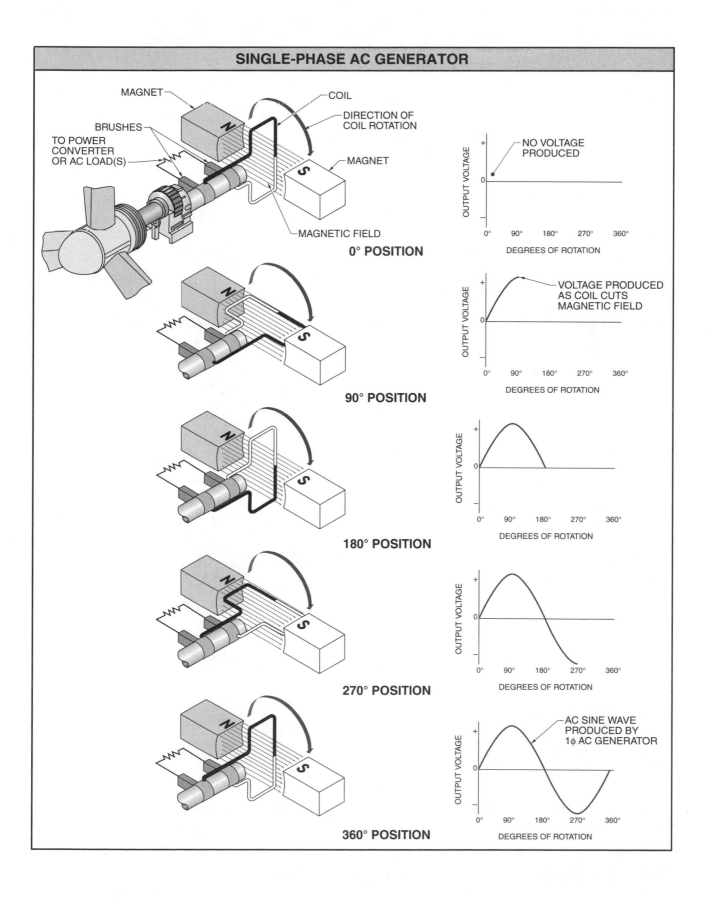

SINGLE-PHASE AC GENERATOR

THREE-PHASE AC GENERATORS

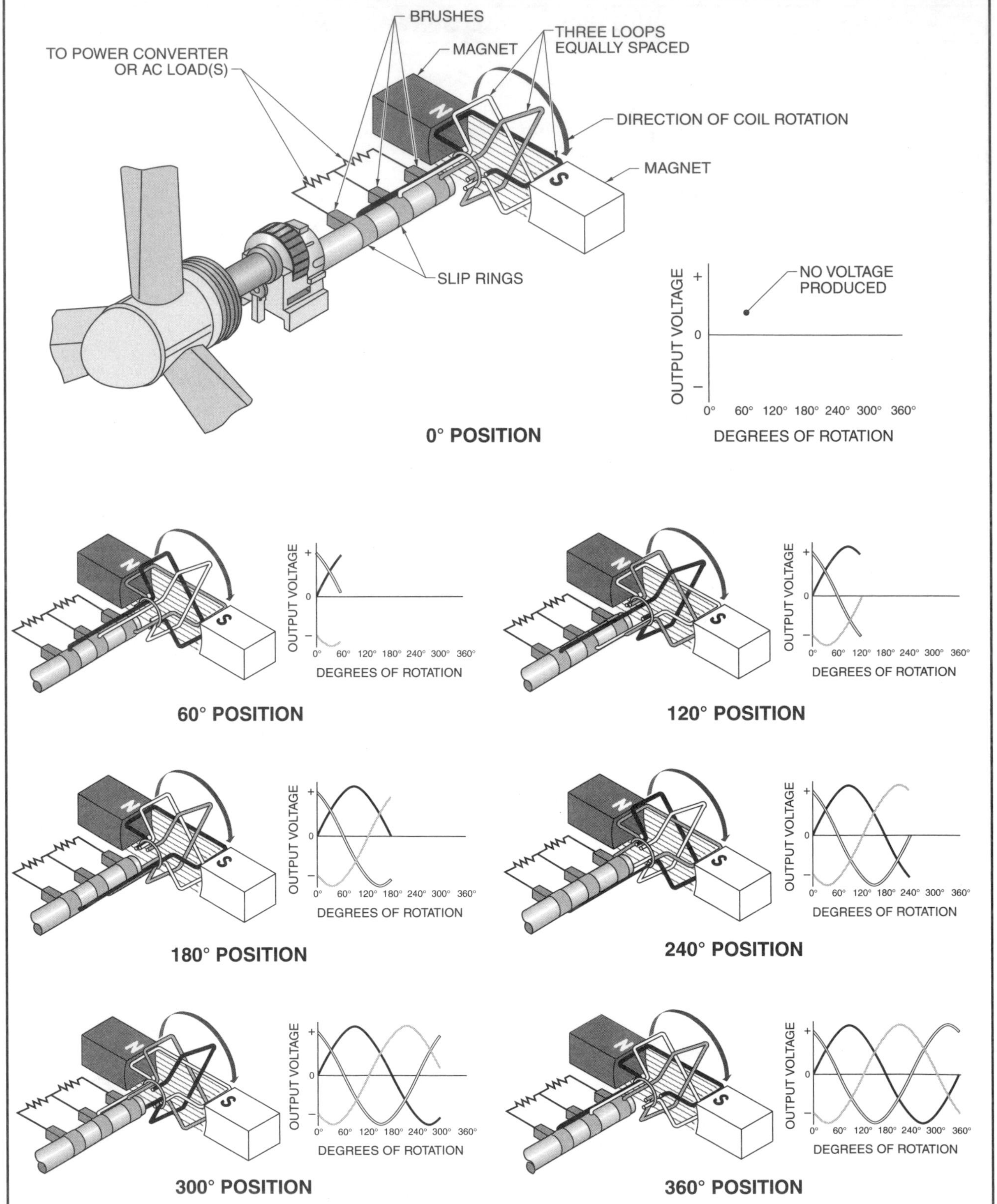

0° POSITION

60° POSITION

120° POSITION

180° POSITION

240° POSITION

300° POSITION

360° POSITION

Power Converter

Electrical loads and the utility power grid are designed to operate at a constant frequency of 60 Hz (50 Hz in some countries). The output of an AC variable speed wind turbine generator is an AC voltage at a varying AC frequency (Hz). In order to use the AC voltage on electrical loads designed to operate at a constant 60 Hz (or 50 Hz) frequency, or to add the voltage to the utility power grid, the voltage must be set to a constant frequency regardless of the wind or generator speed. For this reason, the output of an AC generator is first converted (changed from AC to DC), filtered, inverted to a constant frequency AC voltage (60 Hz or 50 Hz), filtered (conditioned), and sent to transformers to step up the low voltage to the high voltage of the utility grid. **See Electrical Power Production and Conversion.**

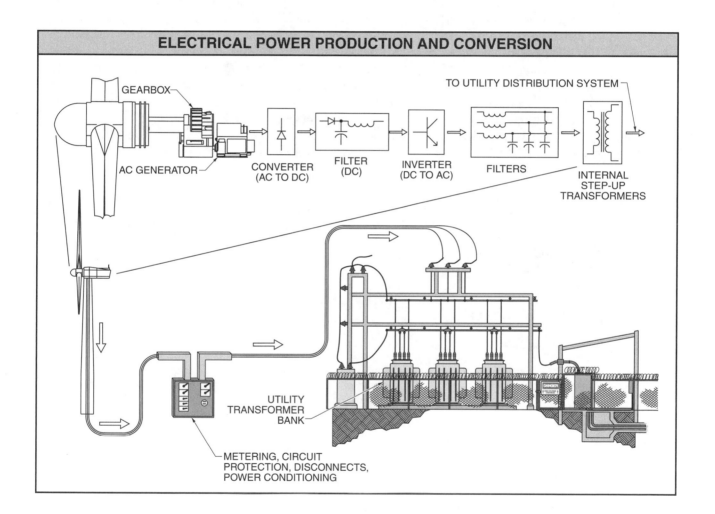

ELECTRICAL POWER PRODUCTION AND CONVERSION

A *converter* is an electronic device that changes AC voltage to DC voltage. The power converter of a wind turbine uses diodes to rectify (change) AC to DC. A *diode* is an electronic device that allows current to pass through in only one direction. Diodes are also referred to as rectifiers because diodes rectify (change) AC voltage to DC voltage. Because the DC voltage output of a rectifier is a pulsating DC, a filter (capacitors and/or inductors) is normally added to smooth the pulsating DC voltage.

Capacitors and inductors (coils) are used together in DC filter circuits. Working together, capacitors and inductors maintain a smooth waveform because capacitors oppose a change in voltage, and coils oppose a change in current. Capacitors oppose a change in voltage by holding a voltage charge that is discharged into the circuit any time the circuit voltage decreases. As current flows through a coil, a magnetic field is produced. The magnetic field remains at maximum potential until the circuit current is reduced. As the circuit current is reduced, the collapsing magnetic field around the coil induces current back into the circuit. The coil smoothes or filters the power by building a magnetic field as current is applied and adding current into the circuit as the magnetic field collapses.

An *inverter* is an electronic device that changes DC voltage into AC voltage. Inverters used in wind turbines change the DC voltage into an AC voltage at a fixed frequency and a set voltage level. Inverters in wind turbine systems use fast-acting solid-state switches that can control high current. The fast-acting solid-state switches must also have the lowest amount of power loss (voltage drop) possible to reduce the amount of heat produced and power loss. The solid-state switches turn the DC voltage on and off to remake AC voltage. High-power transistors are used to switch the voltage on and off. The inverter changes the DC voltage into an AC voltage using pulse width modulation (PWM) technology.

Pulse width modulation (PWM) controls the amount of voltage output by converting the DC voltage into fixed values of individual DC pulses. **See Pulse Width Modulation.** The fixed-value pulses are produced by the high-speed switching of transistors. By varying the width of each pulse (time ON), the output voltage can be increased or decreased. The wider the individual pulses, the higher the DC output voltage. By varying the DC output levels, an equivalent AC voltage is produced. The inverter circuit reverses the DC voltage to produce the negative side of the equivalent AC output. Filters (inductors and capacitors) can be added to help smooth the AC voltage.

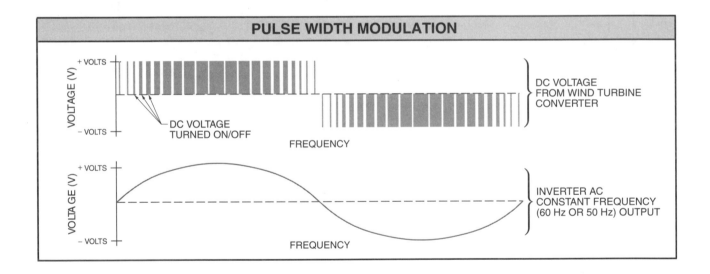

PULSE WIDTH MODULATION

Controller

Wind turbines are designed to generate electrical power to be used locally and/or to be added to the utility grid. In order to have a safe, economical, and efficient system, the electrical, mechanical, and environmental operating conditions must be measured and the information input into the control system. The wind turbine controller is the central electronic control device that receives signals based on operating conditions such as voltage level, current level, power output, frequency, wind speed, wind direction, rotor speed, pitch position, vibration, operating temperature, and faults. This information is used to determine when the generated power can be connected or disconnected from the utility grid. Controllers can be standard microprocessors, PLCs, or manufacturer-built proprietary electronic control systems. The controller can be limited to the local wind turbine system only or connected to an external system that can monitor the wind turbine system from a remote location.

Tower

The tower is the main structural component that holds the wind turbine rotor, nacelle, and other elevated components securely in place. Towers are normally made of tubular steel or concrete. Towers also provide space/support to allow the generated power to be brought down to ground level. On small wind turbines, reaching the nacelle is accomplished by climbing an external ladder. On large wind turbines, reaching the nacelle is accomplished by climbing an internal ladder. The higher the tower, the greater the power generated because wind speed increases with height.

 Refer to **Activity 26-2—Wind Turbine System** ....

APPLICATION 26-3—Wind Turbine Supporting Components

In addition to the wind turbine primary components, supporting components may be included as part of a wind turbine system. Wind turbine supporting components include brakes, fall arrest systems, anemometers, wind vanes, capacitors, heat exchangers, transformers, yaw motors/drives, and pitch control. **See Wind Turbine Supporting Components.**

Brakes

For optimum generating efficiency and safety, all wind turbines require a braking method to limit the speed of the rotor and must have a means of preventing the rotor from rotating during certain conditions. For example, during installation, inspection, and maintenance, the rotor must be locked to prevent rotation. Likewise, when high winds can rotate the rotor at dangerous speeds, the rotor speed must be reduced or the rotor stopped.

Most wind turbine brakes require power to release them so that the brakes are automatically applied during a power failure. Brakes can be controlled mechanically, hydraulically, or electrically. Brakes can be mounted on the low-speed shaft (rotor) or high-speed shaft (generator).

Fall Arrest Systems

A *fall arrest system* is a system of body harnesses/lanyards and attachments that limit the distance of a fall. A fall arrest system is designed to protect individuals against falls by limiting a fall to a total of 6' or less. Fall arrest systems are not designed to be used for support during construction, inspection, or maintenance. Properly designed harnesses should provide freedom of movement, minimize stress forces on the body in the event of an accidental fall requirements, and not interfere with normal work.

Anemometers

An *anemometer* is an instrument that measures wind speed. An anemometer is connected to the controller and is used to provide information to help determine when the wind turbine should rotate and when it should be stopped, such as during high winds and maintenance. The output signal from an anemometer may also be used to track wind speeds at all times and record, store, or send the wind speed information to a centralized location.

Wind Vanes

A *wind vane* is an instrument that measures wind direction. Like an anemometer, a wind vane is connected to the controller and used to provide information to help determine when the wind turbine should rotate and when it should be stopped. The output signal of a wind vane may also be used to track wind direction and record, store, or send the wind direction information to a centralized location.

Capacitors

A *capacitor* is an electrical device that can store an electrical charge. DC capacitors are used in wind turbine systems to help filter (smooth) the voltage coming out of the converter (rectifier) so the DC voltage is at a constant level before it travels to the inverter.

WIND TURBINE SUPPORTING COMPONENTS

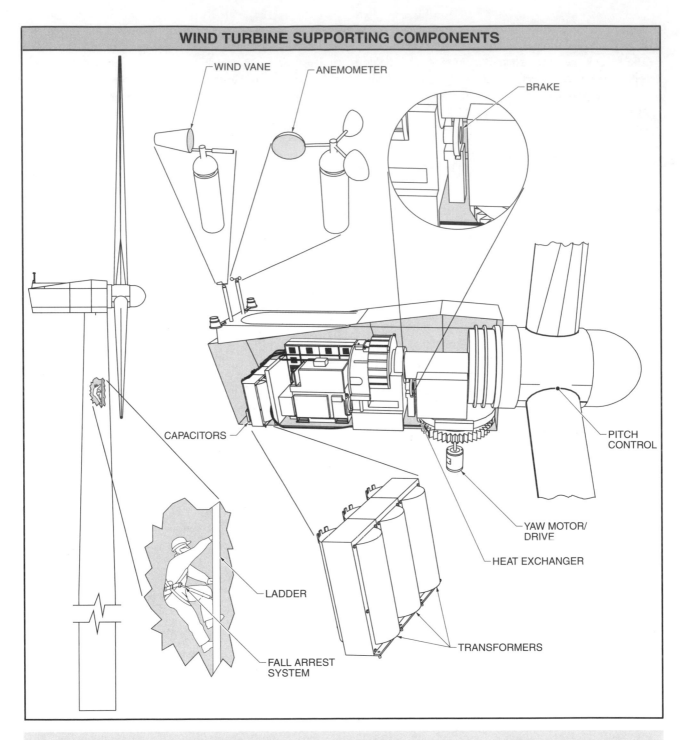

WIND VANE

ANEMOMETER

BRAKE

PITCH CONTROL

CAPACITORS

YAW MOTOR/ DRIVE

HEAT EXCHANGER

LADDER

TRANSFORMERS

FALL ARREST SYSTEM

AC capacitors are used in wind turbine systems to help correct the circuit power factor (PF). A circuit power factor is 1 (100%) when both voltage and current are in phase, as they are in a circuit that contains only resistance. However, in circuits that contain inductive reactance (motors, transformers, solenoids, and other coils), voltage and current are out of phase and the circuit power factor is less than 1. For example, a circuit that contains a motor may have a power factor of 0.75 (75%) because the circuit reactance has shifted the circuit voltage and current out of phase. A 0.75 PF means that the circuit is operating at 75% efficiency. Adding capacitance to an inductive circuit raises the circuit power factor and thus the efficiency.

Heat Exchangers

Electrical generators, components, and gearboxes all produce heat. Because the components that produce heat are also enclosed inside the nacelle, a heat exchanger is used in wind turbines to help cool air, oil, and/or water. A heat exchange system may consist of a fan motor blowing air over system components, or a dedicated heat exchanger that circulates and cools oil or water that is used to indirectly cool other components.

Transformers

A *transformer* is an electrical device that uses electromagnetism to change voltage from one level to another. Both step-up and step-down transformers are used in wind turbine systems. Step-up transformers are used in wind turbine systems to increase the generator AC output voltage to the high voltage of the utility grid. For example, the output of a generator may be 690 VAC (common output voltage on 50 Hz turbines) or 600 VAC (common output voltage on 60 Hz turbines), but the utility grid voltage can be thousands or tens of thousands of volts. Transformers are the system interface that allows the generated power of wind turbines to be added to the utility grid.

Step-down transformers are used in wind turbine systems to reduce the generator AC output voltage to a low voltage level required by many of the system control and monitoring devices. Some of the low-voltage AC is also rectified to operate low-voltage DC loads and controls. The DC voltage can also be used to store power in, or recharge, DC batteries. Batteries in wind turbine systems are used for storing power (large battery systems) and also as emergency power for automatically stopping the blades, sounding an alarm, and/or sending a system failure signal to operators when a power failure occurs.

Yaw Motors/Drives

Yawing is a side-to-side movement. In wind turbine systems, yaw control refers to aligning the blades (and nacelle) in the direction of the wind. Aligning the blades for maximum wind pressure is required for maximum power generation. Aligning the blades to reduce wind pressure during high winds is also required to prevent overspeed and damage. **See Yaw Control.**

Downwind turbines do not require a yaw drive motor because the wind blows the rotor downwind (as in a windmill). Upwind turbines face into the wind and require a yaw control to keep the rotor facing into the wind as the wind direction changes. Since most wind turbines are of the upwind design, they require some type of yaw control.

Yaw control can be either passive or active. Passive yaw control uses the wind acting on a tail vane to position the blades into the wind. Passive yaw control is used on all windmills and small wind turbines. Active yaw control uses sensors and drive motors (electric and/or hydraulic) to position the blades into the wind. Active yaw control is used on large wind turbines to keep the blades in the optimum operating position for a given wind speed and direction.

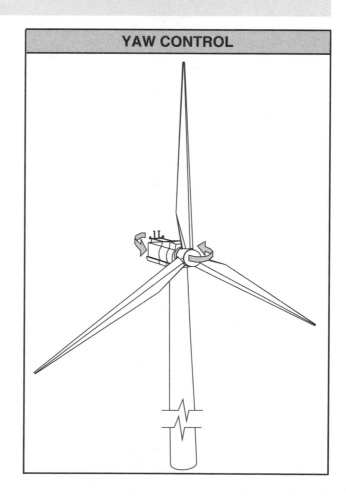

YAW CONTROL

Pitch Control

Wind turbine blades are designed to capture the greatest amount of wind possible. However, during times of high winds, the blades can rotate the rotor at unsafe speeds. Wind turbine systems include brake systems, but other methods are required to reduce rotor speed during excessively high winds because brake systems can fail. To reduce the speed of rotor rotation, the blades are pitched (turned) out of the wind to reduce rotational speed. Active pitch control uses a method of changing the angle of the blades and/or angle at which the wind hits the blades to reduce the force applied to the blades and thus the rotor speed. **See Pitch Control.**

The electrical power output of a wind turbine increases with wind speed. Since the wind speed varies, the electrical power output also varies. Electronic pitch controls monitor the power output of the turbine several times per second and sends a signal to the blade pitch system, which automatically pitches the rotor blades slightly out of the wind when the power output becomes excessively high. Likewise, the blades are pitched back into the wind when the wind speed slows. Pitch control is also used to help smooth the electrical power output (or reduce it if needed) by changing the blade pitch to slow down or speed up the rotor to help maintain a desired power output.

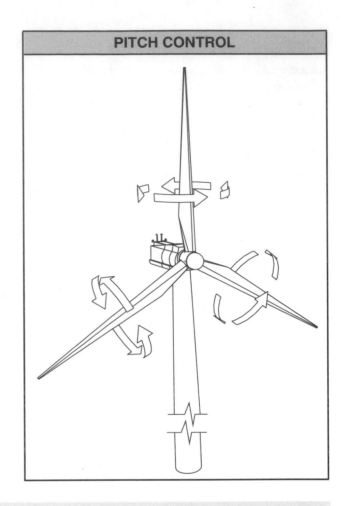

PITCH CONTROL

Wind Turbine System Safety

Wind turbines can safely produce electrical power if they are installed properly and are routinely inspected and maintained, and if personnel working on and around them are properly trained. However, since wind turbines are mounted at heights, include both internal and external rotating parts, and produce electricity, the dangers must be understood and steps taken to prevent accidents. Wind turbine hazards include falling (working at heights and in high winds); being caught in rotating equipment; battery hazards; and electrical shock, burns, and electrocution.

Falls can be prevented by using proper climbing equipment and proper fall arrest equipment, and by following all safety rules. These include wearing proper body harnesses and anchors during ascending and descending and when working on the nacelle or tower. Large wind turbines incorporate a fall-arrest system that includes a cable running the length of the tower to which a person's body harness can be connected to prevent falls and reduce the distance of a fall to a short distance.

Prevention against being caught in rotating equipment requires first guarding the rotating equipment as much as possible and educating workers about the dangers and steps to follow to prevent contact with dangerous parts of the system. This includes stopping the blades (applying system brakes) whenever possible before working on the system and avoiding rotating internal or external parts.

If batteries are used to store electrical power, steps must be taken to prevent an explosion. Batteries produce hydrogen gas when fully charged. Hydrogen gas can cause an explosion if a spark is present and areas are not ventilated. All battery storage containers must be properly vented to prevent gas buildup. Care must be taken to prevent uninsulated tools or other metal parts from causing a short (spark) by contacting the positive side of the battery circuit and the negative/ground side. The appropriate personal protective equipment (PPE) must be worn at all times.

Preventing electrical shock requires proper installation and testing of the system. All non-current-carrying metal parts should be checked for proper grounding before any work is done on the mechanical or electrical systems. Wind turbine metal parts are tested for proper grounding by applying the following procedure:

1. Wear and use proper PPE including fall restraints, insulated gloves, eye protection, and flame-resistant clothing.

2. Set the DMM to measure AC voltage. Start with the highest measurement range and use a high-voltage attachment probe if the voltage may be higher than the highest meter range.

3. Connect the black meter lead to the unpainted (bare) metal part under test.

4. Connect the red meter lead to a known ungrounded (hot/energized) terminal. The DMM should read the hot-to-ground AC voltage. If the DMM does not read the hot-to-ground voltage, the metal under test may not be grounded and/or the red test lead is not on a hot terminal or the black lead is not securely connected.

If the metal under test does not seem to be grounded, test the circuit again using the DC voltage setting on the DMM because the voltage at the test point may be DC. The DMM should read the hot-to-ground DC voltage. If the DMM does not read the hot-to-neutral (ground on DC) voltage, the metal under test may not be grounded and/or the red test lead is not on a hot (positive on DC) terminal. If the non-current-carrying metal part under test is not grounded, the system can cause severe/fatal electrical shock and the entire grounding system must be checked.

 Refer to **Activity 26-3—Wind Turbine Supporting Components** ....

APPLICATION 26-4—Wind Turbine Power Output

All wind turbines are rated for the amount of power they can produce under normal operating conditions. The amount of power produced by a wind turbine depends on the following fixed and variable factors:

- Size (area) of the blades. The larger the surface area of the blades is, the greater the force is applied to the rotor.
- Design of the blades. The more aerodynamic (less resistant) the blades are, the easier and faster the blades rotate.
- Wind speed. The faster the wind speed is, the greater is the force pushing on the blades and the higher is the force produced to rotate the gearbox/generator shaft.
- System control circuit. All systems have losses. Wind turbine power loss comes from heat loss, friction loss, electrical power loss, and wind energy loss. In addition to power loss, wind turbine components fail. Power loss and component failures can be kept to a minimum when the control system is used to monitor the system operating conditions, make corrections, and identify (and report) potential problems before major damage occurs.
- Generator electrical power rating. Generators are rated for the amount of power they can output, such as 300 kW, 2 MW, etc. The higher the generator rating is, the greater the power output is. However, the blades/rotor must be sized to produce enough force so that the generator can output its rated power at an expected normal wind speed. Wind turbine power outputs range from as low as 50 W to over 5 MW. **See Wind Turbine Size and Power Ratings.**

WIND TURBINE SIZE AND POWER RATINGS		
Rotor*	Diameter†	Approximate Generator Output
0.82	0.25	50 W
7.87	2.4	900 W (0.9 kW)
13.45	4.1	2 kW
41	12.5	50 kW
59	18	100 kW
88.5	27	225 kW
108	33	300 kW
131	40	500 kW
144	44	600 kW
157	48	750 kW
177	54	1 MW (1000 kW)
209	64	1.5 MW
236	72	2 MW
262	80	2.5 MW
328	100	3 MW

* in ft
† in meters

The required power output depends on the intended use of a wind turbine. Wind turbines designed to feed all of their power to the utility grid are useful regardless of the amount of power they produce at any given time. Wind turbines designed as a primary power supply for individual homes, farms, and small businesses should be sized large enough to deliver enough power under normal operating conditions. For example, a wind turbine designed for supplying power to a home can range from 3 kW to 10 kW or higher, depending on the home size and electrical usage. Wind turbines designed for home use are selected based on the desired amount of power from the wind turbine, if there are other power sources (e.g., photovoltaic), and if the home is on or off the utility grid. If the home is on the utility grid, a small wind turbine can be used because power can always be bought during high power usage or low-wind-speed conditions. If the home is off the utility grid, a large wind turbine should be used and/or battery storage provided.

In order for the generator to produce power, the rotor must rotate the generator shaft. The faster the shaft rotates, the higher the produced power output. Wind speed is greater at high altitudes. Thus, increasing the height of the tower/blades produces a greater force on the rotor. Wind turbine power ratings are based on installing the blades/rotor at a specified height. **See Wind Speed and Power Output.** Wind speed is stated in meters per second (m/s), miles per hour (mph), or knots. One m/s equals 2.24 mph and 1 mph equals 0.870 knots.

WIND SPEED AND POWER OUTPUT*

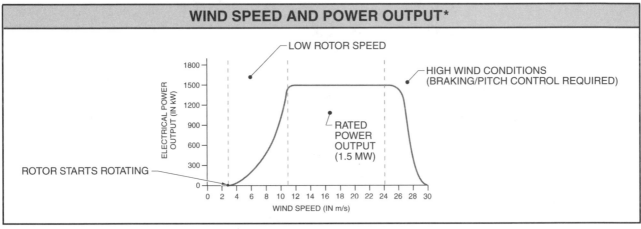

* FOR 1.5 MW RATED WIND TURBINE

For any given power rating, increasing voltage lowers current by the same percent, and decreasing voltage increases current by the same percent. For example, a 50 kW wind turbine produces 434.7 A at 115 V, 217.3 A at 230 V, and 108.7 A at 460 V. This same wind turbine produces 83.3 A at 600 V, 72.42 A at 690 V, 10 A at 5000 V, and 5 A at 10,000 V. Because conductor (wire) size is based on the amount of current it can safely handle, in theory, the higher the voltage the better because current is less for any given amount of power. Theoretically, wind turbines should generate an output voltage as high as possible to reduce the amount of current, and thus the conductor size and the size of the other electrical components and switches within the system. Another advantage to increasing the voltage is that the higher the voltage is transmitted over conductors, the lower the power loss. However, wind turbines do not output high voltages for the following reasons:

- Voltages over 1000 V require special installation, circuit and worker protection, test instruments, etc. and normally fall under stronger regulations and codes.
- High voltages can produce more sparking and require greater distances between conducting parts, increasing size and weight.
- Working around high voltages requires additional training and tools.

For these reasons, wind turbine system output voltage is normally kept below 1000 V (normally 600 V or 690 V). However, the low output voltage must be limited to as short a distance as possible to minimize power loss. In a wind turbine system, if the generated voltage (600 V/690 V) is brought down from the nacelle to ground level before increasing the voltage, there is a much higher power loss than if the voltage is first increased by transformers located in the nacelle and then brought down to ground level. The high voltage is connected to a step-up transformer on the ground that steps the voltage up to the level of the utility grid.

Between the power output of a wind turbine and the utility grid, and at other points in the electrical system, there must be disconnect switches and overcurrent/short circuit protection (circuit breakers and/or fuses). The disconnect switches provide a safe place to turn the power off (lockout/tagout) and on. The circuit breakers or fuses provide a point at which power is automatically removed when the system is overloaded or shorted and provide a known point for troubleshooting.

Electric power meters are connected to the wind turbine system to measure and record the amount of power output. Other meters, such as voltmeters, ammeters, frequency meters, and power factor meters are connected to the system to monitor electrical quantities. Metering the system operating parameters helps in determining any problems, provides data for system improvements, and provides documentation for billing power output and meeting regulatory requirements.

 Refer to **Activity 26-4—Wind Turbine Power Output** ....

APPLICATION 26-5—Photovoltaic Systems

A *photovoltaic (PV) system* is an electrical system that converts light energy into electrical energy. Since PV systems produce electricity using light energy, they are not affected by cold weather but are affected by the amount of light that falls on them. PV systems have an electrical output between 0% (no light) to full-rated power output in direct sunlight. Diminished light (cloudy weather and sun at indirect angles) reduces the power output. The primary parts of a PV system are the photovoltaic cells that convert light energy into direct current (DC), a DC/DC controller that controls and regulates the flow of DC, and a DC/AC controller (inverter) that changes DC into 60 Hz AC. **See Photovoltaic System.**

PV systems may be stand-alone, hybrid, or utility grid tied systems. Stand-alone PV systems are systems that are the only source of electrical power and are not connected to the utility grid. Stand-alone systems can be small for producing power for a few loads and/or a battery bank, such as the systems used on remote traffic warning/safety signs. Other stand-alone systems can be large enough to supply all the power for a remote residential or commercial building. Although both small and large systems produce DC voltage, a DC to AC controller (inverter) is used to output a standard 60 Hz AC voltage.

Hybrid systems are systems that combine PV systems with additional electrical power supplied from the utility grid or other source. Hybrid systems are the most commonly used systems because photovoltaic cells only produce electrical power when light shines on them, and battery storage is generally limited and expensive. Normally, the PV system is backed up with power from the utility grid. The utility grid can supply part or all of the required power as needed. The PV system can also be used to feed extra power to the utility grid during times of high power output and low power usage.

Utility grid tied systems are systems in which the primary purpose of the PV system is to supply electrical power to the utility grid for general power distribution. Utility grid tied systems are normally very large, so they are cost effective. Utility grid tied systems can be stand-alone systems in a remote location, such as an open field that is close enough to be connected to the utility lines. Large stand-alone PV plants can produce power outputs in the megawatt range. However, more practical systems are placed on roofs of large buildings, such as warehouses, to produce power for the utility grid that supplies power to the building.

PHOTOVOLTAIC SYSTEM

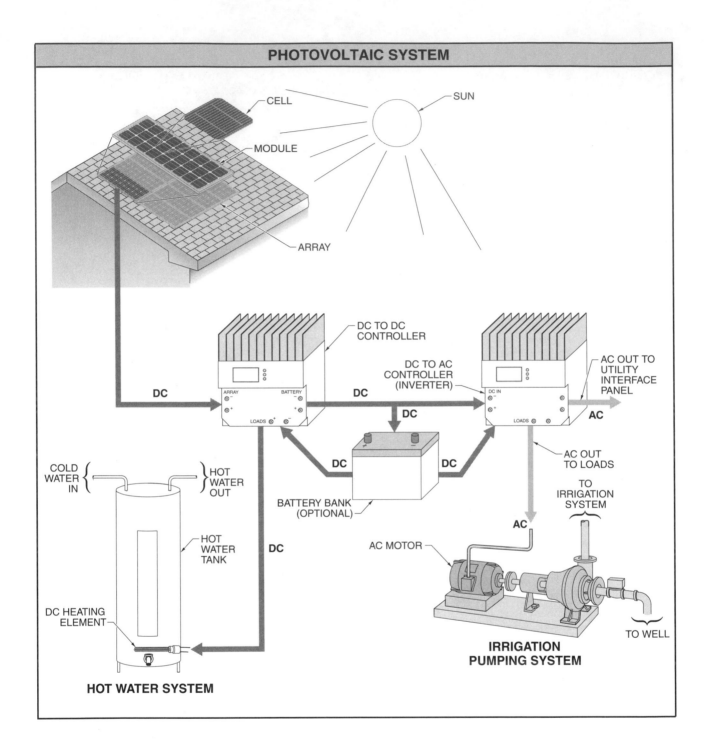

When designing, selecting, installing, and troubleshooting a PV system, each component and its relationship to the other components in the system must be understood. Understanding the individual components and systems helps in determining the test instruments to use when testing or troubleshooting the system and the measurements to take that indicate a working component or system or a problem or fault.

Photovoltaic Arrays

A *photovoltaic cell* is a solid-state semiconductor that converts light energy into electrical energy. Individual photovoltaic cells produce little power and several must be interconnected into modules and arrays to increase power output. A *photovoltaic array* is a group of modules that include many individual photovoltaic cells interconnected in series and/or parallel combinations to produce a DC power output. Cells/modules/arrays are connected in series to increase the total DC voltage output. Cells/modules/arrays are connected in parallel to increase the total DC current output. Cells/modules/arrays are connected in series/parallel combinations to produce the required DC voltage output at a rated DC current output.

A DMM set to measure DC voltage can be used to measure the DC voltage out of a PV cell, module, or array. A DMM set to measure DC current can be used to measure the DC current out of a PV cell, module, or array. Since individual cells produce small amounts of current, an in-line current measurement is the most accurate method of measuring current values below 10 A. When individual cells are connected to form modules or arrays, their total current and voltage levels increase and a DC clamp-on ammeter can be used to measure the high current. Current measurements can be taken when the system is connected to the load (operating current flow), or when the load is removed and the ammeter is placed directly across the cell/module/array (short-circuit current). Measuring the short-circuit current indicates the maximum current output under the given light conditions. **See Measuring PV System Voltage and Current Output.**

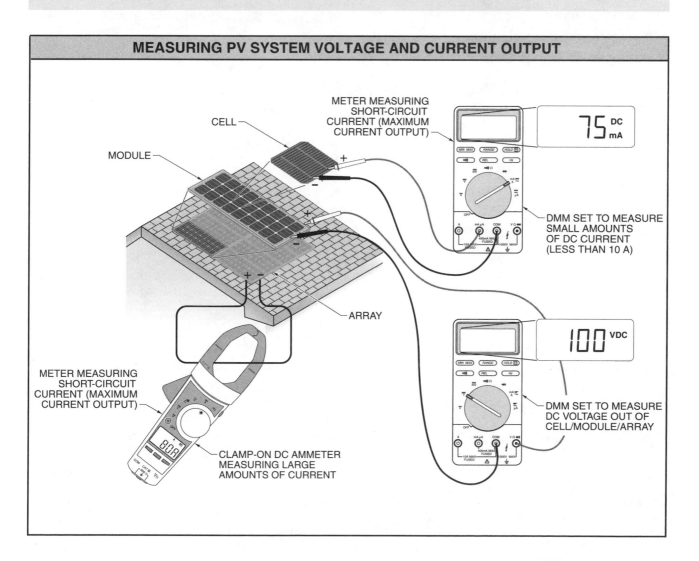

MEASURING PV SYSTEM VOLTAGE AND CURRENT OUTPUT

DC/DC Controller

The DC power output of a PV system can be used to directly control some DC loads. For example, DC heating elements produce heat at any applied voltage below their rated DC voltage. The higher the voltage is, the higher the power is and hotter the heating elements are. Such heating elements can be used to directly heat air or water. However, in a PV system, since the DC voltage and power output varies based on the changing amount of light, and the output power of a PV system is normally used for different types of electrical loads, the DC output of a PV system is normally connected to a DC/DC controller that controls the power to the loads and/or batteries. The controller must control the amount of voltage/power into and out of the batteries to prevent them from being overcharged or discharged too fast. A DMM set to measure DC voltage can be used to measure the DC voltage into and out of a DC/DC controller. A DMM set to measure DC current can be used to measure the current into and out of a DC/DC controller. **See Measuring DC/DC Controller Voltage and Current.**

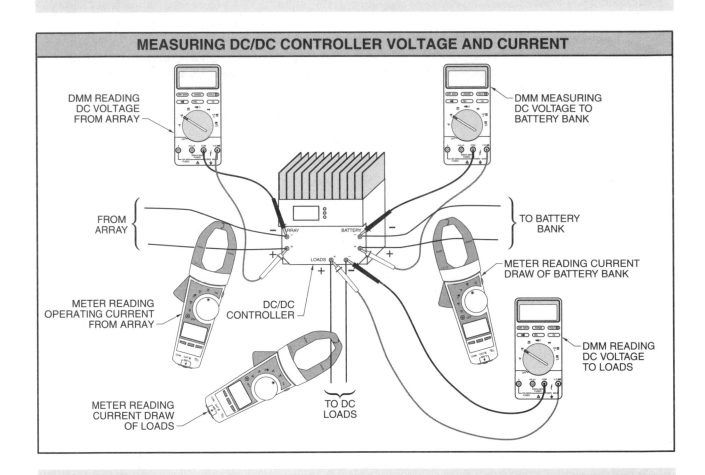

MEASURING DC/DC CONTROLLER VOLTAGE AND CURRENT

Batteries

PV systems can only produce power when light shines on them. Thus, most PV systems include batteries for storing energy, and/or are connected to an alterative source of power. Batteries must be checked, cleaned, and replaced as required. Although sealed batteries require less maintenance, open-vent, lead-acid batteries are normally used in PV systems because they are a more economical way of storing large amounts of electrical power and are designed for long charging/discharging cycles. Open-vent batteries require checking and replacing water and require scheduled checking and cleaning.

Checking a battery open-circuit (no load) voltage and specific gravity gives an approximation of the battery state of charge. A *hydrometer* is a test instrument that measures the specific gravity of a liquid compared to water, which has a specific gravity of 1. Any liquid thinner than water has a specific gravity of less than 1. A liquid thicker than water has a specific gravity greater than 1. Any reading over 1 indicates that the electrolyte in the battery is thicker than water. As a battery discharges, the electrolyte thins (less specific gravity) and as the battery charges, the electrolyte thickens (higher specific gravity). Battery testing should be done when the battery has had time to properly charge and has been disconnected for several minutes. **See Battery-Specific Gravity and Open-Circuit Voltage Test.**

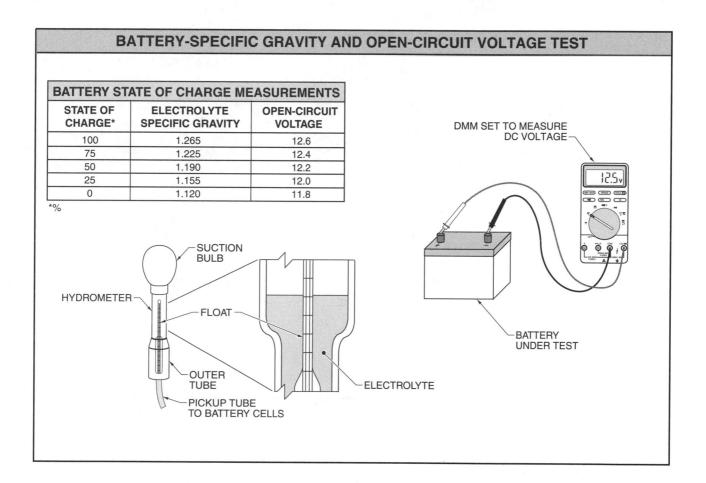

BATTERY-SPECIFIC GRAVITY AND OPEN-CIRCUIT VOLTAGE TEST

BATTERY STATE OF CHARGE MEASUREMENTS

STATE OF CHARGE*	ELECTROLYTE SPECIFIC GRAVITY	OPEN-CIRCUIT VOLTAGE
100	1.265	12.6
75	1.225	12.4
50	1.190	12.2
25	1.155	12.0
0	1.120	11.8

*%

DC/AC Controller

Since most electrical loads (except boats, recreational vehicles, and other DC loads) are designed to be supplied with AC voltage, a DC/AC controller is normally used to convert DC into 60 Hz AC. Inverters used in PV systems change DC voltage from a PV system to AC voltage at a fixed frequency and set voltage level. A DMM set to measure DC voltage can be used to measure the DC voltage into an inverter and a DMM set to measure AC voltage can be used to measure the voltage out of an inverter. Likewise, a DMM set to measure DC current can be used to measure the DC current into an inverter, and a DMM set to measure AC current can be used to measure the current out of an inverter. **See Measuring DC/AC Controller Voltage and Current.**

MEASURING DC/AC CONTROLLER VOLTAGE AND CURRENT

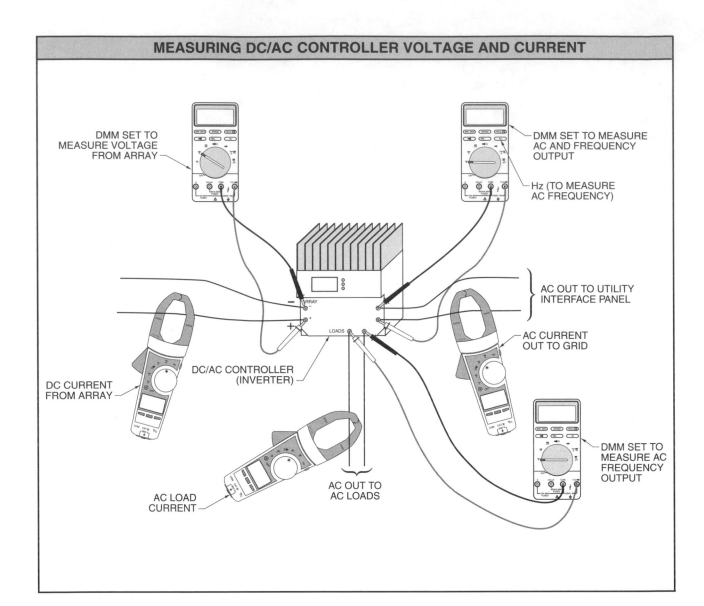

DMM SET TO MEASURE VOLTAGE FROM ARRAY

DMM SET TO MEASURE AC AND FREQUENCY OUTPUT

Hz (TO MEASURE AC FREQUENCY)

AC OUT TO UTILITY INTERFACE PANEL

AC CURRENT OUT TO GRID

DC CURRENT FROM ARRAY

DC/AC CONTROLLER (INVERTER)

DMM SET TO MEASURE AC FREQUENCY OUTPUT

AC LOAD CURRENT

AC OUT TO AC LOADS

ARRAY

LOADS

Fuses/Circuit Breakers

Fuses and/or circuit breakers are installed at several points within a PV system to provide protection from short circuits and overloads. In addition to fuses/circuit breakers, PV systems should also include surge protection to prevent damage from high-voltage surges on the power lines. A *surge protector (suppressor)* is an electronic device that provides protection from high voltage levels by limiting the voltage level downstream from the surge protector. In PV systems, a surge protector is placed on the output of the PV array to prevent high-voltage surges from feeding back into the cells/modules/arrays. **See Photovoltaic System Protection.**

PV System Conductors

All outdoor conductors should be rated for wet locations with a minimum 90°C rating. All outdoor and indoor conductors must be rated to safely carry the maximum amount of current produced plus a safety factor, which rates the conductor at least 125% higher than the maximum PV-system-rated output current.

PHOTOVOLTAIC SYSTEM PROTECTION

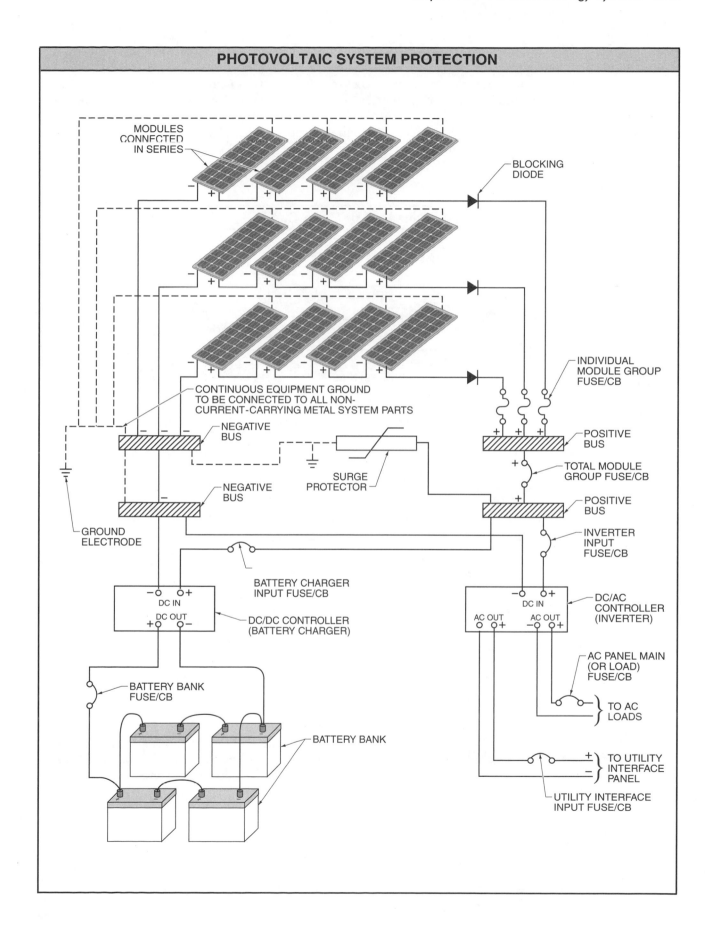

MODULES CONNECTED IN SERIES

BLOCKING DIODE

INDIVIDUAL MODULE GROUP FUSE/CB

CONTINUOUS EQUIPMENT GROUND TO BE CONNECTED TO ALL NON-CURRENT-CARRYING METAL SYSTEM PARTS

NEGATIVE BUS

POSITIVE BUS

SURGE PROTECTOR

TOTAL MODULE GROUP FUSE/CB

NEGATIVE BUS

POSITIVE BUS

GROUND ELECTRODE

INVERTER INPUT FUSE/CB

BATTERY CHARGER INPUT FUSE/CB

DC IN

DC OUT

DC/DC CONTROLLER (BATTERY CHARGER)

DC IN

AC OUT   AC OUT

DC/AC CONTROLLER (INVERTER)

AC PANEL MAIN (OR LOAD) FUSE/CB

TO AC LOADS

BATTERY BANK FUSE/CB

TO UTILITY INTERFACE PANEL

BATTERY BANK

UTILITY INTERFACE INPUT FUSE/CB

Blocking and Bypass Diodes

Diodes are used in PV systems to ensure current flows only out of the PV cells in the correct direction (blocking diodes) and to allow current to flow around a system fault (bypass diodes or shaded cells). A system fault occurs when there is an open in part of a PV module or when cells develop high resistance that reduces the current flow of all the cells in that part of the module. Diodes can be checked using the diode test function included on most DMMs. DMMs set to a diode test position allow the testing of diodes without having to remove them from the circuit. **See Blocking and Bypass Diodes.**

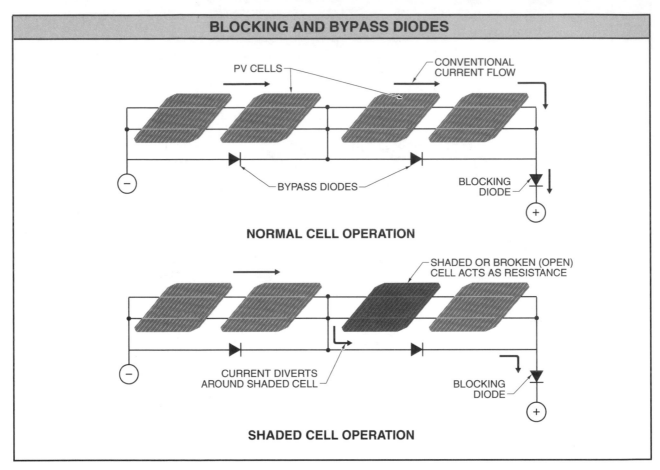

BLOCKING AND BYPASS DIODES

PV CELLS

CONVENTIONAL CURRENT FLOW

BYPASS DIODES

BLOCKING DIODE

NORMAL CELL OPERATION

SHADED OR BROKEN (OPEN) CELL ACTS AS RESISTANCE

CURRENT DIVERTS AROUND SHADED CELL

BLOCKING DIODE

SHADED CELL OPERATION

 Refer to **Activity 26-5—Photovoltacis Systems** ....

 *Refer to Quick Quiz® on CD-ROM*

 Refer to Chapter 26 in the **Troubleshooting Electrical/Electronic Systems Workbook** *for additional questions.*

Name_____ Date _____

ACTIVITY 26-1—Geothermal Systems

Electrical measurements help indicate if electrical components are operating properly. Using the system circulation pump motor nameplate, determine the expected measurements.

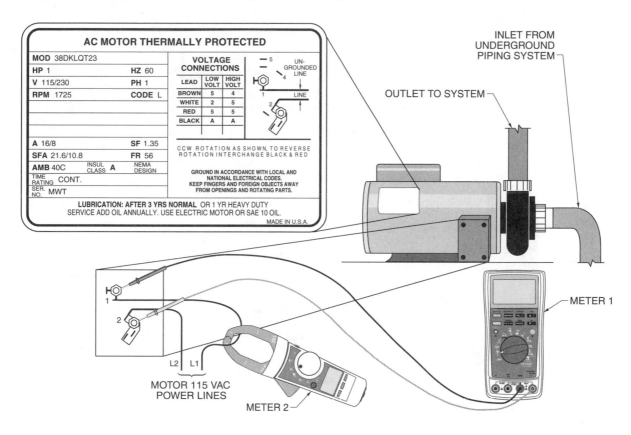

AC MOTOR THERMALLY PROTECTED

MOD 38DKLQT23	
HP 1	HZ 60
V 115/230	PH 1
RPM 1725	CODE L

A 16/8	SF 1.35
SFA 21.6/10.8	FR 56
AMB 40C	INSUL CLASS A	NEMA DESIGN
TIME RATING CONT.	
SER. NO. MWT	

VOLTAGE CONNECTIONS

LEAD	LOW VOLT	HIGH VOLT
BROWN	5	4
WHITE	2	5
RED	5	5
BLACK	A	A

CCW ROTATION AS SHOWN, TO REVERSE ROTATION INTERCHANGE BLACK & RED

GROUND IN ACCORDANCE WITH LOCAL AND NATIONAL ELECTRICAL CODES. KEEP FINGERS AND FOREIGN OBJECTS AWAY FROM OPENINGS AND ROTATING PARTS.

LUBRICATION: AFTER 3 YRS NORMAL OR 1 YR HEAVY DUTY SERVICE ADD OIL ANNUALLY. USE ELECTRIC MOTOR OR SAE 10 OIL.

MADE IN U.S.A.

INLET FROM UNDERGROUND PIPING SYSTEM

OUTLET TO SYSTEM

METER 1

MOTOR 115 VAC POWER LINES

METER 2

_____ **1.** Meter 1 reads ___ VAC if the system is running and the pump motor is operating properly.

_____ **2.** The normal maximum reading of Meter 2 is ___ A if the system is running, the pump motor is operating properly, and the motor not operating in its margin of safety rating (service factor rating).

_____ **3.** The absolute maximum reading of Meter 2 is ___ A if the system is running, the pump motor is operating properly, and the motor is operating up to its margin of safety rating (service factor rating).

_____ **4.** Is the motor current more or less than the motor nameplate rating if the pump motor is operating but there is no flow through the pipes?

ACTIVITY 26-2—Wind Turbine Systems

Understanding meter readings is important for determining whether or not a system is operating properly. In the following wind turbine electrical system, a three-phase power quality meter is connected at five different test points within the system. Match the five waveforms to the expected waveform at each measuring point. Note: Waveforms may be used more than once or not at all.

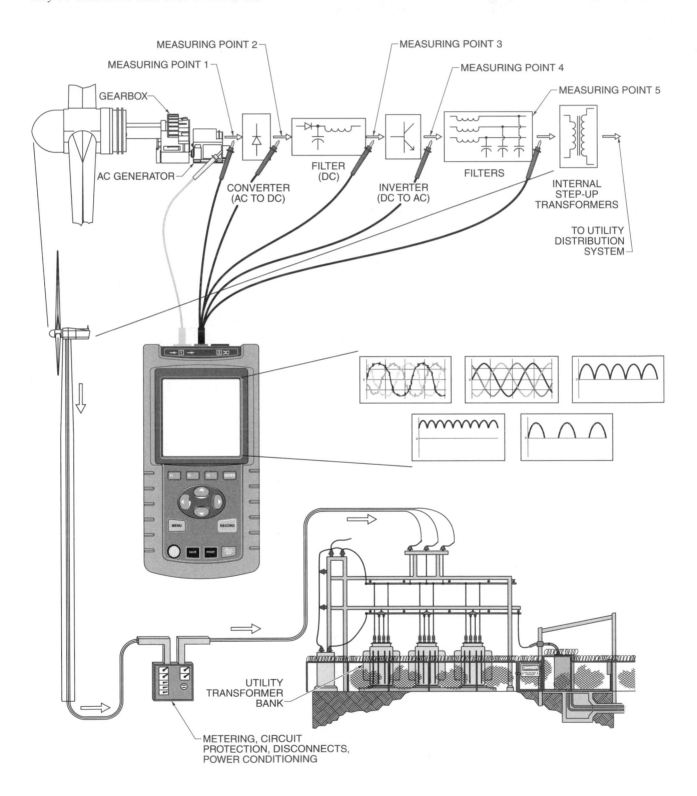

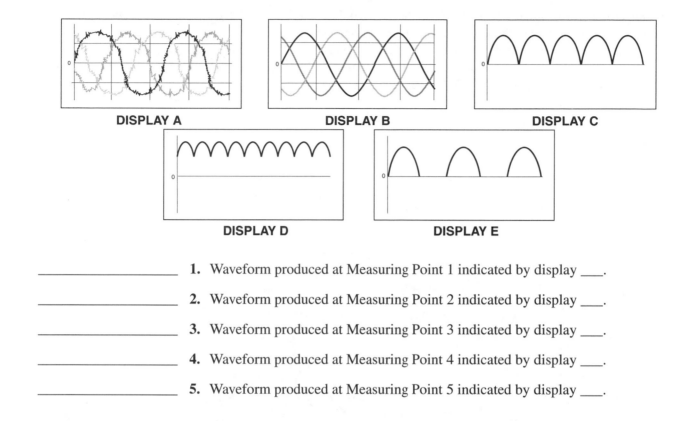

DISPLAY A **DISPLAY B** **DISPLAY C**

DISPLAY D **DISPLAY E**

_____ **1.** Waveform produced at Measuring Point 1 indicated by display ___.

_____ **2.** Waveform produced at Measuring Point 2 indicated by display ___.

_____ **3.** Waveform produced at Measuring Point 3 indicated by display ___.

_____ **4.** Waveform produced at Measuring Point 4 indicated by display ___.

_____ **5.** Waveform produced at Measuring Point 5 indicated by display ___.

ACTIVITY 26-3—Wind Turbine Supporting Components

Wind vanes are subject to the outside elements and must be tested and/or replaced when they fail. Anyone working on a wind turbine system should understand how to connect and test a wind vane. The following sample wind vane outputs both a PNP and an NPN open collector transistor output at approximately 12 mA at 12 VDC to 24 VDC based on the direction of the wind (N, S, E, or W).

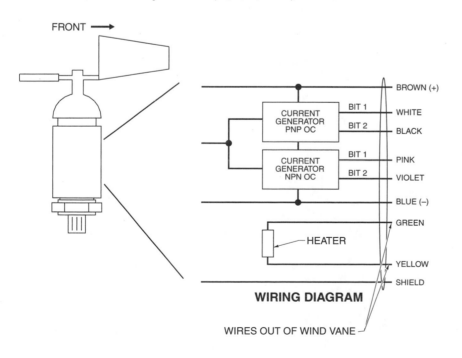

SIGNAL OUTPUT		
WIND DIRECTION	**BIT 1**	**BIT 2**
0° to 90°	0	1
90° to 180°	0	0
180° to 270°	1	0
270° to 360°	1	1

1. Connect the wind vane to the 12 VDC so that the wind vane is powered and its output directional bits can be tested. Also, connect the heater through the two-way switch so that it also can be tested to see if it heats the wind vane. Connect DMM 1 to test the output of the PNP bit 1 transistor. Connect DMM 2 to test the output of the PNP bit 2 transistor. Connect DMM 3 to test the output of the NPN bit 1 transistor. Connect DMM 4 to test the output of the NPN bit 2 transistor.

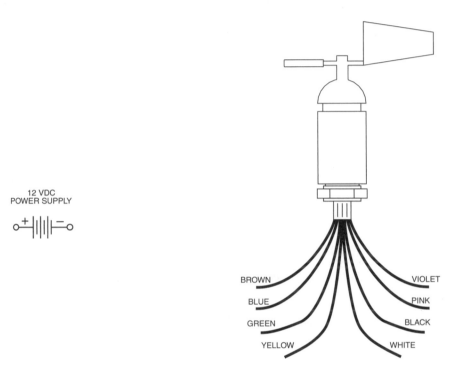

12 VDC
POWER SUPPLY

BROWN   VIOLET

BLUE   PINK

GREEN   BLACK

YELLOW   WHITE

DMM 1   DMM 2   DMM 3   DMM 4

Anemometers are subject to the outside elements and must be tested and/or replaced when they fail. Anyone working on a wind turbine system should understand how to connect and test an anemometer. The following sample anemometer outputs both a PNP and an NPN open collector transistor output at approximately 12 mA at 12 VDC to 24 VDC. The speed at which the output is switched ON/OFF is proportional to the air speed at a rate of 10 pulses per m/s (meter per second). 1 m/s = 2.24 mph and 1 mph = 0.447 m/s

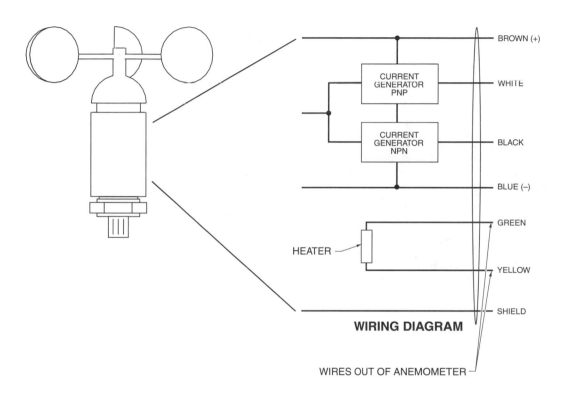

WIRING DIAGRAM

WIRES OUT OF ANEMOMETER

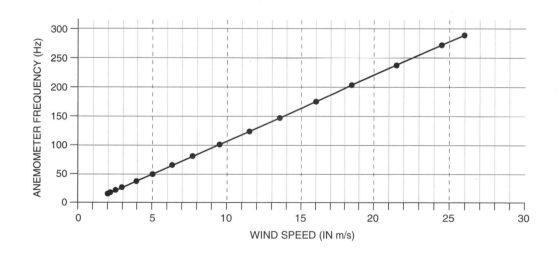

2. Connect the anemometer to the analog input card using input 1 and input 5 so that both inputs 1 and 5 can receive input pulses from the anemometer. Power the anemometer and heater from the power supply provided on the analog input card. Connect DMM 1 to test the output from the NPN anemometer output. Connect DMM 2 to test the output from the PNP anemometer output.

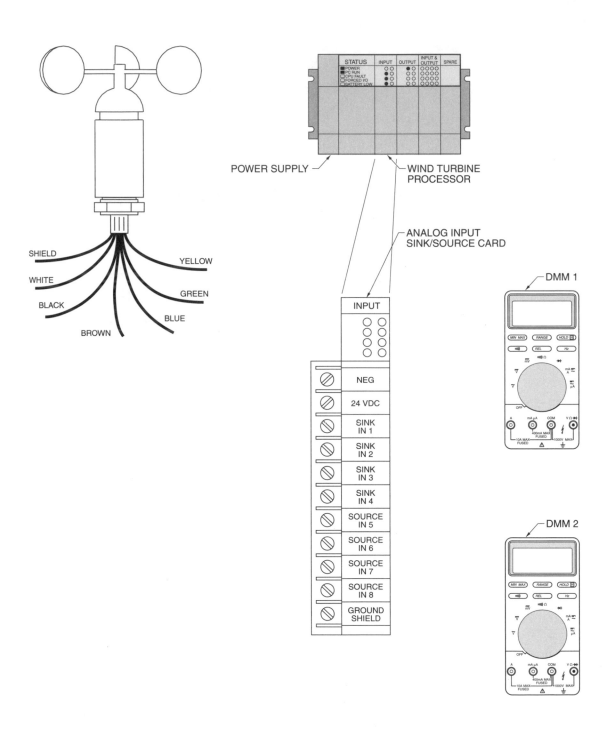

A high voltage probe is used when taking high voltage measurements inside a control panel.

_____ **3.** The high voltage adapter should be connected to which two meter jacks?

_____ **4.** A reading of 692.3 mV on the meter equals ___ V at the tip of the high voltage probe.

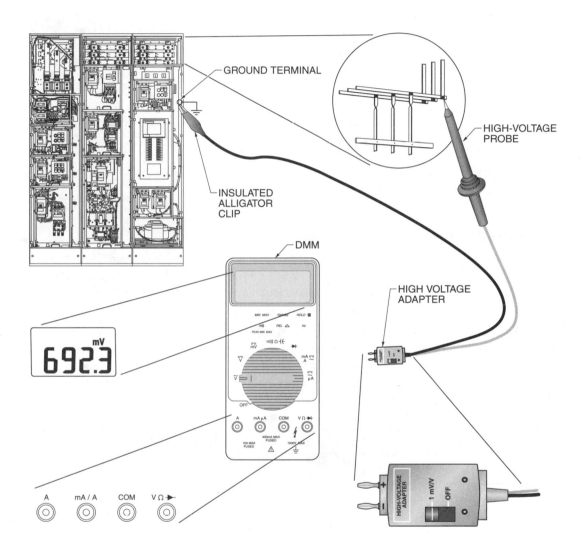

GROUND TERMINAL

HIGH-VOLTAGE PROBE

INSULATED ALLIGATOR CLIP

DMM

692.3 mV

MIN MAX   RANGE   HOLD

PEAK MIN MAX   REL   Hz

mV   )))) Ω

V   mA A

V   μA

OFF

A   mA μA   COM   V Ω

10A MAX FUSED   400mA MAX FUSED   1000V MAX

HIGH VOLTAGE ADAPTER

HIGH-VOLTAGE ADAPTER   1 mV/V   OFF

A   mA / A   COM   V Ω

When a transformer is placed in a circuit, the power/voltage is transformed from the primary (input) side to the secondary (output) side, but the system ground is not. An ungrounded control transformer secondary can cause electrical/electronic problems in the control circuit and thus in the system operation. Check to ensure the secondary of the low-voltage transformer is grounded.

_____ **5.** What is the reading on the meter if the secondary of the transformer is grounded?

_____ **6.** What is the reading on the meter if the secondary of the transformer is not grounded?

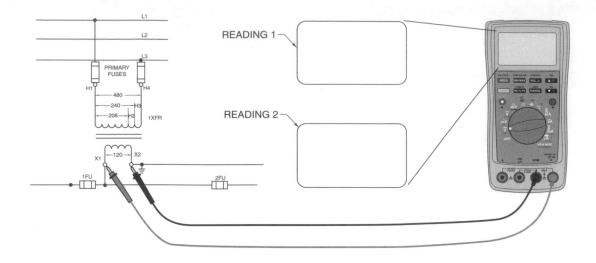

ACTIVITY 26-4—Wind Turbine Power Output

The amount of power required for a residence depends on the size of the residence and the amount of power consumed. A residential power panel is sized for the maximum power usage (200 A service). However, only a fraction of the available power is used. Thus, when utilities determine the amount of generated power required, they factor in the total number of residences and their average size.

Wind farms normally supply power to the utility grid. Wind farms/wind turbines relate their total power output to the number of residences they can supply with power. Often, a power average of 3333 W per residence is used. This accounts for the fact that some residences use more power than others and some use less power than others at any given time. Complete the chart by listing the approximate number of residences the wind turbine could supply with power. Round each answer to the next lowest whole number.

1.

WIND TURBINE SIZE AND POWER RATINGS			
Rotor*	Diameter†	Approximate Generator Output	Approximate Number of Homes Serviced
0.82	0.25	50 W	
7.87	2.4	900 W (0.9 kW)	
13.45	4.1	2 kW	
41	12.5	50 kW	
59	18	100 kW	
88.5	27	225 kW	
108	33	300 kW	
131	40	500 kW	
144	44	600 kW	
157	48	750 kW	
177	54	1 MW (1000 kW)	
209	64	1.5 MW	
236	72	2 MW	
262	80	2.5 MW	
328	100	3 MW	

* in ft
† in m

ACTIVITY 26-5—Photovoltaic Systems

Voltage measurements are taken to determine system and component operating conditions. State each expected meter reading when the system is operating under nominal operating conditions (rated system output conditions).

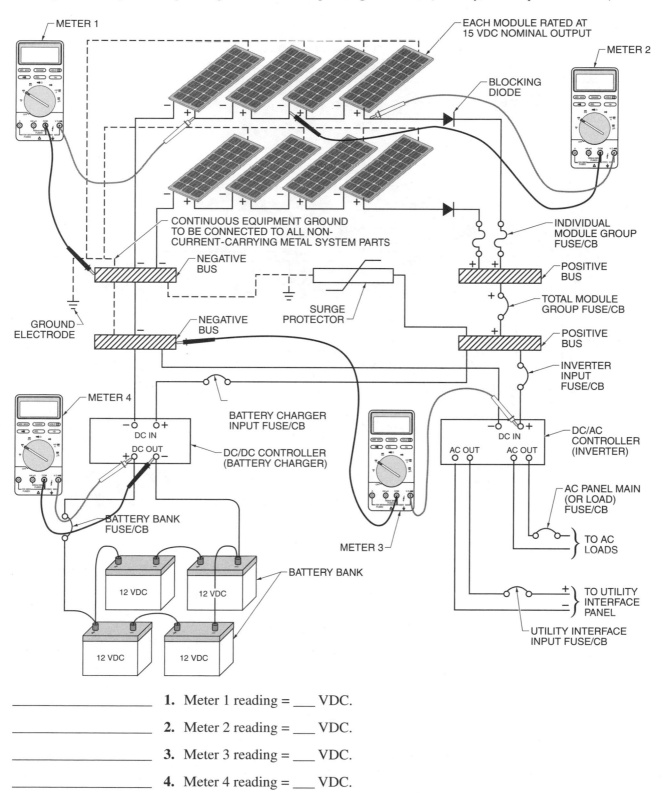

_____ **1.** Meter 1 reading = ___ VDC.

_____ **2.** Meter 2 reading = ___ VDC.

_____ **3.** Meter 3 reading = ___ VDC.

_____ **4.** Meter 4 reading = ___ VDC.

Current measurements are taken to determine system and component operating conditions. State each expected meter reading when the system is operating under nominal operating conditions (rated system output conditions).

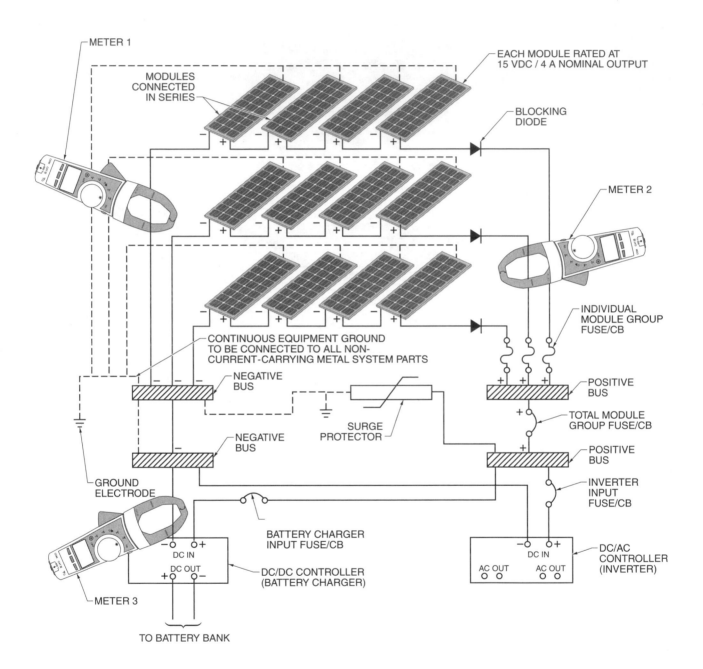

_____ **5.** Meter 1 reading = ___ ADC.

_____ **6.** Meter 2 reading = ___ ADC.

_____ **7.** Meter 3 reading = ___ ADC.

APPLICATION 27-1—General Motor Maintenance

Preventive maintenance is maintenance performed to keep machines, assembly lines, production operations, and plant operations running with little or no downtime. Preventive maintenance is performed on equipment that is not in operation.

Preventive maintenance helps prevent motor problems from occurring. A motor that is well-maintained and used in an application for which it is suited can have a long service life. Maintenance may be performed as needed or as part of a routine scheduled maintenance program.

Water-Soaked Motors

Motors may become water-soaked from flooding, broken water pipes, or lack of proper protection. Normally, water alone does not damage a motor if it is serviced shortly after soaking. Servicing includes a thorough cleaning, drying, and testing.

Motor Cleaning

A water-soaked motor must be cleaned as soon as possible. Cleaning normally requires a cleaning solvent to remove grease, oil, and contaminations. Warm water is normally enough to clean a motor that has been soaked in relatively clean water.

A motor must be thoroughly dried after the motor has been cleaned. Applying power to wet or damp motor windings damages the motor far more than water alone. A motor that has been cleaned may need relubrication because cleaning may remove some of the bearing lubrication. **See Motor Cleaning.**

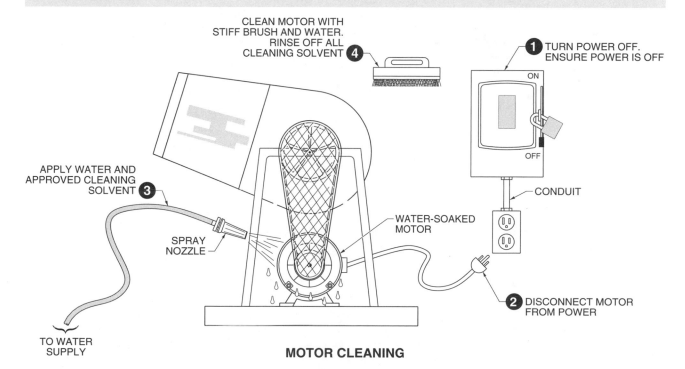

CLEAN MOTOR WITH STIFF BRUSH AND WATER. RINSE OFF ALL CLEANING SOLVENT ❹

❶ TURN POWER OFF. ENSURE POWER IS OFF

ON

OFF

CONDUIT

APPLY WATER AND APPROVED CLEANING SOLVENT ❸

SPRAY NOZZLE

WATER-SOAKED MOTOR

TO WATER SUPPLY

❷ DISCONNECT MOTOR FROM POWER

MOTOR CLEANING

To clean a motor, apply the procedure:
1. Turn power to the motor circuit OFF. Ensure that all power is OFF.
2. Disconnect the motor from the circuit.
3. Apply water and an approved cleaning solvent to the motor. Water pressure should not exceed 25 psi. Water temperature should not exceed 200°F. Allow the cleaning solvent sufficient time to work per manufacturer's recommendations.
4. Clean the motor with a stiff brush and water. Rinse off all the cleaning solvent.

Motor Drying

A motor must be thoroughly dried before any power is applied. Motors may be slow-dried or force-dried. Slow-drying a motor is better than force-drying. A motor is slow-dried by placing it in a warm room or in the sun for two to four days.

Motors are force-dried by heating or by applying low voltage. To force-dry a motor by heating, apply 150°F to 200°F heat for approximately 18 hours to 30 hours depending on the motor type and temperature. To force-dry a motor by applying low voltage, apply approximately 10% of the motor's rated voltage for two to four hours to dry the coils.

Motor Testing

Always check the resistance of a motor before reapplying power. The motor insulation is checked to ensure it has the proper resistance value. A megohmmeter or ohmmeter is used to test the resistance of the insulation.

Although a motor may operate satisfactorily with a reading of less than 1 MΩ, a resistance of 1 MΩ or more is recommended as the minimum resistance. Additional drying or running the motor for a short time increases the resistance.

Check for signs of spitting or smoking when restarting the motor. Spitting indicates moisture in the motor. Smoking indicates an insulation breakdown. In either case, turn the power OFF and completely dry and retest the motor.

Motor Bearing Lubrication

Motors are lubricated at the factory to provide long operation under normal service conditions without relubrication. Excessive and frequent lubrication may damage a motor. The time period between lubrications depends on the motor's service conditions, ambient temperature, and environment. **See Motor Lubrication.**

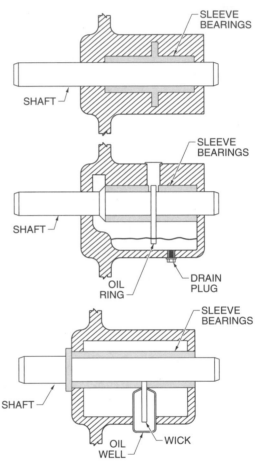

ADD ELECTRIC MOTOR OIL TO SLEEVE BEARINGS AFTER THREE YEARS OF NORMAL OPERATION

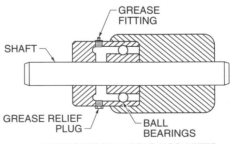

LUBRICATE BALL BEARINGS AFTER FIVE YEARS OF NORMAL OPERATION

MOTOR LUBRICATION

Always follow lubrication instructions provided with the motor. These instructions are normally listed on the nameplate or terminal box cover. There may be separate lubrication instructions furnished with the motor. Relubricate sleeve bearings and ball bearings per schedule if lubrication instructions are not available.

For normal operation, add electric motor oil (or SAE #10 or #20 non-detergent oil) to sleeve bearings after three years. For heavy-duty operation, relubricate once a year. For light operation, relubricate every four years.

Ball bearings are designed for many years of operation without relubrication. The schedule for relubricating ball bearings varies from one year to 10 years, depending on the motor's service conditions, ambient temperature, and environment. When relubricating, use standard, long-life ball-bearing grease.

Motor Mounting/Remounting

Motors other than dripproof types may be mounted in any position or at any angle. Dripproof motors must be mounted in the normal horizontal position to meet the requirements of the enclosure. Motors with sleeve bearings must be mounted with the oil cap up.

Mount a motor securely to the mounting base of equipment or to a rigid, flat surface. Metal surfaces are preferred. For direct-coupled installations, align the motor shaft and coupling carefully. Place shims under the motor base as required to assure proper alignment. Use flexible couplings whenever possible. In belt applications, align pulleys and adjust belt tension so that approximately $\frac{1}{64}''$ of belt deflection occurs per inch of span. Position sleeve-bearing motors so that the belt is away from the oil in the sleeve bearing (located approximately under the oiler of the motor). **See Motor Mounting.**

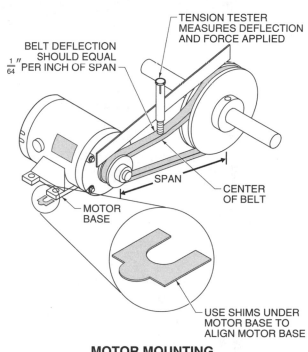

MOTOR MOUNTING

Dirt and Corrosion

Remove dirt and corrosion by brushing, vacuuming, or blowing. After removing obvious dirt, feel for air being discharged from cooling ports of the motor. The internal air passages may be clogged if the air flow is weak.

As time passes, small amounts of corrosion are normal. Large amounts of corrosion may indicate a problem. Check for signs of corrosion. After removing corrosion, use corrosion-resistant, high-temperature paint to repaint the motor.

Heat, Noise, and Vibration

Excessive heat or vibration is determined by touching the motor frame and bearings lightly. Listen closely for any abnormal noise. Problems with heat are normally due to improper ventilation, excessive motor cycling, or overloads. Vibration and noise are caused by misalignment and loose parts. Tighten all mountings connecting the motor and motor load. Realign the motor if excessive vibration remains.

Scheduled Maintenance

A maintenance schedule is essential to the satisfactory operation of electric motors. The frequency of scheduled maintenance on motors varies. Motor maintenance checklists for scheduled semiannual and annual maintenance provide a record of maintenance performed. A motor repair record provides a record of motor repairs. **See Annual Motor Maintenance Checklist** and **Motor Repair and Service Record in Appendix.**

 Refer to *Activity 27-1—General Motor Maintenance*...

APPLICATION 27-2—*Recording Meters*

A *recording meter* is a meter that provides a written record of readings over a time period. Recording meters are used to record quantities such as voltage, current, power, speed, frequency, temperature, humidity, etc.

Recording meters record either analog traces or actual numerical values onto paper. The two basic types of recording meters are the vertical chart recorder and the circular chart recorder. A *vertical chart recorder* is a graphic pen and ink recorder that has measured values drawn on a continuous roll of graph paper. Vertical chart recorders are normally used in permanent installations and are designed to record over long time periods (days or weeks). A *circular chart recorder* is a graphic pen and ink recorder that has measured values drawn on a rotating circular chart. Circular chart recorders are normally used for temporary measurements and normally record over a 24-hour time period. **See Recording Meters.**

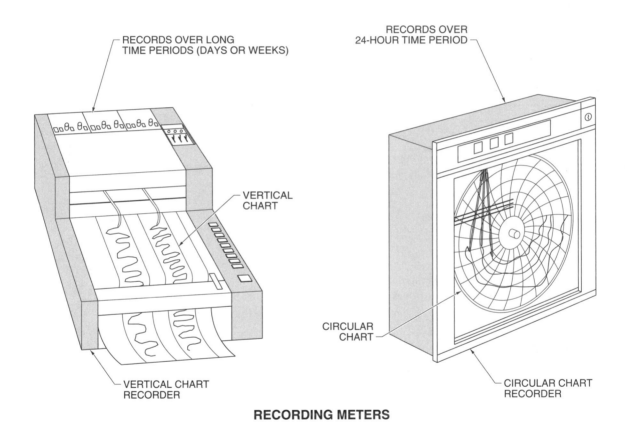

RECORDING METERS

Recording meters are used to record either application or circuit information. Application information, such as temperature and humidity levels, may be used to determine potential process problems. For example, a permanent record of temperature and humidity may be recorded in a food storage area.

Circuit information, such as voltage and power levels, may be used to determine potential control problems. For example, a permanent record of voltage spikes in a computer room may be used to determine intermittent problems.

 Refer to **Activity 27-2—***Recording Meters*..

APPLICATION 27-3—Lockout/Tagout

Per OSHA standards, equipment must be locked out and tagged out before any preventive maintenance or servicing is performed. *Lockout* is the process of removing the source of power and installing a lock which prevents the power from being turned ON. *Tagout* is the process of placing a danger tag on the source of power which indicates that the equipment may not be operated until the danger tag is removed. **See Lockout/Tagout.**

A danger tag has the same importance and purpose as a lock and is used by itself only when a lock does not fit the disconnect device. The danger tag shall be attached at the disconnect device with a tag tie or equivalent and shall have space for the worker's name, craft, and other required information. A danger tag must withstand the elements and expected atmosphere for as long as the tag remains in place.

A lockout/tagout is used when:
• Servicing electrical equipment that does not require power to be ON to perform the service.
• Removing or bypassing a machine guard or other safety device.
• The possibility exists of being injured or caught in moving machinery.
• Clearing jammed equipment.
• The danger exists of being injured if equipment power is turned ON.

Lockouts and tagouts do not by themselves remove power from a circuit. An approved procedure must be followed when applying a lockout/tagout. Lockouts and tagouts are attached only after the equipment is turned OFF and tested to ensure that power is OFF. The lockout/tagout procedure is required for the safety of workers due to modern equipment hazards. OSHA provides a standard procedure for equipment lockout/tagout. OSHA's procedure is:
1. Prepare for machinery shutdown
2. Machinery or equipment shutdown
3. Machinery or equipment isolation
4. Lockout or tagout application
5. Release of stored energy
6. Verification of isolation

Warning: Personnel should consult OSHA Standard 29CFR1910.147 for industry standards on lockout/tagout.

A lockout/tagout shall not be removed by any other person than the person who installed it, except in an emergency. In an emergency, the lockout/tagout may be removed only by authorized personnel. The authorized personnel shall follow approved procedures. A list of company rules and procedures are given to any person that may use a lockout/tagout.

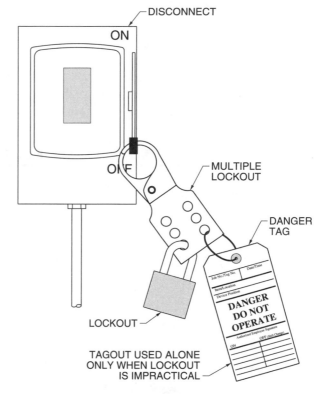

DISCONNECT
ON
OFF
MULTIPLE LOCKOUT
DANGER TAG
LOCKOUT
TAGOUT USED ALONE ONLY WHEN LOCKOUT IS IMPRACTICAL

DANGER DO NOT OPERATE

LOCKOUT/TAGOUT

Always remember:

- Use a lockout and tagout when possible.
- Use a tagout when a lockout is impractical. A tagout is used alone only when a lock does not fit the disconnect device.
- Use a multiple lockout when individual employee lockout of equipment is impractical.
- Notify all employees affected before using a lockout/tagout.
- Remove all power sources including primary and secondary.
- Measure for voltage using a voltmeter to ensure that power is OFF.

Lockout Devices

Lockout devices are lightweight enclosures that allow the lock out of standard control devices. Lockout devices are available in various shapes and sizes that allow for the lock out of ball valves, gate valves, and electrical plugs. Lockout devices resist chemicals, cracking, abrasion, and temperature changes, and are available in colors to match ANSI pipe colors. Lockout devices are sized to fit standard industry control device sizes. **See Lockout Devices.**

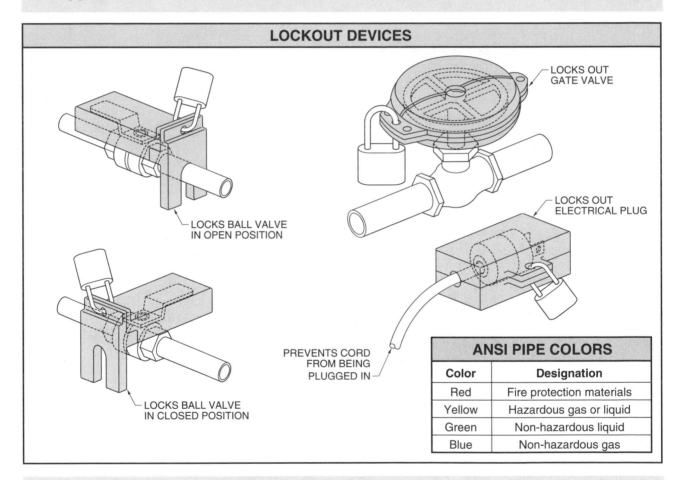

LOCKOUT DEVICES

LOCKS OUT GATE VALVE

LOCKS BALL VALVE IN OPEN POSITION

LOCKS OUT ELECTRICAL PLUG

LOCKS BALL VALVE IN CLOSED POSITION

PREVENTS CORD FROM BEING PLUGGED IN

ANSI PIPE COLORS	
Color	**Designation**
Red	Fire protection materials
Yellow	Hazardous gas or liquid
Green	Non-hazardous liquid
Blue	Non-hazardous gas

Locks used to lock out a device may be color-coded and individually keyed. The locks are rust-resistant and are available with various size shackles.

Danger tags provide additional lockout and warning information. Various danger tags are available. Danger tags may include warnings such as "Do Not Start," "Do Not Operate," or may provide space to enter worker, date, and lockout reason information. Tag ties must be strong enough to prevent accidental removal and must be self-locking and nonreusable. **See Danger Tags.**

Lockout/tagout kits are also available. A lockout/tagout kit contains items required to comply with the OSHA lockout/tagout standard. Lockout/tagout kits contain reusable danger tags, tag ties, multiple lockouts, locks, magnetic signs, and information on lockout/tagout procedures. **See Lockout/Tagout Kit.**

DANGER TAGS

LOCKOUT/TAGOUT KIT

 Refer to **Activity 27-3—Lockout/Tagout**...

APPLICATION 27-4—Enclosure Selection

An *enclosure* is a housing used to protect wires and equipment and prevent personnel injury by accidental contact with a live circuit. Enclosures are used to protect electrical equipment such as programmable controllers, motor starters, fuses/breakers, and control circuits. An enclosure provides the main protection from atmospheric conditions. Using the proper enclosure helps prevent problems caused by contamination, moisture, and physical damage. Enclosures are categorized by the protection they provide. An enclosure is selected based on the location of the equipment and the NEC® requirements. **See Enclosures.**

The NEC® classifies hazardous locations according to the properties and quantities of the hazardous material that may be present. Hazardous locations are divided into three classes, two divisions, and seven groups.

Class refers to the generic hazardous material present. Class I applies to locations where flammable gases or vapors may be present in the air in quantities sufficient to produce an explosive ignitable mixture. Class II applies to locations where combustible dusts may be present in sufficient quantity to cause an explosion. Class III applies to locations where the hazardous material consists of easily ignitable fibers or flyings that are not normally in suspension in the air in large enough quantities to produce an ignitable mixture.

	ENCLOSURES		
Type	**Service Conditions**	**Sealing Method**	**Cost**
1	No unusual		Base
4	Windblown dust and rain, splashing water, hose-directed water, and ice on enclosure		12 x Base
4X	Corrosion, windblown dust and rain, splashing water, hose-directed water, and ice on enclosure		12 x Base
7	Withstand and contain an internal explosion of specified gases, contain an explosion sufficiently so an explosive gas-air mixture in the atmosphere is not ignited		48 x Base
9	Dust		48 x Base
12	Dust, falling dirt, and dripping noncorrosive liquids		5 x Base

Division applies to the probability that a hazardous material is present. Division 1 applies to locations where ignitable mixtures exist under normal operating conditions found in the process, operation, or during periodic maintenance. Division 2 applies to locations where ignitable mixtures exist only in abnormal situations. Abnormal situations occur as a result of accidents or when equipment fails.

Air mixtures of gases, vapors, and dusts are grouped according to their similar characteristics. The NEC® classifies gases and vapors in Groups A, B, C, and D for Class I locations and combustible dusts in Groups E, F, and G for Class II locations. **See Hazardous Locations.** For example, a Type 7 enclosure is required for an indoor application where gasoline is stored (Class I, Group D).

	HAZARDOUS LOCATIONS	
Class	**Group**	**Material**
I	A	Acetylene
	B	Hydrogen, butadiene, ethylene oxide, propylene oxide
	C	Carbon monoxide, ether, ethylene, hydrogen sulfide, morpholine, cyclopropane
	D	Gasoline, benzene, butane, propane, alcohol, acetone, ammonia, vinyl chloride
II	E	Metal dusts
	F	Carbon black, coke dust, coal
	G	Grain dust, flour, starch, sugar, plastics
III	No groups	Wood chips, cotton, flax, and nylon

 *Refer to **Activity 27-4—Enclosure Selection**.*

APPLICATION 27-5—Conveyor Belt Tracking

Conveyor Terminology

The two drive methods used to move a product along a conveyor belt are the direct drive and roller drive methods. The *direct drive method* has the product riding directly on the driven belt and is used for light loads. The *roller drive method* has the belt-driving rollers on which the product rides. The roller drive method is used for heavy loads. **See Direct Drive** and **Roller Drive.**

The right side of a conveyor is the right side when facing the forward direction of material travel. The two ends of a conveyor are identified by the relationship to the forward direction of material travel. The *tail end* of a conveyor is the end where material is fed. The *head end* of a conveyor is the end from which material is discharged. A drive unit is connected to the tail-end pulley or head-end pulley.

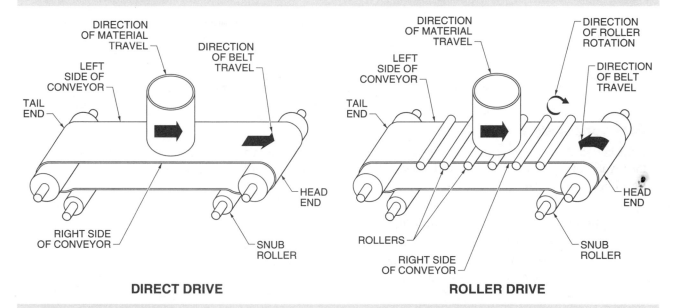

DIRECT DRIVE　　　　　　　　　　　**ROLLER DRIVE**

Conveyor Belt Tension

To prevent slippage, a conveyor belt must have the correct tension. Proper belt tension is accomplished by adjusting the take-up pulley. A *take-up pulley* is a pulley that is used for correcting belt tension and is not connected to a drive.

Conveyor Belt Tracking

Proper belt tracking depends on the alignment of the pulleys and rollers. A pulley is a revolving cylinder connected to a drive. A roller is a revolving cylinder not connected to a drive. As a conveyor belt moves, the belt should remain in the center of the pulleys and rollers. When a belt drifts to one side, the belt edge wears or folds up on the conveyor guard.

A conveyor belt drifts toward the side where pulley and roller centers are not parallel and are the closest to each other. Conveyor belt tracking is adjusted by moving the pulley or snub roller. A *snub roller* is a roller not connected to a drive that is used to correct belt alignment. When aligning a belt, all adjustments should be slight with time allowed for the belt to react to the adjustment.

Conveyor Adjustment—Head-End Pulley

If the belt drifts to the right on the head-end pulley during forward material travel, adjust the right side of the head-end snub roller in the forward direction of material travel and/or the left side of the head-end snub roller in the reverse direction of material travel. **See Head-End Pulley.**

Conveyor Adjustment—Tail-End Pulley

If the belt drifts to the right on the tail-end pulley during forward material travel, adjust the right side of the tail-end snub roller in the reverse direction of material travel and/or the left side of the tail-end snub roller in the forward direction of material travel. **See Tail-End Pulley.**

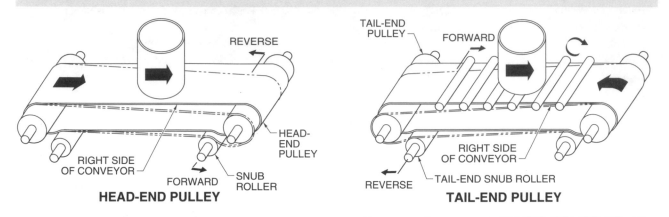

HEAD-END PULLEY

TAIL-END PULLEY

Conveyor Adjustment—Center Drive

If the belt drifts to the right side of the center drive and take-up pulleys, adjust the right side of the snub roller in the reverse direction of material travel and/or the left side of the snub roller in the forward direction of material travel. **See Center Drive Conveyor.**

Conveyor Adjustment—Reverse-Running

If the belt drifts to the right side of the center drive and take-up pulleys, adjust the right side of the snub roller in the forward direction of material travel and/or the left side of the snub roller in the reverse direction. **See Reverse-Running Center Drive Conveyor.**

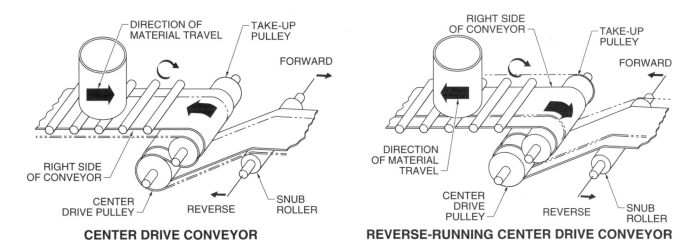

CENTER DRIVE CONVEYOR

REVERSE-RUNNING CENTER DRIVE CONVEYOR

Refer to **Activity 27-5—Conveyor Belt Tracking**..

Refer to **Quick Quiz®** on CD-ROM

Refer to Chapter 27 in the **Troubleshooting Electrical/Electronic Systems Workbook** for additional questions.

Name_____ **Date** _____

ACTIVITY 27-1—General Motor Maintenance

Determine if the belt tension is loose, tight, or correct.

_____ **1.** The belt tension is ___.   _____ **2.** The belt tension is ___.

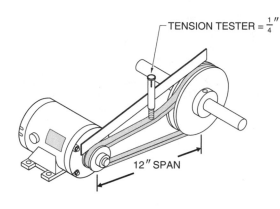

TENSION TESTER = $\frac{1}{4}''$

12" SPAN

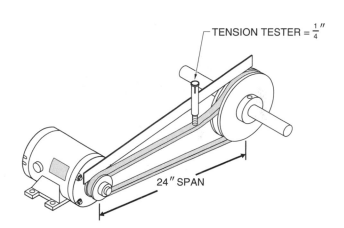

TENSION TESTER = $\frac{1}{4}''$

24" SPAN

_____ **3.** The belt tension is ___.   _____ **4.** The belt tension is ___.

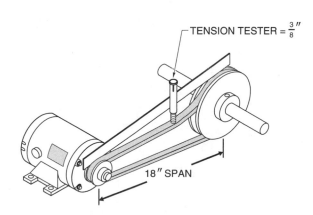

TENSION TESTER = $\frac{3}{8}''$

18" SPAN

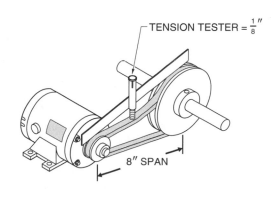

TENSION TESTER = $\frac{1}{8}''$

8" SPAN

ACTIVITY 27-2—Recording Meters

Using the circular chart, answer the questions.

_____ 1. The maximum recorded humidity is ___%.

_____ 2. The maximum humidity was recorded at ___.

_____ 3. The minimum recorded humidity is ___%.

_____ 4. The minimum humidity was recorded at ___.

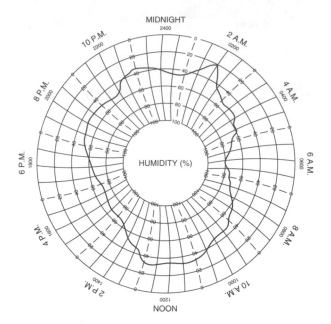

Using the vertical chart, answer the questions.

_____ 5. The maximum recorded temperature is ___°F.

_____ 6. The maximum temperature was recorded at ___.

_____ 7. The minimum recorded temperature is ___°F.

_____ 8. The minimum temperature was recorded at ___.

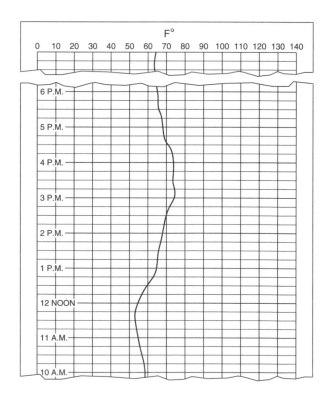

ACTIVITY 27-3—Lockout/Tagout

Determine if a lockout/tagout is required for each service call.

_____ **1.** The service call requires the pump to be disconnected from the motor.

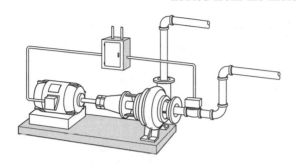

_____ **2.** The service call requires checking the output of an HID ballast.

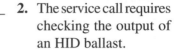
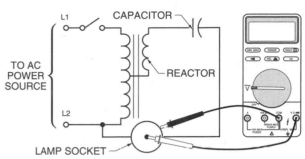

_____ **3.** The service call requires checking belt tension.

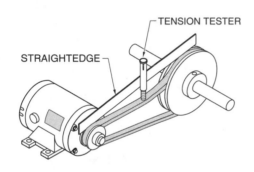

_____ **4.** The service call requires checking for hot spots that may be caused by loose connections.

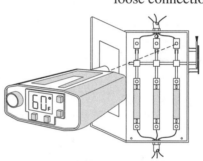

ACTIVITY 27-4—Enclosure Selection

Using Enclosure Types in Appendix, determine the enclosure type required for each location.

_____ **1.** Required enclosure type is ___.

_____ **2.** Required enclosure type is ___.

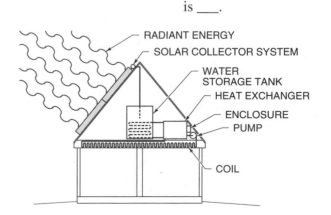

_____ **3.** Required enclosure type is ___.

_____ **4.** Required enclosure type is ___.

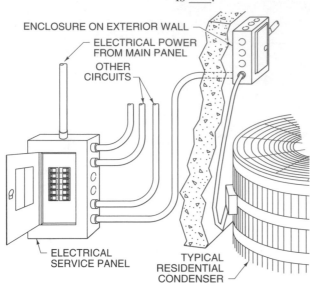

ENCLOSURE ON EXTERIOR WALL
ELECTRICAL POWER FROM MAIN PANEL
OTHER CIRCUITS
ELECTRICAL SERVICE PANEL
TYPICAL RESIDENTIAL CONDENSER

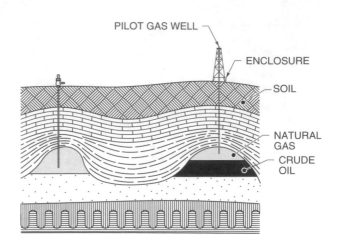

PILOT GAS WELL
ENCLOSURE
SOIL
NATURAL GAS
CRUDE OIL

ACTIVITY 27-5—Conveyor Belt Tracking

Determine if the roller is adjusted properly or improperly to correct the belt tracking.

_____ **1.** The roller is adjusted ___ to correct the belt tracking.

_____ **2.** The roller is adjusted ___ to correct the belt tracking.

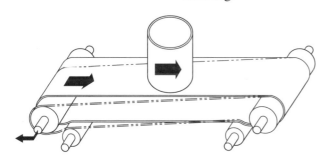

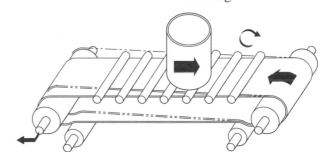

_____ **3.** The roller is adjusted ___ to correct the belt tracking.

_____ **4.** The roller is adjusted ___ to correct the belt tracking.

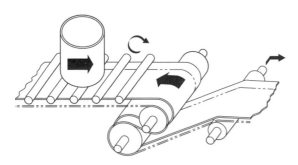

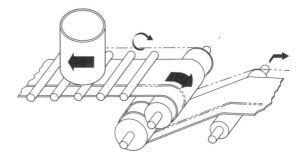

Appendix

Charts and Tables

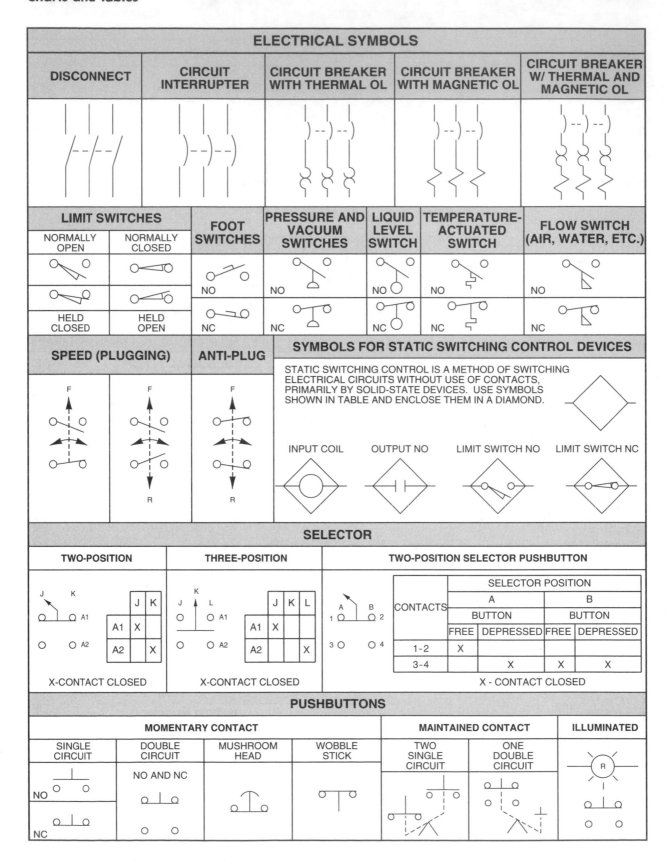

ELECTRICAL SYMBOLS

CONTACTS

INSTANT OPERATING

WITH BLOWOUT		WITHOUT BLOWOUT	
NO	NC	NO	NC

TIMED CONTACTS - CONTACT ACTION RETARDED AFTER COIL IS:

ENERGIZED		DE-ENERGIZED	
NOTC	NCTO	NOTO	NCTC

OVERLOAD RELAYS

THERMAL	MAGNETIC

SUPPLEMENTARY CONTACT SYMBOLS

SPST NO		SPST NC		SPDT		TERMS
SINGLE BREAK	DOUBLE BREAK	SINGLE BREAK	DOUBLE BREAK	SINGLE BREAK	DOUBLE BREAK	

TERMS:

SPST
SINGLE-POLE, SINGLE-THROW

SPDT
SINGLE-POLE, DOUBLE-THROW

DPST
DOUBLE-POLE, SINGLE-THROW

DPDT
DOUBLE-POLE, DOUBLE-THROW

NO
NORMALLY OPEN

NC
NORMALLY CLOSED

DPST, 2NO		DPST, 2NC		DPDT	
SINGLE BREAK	DOUBLE BREAK	SINGLE BREAK	DOUBLE BREAK	SINGLE BREAK	DOUBLE BREAK

METER (INSTRUMENT)

INDICATE TYPE BY LETTER

TO INDICATE FUNCTION OF METER OR INSTRUMENT, PLACE SPECIFIED LETTER OR LETTERS WITHIN SYMBOL.

AM or A	AMMETER	VA	VOLTMETER
AH	AMPERE HOUR	VAR	VARMETER
μA	MICROAMMETER	VARH	VARHOUR METER
mA	MILLAMMETER	W	WATTMETER
PF	POWER FACTOR	WH	WATTHOUR METER
V	VOLTMETER		

PILOT LIGHTS

INDICATE COLOR BY LETTER

NON PUSH-TO-TEST	PUSH-TO-TEST

INDUCTORS

IRON CORE

AIR CORE

COILS

DUAL-VOLTAGE MAGNET COILS

HIGH-VOLTAGE	LOW-VOLTAGE
LINK	LINKS
1 2 3 4	1 2 3 4

BLOWOUT COIL

ELECTRICAL SYMBOLS

TRANSFORMERS

AUTO	AIR CORE	CURRENT	CONTROL TRANSFORMER		AUTOTRANSFORMER FOR REDUCED-VOLTAGE STARTING
			SINGLE-VOLTAGE	DUAL-VOLTAGE	

CONTROL TRANSFORMER — SINGLE-VOLTAGE: H1 H2 / X2 X1

CONTROL TRANSFORMER — DUAL-VOLTAGE: H1 H3 H2 H4 / X2 X1

AUTOTRANSFORMER FOR REDUCED-VOLTAGE STARTING: % 50 65 80 100 0

AC MOTORS

SINGLE-PHASE	SEPARATE PHASE, TWO-SPEED	THREE-PHASE	SEPARATE WINDING, TWO-SPEED	CONSTANT-TORQUE, TWO-SPEED
T1 T2	HIGH COM LOW / T1 T2 T3	T1 T2 T3	T1 T11 / T3 T2 T13 T12	T4 / T3 T1 / T5 T2 T6

VARIABLE-TORQUE, TWO-SPEED	CONSTANT-HORSEPOWER, TWO-SPEED	WYE/DELTA, REDUCED-VOLTAGE	WYE-CONNECTED, PART WINDING, REDUCED-VOLTAGE
T4 T1 / T3 / T5 T2 T6	T4 / T3 T1 / T5 T2 T6	T6 T1 / T3 T4 / T5 T2	T1 T2 T3 T5 T7 T8 T9 / T4 T6

DC MOTORS / WIRING / CONNECTIONS

DC MOTORS				WIRING			CONNECTIONS
ARMATURE	SHUNT FIELD	SERIES FIELD	COMM OR COMPENS FIELD	NOT CONNECTED	POWER	WIRING TERMINAL	MECHANICAL
ARM	SHOW 4 LOOPS	SHOW 3 LOOPS	SHOW 2 LOOPS	CONNECTED	CONTROL	GROUND	MECHANICAL INTERLOCK

CONTROL AND POWER CONNECTIONS-600 V OR LESS ACROSS-THE-LINE STARTERS

		1φ	2φ, 4-WIRE	3φ
LINE MARKINGS		L1, L2	L1, L3 PHASE 1 L2, L4 PHASE 2	L1, L2, L3
GROUND WHEN USED		L1 IS ALWAYS UNGROUNDED	—	L2
MOTOR RUNNING OVERCURRENT UNITS IN	1 ELEMENT	L1	—	—
	2 ELEMENT	—	L1, L4	—
	3 ELEMENT	—	—	L1, L2, L3
CONTROL CIRCUIT CONNECTED TO		L1, L2	L1, L3	L1, L2
FOR REVERSING INTERCHANGE LINES		—	L1, L3	L1, L3

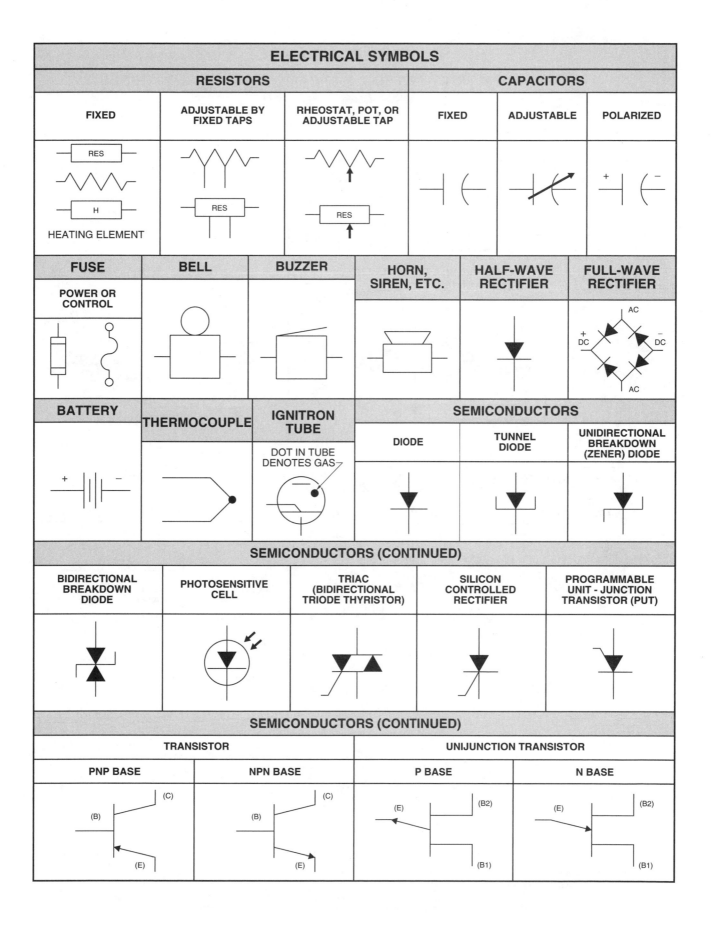

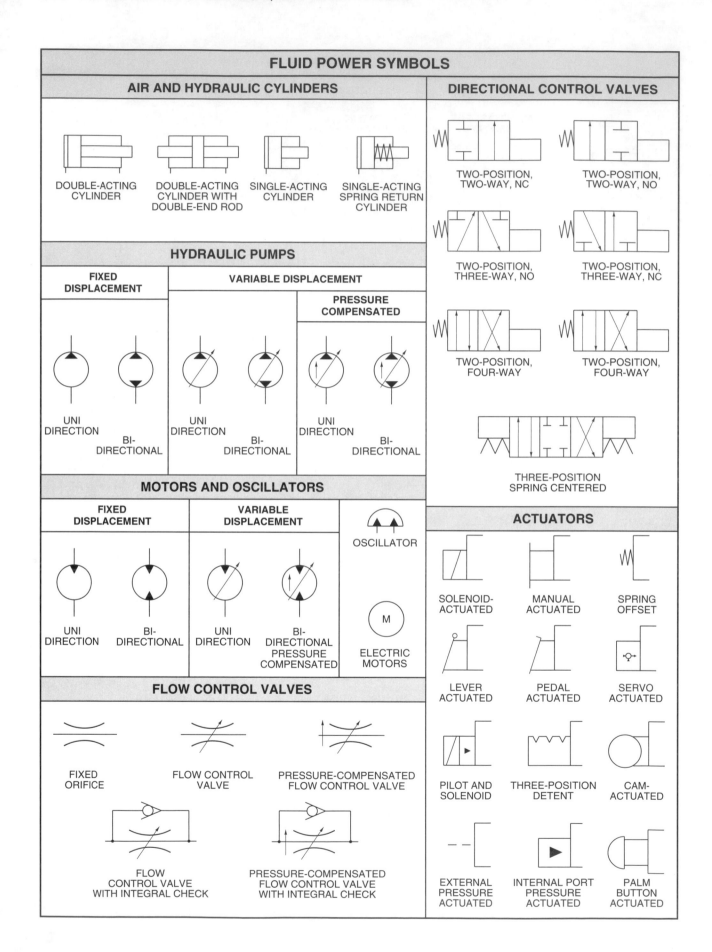

FLUID POWER SYMBOLS

AIR AND HYDRAULIC CYLINDERS

DOUBLE-ACTING CYLINDER

DOUBLE-ACTING CYLINDER WITH DOUBLE-END ROD

SINGLE-ACTING CYLINDER

SINGLE-ACTING SPRING RETURN CYLINDER

HYDRAULIC PUMPS

FIXED DISPLACEMENT

UNI DIRECTION

BI-DIRECTIONAL

VARIABLE DISPLACEMENT

UNI DIRECTION

BI-DIRECTIONAL

PRESSURE COMPENSATED

UNI DIRECTION

BI-DIRECTIONAL

MOTORS AND OSCILLATORS

FIXED DISPLACEMENT

UNI DIRECTION

BI-DIRECTIONAL

VARIABLE DISPLACEMENT

UNI DIRECTION

BI-DIRECTIONAL PRESSURE COMPENSATED

OSCILLATOR

ELECTRIC MOTORS

FLOW CONTROL VALVES

FIXED ORIFICE

FLOW CONTROL VALVE

PRESSURE-COMPENSATED FLOW CONTROL VALVE

FLOW CONTROL VALVE WITH INTEGRAL CHECK

PRESSURE-COMPENSATED FLOW CONTROL VALVE WITH INTEGRAL CHECK

DIRECTIONAL CONTROL VALVES

TWO-POSITION, TWO-WAY, NC

TWO-POSITION, TWO-WAY, NO

TWO-POSITION, THREE-WAY, NO

TWO-POSITION, THREE-WAY, NC

TWO-POSITION, FOUR-WAY

TWO-POSITION, FOUR-WAY

THREE-POSITION SPRING CENTERED

ACTUATORS

SOLENOID-ACTUATED

MANUAL ACTUATED

SPRING OFFSET

LEVER ACTUATED

PEDAL ACTUATED

SERVO ACTUATED

PILOT AND SOLENOID

THREE-POSITION DETENT

CAM-ACTUATED

EXTERNAL PRESSURE ACTUATED

INTERNAL PORT PRESSURE ACTUATED

PALM BUTTON ACTUATED

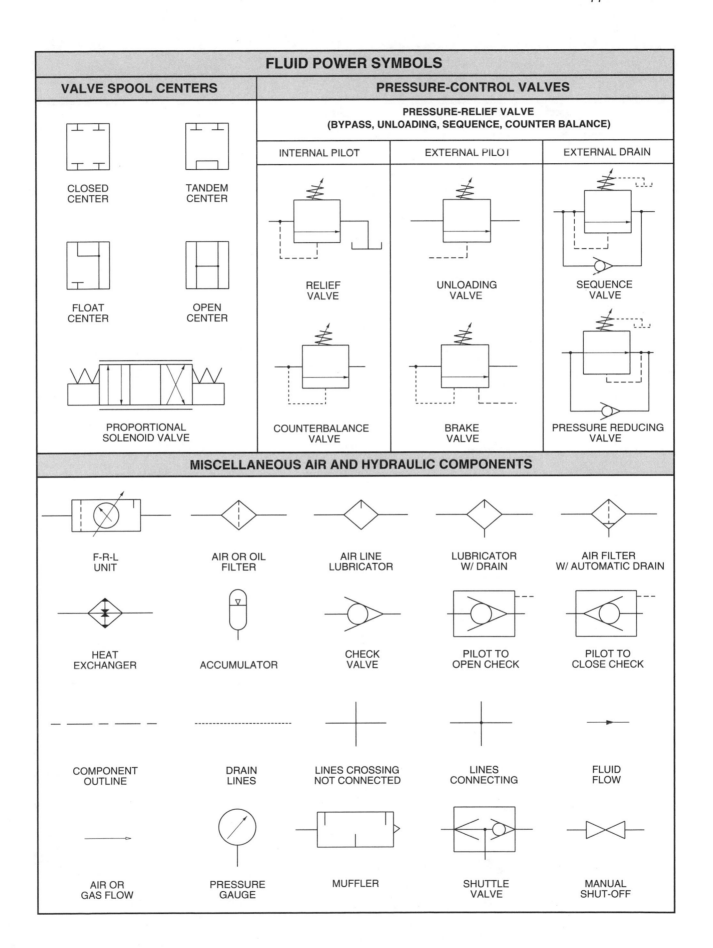

ELECTRICAL/ELECTRONIC ABBREVIATIONS/ACRONYMS

Abbr/Acronym	Meaning	Abbr/Acronym	Meaning	Abbr/Acronym	Meaning
A	Ammeter; Ampere; Anode; Armature	FU	Fuse	PNP	Positive-Negative-Positive
AC	Alternating Current	FWD	Forward	POS	Positive
AC/DC	Alternating Current; Direct Current	G	Gate; Giga; Green; Conductance	POT.	Potentiometer
A/D	Analog to Digital	GEN	Generator	P-P	Peak-to-Peak
AF	Audio Frequency	GRD	Ground	PRI	Primary Switch
AFC	Automatic Frequency Control	GY	Gray	PS	Pressure Switch
Ag	Silver	H	Henry; High Side of Transformer; Magnetic Flux	PSI	Pounds Per Square Inch
ALM	Alarm			PUT	Pull-Up Torque
AM	Ammeter; Amplitude Modulation	HF	High Frequency	Q	Transistor
AM/FM	Amplitude Modulation; Frequency Modulation	HP	Horsepower	R	Radius; Red; Resistance; Reverse
		Hz	Hertz	RAM	Random-Access Memory
ARM.	Armature	I	Current	RC	Resistance-Capacitiance
Au	Gold	IC	Integrated Circuit	RCL	Resistance-Inductance-Capacitance
AU	Automatic	INT	Intermediate; Interrupt	REC	Rectifier
AVC	Automatic Volume Control	INTLK	Interlock	RES	Resistor
AWG	American Wire Gauge	IOL	Instantaneous Overload	REV	Reverse
BAT.	Battery (electric)	IR	Infrared	RF	Radio Frequency
BCD	Binary Coded Decimal	ITB	Inverse Time Breaker	RH	Rheostat
BJT	Bipolar Junction Transistor	ITCB	Instantaneous Trip Circuit Breaker	rms	Root Mean Square
BK	Black	JB	Junction Box	ROM	Read-Only Memory
BL	Blue	JFET	Junction Field-Effect Transistor	rpm	Revolutions Per Minute
BR	Brake Relay; Brown	K	Kilo; Cathode	RPS	Revolutions Per Second
C	Celsius; Capacitiance; Capacitor	L	Line; Load; Coil; Inductance	S	Series; Slow; South; Switch
CAP.	Capacitor	LB-FT	Pounds Per Foot	SCR	Silicon Controlled Rectifier
CB	Circuit Breaker; Citizen's Band	LB-IN.	Pounds Per Inch	SEC	Secondary
CC	Common-Collector Configuration	LC	Inductance-Capacitance	SF	Service Factor
CCW	Counterclockwise	LCD	Liquid Crystal Display	1 PH; 1φ	Single-Phase
CE	Common-Emitter Configuration	LCR	Inductance-Capacitance-Resistance	SOC	Socket
CEMF	Counter Electromotive Force	LED	Light Emitting Diode	SOL	Solenoid
CKT	Circuit	LRC	Locked Rotor Current	SP	Single-Pole
CONT	Continuous; Control	LS	Limit Switch	SPDT	Single-Pole, Double-Throw
CPS	Cycles Per Second	LT	Lamp	SPST	Single-Pole, Single-Throw
CPU	Central Processing Unit	M	Motor; Motor Starter; Motor Starter Contacts	SS	Selector Switch
CR	Control Relay			SSW	Safety Switch
CRM	Control Relay Master	MAX.	Maximum	SW	Switch
CT	Current Transformer	MB	Magnetic Brake	T	Tera; Terminal; Torque; Transformer
CW	Clockwise	MCS	Motor Circuit Switch	TB	Terminal Board
D	Diameter; Diode; Down	MEM	Memory	3 PH; 3φ?	Three-Phase
D/A	Digital to Analog	MED	Medium	TD	Time Delay
DB	Dynamic Braking Contactor; Relay	MIN	Minimum	TDF	Time Delay Fuse
DC	Direct Current	MN	Manual	TEMP	Temperature
DIO	Diode	MOS	Metal-Oxide Semiconductor	THS	Thermostat Switch
DISC.	Disconnect Switch	MOSFET	Metal-Oxide Semiconductor Field-Effect Transistor	TR	Time Delay Relay
DMM	Digital Multimeter			TTL	Transistor-Transistor Logic
DP	Double-Pole	MTR	Motor	U	Up
DPDT	Double-Pole, Double-Throw	N; NEG	North; Negative	UCL	Unclamp
DPST	Double-Pole, Single-Throw	NC	Normally Closed	UHF	Ultrahigh Frequency
DS	Drum Switch	NEUT	Neutral	UJT	Unijunction Transistor
DT	Double-Throw	NO	Normally Open	UV	Ultraviolet; Undervoltage
DVM	Digital Voltmeter	NPN	Negative-Positive-Negative	V	Violet; Volt
EMF	Electromotive Force	NTDF	Nontime-Delay Fuse	VA	Volt Amp
F	Fahrenheit; Fast; Field; Forward; Fuse	O	Orange	VAC	Volts Alternating Current
FET	Field-Effect Transistor	OCPD	Overcurrent Protection Device	VDC	Volts Direct Current
FF	Flip-Flop	OHM	Ohmmeter	VHF	Very High Frequency
FLC	Full-Load Current	OL	Overload Relay	VLF	Very Low Frequency
FLS	Flow Switch	OZ/IN.	Ounces Per Inch	VOM	Volt-Ohm-Milliammeter
FLT	Full-Load Torque	P	Peak; Positive; Power; Power Consumed	W	Watt; White
FM	Fequency Modulation	PB	Pushbutton	w/	With
FREQ	Frequency	PCB	Printed Circuit Board	X	Low Side of Transformer
FS	Float Switch	PH; ?	Phase	Y	Yellow
FTS	Foot Switch	PLS	Plugging Switch	Z	Impedance

LAMP RATINGS

Lamp	Initial Lumen	Mean Lumen
40 W standard incandescent	480	N/A
100 W standard incandescent	1750	N/A
40 W standard fluorescent	3400	3100
100 W tungsten-halogen	1800	1675
100 W mercury-vapor	4000	3000
250 W mercury-vapor	12,000	9800
100 W high-pressure sodium	9500	8500
250 W high-pressure sodium	30,000	27,000
250 W metal-halide	20,000	17,000

LAMP IDENTIFICATION

Lamp	Letter	Number	Rating
Mercury-vapor	H	33	400 W
		34	1000 W/130 V
		35	700 W
		36	1000 W/265 V
		37	250 W
		38	100 W
		39	175 W
		43	75 W
Self-ballasted mercury-vapor	B	78	750 W/120 V
Metal-halide	M	47	1000 W
		48	1500 W
		57	175 W
		58	250 W
		59	400 W
High-pressure sodium	S	50	250 W
		51	400 W
		52	1000 W
		54	100 W
		55	150 W/55 V
		56	150 W/100 V
		62	70 W
		66	200 W
		67	310 W
		68	50 W
		76	35 W

BALLAST LOSS

Lamp	Lamp Rated Wattage (in W)	Ballast Loss (power loss %)
Low-pressure sodium	70	27
	100	25
	150	22
	250	20
	400	15
	1000	7
Mercury-vapor	40	32
	75	25
	100	22
	175	17
	250	16
	400	14
	700	7
	1000	7
Metal-halide	175	20
	250	17
	400	13
	1000	7
	1500	7
High-pressure sodium	50	30
	100	24
	150	22
	250	20
	400	16
	1000	7

RECOMMENDED LIGHT LEVELS

Area	Footcandles
Hospital operating table	2500
Factory assembly area	100-200
Accounting office	150
Major league baseball infield	150
Major league baseball outfield	100
School classroom	60-90
Home kitchen	50-70
Bank lobby	50
Active warehouse storage area	20-30
Inactive warehouse storage area	5
Auditorium	15
Elevator	10
Parking lot	5
Interstate roadway	1.4
Street	.9

RECOMMENDED BALLAST OUTPUT VOLTAGE LIMITS

Ballast	Lamp Size		rms Voltage (Volts)
	Wattage	ANSI Number	
Low-pressure sodium	18	L69	300-325
	35	L70	455-505
	55	L71	455-505
	90	L72	455-525
	135	L73	645-715
	180	L74	645-715
Mercury-vapor	50	H46	225-255
	75	H43	225-255
	100	H38	225-255
	175	H39	225-255
	250	H37	225-255
	400	H33	225-255
	700	H35	405-455
	1000	H36	405-455
Metal-halide	70	M85	210-250
	100	M90	250-300
	150	M81	220-260
	175	M57	285-320
	250	M80	230-270
	250	M58	285-320
	400	M59	285-320
	1000	M47	400-445
	1500	M48	400-445
High-pressure sodium	35	S76	110-130
	50	S68	110-130
	70	S62	110-130
	100	S54	110-130
	150	S55	110-130
	150	S56	200-250
	200	S66	200-230
	250	S50	175-225
	310	S67	155-190
	400	S51	175-225
	1000	S52	420-480

RECOMMENDED SHORT-CIRCUIT CURRENT TEST LIMITS

Ballast	Lamp Size		Short-Circuit Current (Amps)
	Wattage	ANSI Number	
Low-pressure sodium	18	L69	.30-.40
	35	L70	.52-.78
	55	L71	.52-.78
	90	L72	.8-1.2
	135	L73	.8-1.2
	180	L74	.8-1.2
Mercury-vapor	50	H46	.85-1.15
	75	H43	.95-1.70
	100	H38	1.10-2.00
	175	H39	2.0-3.6
	250	H37	3.0-3.8
	400	H33	4.4-7.9
	700	H35	3.9-5.85
	1000	H36	5.7-9.0
Metal-halide	70	M85	.85-1.30
	100	M90	1.15-1.76
	150	M81	1.75-2.60
	175	M57	1.5-1.90
	250	M80	2.9-4.3
	250	M58	2.2-2.85
	400	M59	3.5-4.5
	1000	M47	4.8-6.15
	1500	M48	7.4-9.6
High-pressure sodium	35	S76	.85-1.45
	50	S68	1.5-2.3
	70	S62	1.6-2.9
	100	S54	2.45-3.8
	150	S55	3.5-5.4
	150	S56	2.0-3.0
	200	S66	2.50-3.7
	250	S50	3.0-5.3
	310	S67	3.8-5.7
	400	S51	5.0-7.6
	1000	S52	5.5-8.1

VOLTAGE CONVERSIONS

To Convert	To	Multiply By
rms	Average	.9
rms	Peak	1.414
Average	rms	1.111
Average	Peak	1.567
Peak	rms	.707
Peak	Average	.637
Peak	Peak-to-peak	2

MOTOR REPAIR AND SERVICE RECORD

Motor File #: _____ Serial #: _____

Date Installed: _____ Motor Location: _____

_____

MFR: _____ Type: _____ Frame: _____

HP: _____ Volts: _____ Amps: _____

RPM: _____ Filter Sizes: _____

Date	Operation	Mechanic

SEMIANNUAL MOTOR MAINTENANCE CHECKLIST

Motor File #: _____ Serial #: _____

Date Installed: _____ Motor Location: _____

_____

MFR: _____ Type: _____ Frame: _____

HP: _____ Volts: _____ Amps: _____

RPM: _____ Date Serviced: _____

Step	Operation	Mechanic
1	Turn OFF and lock out all power to the motor and its control circuit.	
2	Clean motor exterior and all ventilation ducts.	
3	Check motor's wire raceway.	
4	Check and lubricate bearings as needed.	
5	Check drive mechanism.	
6	Check brushes and commutator.	
7	Check slip rings.	
8	Check motor terminations.	
9	Check capacitors.	
10	Check all mounting bolts.	
11	Check and record line-to-line resistance.	
12	Check and record megohmmeter resistance from L1 to ground.	
13	Check motor controls.	
14	Reconnect motor and control circuit power supplies.	
15	Check line-to-line voltage for balance and level.	
16	Check line current draw against nameplate rating.	
17	Check and record inboard and outboard bearing temperatures.	

ANNUAL MOTOR MAINTENANCE CHECKLIST

Motor File #: _____   Serial #: _____

Date Installed: _____   Motor Location: _____

_____

MFR: _____   Type: _____   Frame: _____

HP: _____   Volts: _____   Amps: _____

RPM: _____   Date Serviced: _____

Step	Operation	Mechanic
1	Turn OFF and lock out all power to the motor and its control circuit.	
2	Clean motor exterior and all ventilation ducts.	
3	Uncouple motor from load and disassemble.	
4	Clean inside of motor.	
5	Check centrifugal switch assemblies.	
6	Check rotors, armatures, and field windings.	
7	Check all peripheral equipment.	
8	Check bearings.	
9	Check brushes and commutator.	
10	Check slip rings.	
11	Reassemble motor and couple to load.	
12	Flush old bearing lubricant and replace.	
13	Check motor's wire raceway.	
14	Check drive mechanism.	
15	Check motor terminations.	
16	Check capacitors.	
17	Check all mounting bolts.	
18	Check and record line-to-line resistance.	
19	Check and record megohmmeter resistance from T1 to ground.	
20	Check and record insulation polarization index.	
21	Check motor controls.	
22	Reconnect motor and control circuit power supplies.	
23	Check line-to-line voltage for balance and level.	
24	Check line current draw against nameplate rating.	
25	Check and record inboard and outboard bearing temperatures.	

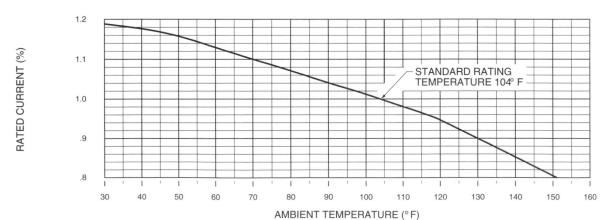

HEATER AMBIENT TEMPERATURE CORRECTION

FULL-LOAD CURRENTS — DC MOTORS

Motor rating (HP)	Current (A)	
	120 V	240 V
¼	3.1	1.6
⅓	4.1	2.0
½	5.4	2.7
¾	7.6	3.8
1	9.5	4.7
1½	13.2	6.6
2	17	8.5
3	25	12.2
5	40	20
7½	48	29
10	76	38

FULL-LOAD CURRENTS — 1ϕ, AC MOTORS

Motor rating (HP)	Current (A)	
	115 V	230 V
⅙	4.4	2.2
¼	5.8	2.9
⅓	7.2	3.6
½	9.8	4.9
¾	13.8	6.9
1	16	8
1½	20	10
2	24	12
3	34	17
5	56	28
7½	80	40

FULL-LOAD CURRENTS — 3ϕ, AC INDUCTION MOTORS

Motor rating (HP)	Current (A)			
	208 V	230 V	460 V	575 V
¼	1.11	.96	.48	.38
⅓	1.34	1.18	.59	.47
½	2.2	2.0	1.0	.8
¾	3.1	2.8	1.4	1.1
1	4.0	3.6	1.8	1.4
1½	5.7	5.2	2.6	2.1
2	7.5	6.8	3.4	2.7
3	10.6	9.6	4.8	3.9
5	16.7	15.2	7.6	6.1
7½	24.0	22.0	11.0	9.0
10	31.0	28.0	14.0	11.0
15	46.0	42.0	21.0	17.0
20	59	54	27	22
25	75	68	34	27
30	88	80	40	32
40	114	104	52	41
50	143	130	65	52
60	169	154	77	62
75	211	192	96	77
100	273	248	124	99
125	343	312	156	125
150	396	360	180	144
200	—	480	240	192
250	—	602	301	242
300	—	—	362	288
350	—	—	413	337
400	—	—	477	382
500	—	—	590	472

TYPICAL MOTOR EFFICIENCIES

HP	Standard Motor (%)	Energy-Effiecient Motor (%)	HP	Standard Motor (%)	Energy-Efficient Motor (%)
1	76.5	84.0	30	88.1	93.1
1.5	78.5	85.5	40	89.3	93.6
2	79.9	86.5	50	90.4	93.7
3	80.8	88.5	75	90.8	95.0
5	83.1	88.6	100	91.6	95.4
7.5	83.8	90.2	125	91.8	95.8
10	85.0	90.3	150	92.3	96.0
15	86.5	91.7	200	93.3	96.1
20	87.5	92.4	250	93.6	96.2
25	88.0	93.0	300	93.8	96.5

Frame No. Series	\multicolumn{8}{c}{Third/Fourth Digit of Frame No.}
	D	**1**	**2**	**3**	**4**	**5**	**6**	**7**
140	3.50	3.00	3.50	4.00	4.50	5.00	5.50	6.25
160	4.00	3.50	4.00	4.50	5.00	5.50	6.25	7.00
180	4.50	4.00	4.50	5.00	5.50	6.25	7.00	8.00
200	5.00	4.50	5.00	5.50	6.50	7.00	8.00	9.00
210	5.25	4.50	5.00	5.50	6.25	7.00	8.00	9.00
220	5.50	5.00	5.50	6.25	6.75	7.50	9.00	10.00
250	6.25	5.50	6.25	7.00	8.25	9.00	10.00	11.00
280	7.00	6.25	7.00	8.00	9.50	10.00	11.00	12.50
320	8.00	7.00	8.00	9.00	10.50	11.00	12.00	14.00
360	9.00	8.00	9.00	10.00	11.25	12.25	14.00	16.00
400	10.00	9.00	10.00	11.00	12.25	13.75	16.00	18.00
440	11.00	10.00	11.00	12.50	14.50	16.50	18.00	20.00
500	12.50	11.00	12.50	14.00	16.00	18.00	20.00	22.00
580	14.50	12.50	14.00	16.00	18.00	20.00	22.00	25.00
680	17.00	16.00	18.00	20.00	22.00	25.00	28.00	32.00

MOTOR FRAME TABLE

Frame No. Series	\multicolumn{8}{c}{Third/Fourth Digit of Frame No.}
	D	**8**	**9**	**10**	**11**	**12**	**13**	**14**	**15**
140	3.50	7.00	8.00	9.00	10.00	11.00	12.50	14.00	16.00
160	4.00	8.00	9.00	10.00	11.00	12.50	14.00	16.00	18.00
180	4.50	9.00	10.00	11.00	12.50	14.00	16.00	18.00	20.00
200	5.00	10.00	11.00	—	—	—	—	—	—
210	5.25	10.00	11.00	12.50	14.00	16.00	18.00	20.00	22.00
220	5.50	11.00	12.50	—	—	—	—	—	—
250	6.25	12.50	14.00	16.00	18.00	20.00	22.00	25.00	28.00
280	7.00	14.00	16.00	18.00	20.00	22.00	25.00	28.00	32.00
320	8.00	16.00	18.00	20.00	22.00	25.00	28.00	32.00	36.00
360	9.00	18.00	20.00	22.00	25.00	28.00	32.00	36.00	40.00
400	10.00	20.00	22.00	25.00	28.00	32.00	36.00	40.00	45.00
440	11.00	22.00	25.00	28.00	32.00	36.00	40.00	45.00	50.00
500	12.50	25.00	28.00	32.00	36.00	40.00	45.00	50.00	56.00
580	14.50	28.00	32.00	36.00	40.00	45.00	50.00	56.00	63.00
680	17.00	36.00	40.00	45.00	50.00	56.00	63.00	71.00	80.00

MOTOR FRAME LETTERS

LETTER	DESIGNATION
G	Gasoline pump motor
K	Sump pump motor
M and N	Oil burner motor
S	Standard short shaft for direct connection
T	Standard dimensions established
U	Previously used as frame designation for which standard dimensions are established
Y	Special mounting dimensions required from manufacturer
Z	Standard mounting dimensions except shaft extension

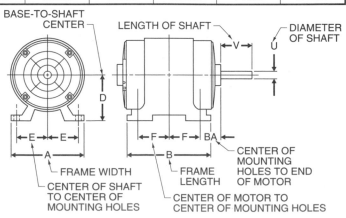

BASE-TO-SHAFT CENTER
LENGTH OF SHAFT
DIAMETER OF SHAFT
FRAME WIDTH
CENTER OF SHAFT TO CENTER OF MOUNTING HOLES
FRAME LENGTH
CENTER OF MOTOR TO CENTER OF MOUNTING HOLES
CENTER OF MOUNTING HOLES TO END OF MOTOR

MOTOR FRAME DIMENSIONS...

Frame No.	Shaft U	Shaft V	Key W	Key T	Key L	A	B	D	E	F	BA
48	1/2	1 1/2*	flat	3/64	—	5 5/8*	3 1/2*	3	2 1/8	1 3/8	2 1/2
56	5/8	1 7/8*	3/16	3/16	1 3/8	6 1/2*	4 1/4*	3 1/2	2 7/16	1 1/2	2 3/4
143T	7/8	2	3/16	3/16	1 3/8	7	6	3 1/2	2 3/4	2	2 1/4
145T	7/8	2	3/16	3/16	1 3/8	7	6	3 1/2	2 3/4	2 1/2	2 1/4
182	7/8	2	3/16	3/16	1 3/8	9	6 1/2	4 1/2	3 3/4	2 1/4	2 3/4
182T	1 1/8	2 1/2	1/4	1/4	1 3/4	9	6 1/2	4 1/2	3 3/4	2 1/4	2 3/4
184	7/8	2	3/16	3/16	1 3/8	9	7 1/2	4 1/2	3 3/4	2 3/4	2 3/4
184T	1 1/8	2 1/2	1/4	1/4	1 3/4	9	7 1/2	4 1/2	3 3/4	2 3/4	2 3/4
203	3/4	2	3/16	3/16	1 3/8	10	7 1/2	5	4	2 3/4	3 1/8
204	3/4	2	3/16	3/16	1 3/8	10	8 1/2	5	4	3 1/4	3 1/8
213	1 1/8	2 3/4	1/4	1/4	2	10 1/2	7 1/2	5 1/4	4 1/4	2 3/4	3 1/2
213T	1 3/8	3 1/8	5/16	5/16	2 3/8	10 1/2	7 1/2	5 1/4	4 1/4	2 3/4	3 1/2
215	1 1/8	2 3/4	1/4	1/4	2	10 1/2	9	5 1/4	4 1/4	3 1/2	3 1/2
215T	1 3/8	3 1/8	5/16	5/16	2 3/8	10 1/2	9	5 1/4	4 1/4	3 1/2	3 1/2
224	1	2 3/4	1/4	1/4	2	11	8 3/4	5 1/2	4 1/2	3 3/8	3 1/2
225	1	2 3/4	1/4	1/4	2	11	9 1/2	5 1/2	4 1/2	3 3/4	3 1/2
254	1 1/8	3 1/8	1/4	1/4	2 3/8	12 1/2	10 3/4	6 1/4	5	4 1/8	4 1/4
254U	1 3/8	3 1/2	5/16	5/16	2 3/4	12 1/2	10 3/4	6 1/4	5	4 1/8	4 1/4
254T	1 5/8	3 3/4	3/8	3/8	3 1/4	12 1/2	10 3/4	6 1/4	5	4 1/8	4 1/4
256U	1 3/8	3 1/2	5/16	5/16	2 3/4	12 1/2	12 1/2	6 1/4	5	5	4 1/4
256T	1 5/8	3 3/4	3/8	3/8	2 7/8	12 1/2	12 1/2	6 1/4	5	5	4 1/4
284	1 1/4	3 1/2	1/4	1/4	2 3/4	14	12 1/2	7	5 1/2	4 3/4	4 3/4
284U	1 5/8	4 5/8	3/8	3/8	3 3/4	14	12 1/2	7	5 1/2	4 3/4	4 3/4
284T	1 7/8	4 3/8	1/2	1/2	3 1/4	14	12 1/2	7	5 1/2	4 3/4	4 3/4
284TS	1 5/8	3	3/8	3/8	1 7/8	14	12 1/2	7	5 1/2	4 3/4	4 3/4
286U	1 5/8	4 5/8	3/8	3/8	3 3/4	14	14	7	5 1/2	5 1/2	4 3/4
286T	1 7/8	4 3/8	1/2	1/2	3 1/4	14	14	7	5 1/2	5 1/2	4 3/4
286TS	1 5/8	3	3/8	3/8	1 7/8	14	14	7	5 1/2	5 1/2	4 3/4
324	1 5/8	4 5/8	3/8	3/8	3 3/4	16	14	8	6 1/4	5 1/4	5 1/4
324U	1 7/8	5 3/8	1/2	1/2	4 1/4	16	14	8	6 1/4	5 1/4	5 1/4
324S	1 5/8	3	3/8	3/8	1 7/8	16	14	8	6 1/4	5 1/4	5 1/4
324T	2 1/8	5	1/2	1/2	3 7/8	16	14	8	6 1/4	5 1/4	5 1/4
324TS	1 7/8	3 1/2	1/2	1/2	2	16	14	8	6 1/4	5 1/4	5 1/4
326	1 5/8	4 5/8	3/8	3/8	3 3/4	16	15 1/2	8	6 1/4	6	5 1/4
326U	1 7/8	5 3/8	1/2	1/2	4 1/4	16	15 1/2	8	6 1/4	6	5 1/4
326S	1 5/8	3	3/8	3/8	1 7/8	16	15 1/2	8	6 1/4	6	5 1/4
326T	2 1/8	5	1/2	1/2	3 7/8	16	15 1/2	8	6 1/4	6	5 1/4
326TS	1 7/8	3 1/2	1/2	1/2	2	16	15 1/2	8	6 1/4	6	5 1/4
364	1 7/8	5 3/8	1/2	1/2	4 1/4	18	15 1/4	9	7	5 5/8	5 7/8

...MOTOR FRAME DIMENSIONS

Frame No.	Shaft U	Shaft V	Key W	Key T	Key L	A	B	D	E	F	BD
364S	1 5/8	3	3/8	3/8	1 7/8	18	15 1/4	9	7	5 5/8	5 7/8
364U	2 1/8	6 1/8	1/2	1/2	5	18	15 1/4	9	7	5 5/8	5 7/8
364US	1 7/8	3 1/2	1/2	1/2	2	18	15 1/4	9	7	5 5/8	5 7/8
364T	2 3/8	5 5/8	5/8	5/8	4 1/4	18	15 1/4	9	7	5 5/8	5 7/8
364TS	1 7/8	3 1/2	1/2	1/2	2	18	15 1/4	9	7	5 5/8	5 7/8
365	1 7/8	5 3/8	3/8	3/8	4 1/4	18	16 1/4	9	7	6 1/8	5 7/8
365S	1 5/8	3	3/8	3/8	1 7/8	18	16 1/4	9	7	6 1/8	5 7/8
365U	2 1/8	6 1/8	1/2	1/2	5	18	16 1/4	9	7	6 1/8	5 7/8
365US	1 7/8	3 1/2	1/2	1/2	2	18	16 1/4	9	7	6 1/8	5 7/8
365T	2 3/8	5 5/8	5/8	5/8	4 1/4	18	16 1/4	9	7	6 1/8	5 7/8
365TS	1 7/8	3 1/2	1/2	1/2	2	18	16 1/4	9	7	6 1/8	5 7/8
404	2 1/8	6 1/8	1/2	1/2	5	20	16 1/4	10	8	6 1/8	6 5/8
404S	1 7/8	3 1/2	1/2	1/2	2	20	16 1/4	10	8	6 1/8	6 5/8
404U	2 3/8	6 7/8	5/8	5/8	5 1/2	20	16 1/4	10	8	6 1/8	6 5/8
404US	2 1/8	4	1/2	1/2	2 3/4	20	16 1/4	10	8	6 1/8	6 5/8
404T	2 7/8	7	3/4	3/4	5 5/8	20	16 1/4	10	8	6 1/8	6 5/8
404TS	2 1/8	4	1/2	1/2	2 3/4	20	16 1/4	10	8	6 1/8	6 5/8
405	2 1/8	6 1/8	1/2	1/2	5	20	17 3/4	10	8	6 7/8	6 5/8
405S	1 7/8	3 1/2	1/2	1/2	2	20	17 3/4	10	8	6 7/8	6 5/8
405U	2 3/8	6 7/8	5/8	5/8	5 1/2	20	17 3/4	10	8	6 7/8	6 5/8
405US	2 1/8	4	1/2	1/2	2 3/4	20	17 3/4	10	8	6 7/8	6 5/8
405T	2 7/8	7	3/4	3/4	5 5/8	20	17 3/4	10	8	6 7/8	6 5/8
405TS	2 1/8	4	1/2	1/2	2 3/4	20	17 3/4	10	8	6 7/8	6 5/8
444	2 3/8	6 7/8	5/8	5/8	5 1/2	22	18 1/2	11	9	7 1/4	7 1/2
444S	2 1/8	4	1/2	1/2	2 3/4	22	18 1/2	11	9	7 1/4	7 1/2
444U	2 7/8	8 3/8	3/4	3/4	7	22	18 1/2	11	9	7 1/4	7 1/2
444US	2 1/8	4	1/2	1/2	2 3/4	22	18 1/2	11	9	7 1/4	7 1/2
444T	3 3/8	8 1/4	7/8	7/8	6 7/8	22	18 1/2	11	9	7 1/4	7 1/2
444TS	2 3/8	4 1/2	5/8	5/8	3	22	18 1/2	11	9	7 1/4	7 1/2
445	2 3/8	6 7/8	5/8	5/8	5 1/2	22	20 1/2	11	9	8 1/4	7 1/2
445S	2 1/8	4	1/2	1/2	2 3/4	22	20 1/2	11	9	8 1/4	7 1/2
445U	2 7/8	8 3/8	3/4	3/4	7	22	20 1/2	11	9	8 1/4	7 1/2
445US	2 1/8	4	1/2	1/2	2 3/4	22	20 1/2	11	9	8 1/4	7 1/2
445T	3 3/8	8 1/4	7/8	7/8	6 7/8	22	20 1/2	11	9	8 1/4	7 1/2
445TS	2 3/8	4 1/2	5/8	5/8	3	22	20 1/2	11	9	8 1/4	7 1/2
504U	2 7/8	8 3/8	3/4	3/4	7 1/4	25	21	12 1/2	10	8	8 1/2
504S	2 1/8	4	1/2	1/2	2 3/4	25	21	12 1/2	10	8	8 1/2
505	2 7/8	8 3/8	3/4	3/4	7 1/4	25	23	12 1/2	10	9	8 1/2
505S	2 1/8	4	1/2	1/2	2 3/4	25	23	12 1/2	10	9	8 1/2

* not NEMA standard dimensions

INCOMING POWER TROUBLESHOOTING MATRIX			
SYMPTOM/FAULT CODE	**PROBLEM**	**CAUSE**	**SOLUTION**
Electric motor drive overvoltage faults. Blown converter (rectifier) semiconductor	High input voltage/voltage swell	Switching of power factor correction capacitors	Stop switching power factor correction capacitors. Install electric motor drive on another feeder
		Utility switching transformer taps for load adjustment	Install line reactor, or install electric motor drive on another feeder
		Proximity to low impedance voltage source	Install line reactor
		Transformer secondary voltage is high	Adjust taps on transformer
Electric motor drive overvoltage faults. Difference between no load and full load voltage is greater than 3%	Low input voltage/voltage sag	Power source for electric motor drive unable to deliver enough current	Increase kVA source rating, or install electric motor drive on another feeder
		Low transformer secondary voltage	Adjust secondary taps on supply transformer
		Large starting load(s)	Increase kVA source rating, or install electric motor drive on another feeder
		Electric motor drive is overloaded	See *Motor and Load Troubleshooting Matrix*
	Voltage sine wave flat topping at electric motor drive	Power source for electric motor drive unable to deliver enough current	Increase kVA source rating, or install electric motor drive on another feeder
Electric motor drive does not turn on. Blown fuse or tripped circuit breaker	No input voltage	Incorrect fuse or circuit breaker	Install correct fuse or circuit breaker
		Conductors feeding electric motor drive shorted or have ground fault	Repair or replace conductors
		Problem with electric motor drive	See *Electric Motor Drive Troubleshooting Matrix*
Electric motor drive overload faults	Voltage unbalance greater than 2%	Unbalance from utility	Contact utility
			Install electric motor drive on separate feeder
Intermittent electric motor drive faults and/or electric motor drive intermittently does not operate per design	Improper grounding	Undersized grounds or no grounds	Install proper size ground
		Loose ground connections	Tighten ground connections
	Intermittent fault/erratic operation	Electric noise, EMI/RFI	See *Electric Motor Drive Troubleshooting Matrix*
Harmonics	Harmonics present on electrical distribution system	Electric motor drive or existing nonlinear loads can be source	Install input reactor

ELECTRIC MOTOR DRIVE TROUBLESHOOTING MATRIX			
ELECTRIC MOTOR DRIVE FAULTS			
SYMPTOM/FAULT CODE	**PROBLEM**	**CAUSE**	**SOLUTION**
Electric motor drive overvoltage fault	Electric motor drive overvoltage	Deceleration time is too short	Increase deceleration time
		High input voltage/voltage swell	See *Incoming Power Troubleshooting Matrix*
		Load is overhauling motor	Add dynamic braking resistors and/or increase decerlation time. Note: also see *Motor and Load Troubleshooting Matrix*
Electric motor drive overcurrent fault	Electric motor drive overcurrent	Motor nameplate data incorrectly programmed	Check motor nameplate and program data correctly
		Acceleration time is too short	Increase acceleration time
		Start boost or continuous boost set too high	Lower start boost or continuous boost
		Short in inverter semiconductor	Replace inverter semiconductor or replace electric motor drive
		Contactor between electric motor drive and motor is changing state while electric motor drive outputs more than 0 Hz	Wire contactor to change state only when electric motor drive outputs 0 Hz
		Motor trying to start in a spinning mode	Enable flying start
Electric motor drive overcurrent fault. Load conductor current readings not equal	Electric motor drive overcurrent	Motor or conductors feeding motor are shorted or have ground fault	See *Motor and Load Troubleshooting Matrix*
Electric motor drive overcurrent fault. Current readings for load conductors are 105% or greater than motor nameplate current when motor powers driven load	Electric motor drive overcurrent	Problem with motor	See *Motor and Load Troubleshooting Matrix*
Electric motor drive overload fault. Current readings for load conductors are 105% or greater than motor nameplate current when motor powers driven load	Electric motor drive overload	Problem with motor and/or load	See *Motor and Load Troubleshooting Matrix*
Electric motor drive undervoltage fault	Electric motor drive undervoltage	Low input voltage/ voltage sag	See *Incoming Power Troubleshooting Matrix*
Electric motor drive overtemperature fault	Electric motor drive overtemperature	High ambient temperature	Add cooling to electric motor drive enclosure, or relocate electric motor drive enclosure
		Drive cooling fan defective	Replace electric motor drive cooling fan, or replace electric motor drive
		Heat sink dirty, or air intake clogged	Clean heatsink or air intake
		Problem with motor and/or load	See *Motor and Load Troubleshooting Matrix*

MOTOR AND LOAD TROUBLESHOOTING MATRIX			
MOTOR AND LOAD PROBLEMS . . .			
SYMPTOM/FAULT CODE	PROBLEM	CAUSE	SOLUTION
Electric motor drive overvoltage fault	Electric motor drive overvoltage	Load is overhauling motor	Contact drive manufacturer for assistance. Note: Also see *Electric Motor Drive Matrix*
Electric motor drive overcurrent fault. Current readings for load conductors/motor not equal. Low megohmmeter readings for conductors feeding motor	Electric motor drive overcurrent	Conductors feeding motor are shorted or have ground fault	Repair or replace conductors
Electric motor drive overcurrent fault. Current readings for load conductors/motor not equal. Low megohmmeter readings for motor	Electric motor drive overcurrent	Motor is shorted or has ground fault	Repair or replace motor
Electric motor drive overcurrent fault, electric motor drive overload fault, or electric motor drive overtemperature fault. Current readings for load conductors/motor greater than 105% of motor nameplate current when motor powers driven load	Electric motor drive overload	Misalignment between motor and driven load	Align motor and load
		Motor and/or driven load not securely fastened	Securely fasten motor and/or driven load
		Defective bearing(s) in motor and/or driven load	Replace defective defective bearing(s). Replace motor
		Object preventing motor or load from turning	Remove object
		Motor and/or electric motor drive not properly sized for load	Contact drive manufacturer for assistance
Unusual noise or vibration when electric motor drive powers load	Problem with motor and/or load	Misalignment between motor and driven load	Align motor and load
		Motor and/or driven load not securely fastened	Securely fasten motor and/or driven load
		Defective bearings(s) in motor and/or driven load	Replace defective bearing(s). Replace motor
		Motor and/or electric motor drive not properly sized for load	Contact drive manufacturer for assistance
	Parameter(s) incorrect	Parameter(s) incorrectly programmed	See *Electric Motor Drive Troubleshooting Matrix*
Bind in rotation of motor shaft or noise when shaft is rotated by hand	Problem with motor	Defective bearing(s) in motor	Replace defective bearing(s). Replace motor

MOTOR AND LOAD TROUBLESHOOTING MATRIX			
. . . MOTOR AND LOAD PROBLEMS			
SYMPTOM/FAULT CODE	PROBLEM	CAUSE	SOLUTION
Bind in rotation of load or noise when load is rotated by hand	Problem with load	Defective bearing(s) in load. Mechanical problem with load	Replace defective bearing(s). Correct mechanical problem
Current readings equal to or greater than motor nameplate current when motor run with load disconnected	Problem with motor	Defective bearing(s) in motor	Replace defective bearing(s). Replace motor
RELATED PROBLEMS			
Electric motor drive operates correctly when set to default parameters, Input Mode is keypad, and motor/load are disconnected. Drive does not operate correctly when set to default parameters (excluding Motor Nameplate Data, and Control Mode), Input Mode is keypad, and motor/load are connected	Problem with motor and/or load Parameters) incorrect	Problem with motor and/or load	See *Motor and Load Problems.*
		Parameter(s) incorrectly programmed	See *Electric Motor Drive Troubleshooting Matrix*

THREE-PHASE VOLTAGE VALUES

For 208 V × 1.732, use 360
For 230 V × 1.732, use 398
For 240 V × 1.732, use 416
For 440 V × 1.732, use 762
For 460 V × 1.732, use 797
For 480 V × 1.732, use 831

CAPACITOR RATINGS

110–125 VAC, 50/60 Hz, Starting Capacitors

Typical Ratings*	Dimensions**		Model Number***
	Diameter	Length	
88 – 106	$1^7/_{16}$	$2^3/_4$	EC8815
108 – 130	$1^7/_{16}$	$2^3/_4$	EC10815
130 – 156	$1^7/_{16}$	$2^3/_4$	EC13015
145 – 174	$1^7/_{16}$	$2^3/_4$	EC14515
161 – 193	$1^7/_{16}$	$2^3/_4$	EC16115
189 – 227	$1^7/_{16}$	$2^3/_4$	EC18915A
216 – 259	$1^7/_{16}$	$3^3/_8$	EC21615
233 – 280	$1^7/_{16}$	$3^3/_8$	EC23315A
243 – 292	$1^7/_{16}$	$3^3/_8$	EC24315A
270 – 324	$1^7/_{16}$	$3^3/_8$	EC27015A
324 – 389	$1^7/_{16}$	$3^3/_8$	EC2R10324N
340 – 408	$1^{13}/_{16}$	$3^3/_8$	EC34015
378 – 454	$1^{13}/_{16}$	$3^3/_8$	EC37815
400 – 480	$1^{13}/_{16}$	$3^3/_8$	EC40015
430 – 516	$1^{13}/_{16}$	$3^3/_8$	EC43015A
460 – 553	$1^{13}/_{16}$	$4^3/_8$	EC5R10460
540 – 648	$1^{13}/_{16}$	$4^3/_8$	EC54015B
590 – 708	$1^{13}/_{16}$	$4^3/_8$	EC59015A
708 – 850	$1^{13}/_{16}$	$4^3/_8$	EC70815
815 – 978	$1^{13}/_{16}$	$4^3/_8$	EC81515
1000 – 1200	$2^1/_{16}$	$4^3/_8$	EC100015A

220–250 VAC, 50/60 Hz, Starting Capacitors

Typical Ratings*	Diameter	Length	Model Number***
53 – 64	$1^7/_{16}$	$3^3/_8$	EC5335
64 – 77	$1^7/_{16}$	$3^3/_8$	EC6435
88 – 106	$1^{13}/_{16}$	$3^3/_8$	EC8835
108 – 130	$1^{13}/_{16}$	$3^3/_8$	EC10835A
124 – 149	$1^{13}/_{16}$	$4^3/_8$	EC12435
130 – 154	$1^{13}/_{16}$	$4^3/_8$	EC13035
145 – 174	$2^1/_{16}$	$3^3/_8$	EC6R22145
161 – 193	$2^1/_{16}$	$3^3/_8$	EC6R2216N
216 – 259	$2^1/_{16}$	$4^3/_8$	EC21635A
233 – 280	$2^1/_{16}$	$4^3/_8$	EC23335A
270 – 324	$2^1/_{16}$	$4^3/_8$	EC27035A

* in μF
** in inches
*** Model numbers vary by manufacturer.

CAPACITOR RATINGS

270 VAC, 50/60 Hz, Running Capacitors

Typical Ratings*	Dimensions**		Model Number***
	Oval	Length	
2	$1^5/_{16} \times 2^5/_{32}$	$2^1/_8$	VH5502
3		$2^1/_8$	VH5503
4		$2^1/_8$	VH5704
5		$2^1/_8$	VH5705
6		$2^5/_8$	VH5706
7.5	$1^5/_{16} \times 2^5/_{32}$	$2^7/_8$	VH9001
10		$2^7/_8$	VH9002
12.5		$3^7/_8$	VH9003
15	$1^{29}/_{32} \times 2^{29}/_{32}$	$2^1/_8$	VH9121
17.5		$2^7/_8$	VH9123
20	$1^{29}/_{32} \times 2^{29}/_{32}$	$2^7/_8$	VH5463
25		$3^7/_8$	VH9069
30		$3^7/_8$	VH5465
35	$1^{29}/_{32} \times 2^{29}/_{32}$	$3^7/_8$	VH9071
40		$3^7/_8$	VH9073
45	$1^{31}/_{32} \times 3^{21}/_{32}$	$3^7/_8$	VH9115
50		$3^7/_8$	VH9075

440 VAC, 50/60 Hz, Running Capacitors

Typical Ratings*	Oval	Length	Model Number***
10	$1^5/_{16} \times 2^5/_{32}$	$3^7/_8$	VH5300
15	$1^{29}/_{32} \times 2^{29}/_{32}$	$2^7/_8$	VH5304
17.5	$1^{29}/_{32} \times 2^{29}/_{32}$	$3^7/_8$	VH9141
20	$1^{29}/_{32} \times 2^{29}/_{32}$	$3^7/_8$	VH9082
25	$1^{29}/_{32} \times 2^{29}/_{32}$	$3^7/_8$	VH5310
30	$1^{29}/_{32} \times 2^{29}/_{32}$	$4^3/_4$	VH9086
35		$4^3/_4$	VH9088
40		$4^3/_4$	VH9641
45	$1^{31}/_{32} \times 3^{21}/_{32}$	$3^7/_8$	VH5351
50		$3^7/_8$	VH5320
55		$4^3/_4$	VH9084

* in μF
** in inches
*** Model numbers vary by manufacturer.

RESISTOR COLOR CODES

Color	Number		Multiplier	Tolerance (%)
	1st	2nd		
Black (BK)	0	0	1	0
Brown (BR)	1	1	10	—
Red (R)	2	2	100	—
Orange (O)	3	3	1000	—
Yellow (Y)	4	4	10,000	—
Green (G)	5	5	100,000	—
Blue (BL)	6	6	1,000,000	—
Violet (V)	7	7	10,000,000	—
Gray (GY)	8	8	100,000,000	—
White (W)	9	9	1,000,000,000	—
Gold (Au)	—	—	0.1	5
Silver (Ag)	—	—	0.01	10
None	—	—	0	20

Formulas

THREE-PHASE VOLTAGE VALUES
For 208 V × 1.732, use 360
For 230 V × 1.732, use 398
For 240 V × 1.732, use 416
For 440 V × 1.732, use 762
For 460 V × 1.732, use 797
For 480 V × 1.732, use 831

Ohm's Law

Ohm's law is the relationship between the voltage, current, and resistance in an electrical circuit. Ohm's law states that current in a circuit is proportional to the voltage and inversely proportional to the resistance. It is written $I = E/R$, $R = E/I$, and $E = R \times I$.

Power Formula

The *power formula* is the relationship between the voltage, current, and power in an electrical circuit. The power formula states that the power in a circuit is equal to the voltage times the current. It is written $P = E \times I$, $E = P/I$, and $I = P/E$. Any value in these relationships is found using Ohm's Law and Power Formula.

P = WATTS
I = AMPS
R = OHMS
E = VOLTS

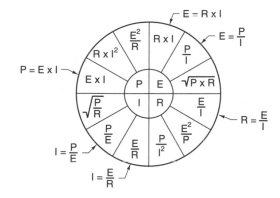

VALUES IN INNER CIRCLE ARE EQUAL TO VALUES IN CORRESPONDING OUTER CIRCLE

OHM'S LAW AND POWER FORMULA

POWER FORMULAS – 1ϕ, 3ϕ					
Phase	To Find	Use Formula	Example		
			Given	Find	Solution
1ϕ	I	$I = \dfrac{VA}{V}$	32,000 VA, 240 V	I	$I = \dfrac{VA}{V}$ $I = \dfrac{32{,}000 \text{ VA}}{240 \text{ V}}$ **I = 133 A**
1ϕ	VA	$VA = I \times V$	100 A, 240 V	VA	$VA = I \times V$ $VA = 100 \text{ A} \times 240 \text{ V}$ **VA = 24,000 VA**
1ϕ	V	$V = \dfrac{VA}{I}$	42,000 VA, 350 A	V	$V = \dfrac{VA}{I}$ $V = \dfrac{42{,}000 \text{ VA}}{350 \text{ A}}$ **V = 120 V**
3ϕ	I	$I = \dfrac{VA}{V \times \sqrt{3}}$	72,000 VA, 208 V	I	$I = \dfrac{VA}{V \times \sqrt{3}}$ $I = \dfrac{72{,}000 \text{ VA}}{360 \text{ V}}$ **I = 200 A**
3ϕ	VA	$VA = I \times V \times \sqrt{3}$	2 A, 240 V	VA	$VA = I \times V \times \sqrt{3}$ $VA = 2 \times 416$ **VA = 832 VA**

AC/DC FORMULAS				
To Find	**DC**	**AC**		
		1ϕ, 115 or 220 V	**1ϕ, 208, 230, or 240 V**	**3ϕ—All Voltages**
I, HP known	$\dfrac{HP \times 746}{E \times E_{ff}}$	$\dfrac{HP \times 746}{E \times E_{ff} \times PF}$	$\dfrac{HP \times 746}{E \times E_{ff} \times PF}$	$\dfrac{HP \times 746}{1.73 \times E \times E_{ff} \times PF}$
I, kW known	$\dfrac{kW \times 1000}{E}$	$\dfrac{kW \times 1000}{E \times PF}$	$\dfrac{kW \times 1000}{E \times PF}$	$\dfrac{kW \times 1000}{1.73 \times E \times PF}$
I, kVA known		$\dfrac{kVA \times 1000}{E}$	$\dfrac{kVA \times 1000}{E}$	$\dfrac{kVA \times 1000}{1.763 \times E}$
kW	$\dfrac{I \times E}{1000}$	$\dfrac{I \times E \times PF}{1000}$	$\dfrac{I \times E \times PF}{1000}$	$\dfrac{I \times E \times 1.73 \times PF}{1000}$
kVA		$\dfrac{I \times E}{1000}$	$\dfrac{I \times E}{1000}$	$\dfrac{I \times E \times 1.73}{1000}$
HP (output)	$\dfrac{I \times E \times E_{ff}}{746}$	$\dfrac{I \times E \times E_{ff} \times PF}{746}$	$\dfrac{I \times E \times E_{ff} \times PF}{746}$	$\dfrac{I \times E \times 1.73 \times E_{ff} \times PF}{746}$

HORSEPOWER FORMULAS				
To Find	**Use Formula**	**Example**		
		Given	**Find**	**Solution**
HP	$HP = \dfrac{I \times E \times E_{ff}}{746}$	240 V, 20 A, 85% E_{ff}	HP	$HP = \dfrac{I \times E \times E_{ff}}{746}$ $HP = \dfrac{20\ A \times 240\ V \times 85\%}{746}$ **HP = 5.5**
I	$I = \dfrac{HP \times 746}{E \times E_{ff} \times PF}$	10 HP, 240 V, 90% E_{ff}, 88% PF	I	$I = \dfrac{HP \times 746}{E \times E_{ff} \times PF}$ $I = \dfrac{10\ HP \times 746}{240\ V \times 90\% \times 88\%}$ **I = 39 A**

VOLTAGE DROP FORMULAS – 1ϕ, 3ϕ					
Phase	**To Find**	**Use Formula**	**Example**		
			Given	**Find**	**Solution**
1ϕ	VD	$VD = \dfrac{2 \times R \times L \times I}{1000}$	240 V, 40 A, 60′ L, .764 R	VD	$VD = \dfrac{2 \times R \times L \times I}{1000}$ $VD = \dfrac{2 \times .764 \times 60 \times 40}{1000}$ **VD = 3.67 V**
3ϕ	VD	$VD = \dfrac{2 \times R \times L \times I}{1000} \times .866$	208 V, 110 A, 75′ L, .194 R, .866 multiplier	VD	$VD = \dfrac{2 \times R \times L \times I}{1000} \times .866$ $VD = \dfrac{2 \times .194 \times 75 \times 110}{1000} \times .866$ **VD = 2.77 V**

$* \dfrac{\sqrt{3}}{2} = .866$

CAPACITORS

Connected in Series		Connected in Parallel	Connected in Series/Parallel
Two Capacitors	**Three or More Capacitors**		
$C_T = \dfrac{C_1 \times C_2}{C_1 + C_2}$ where C_T = total capacitance (in μF) C_1 = capacitance of capacitor 1 (in μF) C_2 = capacitance of capacitor 2 (in μF)	$\dfrac{1}{C_T} = \dfrac{1}{C_1} + \dfrac{1}{C_2} + ...$	$C_T = C_1 + C_2 + ...$	1. Calculate the capacitance of the parallel branch. 2. Calculate the capacitance of the series combination. $C_T = \dfrac{C_1 \times C_2}{C_1 + C_2}$

SINE WAVES

Frequency	Period	Peak-to-Peak Value
$f = \dfrac{1}{T}$ where f = frequency (in hertz) 1 = constant T = period (in seconds)	$T = \dfrac{1}{f}$ where T = period (in seconds) 1 = constant f = frequency (in hertz)	$V_{p-p} = 2 \times V_{max}$ where 2 = *constant* V_{p-p} = peak-to-peak value V_{max} = peak value

Average Value	rms Value
$V_{avg} = V_{max} \times .637$ where V_{avg} = average value (in volts) V_{max} = peak value (in volts) .637 = constant	$V_{rms} = V_{max} \times .707$ where V_{rms} = rms value (in volts) V_{max} = peak value (in volts) .707 = constant

CONDUCTIVE LEAKAGE CURRENT

$I_L = \dfrac{V_A}{R_I}$

where
I_L = leakage current (in microamperes)
V_A = applied voltage (in volts)
R_I = insulation resistance (in megohms)

TEMPERATURE CONVERSIONS

Convert °C to °F	Convert °F to °C
$°F = (1.8 \times °C) + 32$	$°C = \dfrac{(°F - 32)}{1.8}$

FLOW RATE

$Q = \dfrac{N \times V_d}{231}$

where
Q = flow rate (in gpm)
N = pump drive speed (in rpm)
V_d = pump displacement (in cu in./rev)
231 = constant

BRANCH CIRCUIT VOLTAGE DROP

$\%V_D = \dfrac{V_{NL} - V_{FL}}{V_{FL}} \times 100$

where
$\%V_D$ = percent voltage drop (in volts)
V_{NL} = no-load voltage drop (in volts)
V_{FL} = full-load voltage drop (in volts)
100 = constant

A

abbreviation: A letter or combination of letters that represent a word.

AC drive: An electronic device that controls AC motor speed by varying frequency.

AC sine wave: A symmetrical waveform that contains 360 electrical degrees. See *waveform*.

actuator: A device that changes the position of the valve. See *device* and *valve*.

adjustable motor base: A mounting base that allows a motor to be easily moved over a short distance.

alignment chart: A chart that shows the relationship among three variables. See *chart*.

alternating current (AC): Current that reverses its direction of flow at regular intervals. See *current*.

alternating current (AC) motor: A motor that uses AC to produce rotation.

alternation: Half of a cycle. See *cycle*.

alternator: A device that converts mechanical energy (spinning motion) to electrical energy.

ambient temperature: The temperature of the air surrounding a device. See *device*.

amplitude: The height of a waveform. See *waveform*.

analog display: An electromechanical device that indicates readings by the mechanical motion of a pointer.

anemometer: An instrument that measures wind speed.

angular misalignment: Misalignment that occurs when two shafts are not parallel.

anode: The positive lead of a diode. See *diode*.

apparent power: The product of voltage and current in a circuit calculated without considering the phase shift that may be present between total voltage and current in a circuit. See *voltage, current,* and *phase shift*.

application drawing: A drawing that shows the use of a particular piece of equipment or product in an application. See *drawing*.

arc blast hood: An eye and face protection device that covers the entire head with plastic and material.

arcing: The discharge of an electric current across a gap such as when an electric switch is opened.

arc suppressor: A device that dissipates the energy present across opening contacts. See *device* and *contact*.

arc tube: The light-producing element of an HID lamp. See *high-intensity discharge (HID) lamp*.

armature: The rotating part of a DC motor that is mounted on the motor shaft. See *DC motor*.

average value: The mathematical mean of all instantaneous voltage values in a sine wave. See *sine wave*.

B

ballast: A transformer or solid-state circuit that limits current flow and supplies the starting voltage for a fluorescent lamp. See *transformer*.

bar graph: On a digital meter, a graph composed of segments that function as an analog pointer. In drawings and diagrams, a graph that compares statistical data with vertical or horizontal bars. See *graph*.

battery: A DC voltage source that converts chemical energy to electrical energy.

bench testing: Testing performed when equipment under test is brought to a designated service area.

blade: A long slender structure aerodynamically designed to capture the greatest amount of wind and produce the maximum rotating force.

block diagram: A diagram that shows the relationship between individual sections (blocks) of circuits or components. See *diagram* and *component*.

board: A group of electronic components placed on a printed circuit board that performs a set task.

branch circuit: The portion of a distribution system between the final overcurrent protection device and the outlet (receptacle) or load connected to it.

break-in period: The period of time just after the installation of a new piece of equipment in which defects resulting from defective parts, poor manufacturing quality, contamination, or environmental stress appear.

breaks: The number of separate contacts a switch uses to open or close each individual circuit. See *contact*.

brownout: The deliberate reduction of the voltage level by a power company to conserve power during peak usage times.

613

brushes: The sliding contacts that provide contact between the external power source and the commutator in a DC motor. See *commutator* and *direct current (DC) motor.*

bulb: A glass envelope enclosing the light source of an electric lamp.

C

candela (cd): The unit of luminous intensity produced by a light source in a given direction.

capacitance: The property of an electric device that permits the storage of electrically separated charges when potential differences exist between two conductors. See *device* and *conductor.*

capacitive circuit: A circuit in which current leads voltage. See *current* and *voltage.*

capacitive leakage current: Leakage current that flows through conductor insulation due to a capacitive effect. See *leakage current.*

capacitor: An electronic device used to store an electric charge. See *device.*

capacitor motor: A 1ϕ motor with a capacitor connected in series with the stator windings to produce phase displacement in the starting winding. *See capacitor, series,* and *stator.*

capacitor-run motor: A motor that leaves the starting winding and capacitor in the circuit at all times. See *capacitor.*

capacitor start-and-run motor: A motor that uses two capacitors. See *capacitor.*

capacitor-start motor: A motor that has the capacitor connected in series with the starting winding. See *capacitor* and *series.*

carrier frequency: The frequency that controls the number of times the solid state switches in the inverter section of a PWM electric motor drive turn ON and turn OFF.

cathode: The negative lead of a diode. See *diode.*

cell: A unit that produces electricity at a fixed voltage and current level. See *voltage* and *current.*

centrifugal force: The force that moves rotating bodies away from the center of rotation.

centrifugal switch: In motors, a switch that opens to disconnect the starting winding when the rotor reaches a certain preset speed and reconnects the starting winding when the speed falls below a preset value. See *rotor.*

chart: A pictorial representation of data.

circuit analysis method: A method of troubleshooting that uses a logical sequence to determine the reason for device failure. See *troubleshooting* and *device.*

circuit breaker (CB): A reusable overcurrent protective device that opens a circuit automatically at a predetermined overcurrent. See *overcurrent.*

circular chart recorder: A graphic pen and ink recorder that has measures drawn on a rotating circular chart.

clamping diode: A diode that prevents voltage in one part of the circuit from exceeding the voltage in another part. See *diode* and *voltage.*

clamp-on ammeter: A meter that measures current in a circuit by measuring the strength of the magnetic field around a conductor. See *current* and *conductor.*

class: The generic hazardous material present.

closed loop drive: An electric motor drive that operates using a feedback sensor such as an encoder or tachometer connected to the shaft of the motor to send information about motor speed back to the drive.

closed loop system: A system with feedback from the motor sensors to the electric motor drive.

code: A regulation or minimum requirement.

coil: Winding consisting of insulated conductors arranged to produce magnetic flux.

color rendering: The appearance of a color when illuminated by a light source. See *illumination.*

common mode noise: Noise produced between the ground and hot lines, or the ground and neutral lines.

commutator: A series of copper segments connected to the armature. See *armature.*

compact fluorescent lamps (CFLs): Small fluorescent lamps designed to use less energy than incandescent bulbs and screw into standard lamp sockets.

complimentary metal-oxide semiconductor ICs (CMOS ICs): A group of ICs that employ MOS transistors. See *transistor.*

component: An individual device. See *device, board,* and *module.*

compressor: A mechanical device that compresses refrigerant or other fluid. When refrigerant is compressed, its heat increases.

conduction angle: The time during which a thyristor is ON.

conductive leakage current: The small amount of current that normally flows through the insulation of a conductor.

conductor: Any material through which current flows easily. See *current.*

constant-wattage autotransformer ballast: A high-reactance autotransformer ballast with a capacitor added to the circuit. See *high-reactance autotransformer ballast* and *capacitor.*

contact: A conducting part that operates with another conducting part to make or break a circuit.

contact life: The number of times a relays contacts switch the load controlled by the relay before malfunctioning. See *relay* and *load.*

contactor: A control device that uses a small control current to energize or de-energize the load connected to it. See *device* and *load.*

contact protection circuit: A circuit that protects contacts by providing a nondestructive path for generated voltage as a switch is opened. See *contact* and *voltage.*

contact tachometer: A device that measures the rotational speed of an object through direct contact of the tachometer tip with the object to be measured.

contact thermometer: An instrument that measures temperature at a single point.

continuity checker: An instrument that indicates an open or closed circuit in a circuit in which all power is OFF.

converter: An electronic device that changes AC voltage to DC voltage.

conveyor: A mechanical system that extends from a receiving point to a discharge point and conveys, transports, or transfers material between the two points.

current: The amount of electrons flowing through an electrical circuit. See *electron.*

current gain: A measure of the ratio of the collector current to the base current in a transistor. See *current* and *transistor.*

cycle: One complete wave of alternating voltage or current. See *voltage* and *current.*

D

data file: A group of data values (inputs, timers, counters, and outputs) that is displayed as a group and whose status may be monitored.

DC drive: An electronic device that controls DC motor speed by varying the voltage.

decibel (dB): An electrical unit used for expressing the ratio of the magnitudes of two electric values such as voltage or current.

delta configuration: A transformer connection that has each transformer coil connected end-to-end to form a closed loop. See *transformer* and *coil.*

delta-connected 3ϕ motor: A motor that has each phase wired end-to-end to form a completely closed-loop circuit. See *phase.*

device: Equipment or mechanism designed to serve a special purpose or perform a special function.

diac: A two-terminal bi-directional device used primarily to control the triggering of triacs. See *device* and *triac.*

diagram: A drawing that represents an object or area. See *drawing.*

dielectric absorption test: A test that checks the absorption characteristics of humid or contaminated insulation. See *insulator.*

dielectric material: A medium in which an electric field is maintained with little or no outside energy supply.

digital display: An electronic device that displays readings as numerical values.

digital logic probe: A special DC voltmeter that detects the presence or absence of a signal.

diode: An electronic device that allows current to flow in only one direction. See *device.*

direct current (DC): Current that flows in one direction. See *current.*

direct current (DC) compound motor: A motor with a field connected in both series and shunt with the armature. See *field, series,* and *armature.*

direct current (DC) motor: A motor that uses direct current connected to a field and armature to produce rotation. See *direct current, field,* and *armature.*

direct current (DC) permanent-magnet motor: A motor that uses magnets for the field winding. See *field.*

direct current (DC) series motor: A motor with a field connected in series with the armature. See *field, series,* and *armature.*

direct current (DC) shunt motor: A motor with a field connected in shunt (parallel) with the armature. See *field, parallel,* and *armature.*

direct drive method: A conveyor drive method in which the product rides directly on the driven belt and is used for light loads. See *conveyor.*

direct light: Light that travels directly from a lamp to the surface being illuminated.

division: The probability that a hazardous material is present.

double-acting solenoid valve: A valve that uses one solenoid to develop a force in one direction and a second solenoid to develop a force in the opposite direction. See *valve.*

drawing: A representation of an object by means of lines.

dry bulb temperature: The temperature of the air taken with a dry temperature probe.

dual-voltage 3ϕ motor: A motor that operates at more than one voltage level. See *voltage.*

E

effective light beam: In a proximity sensor, the area of light that travels directly from the transmitter to the receiver. See *proximity sensor.*

effective value: The value of a sine wave that produces the same amount of heat in a pure resistive circuit as a DC of the same value. See *root-mean-square value* and *sine wave.*

electrical balance: Transformer balance that occurs when loads on a transformer are placed so that each coil of the transformer carries the same amount of current. See *load, transformer, coil,* and *current.*

electric discharge lamp: A lamp that produces light by an arc discharged between two electrodes.

electric motor drive: An electrical device that controls motor speed between 0 rpm and maximum rpm, and controls motor torque between 0 in.-lb and maximum in.-lb.

electric motor drive test: an initial test that verifies if an electric motor drive is operational.

electric motor drive input and output test: A test used to verify that the inputs and outputs of a drive function properly when operated as designed.

electric motor drive, motor, and load test: A test used to verify that a drive and motor function together properly to rotate the driven load.

electrical noise: Unwanted signals that are present on a power line.

electrolyte: A conducting medium in which the current flow occurs by ion migration.

electromagnetic overload relay: An electromechanical relay operated by the current flow in a circuit. See *electromechanical relay* and *current*.

electromechanical: Electrically operated producing mechanical motion.

electromechanical relay: A device that allows the connection of two different components, voltage levels, voltage types (AC/DC), or systems. See *relay*.

electron: An atomic particle containing a negative electrical charge.

enclosed-coil heating element: A heating element that has the resistance wire that produces the heat surrounded by a refractory insulating material. See *heating element*.

enclosure: A housing used to protect wires and equipment and prevent personnel injury by accidental contact with a live circuit.

erratic relay operation: The proper operation of a relay at times, and the improper operation of the same relay at other times. See *relay*.

exact relay replacement method: A method of troubleshooting that replaces a bad relay with a relay of the same type and size. See *relay*.

expansion valve: A device that reduces the pressure on liquid refrigerant by allowing the refrigerant to expand.

F

face shield: An eye and face protection device that covers the entire face with a plastic shield, and is used for protection from flying objects.

fall arrest system: A system of body harnesses/lanyards and attachments that limit the distance of a fall.

field: The stationary windings, or magnets of a DC motor. See *DC motor*.

filament: A conductor that has a resistance high enough to cause the conductor to heat. See *conductor* and *resistance*.

floating input: A digital input that is not high or low at all times.

flow chart: A diagram that shows a logical sequence of steps for a given set of conditions. See *diagram*.

flow switch: A switch designed to detect the movement (flow) of a product through a pipe or duct.

fluorescent lamp: An electric discharge lamp in which ionization of mercury-vapor transforms ultraviolet energy generated by the discharge into light.

footcandle (fc): The amount of light produced by a lamp (lumen) divided by the area that is illuminated. See *illumination*.

forward bias: The condition of a diode while it conducts current. See *diode* and *current*.

frequency (f): The number of cycles per second (cps) in an AC sine wave. See *AC sine wave*.

function: A mathematical, graphical, or tabular statement of the influence one device or component has on another device or component.

fundamental frequency: The frequency of the voltage used to control motor speed.

fuse: An overcurrent protection device with a fusible link that melts and opens the circuit on an overcurrent condition. See *overcurrent*.

G

gain: The ratio of the amplitude of the output signal to the amplitude of the input signal.

gate turn ON: A method of thyristor turn ON that occurs when the proper signal is applied to the gate at the correct time. See *thyristor*.

gearbox: A housing that contains several different sizes of gears that are used to change the rotational speed between its input shaft and output shaft.

gear ratio: The number of revolutions per minute between the low-speed shaft and the high-speed shaft.

generator (alternator): A device that converts mechanical energy into electrical energy by means of electromagnetic induction.

geothermal system: A system that uses the temperature of the ground or water from a lake or ocean to produce temperature changes that can be used for heating or cooling applications.

ghost voltage: A voltage that appears on a meter that is not connected to a circuit. See *voltage*.

graph: A pictorial representation of data.

grounded circuit: A circuit in which current leaves its normal path and travels to the frame of the motor. See *current.*

grounding: The connection of all exposed non-current-carrying metal parts to the earth.

grounding conductor: A conductor that does not normally carry current, except during a fault (short circuit).

H

half-waving: When a relay fails to turn OFF because the current and voltage in the circuit reach zero at different times. See *relay, phase shift,* and *inductive circuit.*

hardware: The physical components in a system. See *component* and *system.*

hard-wired unit: Equipment or unit that has conductors connected to terminal screws. See *conductor.*

harmonic: A frequency that is an integer (whole number) multiple (2nd, 3rd, 4th, 5th, etc.) of the fundamental frequency.

harmonic analyzer: A meter that detects and measures harmonic distortion. See *harmonic distortion.*

harmonic distortion: The voltage and/or current in a power line that is a multiple of the fundamental line frequency. See *frequency.*

harmonic filter: A device used to reduce harmonic frequencies and total harmonic distortion.

head end: The end of a conveyor from which material is discharged. See *conveyor.*

heating element: A heat-producing device that produces heat from the resistance to the flow of electric current. See *device* and *resistance.*

heat pump: A mechanical compression refrigeration system that contains devices and controls that reverse the flow of refrigerant to move heat from one area to another area.

heat sink: A device that conducts and dissipates heat away from a component. See *component.*

high-intensity discharge (HID) lamp: A lamp that produces light from an arc tube. See *arc tube.*

high-pressure sodium lamp: An HID lamp that produces light when current flows through vaporized sodium under pressure and temperature. See *high-intensity discharge (HID) lamp.*

high-reactance autotransformer ballast: A device that uses two coils (primary and secondary) to regulate both voltage and current. See *device, coil, voltage,* and *current.*

holding current: The minimum current that ensures proper operation of a solid-state switch. See *current.*

hub: The center part of a wind turbine rotor to which the blades are attached.

hydrometer: A test instrument that measures the specific gravity of a liquid compared to water, which has a specific gravity of 1.

I

illumination: The effect of light falling on a surface.

improper phase sequence: The changing of the sequence of any two phases (phase reversal) in a 3φ motor control circuit. See *phase.*

impulse transient voltage: A transient voltage commonly caused by lightning strikes that result in a short, unwanted voltage placed on the power distribution system.

incandescent lamp: An electric lamp that produces light by the flow of current through a tungsten filament inside a gas-filled, sealed glass bulb. See *filament,* and *bulb.*

indirect light: Light that is reflected from one or more objects to the surface being illuminated. See *illumination.*

inductance: The property of a circuit that causes it to oppose a change in current due to energy stored in a magnetic field. See *current.*

induction (asynchronous) generator: An AC generator that rotates at a changing speed slightly below synchronous speed.

inductive circuit: A circuit in which current lags voltage. See *current* and *voltage.*

inductive reactance: The opposition to the flow of alternating current in a circuit due to inductance. See *alternating current* and *inductance.*

infrared meter: A meter that measures heat energy by measuring the infrared energy that a material emits.

initial lumen: The amount of light produced when a lamp is new.

in-line ammeter: A meter that measures current in a circuit by inserting the meter in series with the component under test. See *series* and *component.*

in-phase: The state when voltage and current reach their maximum amplitude and zero level simultaneously. See *amplitude.*

input module: The part of a programmable controller that receives information from pushbuttons, temperature switches, pressure switches, photo-electric and proximity switches, and other input devices. See *programmable controller.*

instant-start circuit: A fluorescent lamp-starting circuit that provides sufficient voltage to strike an arc instantly. See *fluorescent lamp.*

insulated gate bipolar transistor (IGBT): A transistor with fast switching capabilities.

insulation spot test: A test that checks motor insulation over the life of the motor.

insulation step voltage test: A test that creates electrical stress on internal insulation cracks to reveal aging or damage not found during other insulation tests.

insulator: Material that current cannot flow through easily. See *current*.

intensity: The level of brightness.

interconnection diagram: A diagram that shows the external connection between unit assemblies or equipment. See *diagram*.

interface: An electronic device that allows different levels or types of components to be used together in the same circuit. See *component*.

interference: In proximity sensors, any object other than the object to be detected that is sensed by the sensor. See *proximity sensor*.

interlock: A mechanical or electrical method used to prevent two different circuits from being energized at the same time.

initial test: A test that verifies if an electric motor drive is operational.

inverter: An electronic device that changes DC voltage into AC voltage.

isolated grounded receptacle: A receptacle that minimizes electrical noise by providing a separate grounding path. See *electrical noise* and *grounding*.

isometric drawing: A pictorial drawing with a 120° included axis. See *pictorial drawing*.

J

jumper link: A dual-voltage unit used to set the power rating of a programmable controller. See *programmable controller*.

L

leakage current: Current that flows through insulation. See *current*.

let-through voltage: The maximum voltage surge allowed to pass through a surge suppressor. See *voltage surge* and *surge suppressor*.

level switch: A proximity sensor used to detect the presence of a liquid or solid in a tank. See *proximity sensor*.

light emitting diode (LED) lamp: A solid-state semiconductor device that produces light when a DC current passes through it.

light meter: An instrument that measures light levels in footcandles (fc) or lumens (lm).

linear load: Any load in which current increases proportionately as the voltage increases and current decreases proportionately as voltage decreases.

linear scale: A scale divided into equally spaced segments.

linear speed: Distance traveled per unit of time in a straight line.

line (one line) diagram: A diagram that shows with single lines and graphic symbols the logic of an electrical circuit or system of circuits and components. See *diagram* and *component*.

line graph: A graph that shows two related variables plotted on coordinate paper.

line voltage regulator: A power conditioner that maintains a specified output voltage for a given voltage input fluctuation.

load: A device that converts electrical energy to mechanical energy, heat, light, or sound. See *device*.

locked rotor torque: The torque that a motor produces when the rotor is stationary and full power is applied to the motor. See *rotor*.

lockout: The process of removing the source of power and installing a lock which prevents the power from being turned ON.

lockout devices: Lightweight enclosures that allow the lockout of standard control devices.

low-pressure sodium lamp: An HID lamp that operates at a low vapor pressure and uses sodium as the vapor. See *high-intensity discharge (HID) lamp*.

lumen (lm): The unit used to measure the total amount of light produced by a light source.

lumen per watt (lm/W): The number of lumen produced per watt of electrical power.

M

malfunction: The failure of a system, equipment, or part to operate properly.

mean lumen: The average light produced after a lamp has operated about 40% of its rated life.

mechanical life: The number of times a relay's mechanical parts operate before malfunctioning.

mechanical switch: Any switch that uses silver contacts to start and stop the flow of current in a circuit. See *contact* and *current*.

megohmmeter: A device that detects insulation deterioration by measuring high resistance values under high test voltage conditions.

mercury-vapor lamp: An HID lamp that produces light by an electric discharge through mercury-vapor. See *high-intensity discharge (HID) lamp*.

metal-halide: An element (normally sodium and scandium iodide) which is added to mercury in small amounts.

metal-halide lamp: An HID lamp that produces light by an electric discharge through mercury-vapor and metal-halide in the arc tube. See *high-intensity discharge (HID) lamp, metal-halide,* and *arc tube.*

meter loading effect: The additional resistance a meter adds to the total resistance in a circuit. See *resistance.*

minimum holding current: The minimum current that ensures proper operation of a solid-state switch.

module: A group of electronic or electrical components housed in an enclosure that performs a set task. See *component* and *enclosure.*

momentary power interruption: A decrease to 0 V on one or more power lines lasting from .5 cycles up to 3 sec.

motor current test: A test used to find hidden motor problems not found with the motor mechanical test.

motor damage: Any damage that occurs to a properly manufactured motor.

motor defect: An imperfection created during the manufacture of a motor.

motor efficiency: The measure of the effectiveness with which a motor converts electrical energy to mechanical energy.

motor mechanical test: A test that checks the mechanical operation of a motor.

motor starter: A contactor with overload protection added. See *contactor* and *overload.*

multimeter: A meter that is capable of measuring two or more electrical quantities.

N

nacelle: A housing that includes the gearbox, generator, power converter, electrical controls, brakes, and other components required to convert rotating mechanical force into electricity.

neon light: A light that is filled with neon gas and uses two electrodes to ignite the gas.

neutron: An atomic particle containing a neutral electrical charge.

nonlinear load: Any load in which the instantaneous load current is not proportional to the instantaneous voltage.

nonlinear scale: A scale that is divided into unequally spaced segments.

normally-closed (NC) valve: A valve that does not allow pressurized fluid to flow out of the valve in the spring-actuated position. See *valve* and *position.*

normally-open (NO) valve: A valve that allows pressurized fluid to flow out of the valve in the spring-actuated position. See *valve* and *position.*

NPN transistor: A transistor that has a thin layer of P-type material placed between two pieces of N-type material. See *transistor.*

O

off-site troubleshooting: Troubleshooting at a location other than where hardware is installed. See *troubleshooting* and *hardware.*

off-state leakage current: The amount of current that leaks through a solid-state switch when the switch is turned OFF. See *current.*

ohmmeter: An instrument that measures resistance. See *resistance.*

Ohm's law: The relationship between voltage, current, and resistance in an electrical circuit. See *voltage, current,* and *resistance.*

one-level registration: Determining the product level between the ON and OFF points of an operation using one sensor.

one-time fuse: A fuse that cannot be reused after it has opened. See *fuse.*

on-site troubleshooting: Troubleshooting at the location where hardware is installed. See *troubleshooting* and *hardware.*

open circuit: A circuit that no longer provides a path for current to flow. See *current.*

open circuit transition switching: A process in which power is momentarily disconnected when switching a circuit from one voltage supply (or level) to another.

open loop drive: An electric motor drive that operates without any feedback to the drive about motor speed.

open loop system: A system that has no feedback method.

operational diagram: A diagram that shows the operation of individual components or circuits. See *diagram* and *component.*

oscillatory transient voltage: Transient voltage commonly caused by turning OFF high inductive loads and by switching large utility power factor correction capacitors.

oscilloscope: An instrument that displays an instantaneous voltage. See *voltage.*

output module: The part of a programmable controller that delivers the output voltage required to control alarms, lights, solenoids, motor starters, and other output devices. See *programmable controller.*

overcurrent: Any current above the normal current level. See *current.*

overcycling: The process of turning a motor ON and OFF repeatedly.

overload: The application of excessive load to a motor. See *load.*

overload relay: A time-delay device that allows temporary overloads without disconnecting the load. See *device, overload,* and *load.*

overloaded motor: A motor with a current reading greater than 105% of nameplate current rating.

overvoltage: More than a 10% increase above the normal rated line voltage for a period of longer than 1 min.

P

parallel: Two or more components connected creating more than one path for current flow. See *component* and *current*.

parallel circuit: A circuit that contains two or more loads and has more than one path through which current flows.

parallel misalignment: Misalignment that occurs when two shafts are parallel but not on the same line.

peak value: The maximum value of either the positive or negative alternation of a sine wave. See *alternation* and *sine wave*.

peak-to-peak value: The value measured from the maximum positive alternation to the maximum negative alternation of a sine wave. See *alternation* and *sine wave*.

period: The time required to produce one complete cycle of a waveform. See *cycle* and *waveform*.

personal protective equipment (PPE): Clothing and/or equipment worn by a technician to reduce the possibility of injury in the work area.

phase: The fractional part of a period through which the time variable of a periodic quantity has moved. See *period*.

phase shift: The state when voltage and current in a circuit do not reach their maximum amplitude and zero level simultaneously. See *voltage*, *current*, and *amplitude*.

phase unbalance: The unbalance that occurs when power lines are out of phase. See *phase*.

phasor rotation: The order in which waveforms from each phase (phase A, phase B, and phase C) cross zero.

photo tachometer: A device that measures the speed of an object without direct contact with the object.

photovoltaic array: A group of modules that include many individual photovoltaic cells interconnected in series and/or parallel combinations to produce a DC power output.

photovoltaic cell: A solid-state semiconductor that converts light energy into electrical energy.

photovoltaic (PV) system: An electrical system that converts light energy into electrical energy.

pie chart: A chart that compares parts of a whole in relation to their total use.

pictorial drawing: A drawing that shows the length, height, and depth of an object in one view.

PNP transistor: A transistor that has a thin layer of N-type material placed between two pieces of P-type material. See *transistor*.

polarity: The positive (+) or negative (−) state of an object.

poles: The number of completely isolated circuits that can pass through a switch at one time.

position: In valves, the number of positions within the valve that the spool is placed in to direct fluid-flow through the valve. See *valve, position,* and *valve spool*.

power: The rate of doing work or using energy.

power circuit: The part of an electrical circuit that connects the loads to the main power lines. See *load*.

power factor (PF): The ratio of true power used in an AC circuit to apparent power delivered to the circuit. See *true power* and *apparent power*.

power supply: The device that provides all the voltages required for the internal operation of a device or system. See *device* and *system*.

preheat circuit: A fluorescent lamp-starting circuit that heats the cathode before an arc is created. See *fluorescent lamp* and *cathode*.

pressure switch: A switch that is designed to activate its contacts at a preset pressure.

preventive maintenance: Maintenance performed to keep machines, assembly lines, production operations, and plant operations running with little or no downtime.

primary division: On meter displays, a division with a listed value.

process: A sequence of operations that accomplishes desired results.

processor (CPU): Part of the programmable controller that performs logic functions, program storage, and supervisory functions. See *programmable controller*.

programmable controller: A solid-state control device that is programmed and reprogrammed to automatically control machines and industrial processes.

programming terminal: A keyboard-operated device that is used to insert the control program into memory, make changes to the stored program, and display the program circuit and data. See *device*.

proton: An atomic particle containing a positive electrical charge.

proximity: The quality or state of being close.

proximity sensor: A device that reacts to the proximity of an actuating means without physical contact.

psychrometric chart: A graph that is used to find various properties of air. See *graph*.

pulley: 1. A shaft-mounted wheel used to transmit power by means of a belt, chain, etc. 2. In conveyor systems, a revolving cylinder connected to a drive. See *conveyor*.

pull-up resistor: A resistor that has one side connected to the power supply at all times and the other side connected to one side of an input switch.

pulse width modulation (PWM): The varying of the voltage pulse width to control the amount of voltage supplied to a motor.

pure DC power: Power obtained from a battery or DC generator. See *battery.*

R

rapid-start circuit: A fluorescent lamp-starting circuit that brings the lamp to full brightness in about two seconds. See *fluorescent lamp.*

RC circuit: A circuit in which resistance and capacitance are used to help filter the power. See *resistance* and *capacitance.*

reactor ballast: A ballast that connects a coil (reactor) in series with the power line leading to the lamp. See *ballast, coil,* and *series.*

recording meter: A meter that provides a written record of readings over a time period.

relay: An interface that controls one electrical circuit by opening and closing contacts in another circuit. See *interface* and *contact.*

renewable fuse: A fuse that is designed so that the fusible link can be replaced.

resistance: Opposition to the flow of electrons.

resistive circuit: A circuit that contains only resistive components, such as heating elements and incandescent lamps. See *component, heating element,* and *incandescent lamp.*

reverse bias: The condition of a diode when it acts as an insulator. See *diode* and *insulator.*

reversing valve: A four-way valve that reverses the flow of refrigerant in a heat pump.

roller: A revolving cylinder not connected to a drive.

roller drive method: A conveyor drive method in which the product rides on the belt-driving rollers. See *conveyor.*

root-mean-square value (rms): The value that produces the same amount of heat in a pure resistive circuit as a DC of the same value. See *effective value.*

rotational speed: Distance traveled per unit of time in a circular direction.

rotor: The central rotating shaft of a wind turbine that includes the blades and hub. See *alternating current (AC) motor.*

rough service lamp: A lamp made of a specially processed tungsten filament that has extra supports to resist breakage. See *tungsten* and *filament.*

rubber insulating matting: A personal protective device placed on the floor to protect technicians from electrical shock when working on energized electrical circuits.

S

safety glasses: An eye protection device with special impact-resistant glass or plastic lenses, reinforced frames, and side shields.

schematic (elementary) diagram: A diagram that shows with graphic symbols the electrical connections and functions of a specific circuit arrangement. See *diagram.*

scope: A device that gives a visual display of voltages. See *device.*

scopemeter: A combination oscilloscope and digital multimeter. See *oscilloscope.*

secondary division: On meter displays, a division that divides primary divisions in halves, thirds, fourths, fifths, etc.

sectional view drawing: A drawing that shows the internal features of an object. See *drawing.*

series: Two or more components connected creating one path for current flow. See *component.*

series circuit: A circuit that contains two or more loads and one path through which current flows. See *load.*

series/parallel: A combination of series- and parallel-connected components. See *series, parallel,* and *component.*

shaded-pole motor: An AC motor that uses a shaded stator pole for starting. See *alternating current (AC) motor* and *stator.*

short circuit: A circuit in which current takes a shortcut around the normal path of current flow.

signal (function) generator: A test instrument that provides a known test signal to a circuit, device, or system for testing purposes.

silicon controlled rectifier (SCR): A thyristor that is triggered into conduction in only one direction. See *thyristor.*

sine wave: A wave that can be expressed as the sine of a linear function. See *function.*

single phasing: The operation of a motor designed to operate on three phases operating on two phases because one phase is lost. See *phase.*

single-voltage 3ϕ motor: A motor that operates at only one voltage level. *See voltage.*

slip: The difference between the synchronous speed and actual speed of a motor. See *synchronous speed.*

snub roller: In conveyor systems, a roller not connected to a drive that is used to correct belt alignment. See *conveyor.*

software: The computer programs and procedures that operate hardware. See *hardware.*

solarcell: A voltage source that converts light energy to electrical energy.

solenoid: An electromechanical device that converts electrical energy to mechanical energy. See *device*.

solenoid-operated valve: A valve that uses electrical force to directly or indirectly change the position of a valve spool. See *valve, position,* and *valve spool*.

solid-state relay: An electronic switching device that has no moving parts. See *device*.

split-phase motor: An AC motor that can run on one or more phases. See *alternating current (AC) motor* and *phase*.

standard: An accepted reference or practice.

standard ambient temperature: In transformers, the average temperature of the air that cools a transformer over a 24-hour period. See *transformer*.

startup procedure: A logical, systematic procedure that isolates and eliminates potential problems that may cause personal injury and equipment damage.

static turn ON: A method of thyristor turn ON that occurs when a fast rising voltage is applied across the anode and cathode terminals of an SCR or the main terminals of a triac. See *thyristor, anode, cathode, SCR,* and *triac*.

stator: The stationary part of an AC motor to which power lines are connected. See *alternating current (AC) motor*.

status indicator: A light that shows the condition of the components in a programmable controller system. See *programmable controller*.

strobe tachometer: A device that uses a flashing light to measure the speed of a moving object. See *device*.

subdivision: On meter displays, a division that divides secondary divisions in halves, thirds, fourths, fifths, etc. See *secondary division*.

substation: An assemblage of equipment installed for switching, changing, or regulating the voltage of electricity.

substitution: The replacement of malfunctioning equipment or units. See *unit*.

sulfidation: The formation of film on a contact surface. See *contact*.

surface leakage current: Current that flows from areas on conductors where insulation is removed to allow electrical connections. See *current* and *conductor*.

surge protection device: A device that limits voltage surges that may be present on the power lines.

surge suppressor: An electrical device that provides protection from high-level transients by limiting the level of voltage allowed downstream from the surge suppressor. See *device* and *voltage surge*.

sustained power interruption: A decrease to 0 V on all power lines for a period of more than 1 min.

sweep: On oscilloscopes, the movement of the displayed trace across a scope screen. See *oscilloscope, trace,* and *scope*.

symbol: A graphic element that represents a quantity or unit.

synchronous generator: An AC generator that rotates at a constant speed.

synchronous speed: The theoretical speed of a motor based on the motor's number of poles and the line frequency. See *frequency*.

system: A combination of components, units, or modules that are connected to perform work or meet a specific need. See *component, unit,* and *module*.

system analysis: The breakdown of system requirements and components performed when designing, implementing, maintaining, or troubleshooting a system. See *system, component,* and *troubleshooting*.

systems analyst: An individual who troubleshoots at the system level. See *troubleshooting* and *system*.

T

tabular chart: A chart that presents data in columnar form. See *chart*.

tachometer: A device that measures the speed of a moving object. See *device*.

tagout: The process of placing a danger tag on the source of power, which indicates that the equipment may not be operated until the danger tag is removed.

tail end: In conveyor systems, the end of a conveyor where material is fed. See *conveyor*.

take-up pulley: In conveyor systems, a pulley that is used for correcting belt tension and is not connected to a drive.

taps: Connecting points that are provided along a transformer coil. See *transformer* and *coil*.

temperature: A measurement of the intensity of heat.

temperature rise: The difference between the motor winding temperature when running and the ambient temperature. See *ambient temperature*.

temporary power interruption: A decrease to 0 V on one or more power lines lasting for more than 3 sec up to 1 min.

terminal: A point of connection in an electrical circuit.

terminal diagram: A diagram that relates the internal circuit of a piece of equipment to its terminal physical configuration. See *diagram*.

test light: A light that is connected to two test leads and gives a visual indication when voltage is present in a circuit. See *voltage*.

thermal overload relay: An electromechanical relay that operates by heat developed in the relay. See *relay*.

thermal resistance (R_{TH}): The ability of a device to impede the flow of heat. See *device*.

thermal switch: A switch that operates its contacts when a preset temperature is reached.

thermal turn ON: A method of thyristor turn ON that occurs when the heat level exceeds the limit of a thyristor (normally 110°C). See *thyristor*.

three-wire solid-state switch: A switch that has three connecting terminals or wires (exclusive of ground). See *terminal*.

throws: The number of different closed contact positions per pole that are available on a switch. See *contact* and *pole*.

thyristor: A solid-state switching device that switches current ON by a quick pulse of control current. See *device*.

time/division (time per division) control: An oscilloscope or scopemeter control that selects the width of the displayed waveform. See *oscilloscope*, *scopemeter*, and *waveform*.

trace: A reference point/line that is visually displayed on the face of a scope screen. See *scope*.

transformer: An electrical device that uses electromagnetism to change voltage from one level to another. See *device*, *voltage*, and *alternating current*.

transient voltage: Temporary, unwanted voltage in an electrical circuit.

transistor: A three-element device made of semiconductor material. See *device*.

transistor-transistor logic ICs (TTL ICs): A broad family of ICs that employ a two-transistor arrangement. See *transistor*.

transverse mode noise: Noise produced between the hot and neutral lines.

triac: A thyristor that is triggered into conduction in either direction. See *thyristor*.

troubleshooting: The systematic elimination of the various parts of a system or process to locate a malfunctioning part. See *system*, *malfunction*, and *process*.

troubleshooting procedure: A logical step-by-step process used to find a malfunction in a system or process as quickly and easily as possible. See *malfunction*, *system*, and *process*.

true power: Actual power used in an electrical circuit.

tungsten: A heavy high-melting ductile metal that is used for electrical purposes.

tungsten-halogen lamp: An incandescent lamp that is filled with a halogen gas (iodine or bromine). See *incandescent lamp*.

two-level registration: Determining the product level between the ON and OFF points of an operation using two sensors.

two-winding constant-wattage ballast: A ballast that uses a transformer that provides isolation between the primary and secondary circuits. See *ballast*.

two-wire solid-state switch: A switch that has two connecting terminals or wires (exclusive of ground). See *terminal*.

U

undervoltage: More than a 10% decrease (but not to 0 V) below the normal rated line voltage for a period of longer than 1 min.

uninterruptible power system (UPS): A power supply that provides constant on-line power when the primary power supply is interrupted.

unit: An individual component that performs a specific task by itself. See *component*.

useful life: Period of time after the break-in period when most equipment operates properly. See *break-in period*.

V

valve: A device that controls fluid flow, direction, pressure, or flow rate. See *device*.

valve spool: The movable part of a valve that opens, shuts, or partially obstructs one or more passageways. See *valve*.

varistor: A resistor whose resistance is inversely proportional to the voltage applied to it. See *resistance* and *voltage*.

vertical chart recorder: A graphic pen and ink recorder that has measured values drawn on a continuous roll of graphic paper.

voltage: The amount of electrical pressure in a circuit.

voltage break-over turn ON: A method of thyristor turn ON that occurs when the voltage across the thyristor terminals exceeds the maximum voltage rating of the device. See *thyristor, voltage,* and *device*.

voltage dip: A momentarily low voltage.

voltage fluctuation: An increase or decease in the normal line voltage within the range of ±10%.

voltage regulator (stabilizer): A device that provides precise voltage regulation to protect equipment from voltage sags (voltage dips) and swells (voltage surges).

voltage sag (voltage dip): More than a 10% decrease (but not to 0 V) below the normal rated line voltage lasting from .5 cycles up to 1 min.

voltage spike: An increase in voltage (typically several thousand volts) that lasts for a very short time (microseconds to milliseconds). See *voltage*.

voltage stabilizer (regulator): A device that provides precise voltage regulation to protect equipment from voltage dips and voltage surges. See *device, voltage dip,* and *voltage surge*.

voltage surge: Any higher-than-normal voltage that temporarily exists on one or more power lines.

voltage swell (voltage surge): More than a 10% increase above the normal rated line voltage lasting from .5 cycles up to 1 min.

voltage tester: A device that indicates approximate voltage level and type (AC or DC) by the movement and vibration of a pointer on a scale. See *device*.

voltage unbalance: The unbalance that occurs when the voltages at different motor terminals are not equal. See *voltage* and *terminal*.

voltage variance: The difference in voltage between a voltage surge and a voltage dip. See *voltage, voltage surge,* and *voltage dip*.

voltmeter: A meter that measures voltage.

volt-ohm-milliammeter (VOM): A meter that measures voltage, resistance, and current. See *voltage, resistance,* and *current*.

volts/division (volts per division) control: An oscilloscope and scopemeter control that selects the height of the displayed waveform. See *oscilloscope, scopemeter,* and *waveform*.

W

wall-bulb temperature: The operating temperature of a fluorescent lamp's wall, that affects the mercury-vapor pressure inside the bulb. See *fluorescent lamp* and *bulb*.

watt density: The watts per square inch (W/sq in.) on the outer cover of a heating element.

wattmeter: An instrument that measures true power in a circuit. See *true power*.

waveform: The geometrical shape obtained by displaying a wave as a function of time when plotted over one period. See *period*.

way: The flow path through a valve. See *valve*.

wear-out period: Period of time after a piece of equipment's useful life terminates when normal equipment failure occurs. See *useful life*.

wet bulb temperature: The temperature of the air taken with a wet sock covering the temperature probe.

wind turbine: A machine that converts the kinetic energy of wind into rotating mechanical energy.

wind vane: An instrument that measures wind direction.

wiring (connection) diagram: A diagram that shows the connection of an installation or its component devices or parts. See *diagram, component,* and *device*.

wrap-around bar graph: On a digital meter, a bar graph that displays a fraction of the full range on the bar graph. See *bar graph*.

wye configuration: A transformer connection that has one end of each transformer coil connected together. See *transformer* and *coil*.

wye-connected 3φ motor: A motor that has one end of each of the three phases internally connected to the other phases. See *phase*.

Y

yawing: A side-to-side movement. In wind turbine systems, yaw control refers to aligning the blades (and nacelle) in the direction of the wind.

Page numbers in italic refer to figures.

USING THE *TROUBLESHOOTING ELECTRICAL/ELECTRONIC SYSTEMS* INTERACTIVE CD-ROM

Before removing the Interactive CD-ROM from the protective sleeve, please note that the book cannot be returned for refund or credit if the CD-ROM sleeve seal is broken.

System Requirements

To use this Windows®-compatible CD-ROM, your computer must meet the following minimum system requirements:

- Microsoft® Windows Vista™, Windows 2000®, or Windows NT® operating system
- Intel® 1.3 GHz processor (or equivalent)
- 128MB of available RAM (256MB recommended)
- 335MB of available hard disc space
- 1024 x 768 monitor resolution
- CD-ROM drive (or equivalent optical drive)
- Sound output capability and speakers
- Microsoft® Internet Explorer® 6.0 or Firefox® 2.0 web browser
- Active Internet connection required for Internet links

Opening Files

Insert the Interactive CD-ROM into the computer CD-ROM drive. Within a few seconds, the home screen will be displayed allowing access to all features of the CD-ROM. Information about the usage of the CD-ROM can be accessed by clicking on Using This Interactive CD-ROM. The Quick Quizzes®, Illustrated Glossary, Flash Cards, Virtual Meters, Media Clips, and ATPeResources.com can be accessed by clicking on the appropriate button on the home screen. Clicking on the American Tech web site button (www.go2atp.com) accesses information on related educational products. Unauthorized reproduction of the material on this CD-ROM is strictly prohibited.

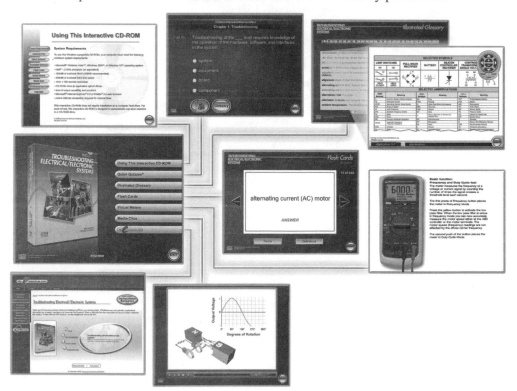